# Green Consumerism

# Green Consumerism

## An A-to-Z Guide

JULIANA MANSVELT, GENERAL EDITOR
*Massey University, Palmerston North, New Zealand*

PAUL ROBBINS, SERIES EDITOR
*University of Arizona*

Los Angeles | London | New Delhi
Singapore | Washington DC

Los Angeles | London | New Delhi
Singapore | Washington DC

FOR INFORMATION:

SAGE Publications, Inc.
2455 Teller Road
Thousand Oaks, California 91320
E-mail: order@sagepub.com

SAGE Publications Ltd.
1 Oliver's Yard
55 City Road
London EC1Y 1SP
United Kingdom

SAGE Publications India Pvt. Ltd.
B 1/I 1 Mohan Cooperative Industrial Area
Mathura Road, New Delhi 110 044
India

SAGE Publications Asia-Pacific Pte. Ltd.
33 Pekin Street #02-01
Far East Square
Singapore 048763

Publisher: Rolf A. Janke
Assistant to the Publisher: Michele Thompson
Senior Editor: Jim Brace-Thompson
Production Editors: Kate Schroeder, Tracy Buyan
Reference Systems Manager: Leticia Gutierrez
Reference Systems Coordinator: Laura Notton
Typesetter: C&M Digitals (P) Ltd.
Proofreader: Christina West
Indexer: Julie Sherman Grayson
Cover Designer: Gail Buschman
Marketing Manager: Kristi Ward

Golson Media
President and Editor: J. Geoffrey Golson
Author Manager: Ellen Ingber
Editors: Kenneth Heller, Mary Jo Scibetta
Copy Editors: Anne Hicks, Barbara Paris



Printed in the United States of America

Library of Congress Cataloging-in-Publication Data

Green consumerism : an A-to-Z guide / general editor Juliana Mansvelt.

p. cm. — (The Sage references series on green society: toward a sustainable future)
Includes bibliographical references and index.

ISBN 978-1-4129-9685-3 (cloth)—ISBN 978-1-4129-7380-9 (ebk)

1. Consumption (Economics)—Environmental aspects. 2. Environmental responsibility. I. Mansvelt, Juliana.

HB801.G785 2010 339.4'7—dc22 2011002874

11 12 13 14 15 10 9 8 7 6 5 4 3 2 1

# Contents

# About the Editors

## Green Series Editor: Paul Robbins

**Paul Robbins** is a professor and the director of the University of Arizona School of Geography and Development. He earned his Ph.D. in Geography in 1996 from Clark University. He is General Editor of the *Encyclopedia of Environment and Society* (2007) and author of several books, including *Environment and Society: A Critical Introduction* (2010), *Lawn People: How Grasses, Weeds, and Chemicals Make Us Who We Are* (2007), and *Political Ecology: A Critical Introduction* (2004).

Robbins's research centers on the relationships between individuals (homeowners, hunters, professional foresters), environmental actors (lawns, elk, mesquite trees), and the institutions that connect them. He and his students seek to explain human environmental practices and knowledge, the influence nonhumans have on human behavior and organization, and the implications these interactions hold for ecosystem health, local community, and social justice. Past projects have examined chemical use in the suburban United States, elk management in Montana, forest product collection in New England, and wolf conservation in India.

## *Green Consumerism* General Editor: Juliana Mansvelt

**Juliana Mansvelt**, Ph.D., is a senior lecturer in human geography at Massey University, Palmerston North, New Zealand. She is author of *Geographies of Consumption* (2005) in which she argued a consideration of issues of power, ethics, and sustainability are essential to understanding how people, places, and consuming are connected in a globalizing world. She has recently completed a series of review articles on consumption for the journal *Progress in Human Geography* and has published on consumption-related topics in a variety of books, journals, and encyclopedias. Her teaching has primarily been in the field of geographic theory and qualitative research techniques, geographies of consumption, and globalization. She has a received a number of teaching awards, including a National Tertiary Teaching Award for sustained excellence in 2006. Mansvelt is a qualitative researcher and her recent research has focused on catalog shopping, ethnic food marketing, and also aging. She is currently involved in several research projects that endeavor to explore relationships between aging, consumption, and place, with a view to investigating purchase, use, and divestment of commodities across the lifespan; the interactions of older consumers with organizations; and factors influencing standards of living for older people.

# Introduction

Perhaps no change has been more evident over the last century than the growth of consumer society. Though the chronology of the emergence of consumer society is much debated, it is generally accepted that its rapid expansion has been a 20th-century phenomenon. The rise of the advertising and marketing industries; the development of Fordist and post-Fordist industrial processes enabling swift production of different commodities; improvements in transportation, logistics, and information technologies; the emergence of communicative means such as the Internet, television, and mobile technologies; and political, cultural, and economic changes associated with globalization have meant few places in contemporary society are exempt from the impact of consumer culture. Processes of commodification have resulted in an increasing volume and variety of goods and services sold in the marketplace and the emergence of new forms of markets, particularly with regard to digital technologies. Such changes have led some commentators to suggest that consumption, rather than production, is now a key engine driving societal change. The incursion of more and more commodities and the images surrounding them into the spaces and practices of everyday life has encouraged consumerism, that is, the process whereby individuals' work and private lives are intricately connected to the acquisition of commodities, and where goals surrounding these become a part of life course trajectories.

Whether people are in situations of material lack or plenty, consumer culture and consumption processes have become a pervasive and visible part of material and imagined landscapes, and of the social relations and practices that comprise everyday life. Commodities provide a means of mediating human relations, and they may offer liberatory, narcissistic, and even hedonistic possibilities for self-fulfillment and self-expression. Consumption can be a medium for processes of identity formation and subjectivity, with commodities and their meanings facilitating consumers' reflexive constructions of multiple and changing selves. Yet commodity relationships cannot be reduced to possibilities for identity formation, as commodities themselves form an important part of material cultures, social relationships, and meaningful places. Capitalist systems of accumulation rely on consumption as a mechanism for realizing the value inherent in the production of commodities. Consumption is critical for production systems to thrive, and for reinvestment of profit from commodities. Though consumerism plays a role in securing the economic and individual well-being, it does so in ways that both enhance and diminish this. Consumer choices and actions may influence social and environmental relations at the time of purchase and use, or in the future. Similarly, consumer actions may impact environments positively or negatively in local contexts (such as in households and communities), or at a distance (for example, via connections through production or disposal processes).

The global proliferation of commodities and the extent of commodification and consumerism have become a source of environmental and social concern. Changes in consumption have occurred unevenly, reflecting and producing social division. Wide material disparities exist between both First and Third Worlds, and within many nations, regions, and local spaces. Concerns about carbon dioxide and other greenhouse gas emissions and their contribution to global warming have coincided with a continuing recognition of limits to growth based on the use of nonrenewable resources, and evidence of the negative impacts of commodity consumption on the environment. The consequences of inputs to production processes can be detrimental to society and space (for example, unsustainable farming or horticultural practices, land degradation, and pollution), and problems may also occur as a consequence of the characteristics of commodities produced and their consumption (the packaging, durability, energy efficiency, and biodegradability of commodities). The use and disposal of commodities can have harmful consequences for both current and future generations through such activities as the leaching of chemicals into soil and water, the burning of fossil fuels, and the creation of stockpiles of inorganic waste. Consequently, the kinds of commodity practices people engage in, the types and amounts of commodities consumed, and the potential reuse and recycling of commodities must be considered with regard to effects on people and environments.

The 20th and 21st centuries have not only seen a proliferation of commodities and consumer choices, but also efforts by firms, marketers, organizations, and governments to construct consumers in multiple ways. Governments informed by neo-liberal ideologies shape notions of citizens as responsible, knowledgeable, and sovereign consumers purchasing everything from food to health care and education through the market mechanism. Advertising and marketing industries appeal not only to capacities of goods and services to fulfill material needs, but also to a multitude of symbolic needs. Advanced information and technology systems and forms of consumer surveillance enable the construction of sophisticated consumer profiles, yet the diversity of people, orientations, and products mean the notion of "the consumer" is impossible to sustain. It is in the context of multifaceted relationships involving the state, trans-state institutions, private sector firms and organizations, and civil society that green consumerism has emerged as a distinctive sphere of consumer practice, politics, and identification.

Despite the promotion of green commodities as a distinct alternative to other goods and services available on the market, the range of products and practices associated with green consumerism are diverse, with multiple characteristics. For example, green adhesives may be made from natural rather than synthetic substances, may have chemical compositions that mean they emit less hazardous vapors, or may be specifically designed to reduce energy loss in buildings. Green consumerism encompasses a range of practices centered on lowering consumption, consuming more sustainably, or ameliorating the negative social and environmental effects of consumption (such as reducing carbon footprints, reusing, downshifting, dumpster diving, and recycling). The concept also applies to commodities and services that are intended to reduce harmful effects on people and the environment, and that may be framed in relation to their role in production or consumption processes, or the ways these might be regulated. For example, in relation to their production, green products may be derived from renewable or recycled resources, may be natural rather than synthetically produced, involve fairly traded labor or sustainably produced material inputs, and engage with humane animal practices. With regard to their consumption (involving purchase, use, and disposal) green commodities may aid energy efficiency, be commodities that are more durable, reusable, or recyclable, or may be products that avoid toxic or other emissions and reduce waste. Though green consumerism is associated with minimizing overall environmental impact, this often involves consideration of the human and health impacts of commodities sold, with many green products promoted on the basis of securing more sustainable futures.

The construction of green consumerism as a mode of being, a way of thinking and acting on the world, and a form of consumption is both influenced by and an outcome of wider social, economic, political, and environmental processes that are formed relationally in place and across time. Much of the visible work of shaping green consumerism is undertaken by agents (everything from firms to charities, activist groups, and individuals) keen to reveal the connections between producers and consumers by focusing on the material connections along commodity chains and networks, and in linking the symbolic meaning of commodity production and consumption to notions of knowledge, ethicality, fairness, and transparency. The revealing of the commodity fetish can rest on the assumption that knowledgeable and informed consumers will behave in an ethical and socially responsible manner with regard to their consumption, and that this will influence the firms making and marketing the goods and services to produce more green products and to further green their production. There are numerous examples of where producers, manufacturers, distributors, and marketers have responded to calls for more green products or have altered their production processes. Consumer activism and a knowledge of consumer demand and preferences has prompted firms to formulate codes of corporate responsibility, create environmental profiles, and to develop systems for tracking, monitoring, and auditing aspects of green production. In addition, green, eco, and fair trade labeling and quality assurance systems have emerged to assure that production is green; enabling customers to make trusted and informed purchases. However, there is unease surrounding green washing, where firms' professed changes may be more cosmetic than substantive, and about the ways in which green politics and practice might be subsumed within neo-liberal agendas or forms of corporate and institutional power.

Negotiating the complex terrain of consumption may not be easy for individuals, groups, or organizations. The capacity of people and groups to effect change and to engage in green consumerism is uneven given disparities in income, material circumstances, life chances, and differential enrollment in and access to social and spatial networks. Individuals and collective actors also differ in their capacity to take action and to have their views and voices heard, acknowledged, and acted upon. Commodity and consumption practices arise from multiple motivations and moralities and are undertaken in the context of social, political, and economic structures that may both limit and enable possibilities for action. Consuming less, boycotting (avoiding some products completely), and buying or substituting greener alternatives, conservation/careful use, reusing (through practices such as dumpster diving, buying second-hand, composting, and recycling) are all strategies that have been employed by consumers. However, the unveiling of the commodity fetish with the aim of informing consumers about the environmental and social effects of their consumption decisions, or even consumers' professed beliefs in green values, may not be expressed in altered consumption behaviors.

In an article in the journal *Antipode*, "Consuming Ethics: Articulating the Subjects and Spaces of Ethical Consumption," Clive Barnett, Paul Cloke, Nick Clarke, and Alice Malpass argue that it is important to understand everyday consumption as ordinarily ethical. A view of green consumers as those who respond ethically only in relation to knowledge about products or production processes can tend to ignore the ways in which consumption practice is already constituted ethically, thereby obscuring possibilities for how everyday practices and moral dispositions might be enrolled, reworked, and reproduced through policies, campaigns, and collective actions to secure green consumerism. Thus as Peter Jackson, Neil Ward, and Polly Russell argue in their recent contribution to the journal *Transactions of the Institute of British Geographers*, moral and political economies are always co-constituted. In addition, the purchase and use of green products by individual consumers does not necessarily guarantee a reduction in overall levels of consumption, nor an affiliation with green politics, and there are questions yet to be answered about how green consumption can challenge or change issues of power, control, and inequality. As a result, how states, firms, organizations, and activist groups can engage meaningfully with individuals and vice versa becomes a critical issue in progressing green agendas and securing positive social and environmental outcomes.

This encyclopedia endeavors to present a wide-ranging examination of green consumerism, one that reflects the complexity of the subject, and the diversity of views and debates surrounding the concept. The multiplicity of topics and disciplinary perspectives provides a useful survey of the nature of green consumerism, the forms that it takes, the issues that impact it, and the practices it involves. The entries demonstrate in numerous ways how consumers' decisions about what might constitute green practices and products are not always straightforward. There are multiple social and environmental dimensions of commodity production, use, and disposal to address, as well as the costs and values to both the individual consumer and society to consider. For example, simply buying local does not necessarily mean the products have been ethically produced, or have fewer environmental impacts: purchasing food with lower food miles may not actually be more energy-efficient if the energy invested in production processes are taken into account; and the provision of alternative products and spaces of consumption, while less hazardous to the environment, must also meet the needs and values of the communities in which they are located. Leaving issues of the production, construction, and availability of green products and services aside, for consumers negotiating and engaging with a range of commodity choices and potential impacts, consumption is likely to be influenced by numerous factors, including: material and economic circumstances, one's enrollment in particular cultural and social contexts and networks, as well as knowledge, attitudes, beliefs, and preferences regarding green products.

Together the contributors highlight the complexity of consumption, and in doing so provide insights into the social and spatial constitution of green consumerism. Many authors have provided specific examples from the particular geographical contexts in which they are situated. Beyond that, the entries illuminate the multifaceted and sometimes contested contours of green consumerism and the ways it is embedded and shaped in relation to wider cultural, economic, political, and environmental processes. Ultimately I hope readers will derive from the entries a sense not only of what green consumerism involves, but more critically, how it might evolve, addressing both limitations and possibilities for real and meaningful change.

*Juliana Mansvelt*
*General Editor*

# Reader's Guide

## Green Consumer Challenges

Affluenza
Air Travel
Carbon Emissions
Commuting
Conspicuous Consumption
Disparities in Consumption
Dumpster Diving
Durability
Electricity Usage
Energy Efficiency of Products and Appliances
E-Waste
Food Additives
Food Miles
Genetically Modified Products
Greenwashing
Healthcare
Insulation
Lawns and Landscaping
Materialism
Needs and Wants
Overconsumption
Pesticides and Fertilizers
Pets
Pharmaceuticals
Positional Goods
Poverty
Pricing
Quality of Life
Resource Consumption and Usage
Solid and Human Waste
Super Rich
Symbolic Consumption
Waste Disposal
Windows

## Green Consumer Consumables

Beverages
Bottled Beverages (Water)
Coffee
Confections
Dairy Products
Fish
Meat
Poultry and Eggs
Slow Food
Tea
Vegetables and Fruits
Water

## Green Consumer Products and Services

Adhesives
Apparel
Audio Equipment
Automobiles
Baby Products
Books
Car Washing
Certified Products (Fair Trade or Organic)
Cleaning Products
Computers and Printers
Cosmetics
Disposable Plates and Plastic Implements
Floor and Wall Coverings

## Green Consumer Solutions



## Green Consumerism Organizations, Movements, and Planning

# List of Articles

# List of Contributors


Andrews, Mitchell
*University of Sunderland*

Arney, Jo A.
*University of Wisconsin–La Crosse*

Barr, Stewart
*University of Exeter*

Beder, Sharon
*University of Wollongong*

Blaylock, Craig
*University of Houston-Downtown*

Bled, Amandine J.
*Université Libre de Bruxelles*

Boslaugh, Sarah
*Washington University in St. Louis*

Boulanger, Paul-Marie
*Institut pour un Développement Durable*

Bourmorck, Amandine
*Université Libre de Bruxelles*

Bristol, Kelli
*University of Houston-Downtown*

Brohé, Arnaud
*Université Libre de Bruxelles*

Caracciolo di Brienza, Michele
*Graduate Institute of International and Development Studies, Geneva*

Chatterjee, Amitava
*University of California, Riverside*

Chiaviello, Anthony R. S.
*University of Houston-Downtown*

Cliath, Alison Grace
*California State University, Fullerton*

Dale, Gareth
*Brunel University*

Davidsen, Conny
*University of Calgary*

Dermody, Janine
*University of Gloucestershire*

Dorney, Erin E.
*Millersville University*

Emery, Barry
*University of Northampton*

Evans, David
*University of Manchester*

Fenner, Charles R., Jr.
*State University of New York at Canton*

Finley-Brook, Mary
*University of Richmond*

Andrew Fox
*University of the West of England, Bristol*

Gonshorek, Daniel O.
*Knox College*

Gonzalez-Perez, Maria-Alejandra
*Universidad EAFIT*

Good, Ryan Zachary
*University of Florida*

Graddy, Garrett
*University of Kentucky*

Harper, Gavin D. J.
*Cardiff University*

Helfer, Jason A.
*Knox College*

Hoffmann, Sabine H.
*American University of the Middle East*

Hostovsky, Chuck
*University of Toronto*

Irvin, George William
*University of London (SOAS)*

Kaur, Meera
*University of Manitoba*

Keane, Timothy P.
*Saint Louis University*

Kirwan, James
*Countryside and Community Research Institute*

Krogman, Naomi T.
*University of Alberta*

Kte'pi, Bill
*Independent Scholar*

Lang, Steven
*LaGuardia Community College*

Leinaweaver, Jeff
*Fielding Graduate University*

LeVasseur, Todd J.
*University of Florida*

Lowrey, Tina M.
*University of Texas at San Antonio*

Lubitow, Amy
*Northeastern University*

Lynes, Jennifer K.
*University of Waterloo*

Mangone, Giancarlo
*University of Virginia*

Maycroft, Neil
*University of Lincoln*

Maye, Damian
*Countryside and Community Research Institute*

McCarty, John A.
*College of New Jersey*

Merskin, Debra
*University of Oregon*

Mukhopadhyay, Kausiki
*University of Denver*

Murray, Daniel
*Independent Scholar*

Nash, Hazel
*Cardiff University*

Nieuwenhuis, Paul
*Cardiff University*

Nussinow, Jill
*Santa Rosa Junior College*

Orsini, Marco
*Institut de Conseil et d'Etudes en Développement Durable*

Paolini, Federico
*University of Siena*

Patnaik, Rasmi
*Pondicherry University*

Paul, Pallab
*University of Denver*

Pollack, Jeffrey
*Independent Scholar*

Ponting, Cerys Anne
*Cardiff University*

Powell, Marie-Alix
*Knox College*

Poyyamoli, Gopalsamy
*Pondicherry University*

Putnam, Heather R.
*University of Kansas*

Rands, Gordon P.
*Western Illinois University*

Rands, Pamela J.
*Western Illinois University*

Reed, Matt
*Countryside and Community Research Institute*

Robles, Laura Joanne
*University of Houston–Downtown*

Roka, Krishna
*Penn State University*

Ross-Rodgers, Martha J.
*University of Phoenix*

Roy, Abhijit
*University of Scranton*

Roy, Mousumi
*Penn State University, Worthington Scranton*

Salsedo, Carl
*University of Connecticut*

Saravia, Luis
*University of Houston–Downtown*

Schroth, Stephen T.
*Knox College*

Scott, Austin Elizabeth
*University of Florida*

Shrum, L. J.
*University of Texas at San Antonio*

Singh, Satyendra
*University of Winnipeg*

Slinger-Friedman, Vanessa
*Kennesaw State University*

Smith, Alastair M.
*ESRC Centre for Business Relationships, Accountability, Sustainability & Society*

Staddon, Chad
*University of the West of England, Bristol*

Stancil, John L.
*Florida Southern College*

Tam, Michael
*Knox College*

Curtis Thomas
*University of Richmond*

Thompson, Adeboyejo Aina
*Ladoke Akintola University of Technology, Ogbomoso*

Tsoi, Joyce
*Brunel University*

Ugarte, Marco
*Arizona State University*

Wagner, Ralf
*University of Kassel*

Wallenborn, Grégoire
*Université Libre de Bruxelles*

Waskey, Andrew Jackson
*Dalton State College*

Watson, Derek
*University of Sunderland*

Wickstrom, Stefanie
*Independent Scholar*

Woodward, David G.
*University of Southampton School of Management*

Zaccai, Edwin
*Université Libre de Bruxelles*

Zamora, Michael A.
*University of Houston–Downtown*

# Green Consumerism Chronology

**1880s:** C. V. Riley, a scientist working for the U.S. Department of Agriculture, advocates the use of biological pest controls for farming, a method used today in organic farming.

**1899:** In *Theory of the Leisure Class,* economist Thorsten Veblen introduces the notion of "conspicuous consumption," and argues that people acquire unnecessary material goods in order to claim higher status among their peers.

**1906:** The Pure Food and Drugs Act and the Meat Inspection Act establish guidelines and procedures to inspect food and drugs sold in the United States for safety and purity.

**1907:** Petrol fumes are identified as damaging to health at an international hygiene conference in Berlin.

**1921:** Most European countries sign a treaty that prohibits the use of interior house paint (but not exterior house paint) containing lead.

**1939:** Swiss chemist Paul Müller discovers that DDT is effective at killing insects, yet harmless to humans in powder form. The chemical is widely used in the armed forces to control mosquitoes and lice, and after World War II becomes popular in agriculture.

**1940s:** Photochemical smog, created when volatile organic compounds (including hydrocarbons present in automobile exhaust) react with sunlight and oxygen, is first observed in the Los Angeles area. It is also discovered that ozone, a product of the photochemical reaction, is irritating to humans, kills plants, and damages materials such as rubber, fabric, and paint.

**1942:** Jerome Irving Rodale begins publication of *Organic Farming and Gardening,* popularizing the concept of organic food production as advocated by British writers Sir Albert Howard and Lord Northbourne.

**1948:** The Nobel Prize in Physiology or Medicine is awarded to Paul Müller for his discovery of the usefulness of DDT as an insecticide.

**1946**: Edna Ruth Byler founds the fair trade organization Ten Thousand Villages, a nonprofit program of the Mennonite Central Committee, which markets products made by artisans in Africa, Asia, Latin America, and the Middle East.

**1952**: An atmospheric inversion in London, coupled with particulate matter in the air from motor vehicles and coal-burning stoves and factories, caused nearly 3,000 excessive deaths in a single week, and highlights the importance of controlling man-made sources of air pollution.

**1954**: *Let's Eat Right to Keep Fit,* by Adelle Davis, promotes the consumption of natural foods and vitamin supplements.

**1955**: Economist Victor Lebow identifies how much consumerism has become part of American culture in an article in the *Journal of Retailing*, saying both social status and personal satisfaction were closely allied with consumption, while continued high consumption was required to make use of the ever-increasing number of products created by the American economy.

**1958**: The Delaney Clause is added to the Food, Drug, and Cosmetic Act of 1938, banning the use of food additives that have been shown to cause cancer in lab animals. This clause also shifts the burden of proof by requiring that the manufacturer must establish the safety of an additive, rather than requiring the FDA to prove that it is harmful.

**1962**: Rachel Carson's *Silent Spring* calls attention to the harmful effects of human activity on the environment, including air pollution and the use of harmful chemicals in agriculture.

**1965**: Local homemakers in Kanagawa prefecture, Japan, organize a cooperative in which they pay a local family farm in advance for milk that is then provided to them at reduced prices. This cooperative grows into the Seikatsu Club, which currently connects over 20 million Japanese consumers with local producers.

**1965**: The United States passes the first emissions standards for automobiles, regulating carbon monoxide and hydrocarbons, which takes effect beginning with 1968 model year cars.

**1970**: The first Earth Day is celebrated internationally, drawing attention to worldwide interest in environmental protection and reform.

**1970s**: Witkar, the world's first car-sharing scheme, is introduced in Amsterdam using battery electric vehicles.

**1971**: Michael Jacobson, Albert Fritsch, and Jim Sullivan found the Center for Science in the Public Interest. The center becomes famous for its attention-grabbing investigations into the nutritional qualities of popular foods, labeling Fettuccine Alfredo "heart attack on a plate," and demonstrating that a typical takeout order of Kung Pao chicken has more fat content than a McDonald's Quarter Pounder.

**1971**: Frances Moore Lappé publishes *Diet for a Small Planet*, introducing the concept of "complementary proteins" (now considered by some scientists as fallacious). Lappé advocates the adoption of a vegetarian diet, both for reasons of health and because of the much greater resources required to produce meat rather than vegetables and grains.

**1971**: Oregon passes a law requiring that beverages in bottles and cans be sold with a deposit that is refunded to the customer when the contained is returned. The Oregon Bottle Bill is estimated to motivate return of 90 percent of beverage containers, reducing roadside litter and increasing recycling.

**1972**: 459 residents of a village in Iraq die and over 6,500 are sickened after consuming bread made from wheat treated with a fungicide containing mercury, highlighting the dangers of this chemical for humans.

**1972**: The Environmental Protection Agency bans the use of DDT in the United States.

**1973**: Due to increasing evidence of the health risk posed by exposure to asbestos, the U.S. Environmental Protection Agency bans its use in building materials, but does not require inspection or removal of asbestos materials already in place.

**1973**: The United States begins phasing out the use of leaded gasoline, resulting in substantial reduction in lead exposure for most Americans.

**1973/1974, 1979**: Energy crises draw attention to the dependency of the United States on Middle Eastern oil, leading to increased interest in fuel-efficient European and Japanese cars and the introduction of smaller, lighter cars by U.S. manufacturers.

**1975**: Jim Hightower coins the term "McDonaldization" in his book *Eat Your Heart Out,* which warns of the danger of international corporations such as McDonald's destroying local cuisines, and driving small farms and restaurants out of business.

**1975**: *Gourmet Magazine* runs a feature story on Chez Panisse, the California restaurant run by Alice Waters, giving a huge boost to the Waters's ethos of cooking with seasonal, locally grown, organic produce.

**1975**: The United States requires that all nondiesel cars include a catalytic converter, which greatly reduces air pollution by burning carbon monoxide, hydrocarbons, and nitrogen oxides at extremely high temperatures (2,000–3,000 degrees Fahrenheit) so that they are reduced to water and carbon dioxide.

**1977**: A boycott against Nestlé products begins in protest of the company's aggressive marketing of infant formula in the Third World to the detriment of the health and welfare of the citizens of those countries. The boycott costs Nestlé millions of dollars, and is successful in forcing them to follow guidelines published by the World Health Organization regarding the marketing of infant formula.

**1977**: The U.S. Consumer Product Safety Commission bans the sale of house paint (interior or exterior) containing more than 0.06 percent lead by weight, as well as the use of paint containing more than that level of lead on toys or furniture.

**1978:** The United States bans the production of aerosol cans that use CFCs as a propellant because CFCs are found to damage the ozone layer that shields the Earth from ultraviolet radiation.

**1980s:** Green marketing is developed to address consumer desire for products that consume less energy (such as energy-saving light bulbs) or otherwise cause less environmental harm (like recycled paper products).

**1981:** In *Voluntary Simplicity* (revised edition published in 1993) Duane Elgin advocates an alternative way of life that is less tied to career success and the acquisition of status-enriching consumer goods, and focused more on the quality of life and human relationships, as well as a sustainable lifestyle.

**1985:** Robyn Van En coins the term "community-supported agriculture," and establishes the first collective in the United States in Massachusetts. In this system community members buy shares of a farmer's crops in advance and receive their portion of the crops that are delivered as they are harvested. This shifts some of the risk of crop loss, as well as the benefits if a growing season is particularly productive, from the farmer to members of the collective.

**1985:** An estimated 200,000 people in the Midwest are sickened by *Salmonella*-contaminated milk, making it the largest salmonellosis outbreak to date in the United States.

**1986:** Arcigola Slow Food, the forerunner of the international Slow Food movement, is founded in Italy as a protest against standardized food produced by international corporations such as McDonald's, as well as a celebration of local products and traditions.

**1987:** The Montreal Protocol, developed with representatives from 29 nations, marks the first international attempt to control or prohibit the use of chemicals believed to deplete the ozone layer. Industrialized countries agree to ban production of CFCs (chlorofluorocarbons) by 1996, while developing nations must phase them out by 2010.

**1988:** The Dutch development agency Solidaridad introduces fair trade coffee in the Netherlands under the label "Max Havelaar" (referring to a fictional character who opposed exploitation of coffee pickers).

**1989:** Kalle Lasn founds Canadian advocacy group Adbusters after he is unsuccessful in purchasing television time to advertise "Buy Nothing Day," a holiday meant as a counterforce to the consumption-heavy Christmas shopping season.

**1990:** The Organic Foods Production Act, which sets national standards for the production, handling, and labeling of organic food in the United States and established the National Organic Standards Board, is implemented as Title XXI of the Farm Bill.

**1990, 1991:** The U.S. government bans the production of paint containing mercury for interior (1990) and exterior (1991) use.

**1991:** The U.S. government bans the use of lead solder in cans containing food or soft drinks, although not the importation of food housed in cans that use lead solder.

**1992:** The Environmental Protection Agency introduces the voluntary Energy Star program to help consumers evaluate the energy efficiency of common products. Specific standards vary by product, but most Energy Star products represent an improvement of at least 20–30 percent in energy efficiency over the traditional version of the same product.

**1992:** The European Union introduces the Ecolabel to help consumers identify products that have relatively low impact on the environment over the life cycle of the product.

**1992, 1995:** Veganism captures worldwide publicity when American vegan chefs Ken Bergeron and Brother Ron Pickarski win gold medals at the International Culinary Olympics in Berlin.

**1993:** *The McDonaldization of Society* by sociologist George Ritzer describes the effects on American society of the success of McDonald's fast food restaurants: he argues that the principles of efficiency, predictability, and control that are characteristic of McDonald's have come to dominate not just the fast-food industry, but other sections of U.S. and world society.

**1993:** Four children die and hundreds become ill after eating hamburgers contaminated with *Escherichia coli* at several Oregon branches of Jack in the Box fast food restaurants, drawing attention to the dangers of this organism and the lack of effective regulation of the U.S. commercial beef supply.

**1994:** The Nutrition Labeling and Education Act requires all processed foods sold in the United States to include nutrition labels to help shoppers select nutritious foods.

**1994:** The U.S. Dietary and Supplement Health and Education Act places limits on the claims vitamin manufacturers may make about their products' abilities to prevent, treat, or cure disease.

**1995:** The World Resources Institute publishes a report stating that motor vehicles are the primary source of carbon monoxide, nitrous oxides, and hydrocarbons in congested urban areas, highlighting the importance of reducing automobile emissions in order to limit air pollution.

**1997:** The term "affluenza" is introduced in a documentary of the same name to refer to a cycle the filmmakers see in contemporary Western society: the unbounded pursuit of material possessions that provide ever-decreasing satisfaction, but produce the desire to accumulate still more goods.

**1997:** Fairtrade Labelling Organizations International is founded in Germany with the goals of bringing together disparate fair trade organizations and harmonizing standards for fair trade certification.

**1997:** A food poisoning outbreak in Japan is traced to radish sprouts grown from seeds produced in Oregon, highlighting the difficulties of policing the safety of the food supply in a global marketplace.

**1998:** the United Nations issues a report investigating growth in consumption in the 20th century. It reports that the richest fifth of the world's population accounts for 86 percent of the world's total private consumption expenditures, while the poorest fifth accounts for only 1.3 percent.

**1999:** The National Resources Defense Council reports that nearly one in four of the brands of bottled water they tested violated at least one of California's water quality regulations, casting doubt on claims that bottled water is healthier or otherwise superior to ordinary tap water. Most brands of bottled water are not subject to FDA regulation because they are packaged and sold within a single U.S. state.

**2000:** Slow Food U.S.A., the American branch of the international Slow Food movement, is founded to celebrate and preserve local food traditions and to promote biodiversity and sustainable growth.

**2002:** The National Organic Rule goes into effect in the United States. It requires producers and handlers of organic food to be certified, and specifies a number of conditions that food labeled as "organic" must meet.

**2002:** *Food Politics* by Marion Nestle alerts the general public to how extensively the food industry exerts a negative influence on the way food is consumed in the United States.

**2003:** The European Union introduces regulations that govern the disposal of electrical equipment, including establishment of collection centers where consumers can deposit discarded goods (rather than putting it in the trash), with the joint purposes of encouraging recycling and reducing the pollution caused by heavy metals and other hazardous materials.

**2004:** Adbusters Media Foundation introduces the Blackspot sneaker, manufactured from hemp in a Portuguese factory where employees are unionized and paid more than the minimum wage, as a challenge to athletic footwear company Nike and its alleged sweatshop labor practices.

**2006:** WaterSense, a partnership with the U.S. Environmental Protection Agency, identifies products such as toilets and irrigation equipment that maintain high standards of water efficiency.

**2006:** Wal-Mart launches an initiative requiring their suppliers to reduce packaging to the lowest possible levels.

**2007:** California and 16 other states sue the Environmental Protection Agency for prohibiting them from raising fuel economy standards for cars sold or driven in their states.

**2008:** Barbara Kingsolver's *Animal, Vegetable, Miracle: A Year of Food Life* popularizes the concept of the locavore and draws attention to the ecological cost of the modern food industry by living for a year on food grown by her family or by other nearby farmers.

**2009:** *No Impact Man* is published as a book and movie documenting the attempts of Colin Beavan and his family to live for one year in New York City while minimizing their impact on the environment through reduced consumption (such as no purchases of clothing), eschewing motorized transportation, and eating only locally grown food.

*Sarah Boslaugh*
*Washington University in St. Louis*

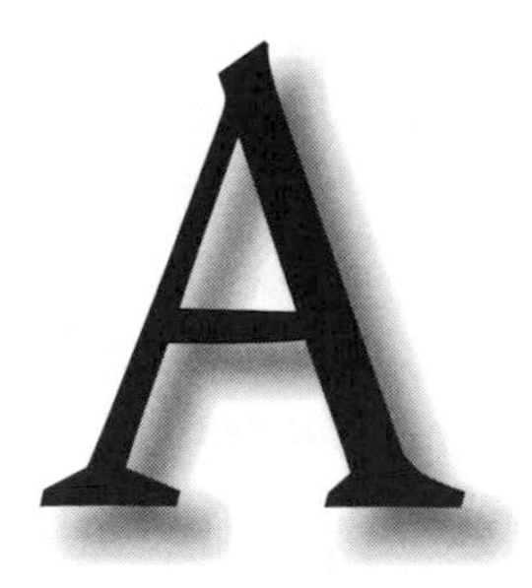

## Adhesives

Bonding agents are called adhesives or glues. They may be made from natural substances or artificial materials. They bind together metals, wood, fabric, ceramics, plastics, composite materials, or virtually any material known. Because of the rising price of oil, scientific investigators have begun a search for natural adhesive materials. At the present time, many are known but are not commercially profitable.

Adhesives have been used by humans for thousands of years. The natural glues have varied from region to region because of the limited materials available locally. One of the simplest adhesives has been made from gluten in flour or cornstarch. Other binding materials have been made from natural resins or from animals. Glue from animals is made by boiling the connective tissues to make protein colloid glues. Horses have often been used in modern times as a source of animal glue. This type of glue is commonly used in the bottle-labeling process.

Starch glues are used to make corrugated cardboard. Huge rolls of paper are threaded though machines that precisely set the folds and use starch glues in a hot mixture to produce a flowing volume of cardboard, which is then cut to make various-sized boxes or other products. Many other products, especially paper products, use natural glues.

Bioadhesives are glues made from naturally occurring organic materials. The term also may be used to describe processed natural materials that have been converted into synthetic adhesives. The bookbinding industry has long used animal glues.

Adhesives are used by dentists to bond crowns to stumps of teeth. The temporary binding agents are adhesives, and the permanent ones are cements. These types of adhesives have to be safe for humans because they are used in the mouth of the patient. Other uses include cavity liners and dental cement.

Actors use adhesives to attach wigs, moustaches, beards, and other objects such as fake noses. These types of adhesives have to be a chemical compound that will not harm the skin or be absorbed as a toxin. Some forms of skin adhesives are an alcohol-resin mixture. Spirit gum remover is used to remove the adhesive. These types of adhesives are not likely to last very long because the body sheds skin cells constantly and the adhesive sheds off with the skin to which it is attached, most likely within a few days or a week.

Bandages with adhesive gums are very common, a variety of which are used in medical treatments and are made from natural or synthetic rubber, acrylic, silicon, or other materials.

Synthetic adhesives are made from chemicals that have been processed through the necessary chemical reactions to produce the adhesive. Adhesives made synthetically include resins, silicon, elastomeric, thermoplastics, acetate, polyvinyl, polyurethane, and others. The artificial adhesives are designed and sold to meet a vast number of needs. Some are for general use, are nontoxic, and are water-soluble. Children and the general public safely use these types of glues for simple crafts and many other projects.

Some types of synthetic glues may be used to bond ceramic materials such as the broken handle of a coffee mug. Properties have been manufactured into some glues of this type that allow the coffee mug handle to remain bonded even after repeated washings in the dishwasher. Other types or glues have such strength that it is unlikely that the materials will separate once glued together.

Large quantities of adhesive materials are used in construction, such as silicon adhesives used to seal bathtubs or the joints in a shower. Other silicon products are used as putty to seal flooring joints, around windows, or in other locations where air may be exchanged in a building with the outside. The sealing adhesive also blocks the entrance of insects into the building or house.

Adhesives are also used for a variety of flooring materials. Linoleum, vinyl, rock, or even wood products are bonded to the rough subflooring to make a floor covering that will be durable and cleanable. Many of the adhesives used in flooring are contact adhesives. Some are composed of two types of solvent glues that require some time to "set up" before bonding can occur. Some types are used to make laminated materials.

Some adhesives are formed by mixing two chemical compounds together. Many epoxy glues use this method to create the chemical adhesive that is needed for the product at hand. These are emulsion adhesives.

Drying adhesives use chemical mixtures that are dissolved in a solvent. When applied, the glue dries because the solvent evaporates. These types of adhesives usually have weaker bonds than those of other glues, so they make safer household glues.

Other adhesives bond when exposed to ultraviolet light. Emulsion adhesives are used for other purposes. Pressure-sensitive adhesives are often used as the coating of a film that protects a surface.

In addition to efforts to develop natural and more sustainable adhesive materials for commercial use, recent years have seen the emergence of a wide variety of "green adhesives." Efforts to produce and market these have been associated not only with the use of more sustainable raw materials but also with minimizing the environmental impacts of adhesives, particularly with regard to reducing harmful compounds and solvents contained in the adhesives. A number of governments have introduced regulations or restrictions on the chemical emissions produced when using adhesives. These regulations have attempted to place limits on the amount of volatile organic compounds contained in adhesive products, as these compounds are thought to release hazardous air pollutants posing both health and environmental risks. Other products are designed to save energy and minimize waste. Low-temperature hot melt glues, for example, require less energy to melt and apply, and new cardboard and foil-based packaging has also been developed to reduce the landfill waste from plastic tube applicators. Adhesives that improve energy efficiency in the building industry (such as sealants that reduce heat leakage) have also been developed.

**See Also:** Environmentally Friendly; Green Consumer; Packaging and Product Containers.

**Further Readings**

Beswick, R. H. and David J. Dunn. *Natural and Synthetic Latex Polymers Market Report.* Shrewsbury, UK: Rapra Technology, 2002.
Dunn, David J. *Adhesives and Sealants: Technology, Applications and Markets.* Shrewsbury, UK: Rapra Technology, 2003.
Petire, Edward M. *Handbook of Adhesives and Sealants.* New York: McGraw-Hill Professional, 2006.
Wilson, Alex, et al., eds. *Green Building Products: The Greenspec Guide to Residential Building Materials,* 3rd Ed. Gabriola Island, British Columbia, Canada: New Society Publishers, 2008.

*Andrew Jackson Waskey*
*Dalton State College*

# Advertising

Advertising may be considered a form of persuasive communication that promotes market goods and services. Advertising companies engage in informing, persuading, and prompting consumer awareness of the value propositions of particular products and services. Institutional advertising in any media demands payment—it is not personal, advertising's task is to identify the sponsoring company and/or the advertised product or services. Political advertising focuses on promoting a party or candidate with the aim of winning votes. Advertising can also be used for social purposes; for example, urging drivers to wear seat belts, promoting antismoking campaigns, and informing consumers about protecting the environment by specifying a green lifestyle (e.g., recycling, being aware of one's carbon footprint). In general, advertising creates as well as responds to new consumer trends, new media, and (sub)cultural contexts. Thus, the development of green advertising can be seen as a means of both shaping new markets and reflecting a change in consumer preferences and desires.

Advertising is not a modern phenomenon. Ancient history reveals that in early times, people employed advertising to announce and promote their products and services. For example, the Romans promoted upcoming gladiator fights on city walls; in Greece, town criers advertised available goods. From the 17th century onward, with the emergence of newspapers, advertising began to emerge as an independent field. The first newspaper advertisement was published in 1704. During the following decades, this promotional tool spread and cumulated. By the end of the 19th century, having passed through the Industrial Revolution, businesses became increasingly interested in advertising their mass products (e.g., soap, canned food). With competition now growing, the need to "shout louder" gained importance. This resulted in the expansion of products and shopping venues advertising (e.g., the newly emerging department stores). The first advertising agency was opened in 1843 by Volney Palmer in Philadelphia, Pennsylvania.

By the 1920s, not only had advertising grown in volume but it was now spreading from print to radio and later to television. Whereas earlier advertising focused on spreading the word about the benefits of new products, new places, or other offers, by the mid-20th century, advertising began to increasingly highlight products and company images.

Today, the advertising industry is big business worldwide. Some estimate that the amount spent on advertising worldwide is in excess of $604 billion. However, many top marketers reduced their advertising spending in 2008 because of the global financial crisis. In the United States, for example, advertisement spending has dropped since the crisis began and is forecast to drop further. *Advertising Age* has identified telecommunications companies Verizon and AT&T as the top two advertisers (based on U.S. ad spending and ranking) of 2008, after Procter & Gamble, which is the world's largest advertiser.

Advertising, being a commercial business, has the potential to generate a lot of profit, but it is not only business firms that are buying advertising. Nonprofit organizations and governments also advertise to promote their social causes and various agendas. The U.S. government, for example, is one of the largest advertisers in the country. Governments frequently advertise to promote a certain mode of (social) behavior; for example, how to cope with traffic situations both locally and when traveling, and antismoking campaigns.

Phillip Kotler, one of the world's leading marketing experts, urges establishing relationships with customers by communicating customer value as a top advertising goal. Kotler stresses that marketing is not the art of finding clever ways to dispose of what companies make. Rather, it is the art of creating genuine customer value. He suggested that marketing is the art of helping customers become better off. The marketer's watchwords are quality, service, and value.

## What Is Advertising?

Advertising, which may be considered an instrument to communicate customer value, is one axiom of the promotion mix or marketing communications mix. Other promotion elements include sales promotion, personal selling, public relations, and direct marketing. All of these promotion tools need to be carefully coordinated and aligned to ensure the high effectiveness and consistency of the communicated message.

Advertising is a nonpersonal form of mass promotion that mostly speaks to a sizable audience or target group. To communicate a marketing message to a target group effectively, the most commonly used method is to first introduce the new product or service, then increase sales and increase consumer awareness. It is of the utmost importance to establish and maintain good customer relationships. The critical question, however, is how and where a company should advertise, taking into consideration highest impact/effectiveness and (often) a limited advertising budget.

Key advertising decisions are determining and aligning communication and sales objectives, setting an advertising budget, developing an advertising strategy, and measuring advertising success. Any advertising campaign needs to start with determining advertising objectives. What is it that the advertising campaign seeks to achieve within the framework of the marketing communications mix—to increase awareness, to persuade consumers to buy, or to simply act as a reminder? Are these objectives in line with sales objectives? The introduction of a new product should be accompanied by an informative advertising campaign that is usually different from one that advertises a product that is already well known. Take the innovative Apple iPhone as an example. When the product was launched, the first advertisements included a demonstration of a few handset functions—functions that were hitherto not widely known. In competitive markets, advertisers usually engage in persuasive marketing: they have to tell the consumers why their product is better (or of more value to the consumers) than others.

The advertising strategy is the advertising blueprint. It contains the advertising messages—aligned with the advertising objectives—and the selected advertising media. Although in recent years companies are spending less money on advertisements, preferring more personal forms of advertising and new media representation, advertisements per se remain a critical element of the communication mix.

As well as understanding the marketing and/or communication department as a profit center, advertising success needs to be carefully considered vis-à-vis the return on advertising investment. John Wanamaker, often considered the father of modern advertising, once said: "I know that half of my advertising is wasted—I just don't know which half." Given that no company can afford to waste advertising dollars, the effectiveness and accountability of advertising of necessity must be measured. There are two key measurement aspects: one is the communication (persuasion) success of the advertising campaign, and the other is the determination surrounding the success of the advertising, as indicated by the sales objectives. Has that particular form of advertising resulted in an increase in sales and profit? The multiple decisions around budget, communication, and viability of response create a complex terrain in which green advertising and the development of green markets is situated. Green consumption can be seen as a niche but rapidly emerging market, with green products, lifestyles, and practices associated with social responsibility and ethical networks of production linked to particular commodities, firms that are associated with their marketing and production, and the moral preferences and practices of consumers.

## Green Ads, Green Marketing

As already suggested, advertising is strongly linked with consumer trends. Today, increased social environmental awareness is reflected in the creation and objectives of advertising. The American Marketing Association defines green marketing as "(1) [t]he marketing of products that are presumed to be environmentally safe . . . ; (2) products designed to minimize negative effects on the physical environment or to improve its quality; [or] (3) efforts by organizations to produce, promote, package, and reclaim products in a manner that is sensitive or responsive to ecological concerns."

Green marketing is a phenomenon that has its genesis in the 1980s, a time when corporate social responsibility reports and the buzzwords *sustainable development* and *green consumerism* gained popularity. One of the biggest challenges of green marketing (green advertising being among them) is the need to address environmental issues while at the same time satisfying core customer needs. This is one of the challenges "green" ad campaigns have to face—convincing consumers that environmental products do not perform lower than regular products. Rather, they often—with regard to specific features—perform better. For example, energy-saving lightbulbs last longer, offer better convenience (they do not have to be replaced as often as regular bulbs), and reduce energy expenses. Convenience and fulfillment of basic human needs are two of the most significant reasons why consumers buy green products—not necessarily for environmental reasons but for better value (safety, money). This underpins the increasing demand for green products. In the case of green marketing, it is very important to understand the value that individual consumers assign to these types of products. Only after gaining an understanding of these needs can an advertising agency successfully promote its products.

Green advertising campaigns often contain an educational communication message; for example, a message that informs the consumer about the added value of using green products, especially for the community and for future generations. The focus is on the conservation of natural resources; for example, "Help Save Our Environment" campaigns. One of the earliest green messages informed the consumer about how long it took the product to reach his/her shopping cart (time and distance). One popular example was the tracking of the worldwide production of a toothbrush—or of food—and the ecological consequences. The World Wildlife Fund Canada, for example, launched a campaign in September 2009 to reduce the individual footprint by changing lifestyles—for instance, by eating locally.

Many companies ride the wave of green marketing without actually marketing products that are environmentally safe or produced according to "green" requirements. To increase transparency and credibility, well-known third parties experienced in environmental testing can enable identification of green products by affixing certificates or special labels; for example, the Energy Star or the Green Seal. At the same time, increasing numbers of questions are being asked about green advertising claims. "Buzz" and "viral" marketing are gaining importance in this respect. When excessive numbers of products are labeled "green," "eco," or "organic," consumers become wary—they tend to listen more to word-of-mouth—to family, friends, and other trusted people—rather than to commercial advertising claims. The Internet also plays an important role in this respect, providing a medium for debate and exchange over the shaping and validity of green claims for products and firms.

**See Also:** Consumer Culture; Green Marketing; Greenwashing.

### Further Readings

*Advertising Age*. "Leading National Advertisers 2009. U.S. Ad Spend Trends: 2008." http://adage.com/datacenter/article?article_id=136308 (Accessed September 2009).

*Advertising Age*. "The Web Version of the 1999 'Advertising Century' Report." 2005. http://adage.com/century/index.html (Accessed September 2009).

de Mooij, Marieke. *Global Marketing and Advertising: Understanding Cultural Paradoxes*. Thousand Oaks, CA: Sage, 2005.

Garfield, Bob. *Top 100 Advertising Campaigns of the Century*. 2005. http://adage.com/century/campaigns.html (Accessed October 2009).

Ginsberg, J. M. and P. N. Bloom. "Choosing the Right Green Marketing Strategy." *MIT Sloan Management Journal* (Fall 2004).

Hoppe, Ralf. "Die Weltbürste" [The World Toothbrush]. *Spiegel Spezial*, 7:136–41 (2005).

Howard, Theresa. "Being Eco-Friendly Can Pay Economically." *USA TODAY* (August 8, 2005).

Kotler, Philip and Gary Armstrong. *Principles of Marketing*, 13th Ed. Boston, MA: Pearson, 2010.

O'Barr, William M. "What Is Advertising?" *Advertising & Society Review Supplement Unit 1*. 2005. http://muse.jhu.edu/journals/asr/v006/6.3unit01.html (Accessed September 2009).

Ottman, Jacquelyn A., et al. "Green Marketing Myopia." *Environment: Science and Policy for Sustainable Development*, 48/5:22–36 (2006).

Stafford, Edwin R. "Energy Efficiency and the New Green Marketing." *Environment*, 45:3 (2003).

World Wildlife Fund Canada. "Have an Appetite for Change? WWF-Canada Launches Localicious." 2009. http://www.wwf.ca (Accessed September 2009).

*Sabine H. Hoffmann*
*American University of the Middle East*

# Affluenza

"Affluenza" refers to a blending of two terms, *affluence* and *influenza*, and is used by critics of consumerism who consider the costs of ownership of material wealth to far outweigh the benefits. The term was first introduced in a documentary by the same name by John de Graf in 1997 for KCTS in Seattle, Washington, and Oregon Public Broadcasting. Later, de Graf, David Wann, and Thomas H. Taylor, in their book *Affluenza: The All Consuming Epidemic*, defined the term as "a painful, contagious, socially transmitted condition of overload, debt, anxiety and waste resulting from the dogged pursuit of more." It is caused by the epidemic of stress, overwork, waste, and indebtedness caused by the pursuit of a utopian dream of living the good life and results in a sluggish and unfulfilled feeling from efforts to "keep up with the Joneses." It is believed to be an unsustainable addiction to economic growth.

The notion of affluenza is based on the claim that the pursuit of wealth results in economic success, yet begets a sense of remaining unfulfilled and a need for more wealth. The condition is particularly severe among those with inherited wealth, who are said to experience guilt, lack of purpose, and dissolute behavior as well as an obsession with holding onto the wealth. This "luxury fever" also causes waste and harm to the environment, causing psychological disorders, distress, and alienation from the rest of society. Mainstream media, such as television, tend to reflect the pervasiveness of this phenomenon, while simultaneously reinforcing these values as well. Affluenza is supposed to be the greatest in the United States and other developed Western nations, where the culture encourages the citizens to measure their worth in terms of material possessions and financial success. Recent studies have further demonstrated the spread of this phenomenon to other regions of the globe, including developing nations.

## Negative Consequences of Affluenza

On the basis of his interviews in several global cities including Auckland, Copenhagen, Moscow, New York, Singapore, Shanghai, and Sydney, British psychologist Oliver James established that there was a correlation between the increasing nature of affluenza and the resulting increase in material inequality: greater inequality in a society led to unhappiness of its citizens. He further attributed the increase of affluenza to the increased use of needs via manipulative tools used by some unethical marketers.

He also demonstrated a higher rate of mental disorder as a consequence of excessive wealth-seeking in consumerist nations based on data from the World Health Organization that showed that English-speaking nations have twice as much mental illness as mainland Europe (23 percent versus 11.5 percent). Placing a higher premium on "selfish capitalism," that is, on money, possessions, and physical and social appearances and fame is a major reason behind a greater incidence of affluenza.

## Proactive Resistance to Affluenza

As noted in Chuck Palahniuk's movie *The Fight Club*, "the things you own end up owning you." As further noted in the movie, "[W]e're the children of history, man. No purpose or place. We have no Great War. No Great Depression. Our Great War's a spiritual war . . . our Great Depression is our lives." Hence, many individuals are "downshifting" by deliberately reducing their hours on the job in exchange for more leisure, time with family, or other pursuits. This is one approach for someone to break through the compulsive spending lifestyle and lack of savings. Societies are increasingly removing the negative consumerist effects by pursuing real needs over perceived wants and by defining themselves by having value independent of their material possessions. Such endeavors have been reflected in the United States through events such as the national Buy Nothing Day on the day after Thanksgiving and public school credit counseling and media literacy educational programs.

In a recent study comparing 200 individuals who practice "voluntary simplicity" with 200 mainstream Americans, the former were much happier and more satisfied with their lives. The simple livers were also more likely to be careful about spending their money, which could translate into increased financial security. More individuals are designing their own homes, growing their own food, and living simply, yet comfortably. They are making their purchases based on their values and finding that practicing sustainable living and being mindful of consumption is more rewarding. Voluntary simplicity groups are also being formed at work, at places of worship, with neighbors, or with friends. The potential for ecoefficiency in design and production and for environmentally conscious consumer choices is being increasingly sought. Furthermore, the significance of civic engagement through avenues such as socially responsible investing, cohousing, community-supported agriculture, and car sharing are also being encouraged.

**See Also:** Advertising; Consumer Culture; Consumer Society; Disparities in Consumption; Downshifting; Ethically Produced Products; Frugality; Materialism; Morality (Consumer Ethics); Needs and Wants; Overconsumption; Poverty; Quality of Life; Simple Living.

### Further Readings

Andrews, C. *Slow Is Beautiful: New Visions of Community, Leisure and Joie de Vivre.* Seattle, WA: New Society Publishers, 2007.

de Graf, J. and V. Boe, producers. "Affluenza!" KCTS Seattle and Oregon Public Broadcasting. Oley, PA: Bullfrog Films, 1997.

de Graf, John and V. Boe, producers. "Escape From Affluenza!" KCTS Seattle and Oregon Public Broadcasting. Oley, PA: Bullfrog Films, 1998.

de Graf, J., et al. *Affluenza: The All Consuming Epidemic.* San Francisco, CA: Berrett Koehler, 2001.

Doherty, D. and A. Etzioni. *Voluntary Simplicity: Responding to Consumer Culture.* Lanham, MD: Rowman & Littlefield, 2004.

Dominguez, J. and V. Robin. *Your Money or Your Life: Transforming Your Relationship with Money and Achieving Financial Independence.* New York: Viking, 1992.

Elgin, D. *Voluntary Simplicity: Toward a Way of Life That Is Outwardly Simple, Inwardly Rich.* New York: Quill, 1998.

Hamilton, C. and R. Denniss. *Affluenza: When Too Much Is Never Enough.* Sydney: Allen & Unwin, 2005.

James, O. *Affluenza: How to Be Successful and Stay Sane.* London: Vermilion, 2007.

Kasser, T. *The High Price of Materialism.* Cambridge, MA: MIT Press, 2003.

O'Neill, J. H. *The Golden Ghetto: The Psychology of Affluence.* Center City, MN: Hazelden, 1996.

Schorr, J. B. *The Overspent American: Why We Want What We Don't Need.* New York: Harper, 1999.

Speck, S. K. S. and A. Roy. "The Interrelationships Between TV Viewing, Values, and Quality of Life: A Global Perspective." *Journal of International Business Studies* (November 2008).

*Abhijit Roy*
*University of Scranton*

# Air Travel

The idea of green air travel at first seems impossible. Though air travel in 2009 accounted for about 2–3 percent of anthropogenic carbon dioxide emissions, some estimates have annual aircraft emissions tripling by 2050, the same time period during which many efforts are scheduled to reduce emissions from all other sources. Air travel not only consumes vast amounts of nonrenewable fossil fuels but is also a major contributor to nitrogen oxide in the troposphere. The cumulative effect is such that the per passenger, per mile effect of air travel is nearly as great as that of each passenger driving the same distance himself and is significantly worse than that of small or hybrid cars, or ground mass transit (whether rail or bus). Aircraft contrails, high-altitude vapor trails formed when atmospheric water vapor condenses around particles of engine exhaust, has a noticeable effect on climate as well—great enough that the three-day grounding of U.S. air traffic after September 11, 2001, had a measurable effect on atmospheric temperatures (1 degree Celsius). Furthermore, a significant amount of air travel is travel that would not otherwise occur—few of the tourists taking three-day vacations to the other side of the country or visiting the other side of the world would make such trips if they were limited to other modes of transport. Much of this polluting transit, in other words, is dispensable in a sense that many other sources of pollution are not.

Aircraft contrails have such an impact on climate that the three-day grounding of U.S. air traffic after September 11, 2001, had a measurable effect (1 degree Celsius) on atmospheric temperatures.

*Source:* iStockphoto

However, these factors also serve to motivate greener forms of air travel. Fuel efficiency is not only good for the environment, it is good for business, and offering a green passenger flight is a value-added incentive that may attract enough customers to offset the expense of the changes necessary to make that flight more energy-efficient or less environmentally damaging.

There are actions consumers can take, regardless of whether an airline offers greener service. Because takeoff and landing are the most fuel-using parts of a flight, and because indirect flights involve more total travel miles, a direct flight, though often more expensive, will consume considerably less fuel. Using an airplane bathroom consumes about as much gasoline as six miles of flight, so avoiding it helps reduce one's effect as well, as does packing less (or, less easy to adjust with short notice, weighing less). As with any activity involving greenhouse gas emissions, there is an option to purchase carbon offsets to balance out one's effect.

However, the real potential lies in "greening" air travel itself, at the airline and manufacturer level rather than through passengers' activities. In April 2008, Boeing tested the first manned plane powered only by fuel cells, using hydrogen, converted through chemical reaction into electricity and water. There is no combustion engine, no carbon emissions, so although water vapor is not a benign substance at high altitudes, the assumption (or hope) is that water vapor emissions would be significantly less harmful than carbon. The test plane was too small for more than two people, but the Wright brothers started with a small craft too. In theory, such planes can someday be built on the same scale as today's commercial passenger planes.

Planes can be made more fuel efficient, too, regardless of the fuel they use. Some carriers have pressed manufacturers to refine and begin producing a more fuel-efficient design that sets open-rotor twin engines above the rear fuselage, which could potentially reduce nitrogen oxide emissions by as much as 75 percent and cut carbon emissions in half. The problem is that such aircraft are much noisier, and noise affects where aircraft can be used and where airfields can be built. In the long term, despite the negative public image of nuclear power, the ability to harvest uranium from seawater makes nuclear-powered clean aircraft a substantive consideration. There has even been some development in the area of solar-powered aircraft, though it seems unlikely they will be used for commercial passenger or freight purposes.

In the meantime, using biodiesel instead of, or in addition to, conventional jet fuel is another measure airlines can take, which Virgin and others have been exploring since 2007. In the "biofuel jet race," Virgin, Continental, and Japan Airlines have been the main contestants, aiming to find a biofuel that can be used by their existing fleets, with emissions of 40 percent less than traditional (kerosene) jet fuel. Such second-generation biofuels use a combination of plant oils, including those of algae, *Jatropha*, and *Camelina*. Widespread usage of such fuels will depend on the ability of agricultural businesses to supply those plants, which is one reason palm oil is not as favored as it once was—there simply is not enough of it to supply significant demand, and the deforestation resulting from its harvest would offset the gains of using it. This has been one reason Greenpeace and other organizations have opposed many biofuel efforts, and why biofuel projects are sometimes accused of being more concerned with economy than ecology.

The appeal of algae is that it can be harvested in conditions that do not compete with other crops—that is, the decision to farm algae need not be the decision not to farm some other crop. *Camelina* can be grown in rotation with wheat crops, and thus benefits grain farmers without incentivizing them away from traditional food crops. The jury is still out on *Jatropha* and some other biofuel sources, and doubts have been raised as to whether air travel could really be considered green if it were supported by a biofuel industry that consumes agricultural resources needed to meet basic food and energy needs. The benefit of biofuel overall is that it does not necessitate significant overhaul of existing aircraft—fuel-cell aircraft may be more efficient still, but they must be built to replace existing fleets, at a significant cost.

**See Also:** Automobiles; Carbon Emissions; Commuting; Fuel.

**Further Readings**

Jenner, Paul and Christine Smith. *The Green Travel Guide*. London: Crimson Publishing, 2008.

Natural Resources Defense Council. *Flying Off Course: Environmental Impacts of America's Airports*. New York: Natural Resources Defense Council, 1995.

Thorpe, Annabelle. "Is Green Air Travel Becoming a Reality?" *The Observer* (April 13, 2008).

Wardle, D. A. "Global Sale of Green Air Travel Supported Using Biodiesel." *Renewable and Sustainable Energy Reviews*, 7/1 (February 2003).

*Bill Kte'pi*
*Independent Scholar*

# Apparel

Traditionally, apparel purchasing decisions have depended predominantly on price, quality, style, brand reputation, and fabric. In the last 10 to 15 years, consumers have become more aware of the realities of the clothing they wear and may believe anything bearing the Fair Trade label is ethically made.

The growth in ethical apparel is being driven by consumer demand for fairly produced and sustainable goods. The data analysis group Mintel reports that the ethical clothing market is now worth $288 million, with increasingly stylish designs of fairly produced apparel and a growing awareness of production. The demand for ethical clothing has been increasingly met by independent stores and a growing number of major upscale retailers and supermarkets. Among the many apparel retailers, Patagonia (an outdoor clothing company) and Timberland were named the world's most ethical apparel companies in 2008, having demonstrated ethical leadership at the Forbes-Ethisphere Ethical Leadership Forum in New York City. The awarded companies were given the highest scores based on the results of a survey in which they were rated in seven distinct categories: corporate citizenship and responsibility; corporate governance; innovations contributing to the public well-being; executive leadership and tone from the top; legal, regulatory, and reputation track record; internal systems; and ethics or compliance programs.

Timberland is among a group of apparel retailers that were recognized as the world's most ethical apparel companies, based on criteria that included corporate responsibility, contributions to the public well-being, and a regulatory track record.

*Source:* Dimpalen/Wikipedia

The term *ethical* means different things to different people and organizations. The term generally means that workers' rights need to be respected throughout the supply chain.

The main ethical concerns are excessive hours, forced overtime, lack of job security, poverty wages, and denial of trade union rights, poor health, exhaustion, sexual harassment, and mental stress. However, it may also mean promoting ecological and sustainable manufacturing practices that use a different model of production or trade. Some alternative brands are making a genuine effort to challenge the way the garment industry currently operates. Taking Patagonia as an example, the company is pushing a more sustainable model by initiating the Organic Exchange with the sole purpose of training other companies to source and manufacture organic goods. In an effort to provide its customers with greater access to information and to promote transparency, Patagonia created the Footprint Chronicles, a study that traces the life cycle of a product from design through to shipment. One of its most exciting goals is the development of not just fully recyclable clothing but completely biodegradable natural garments that will wind up on a compost heap, as opposed to in a landfill.

As the ethical movement has grown, there is a growing diverse use of the following terms; companies designate ethical apparel under different headings to reflect different types of concern:

- Fair Trade and sweatshop-free apparel: Focus on people, human rights, workers' rights, supply chain policy; aims to give producers in poorer countries a "fair deal."
- Vegan apparel: Oppose animal cruelty, animal testing, factory farming, other abrogation of animal rights.
- Organic cotton apparel: Focus on product sustainability, positive environmental features; that is, apparel might be certified to Soil Association standards.
- Natural environmentally friendly, ecofriendly apparel: Focus on pollution and toxics, habitats and resources, such as apparel that uses only high-quality organic and sustainable materials (e.g., ecodyes); its production typically relies on far fewer agrochemicals.

These terms are used interchangeably based on different ethical consideration. Even in factories that on the surface look clean and modern, workers are often deprived of their internationally recognized basic rights.

For over a decade, the process of globalization has affected large numbers of apparel companies, and companies have reallocated or outsourced their manufacturing in part or entirely to low-cost countries. These companies retain the functions of design, sales, and marketing of new products within their own territory. Increasingly, apparel items are imported from developing countries such as Cambodia, China, Bangladesh, India, Vietnam, Sri Lanka, Thailand, Indonesia, and Indonesia, where many global companies such as Levi-Strauss, Wal-Mart, Nike, Disney, and the Gap benefit from cheaper labor markets. Consumers, workers, and campaigners have called on global apparel brands to make sure the apparel that they produce is ethical or, specifically, sweatshop free. From the 1990s on, the antisweatshop movement spread and many activist groups started to form, including activists from the Clean Clothes Campaign, university students and administrators, labor rights experts from the Worker Rights Consortium, and U.S. working men and women from the Solidarity Center. These and many other activist groups campaigned to stop the production and sales of school uniforms and university logo T-shirts that were made by child workers or in sweatshop conditions. This increase in sensitivity of Western consumers to labor issues in developing countries, as well as heightened public scrutiny from nongovernmental organizations, antisweatshop campaigners, and student groups has created enormous media attention on brand-name products from companies such as Nike, Gap, Disney, and Reebok, and pressure to change their practice. As a result, suppliers are chosen for the ethical way in which they produce their goods, their commitment to Fair Trade, and their environmental policies.

These multinational apparel brands are aware of the potential threat of a boycott resulting from extensive media coverage. In addition, both governments and consumers have challenged apparel retailers in their labor practices, with the result being the adoption and implementation of ethical trading principles such as corporate social responsibility: voluntary codes of conduct to improve the labor conditions of their suppliers in developing countries. To a large extent, a code of conduct is formulated as a supply chain policy and is often based on the International Labour Organisation's (ILO's) standards, which define corporate codes as companies' policy statements that define ethical standards for their conduct. The ILO conventions focus on rights of human beings at work, such as freedom of association, abolition of forced and child labor, and equality. The number of companies adopting these conventions rose from 936 in 1999 to 1,218 in 2003. This ILO system consists of the ILO labor standard, which is endorsed, monitored, and enforced by governments. Despite an increasing number of countries ratifying the fundamental ILO conventions in the last five to 10 years, this type of system is often criticized by trade unions and human rights groups for not having enough power to protect basic worker and trade union rights or to prevent violations. Consumers expect companies to disclose their social and environmental performance in addition to their financial results. Most clothing manufacturers now have sections on their Websites with details of their supplier code of conduct and worker and environmental policies, as well as worker and customer feedback mechanisms.

A number of national or international labeling mechanisms are available to assist branded apparel purchase decisions. One of the earliest related developments, which took place in 1997, was the implementation of the Social Accountability International (SA 8000) standards, which offer a certification mark to suppliers to assure consumers of various quality-of-worklife standards met throughout the process of apparel production. Other recent extensions include the "Sweat-Free T" logo and the "No Sweat" logo. However, despite these developments, many leading apparel retailers are yet to endorse such labeling mechanisms. This is in contrast to the Fair Trade logo, which is internationally recognized by consumers, and which subsequently encouraged leading international retailers to formulate their own Fair Trade label.

Pressure and campaign groups such as Oxfam, the Ethical Trading Initiative and Traidcraft Exchange, "Clean Clothes Campaign," "Mind the Gap," "Sweatshop Watch," and "Labour Behind the Label" regularly flag companies of concern to inform and encourage consumers to buy apparel that is manufactured under fair working conditions. The efforts of such groups and the media publicity generated over unfair labor practices have been successful and have resulted in consumers rejecting certain brands because of their unethical reputation, with the use of sweatshops being one example. This reaction by consumers is thought to have cost U.K. companies, including the U.K. apparel industry, 2.8 billion euros, as indicated by the Co-operative Bank.

This ethical apparel movement phenomenon is still a complex work in progress. Ethical trade works from the top down, where it can only guarantee that the raw materials used have been produced under humane conditions, but the complex nature of the garment industry makes it difficult to guarantee that labor practices meet international standards at every stage of the supply chain, from cotton growing, to cutting and sewing, to finishing and packing. Alternatively, ethical consumption can be a powerful tool for change. It stresses the role of the consumer in preventing the exploitation of workers as well as assessing the environmental costs of production.

As a consequence, for the consumer it is important to be aware of who makes a product and how it is made, who profits from it, what local economies are being supported, whether chains of production are based on a fair compensation for labor and time invested, and the sorts of practices involved in the production of raw and manufactured production. Consumers are also being encouraged through green consumption to reflect on the values embodied in apparel, on issues of need and want, on how long a product will last, and on mechanisms for disposal, including recycling merchandise through second-hand consumption, such as donating items to thrift stores.

**See Also:** Certified Products (Fair Trade or Organic); Ethically Produced Products; Fair Trade; Recycling.

### Further Readings

Clean Clothes Campaign. http://www.cleanclothes.org (Accessed May 2009).
Ethical Trading Initiative. http://www.ethicaltrade.org (Accessed May 2009).
Ethisphere. "2008 World's Most Ethical Companies." http://ethisphere.com/wme2008 (Accessed May 2009).
Fairtrade Foundation. http://www.fairtrade.org (Accessed May 2009).
Labour Behind the Label. http://www.labourbehindthelabel.org (Accessed May 2009).
Mintel. "Ethical Clothes Sales Go from Rags to Riches." http://www.mintel.com/press-release/Ethical-clothes-sales-go-from-rags-to-riches?id=341 (Accessed May 2009).
Social Accountability International 8000. http://www.sa-intl.org (Accessed May 2009).
Soil Association. http://www.soilassociation.org/textiles (Accessed May 2009).
Sweatshop Watch. http://www.sweatshopwatch.org (Accessed May 2009).

*Joyce Tsoi*
*Brunel University*

## Audio Equipment

The last several years have seen an increase of green audio equipment. As with other high-end consumer goods, the challenge is to design a product that serves two masters—creating goods that are energy-efficient and environmentally responsible, while also ensuring that performance is at a high enough standard to satisfy demanding customers. Because the initial cost of green audio equipment is likely to be higher than equipment of similar performance, the assumed customer base consists of people who already purchase high-end audio equipment, and those consumers have exacting standards.

As with cell phones, audio equipment can be designed to encourage consumers to recycle rather than dispose of the goods. Components can be built to break down easily in processing so that metals and chemicals do not become trapped in landfills.

Every year, millions of pieces of audio equipment are sold, including MP3 players and other personal media players, stereos and stereo components, speakers, and home entertainment centers, ranging from small 50-Watt devices to powerful 1,000-Watt behemoths. A fully configured high-end system for audiophiles can easily exceed 2,000 Watts. Power consumption varies, but unlike many appliances, audio equipment is often in use for hours every day, which translates into thousands of kilowatt-hours and hundreds of dollars in utility bills. Because high-end audio systems are composed of multiple components—for example, multiple speakers, a CD player, a tuner, woofers and subwoofers, amplifier and preamplifiers, a turntable, and an MP3 player port—there is a great deal of potential for energy inefficiency. One of the first steps toward greener audio is simplifying the system and integrating components. Integrated amplifiers can be used instead of separate amplifiers and preamplifiers. This will often lead to a better-sounding system as well: the more components that are involved in a system, and the more combinations of different brands and models of each component there are, the greater the odds are that the consumer will construct a less-than-ideal combination of those components.

Speaker efficiency is a key part of green audio and a widely misunderstood aspect of audio equipment in general. Speaker efficiency is measured in decibels (dB), a logarithmic unit of measure. A 3-dB increase in efficiency means the speaker needs only half the power. The difference between an 85-dB speaker and a 100-dB speaker, then, is that the latter uses 32 times less power than the former—a staggering difference. Casual consumers, even those spending a good deal of money on their equipment, tend to overlook this. High-efficiency speakers have not historically been valued, partly because audiophile audio systems have typically been power-hungry by definition, involving multiple inefficient components, and partly because wattage is more typically emphasized. But higher-efficiency speakers do not simply use less power, they work better with other components at a wider range of volumes without resulting in distortion. Using more efficient speakers makes for a more efficient sound system.

Another innovation leading to greener audio is the use of class D amplifiers. Once used only for low-end audio equipment like telephone handsets, class Ds have increasingly come into use across the spectrum. Traditionally, class AB amplifiers have been used in most audio equipment. Class AB amplifiers run at 15–40 percent efficiency; much more efficient class Ds run at 40–85 percent. The gain of energy efficiency means more than just a lower electric bill and fewer emissions produced: Because the wasted energy would have been shed as heat, the heat sink can be reduced or eliminated, allowing for an audio component or system with a smaller physical footprint. One of the reasons flat-screen TVs have such crisp audio is because of their use of class D amplifiers, which eliminate the need for the significant amount of space that the audio components took up in older traditional televisions. Smaller systems lead, in turn, to lower shipping costs.

Though lagging somewhat behind green computing and other green electronics, there is a growing amount of Energy Star–certified audio equipment. Energy Star is a standards program maintained by the U.S. Environmental Protection Agency, with energy efficiency standards set according to product type. Less common than the Energy Star–certified audio equipment is equipment that is made without toxic materials; this is a sector that needs expansion. Almost all audio equipment contains amounts of lead, cadmium, mercury, hexavalent chromium, and polybrominated biphenyl—hazardous substances that endanger the soil when the equipment is thrown away. As with many consumer electronics products, wasteful packaging is also a perennial problem.

Components like the Vers 2x iPod dock use wooden casings, which are not only environmentally friendly but help with producing bass tones. The Vers 2x is free of toxic materials, sold in recycled packaging, and uses energy-efficient components. The few energy-efficient amplifiers on the market achieve their efficiency in part by shutting down power to unused channels—a seemingly obvious but rarely implemented technique. Audio equipment consumes a surprising amount of power when turned on but not actively in use—one of the substantial areas of its inefficiency. Properly configured, Energy Star–rated equipment can consume less than a watt in stand-by mode.

A good deal of waste and resources consumption results from physical media: CDs, DVDs, and vinyl albums. There is not only the physical material involved and the energy used in manufacturing them but also the fuel expense of transporting crates of the media across the country. One ecological advantage of digital media, such as MP3 files, is that it bypasses wasted packaging, plastic, and chemicals.

Several MP3 players have been introduced that are powered by alternative methods. Kinetic-powered MP3 players like the Chukka use a kinetic generator, generating power simply through motion, which is ideally suited for an MP3 player used during workouts or walks—particularly a flash-based MP3 player less prone to skipping than hard-drive models. Other MP3 players can be charged up with a crank—one minute of cranking provides varying amounts of power, depending on the model. (A Japanese crank-powered MP3 player has a 1:15 ratio of cranking minutes to playing minutes; a premium crank-powered British player enjoys a large 1:40 ratio.) The same approach has been used to power other devices, such as emergency flashlights and radios, for decades. However, the battery capacity often is not high enough for it to be charged with as much power as a conventional MP3 player, forcing the user to stop and crank more often than they would have to stop and recharge.

Typically sold for outdoor use, solar-powered radios are also produced, some with a hand crank for additional power. Because these are considered only for special use, they are not considered to be strong alternatives to existing standard options.

**See Also:** Mobile Phones; Resource Consumption and Usage; Television and DVD Equipment.

### Further Readings

Brower, Michael and Warren Leon. *The Consumer's Guide to Effective Environmental Choices: Practical Advice from the Union of Concerned Scientists*. New York: Three Rivers, 1999.

Garlough, Donna, et al., eds. *The Green Guide: The Complete Reference for Consuming Wisely*. New York: National Geographic, 2008.

Rogers, Elizabeth and Thomas Kostigen. *The Green Book*. New York: Three Rivers, 2007.

*Bill Kte'pi*
*Independent Scholar*

## Automobiles

The gasoline-powered automobile was introduced in the 1880s, but by the time it reached its first century, it had already become implicated in a range of environmental problems. As the largest, most complex consumer good, the consumption of automobility is also often one of the areas of greatest impact of personal consumption patterns in industrialized nations. Initially, industry saw the environmental debate as yet another temporary fad that could be addressed by technology and then quietly forgotten about. However, increasingly people realized that the environment was not some external entity deserving of our benevolent protection. Instead, at issue was our own living environment—mankind's ability to live on a planet that could survive perfectly well without us. As this realization spread, and with it social and legislative pressure on the car, the debate became incorporated into motor industry strategy such that the car and its use became increasingly shaped by environmental requirements. At first the debate focused on toxic emissions from car exhausts, which was reflected in the legislation that followed. Over the years, the scope of the debate widened to include other issues such as energy use, raw materials use, traffic congestion, and greenhouse gas emissions, as well as end-of-life issues such as recycling, reuse, and vehicle durability. As a result, a more global assessment of the car's impact became possible, leading to a so-called life cycle approach to the problem.

In the car market the success of the Toyota Prius hybrid has shown the power of green consumption; one of the car's strengths is in its distinctive styling.

*Source:* Toyota

## History

Perhaps surprisingly, environmental arguments have surrounded the car from the very beginning. Many welcomed the opportunity of reducing or even eliminating the growing urban problem of horse manure, and dead horses in the street were not uncommon either. It is estimated that at the turn of the century around 15,000 horse carcasses were removed from New York streets annually. Horses also deposited over 1,000 tons of manure and 225,000 liters of urine every day. By 1908, the cost of cleaning up this waste was put at $100 million a year. In London, some 5,000 tons of horse manure had to be removed from the city every day. Although the agricultural sector formed a ready market, it was still regarded as a major problem.

Contemporary environmental arguments centered around the then-fashionable concepts of hygiene and fresh air. The cities, with their deteriorating air and water quality, as well as the side effects of the horse economy, did not have much to offer in this respect and were considered major sources of bad "miasma." Similar to the train and bicycle before it, the car provided a means of escaping the "bad" air of the city in favor of clean country air, with all its perceived health benefits. In fact, some medical authorities at the time argued that the greater speed enjoyed by those traveling in a car allowed the ingestion of larger quantities of clean air. Clearly, the new ability for city dwellers to become aware of the countryside and nature in general did much to promote a love and understanding of the natural environment. This trend started with the bicycle—the first mechanical mode of individual transport—and was picked up by the car. The car was promoted with the same arguments of health and freedom as the bicycle, and many wealthy cyclists quickly transferred their allegiance to the new vehicle, which could take them much farther afield than its nonmotorized counterpart. The sportsman image of the bicycle was also transferred to the car, which meant that it had a head start in being associated with healthy outdoor pursuits and in adopting a positive and healthy aura. If this seems ironic to the modern observer, it must be remembered that environmentalism itself would probably not have developed without the car and its ability to take people to areas of nature they might otherwise have been largely unaware of, let alone appreciative of.

In contrast, even at that time, many saw the potential for negative effects from the car. The most immediate effect was the dust generated by cars on the often-unpaved roads of the period. The Dutch firm of Spyker was one of the first to try to address this problem by offering a fully enclosed undertray for its chassis from 1905, improving airflow under the car and reducing turbulence. The "Dustless Spyker" concept was widely used in the company's advertising in the first decade of the 20th century and may have been one of the first attempts by a carmaker to address an environmental problem through a product engineering solution. In California the solution to dust generation—which many citrus growers blamed for reduced crops—was sought in spreading oil—of which plenty was available in California at that time—on dusty roads.

The more perceptive also saw the potential harm in exhaust emissions from petrol engines, and for this reason, even in the 1890s, some rated the prospects for electric cars more highly than those for cars with gasoline engines, especially in an urban context. At an international hygiene conference in Berlin in 1907, gas fumes were identified as particularly damaging to health. At this time, soot, carbon monoxide, and unburned hydrocarbons were singled out as the harmful substances, and their toxicity had already been shown by animal experiments. However, accidents were identified as the greatest danger to health. As car use grew, the effect on the countryside was further exacerbated by the growing infrastructural requirements of the new technology. Cyclists first started lobbying for improved roads, with organizations such as the Cyclists Touring Club in the United Kingdom and the Algemene Nederlandse Wielrijdersbond (General Netherlands Cyclists Club) in the Netherlands initially regarding lobbying for better roads as their primary objective. The state of California set up its Bureau of Highways in 1895 as a direct response to lobbying by cyclists in the year when the first car was seen in Los Angeles. Nevertheless, the real impetus came with the car and the advent of motoring. The dust problem generated by car use was significantly worse than the environmental impact of cyclists. This meant that it was not just the road users who benefited from road improvement but also those who lived or worked near the roads.

## The Rise of Motoring

Similar to the bicycle, the car was initially marketed as an adventurous machine for the sports enthusiast. Gradually, practical considerations were introduced, and the car also became a means of transport. Various professions envisioned business opportunities offered by the car. Taxi firms were the first users of the car purely as a means of transport, whereas commercial vehicles, often based on a standard car chassis, also developed. Commercial travelers, doctors, and veterinary practitioners were among the first to adopt the car for professional reasons.

By 1907, France, the United Kingdom, and the United States had become the world's most motorized countries, with one car per 640 people in the United Kingdom, one per 608 in the United States, and one per 981 in France. In comparison, Germany had one car per 3,824 people. By 1910, Belgium—another country that motorized early on—had one per 1,180. Japan had fewer than 200 cars in total. Ashleigh Brilliant, in his book *The Great Car Craze,* explained how it was Southern California that first saw the development of a mass motorization phenomenon. In doing so, it established many of the values and problems now associated with automobility as a mass phenomenon. The craze soon spread across the United States, but the rest of the world took longer to follow. By 1930, car ownership in Germany was 10.6 per 1,000 people, compared with 31 in France.

Even by the turn of the 21st century, car ownership in the United States was still higher than anywhere else, although Italy beat Canada and Germany into second place during the middle of the 1990s. Nevertheless, mass motorization was still largely confined to the developed industrialized countries. In most developing countries, the ratio of commercial vehicles to cars is much higher than in the developed world because the need to transport basic goods develops long before a demand for luxuries such as cars. In fact, by the end of the 20th century, it was still the case that more passenger miles were traveled by bicycle than by car worldwide. By 1950, the global car and truck population had reached some 50 million, which worked out at roughly two vehicles for every 100 people. However, by the middle of the 1990s, that figure had risen to over 600 million, or 8.2 vehicles per 100 people. A worldwide car and truck population of one billion was reached by 2010. If this trend continues, then by 2050 there could be over 3 billion cars and trucks worldwide. This equates to around 20 per 100 people and is still well short of the 1990s U.S. ratio of 70 vehicles per 100 people. The question really arises of whether this level of automobility is sustainable.

## Cars and the Environment

After the initial environmental concerns surrounding the motor car, the spread of motorization largely marginalized the critics. Environmental concerns resurfaced in a different form at times of crisis. Shortages of cars and fuel during World War II led to various alternative solutions. Conventional cars were converted to run on gas generated from coal or wood via a heavy apparatus fitted to the front or rear of the vehicle. A limited revival of electric vehicles also occurred, as these were less dependent on imported oil, whereas others opted for human power. Aircraft pioneer and luxury carmaker Gabriel Voisin produced a pedal car for his own use during the German occupation, which even carried a smaller version of the "cocotte" radiator mascot of his cars.

A second wave of renewed interest in the environment came in the wake of the hippy era of the late 1960s. Social movements at this time generally rejected the established value system. Rachel Carson's landmark book *Silent Spring* (1962), an indictment of the overuse of harmful chemicals in agriculture, as well as the growing air pollution problem, started a new concern for the way human activity affects our natural environment, which made a good fit with 1960s philosophy. At the same time, new concerns about traffic congestion led to a series of experimental "city" cars. Many of these, such as Ford's Comuta of 1967, were powered by electricity. This period saw the first wave of environmental legislation affecting the car, with California adopting a pioneering role.

The next wave of environmental concern was probably more of a reinforcement of trends started in the previous decade. The energy crises of 1973–74 and 1979 really concentrated the minds of the industry. The psychological effect was perhaps greatest in the United States, where energy use had never been an issue. Now, real shortages at the pumps, and the realization that America's mobility was largely in the hands of minor Middle

Eastern powers, led to real change. First of all, the market share of more fuel-efficient European and especially Japanese products increased markedly. Second, U.S. carmakers rapidly introduced a product "downsizing" program. Early attempts, such as American Motors' Pacer, turned out smaller, but no lighter, than their predecessors. By the end of the century, though, U.S. cars were markedly smaller and lighter than their ancestors of the early 1970s, which were by then regarded as "dinosaurs." However, U.S. buyers then turned to light trucks, which are much heavier than cars, thus largely negating the gains made in more efficient car design.

The spread of mass motorization created essentially two classes of motorist: the genuine car enthusiasts—carrying on, perhaps, the sportsman motorist tradition—and the mass of car users, who regarded the car as "a means for getting from A to B," a status symbol, a fashion accessory, or a mobile drawing room. This development has marginalized the genuine enthusiast to some extent, prompting the development of cars that, though more reliable, are often less enjoyable and less involving as driving machines. Ironically, environmental pressure may well rectify this—an energy-efficient lightweight car also tends to be environmentally optimized. The Lotus Elise—on which the electric Tesla is based—is a good example.

During the 20th century, the focus of environmental lobbying and regulation has been very much on air pollution and vehicle emissions. Within this narrow perspective, one fact has often been overlooked; namely, that the environmental impact of the car is much greater than emissions alone. Much of this impact can be reduced by making cars last significantly longer, thus minimizing, for example, the 20–30 percent of the lifetime energy use accounted for by the production and dismantling phases. Apart from classic car enthusiasts, the industry has few supporters of this approach, as it would lead to a reduction in new car production—something that the industry still finds difficult to address. This may change in the future as the realization spreads that maintaining and managing cars in the market over a long lifetime is potentially more profitable than making and selling new cars. Profit margins on new cars are painfully thin for both manufacturers and dealers.

At the end of the 20th century, these other aspects began to receive greater attention with the move toward a life cycle approach to environmental impact analysis. This has the potential to dramatically change the car, as the need for drastic weight reduction will force the use of alternative materials, such as plastics, and alternative construction methods. The dwindling oil reserves and continuing concern over air pollution will lead to far-reaching changes in power train by the middle of the 21st century at the latest. During their history, there has been a trend for cars to undergo a transformation from being primarily mechanical devices to being more and more electric and electronic devices. Electric cars are the logical outcome. The main concern of car manufacturers will be the cost of such changes and the risks involved in selling radical new technologies to skeptical consumers. Engineering and product development strategies are already geared toward preparing for this revolution, and prototypes of such environmentally benign and more sustainable vehicles have been built. Several of these prototypes have been shown as concept cars.

## Car Consumption

With the steady rise of consumerism, informing the consumer is increasingly important. From the 1970s on, we can discern the development of a green consumer. Although still rare by the turn of the century, the green consumer is more real in some countries than in others. In practice, a growing number of consumers have started to take environmental considerations into account in their purchasing decisions; however, other elements often override a purely environmental choice. For many green consumers, owning a car is a contradiction in terms. However, in the car market, the success of the Toyota Prius hybrid has shown the symbolic power of green consumption; the car's strength is in its distinctive styling, which means the green consumer does not only know he or she is being green but is seen by others as being green. The car thus enables the displaying of a green, environmentally responsible image.

A problem is the general lack of information available to the consumer. The complex problems surrounding the car's effect on our environment still baffle many within the industry, let alone the car buyer. In response, a growing number of environmental rating systems have appeared; however, few of these take a true life cycle view. In fact, the information needed to carry out a life cycle analysis of a car is rarely available, even within the car industry. The European Commission is trying to introduce such a system gradually during the early years of the 21st century. If a meaningful plan emerges from this process, consumers will for the first time be able to make a truly informed choice. Individual firms have also tried to address this. Volvo was among the first to publish details on the environmental performance of its cars for use by car buyers. These Environmental Product Declarations were available at dealers and were displayed alongside more conventional product catalogs from the 1990s.

Another development involves changes in the nature of car consumption. The phenomenon of car sharing schemes, or car clubs, started in Switzerland in the 1980s. In the 1990s, the Scottish capital, Edinburgh, caused a stir by introducing a centrally located housing scheme for non–car owners only. As part of the package, new residents received membership in a newly formed local car club. The world's largest provider of this type of service is Massachusetts-based Zipcar. In 2007, it merged with Flexcar of Seattle, Washington. By 2008, Zipcar had around 180,000 members and some 5,000 cars and also operated in Europe. In North America, it is particularly successful in the more densely populated East and West Coast cities, such as New York, Boston, San Francisco, Seattle, and Vancouver.

One of the first car sharing schemes was run in the Dutch capital of Amsterdam in the 1970s. This scheme, dubbed Witkar, used unique and rather novel battery electric vehicles. Although the motivation was their zero-emissions nature, it also introduced members to electric vehicle technology. More recent is the "Move About" concept in Norway (http://www.moveabout.no), a car sharing scheme designed to use Th!nk electric vehicles to deliver an urban mobility package. Asian markets feature many large, densely populated cities in which people may aspire to car ownership, but lack of physical space for parking prevents full purchase. However, access to a car sharing scheme here would allow such people access to cars, thereby giving them hands-on experience and extending the number of motorists, while attenuating some of the negative effects of there being too many cars. Car manufacturers would still sell cars, or could even run such schemes themselves. However, the main point is that car sharing schemes, or car clubs, could be used as a reduced risk means to introduce new, low-carbon vehicle technologies to consumers. Taking no action to make the car more sustainable may render it nonviable in a matter of decades.

**See Also:** Carbon Emissions; Consumer Ethics; Fuel; Green Design; Product Sharing.

## Further Readings

Brilliant, A. *The Great Car Craze; How Southern California Collided With the Automobile in the 1920s*. Santa Barbara, CA: Woodbridge, 1989.

Carson, R. *Silent Spring*. New York: Houghton Mifflin, 1962.

Flink, J. *The Automobile Age*. Cambridge, MA: MIT Press, 1988.

Nieuwenhuis, P. and P. Wells. *The Death of Motoring? Car Making and Automobility in the 21st Century*. Hoboken, NJ: John Wiley & Sons, 1997.

*Paul Nieuwenhuis*
*Cardiff University*

# B

## Baby Products

At its core, the green baby-care market merges two larger social and parental concerns: the baby's health and ecological health. These two responsibilities coalesce in a shift from consuming conventional infant and toddler commodities to either more sustainably produced—but still purchased—alternatives, or to homemade equivalents. This phenomenon results in part from a more holistic conception of the environment as not simply the backdrop of life but intricately constitutive of life itself, such that pesticides sprayed on cotton fields not only pollute nearby waterways but also enter into the cotton material itself, and into all its subsequent permutations as pajamas, bibs, and receiving blankets. Recognition by consumers of the consequences of their actions in both the purchasing of and the waste created by baby products, and concerns about product toxicity in the context of the construction of notions of infant care and vulnerability in media discourse and consumer culture, have also influenced the growth of green baby products.

Hence, the recent proliferation of expressly sustainable infant- and children-specific products has formed part of an ecobaby boom. One of the first elements of this new market has been organic cotton baby clothing, bedding, and blankets, advertised alternately as green, natural, low-impact, and safe. Such certified organic cotton fleeces, booties, and sheets also usually boast of natural colors made from nontoxic dyes.

Shortly thereafter, biodegradable diapers appeared on natural food shelves, as did nontoxic baby wipes. The green beauty industry has quickly embraced maternity and infant clothing, with an expanding selection of organic, herbal and/or paraben-free massage oils, stretchmark creams, diaper rash powders, and baby shampoos made without sodium laureth sulfates. Meanwhile, the organic food industry now offers an array of chemical-free, nongenetically modified baby food products and formulas, though new mother magazines and Websites are including more and more recipes for homemade baby food. Some green-minded companies also concentrate on ensuring recycled or recyclable packaging of such products to avoid excessive energy and resource consumption and waste.

Infant mattresses have become another area for green improvements. Conventional baby bedding, made from polyurethane foam, is made by reacting chemicals known as isocyanates and polyols with other chemicals that act as stabilizers, catalysts, surfactants, fire retardants, colorants, stain repellants, and blowing agents, each of which have hazardous ecological and health consequences, particularly for infant immune systems. Baby mattresses made of organic cotton, flax, wool, and even horsehair filling have emerged as alternatives to nonhypoallergenic foam beds, which trap dust as well as moisture, and thus mildew and mold, creating breeding grounds for dust mites and other bugs.

As asthma and allergy cases steadily rise in children, all aspects of the baby's home environment have been reviewed through the green lens. Many parents wanting to engage in green discourses are avoiding products with toxins, choosing ecofriendly furniture, clothing, home decorations, cleaning supplies, bedding, and bathing products. Others opt to purchase and reuse second-hand commodities as a way to reduce their consumption of manufactured goods.

Recent scientific evidence has linked bisphenol A (BPA)—a chemical used in plastic bottles, as well as in canned food and beverage linings (including baby food jar lacquers)—to serious and long-term health problems. Accordingly, the baby bottle industry has since opted to eliminate BPA from its products, though some states have already instigated bans on the sale, manufacture, or distribution of infant formula and baby food in BPA-laced containers. The U.S. Congress has also begun to impose stricter standards on phthalates—a chemical substance found in many conventional toys—in children's products.

Adding to the notoriety of the seemingly innocuous toy market has been a series of recent, international recalls wherein toy manufacturers from China to the United States admit the use of lead paint in toddler playthings. Parents and politicians have begun to demand high safety standards for products, and this is being reflected in legislative changes; for example, the 2009 Safe Baby Products Act was introduced to the U.S. Congress with the intention of directing the Secretary of Health and Human Services to study the presence of impurities and contaminants in cosmetics and personal care products marketed to and used by children.

Reducing the amount of plastics used in baby products has also been a part of green consumption. The sustainable babycare movement and producers of baby products have both been active in this regard, producing natural rubber pacifiers, organic cotton dolls, and BPA-free glass baby bottles.

For consumers, buying green products or services such as "ecobaby proofing," in which contracted public health professionals are hired to decontaminate a house, may be expensive. An alternative way of engaging in green consumption to reduce children's exposure to potentially toxic commodities and to reduce carbon footprints is to consume less. Breastfeeding rather than bottle feeding, thrift store purchases, childcare co-ops, and exchanges of goods and services with other ecominded parents can help promote these goals.

**See Also:** Apparel; Consumer Activism; Green Consumer; Green Marketing; Linen and Bedding; Toys.

### Further Readings

Baby Earth. http://www.babyearth.com (Accessed November 2009).

DK Publishing. *Green Baby*. New York: DK Publishing, 2008.

Fassa, Lynda and Harvey Karp. *Green Babies, Sage Moms: The Ultimate Guide to Raising Your Organic Baby.* New York: NAL Trade, 2008.

Rider, Kimberly. *Organic Baby: Simple Steps for Healthy Living*. New York: Chronicle Books, 2007.

*Garrett Graddy*
*University of Kentucky*

# Beverages

Consumer product businesses are significant users of natural resources, as their output relies on a range of nature's goods, such as water, forestry, and agricultural products. From the high emissions levels, energy consumption, and water usage at production facilities to their packaging and solid waste materials, their environmental footprint is among the largest of any in the global economy. When viewed as a supply chain from raw materials through consumption, their direct and indirect environmental impact continues to grow significantly.

Within consumer products, the beverage industry has arguably the most recognizable brands in any industry. Through billions of dollars invested in marketing over the years, brands like Coke and Budweiser have become icons around the globe. Beverage industry revenue in the United States alone is over $200 billion and growing. Beverage products are deeply rooted in consumer lifestyle activities. Even in difficult economic times, consumers rarely sacrifice their morning coffee, a cold bottle of water to quench their thirst, or a glass of wine over dinner.

The beverage industry is considered a critical player in the environmental sustainability movement. The examples here provide a perspective on the environmental challenges facing the industry.

- Millions of barrels of oil are used each year in beverage industry production and distribution.
- Increased complexity in global supply chains has effectively resulted in higher energy use and emissions output.
- Raw materials used in production, such as the corn needed for corn syrup, disperse significant amounts of chemical fertilizers and pesticides into our rivers through wastewater runoff.
- Millions of beverages are sold in plastic bottles every day, contributing to the issue of waste product disposal.

The beverage industry is challenged as much as any consumer products industry to aggressively pursue sound environmental practices while simultaneously addressing consumer demands and achieving reasonable revenue growth. Heavy reliance on natural resources traditionally considered commodities such as oil, water, and corn creates reduced margins for the beverage companies as raw material input prices increase. Meeting the rapidly increasing demand from consumers in developing countries dictates higher energy and raw material usage, while requiring significant capital investment to expand. Consumer lifestyle activities have increased consumption trends toward convenience and fitness. The beverage industry has seized the opportunity to produce reduced-calorie and reduced-serving size packages, thus increasing the waste disposal problem.

A large percentage of beverage industry products are sold to the consumer through chain retailers. As those retailers attempt to reduce their own costs through supply chain optimization initiatives, certain costs are shifted back to suppliers, further compressing profit margins for the beverage industry. For example, in an attempt to reduce waste disposal issues after consumption, Wal-Mart has initiated efforts to dictate the type of packaging materials suppliers like beverage manufacturers can use in their stores.

In response to consumer and public demand, the beverage industry has initiated programs to reduce its environmental footprint. Select examples illustrate the proenvironment direction the industry seems to have embraced:

- leveraging their aluminum recycling competencies to establish aggressive recycling objectives on all packaging materials, including plastic- and paper-based containers;

- reducing water usage through efficient water management programs in production processes;
- partnering with organizations that certify raw material inputs are grown using sustainable farming practices designed to protect the planet's natural resources; and
- investing in alternative energy sources in production facilities and hybrid vehicles to reduce reliance on fossil fuels and carbon emissions.

**See Also:** Bottled Beverages (Water); Consumerism; Packaging and Product Containers.

### Further Readings

Fuhrman, Elizabeth. "Environmental Approach: From Sustainability to Recycling, Beverage Packaging Keeps Getting Greener." *Beverage Industry* (October 2006).

Roth, Bill. "In Beverage Industry, Sustainability Sells." (January 30, 2009). http://www.entrepreneur.com/management/greencolumnistbillroth/article199840.html (Accessed September 2009).

Theodore, Sarah. "Sustainable Strategies: Beverage Companies Take on Environmental Concerns." *Beverage Industry* (August 2007).

*Timothy P. Keane*
*Saint Louis University*

## Biodegradability

*Biodegradable* generally refers to products that can be broken down by natural processes and biological organisms, including invertebrates, fungi, and bacteria. Until recently, consumer appreciation of the issues behind biodegradability was limited for a number of reasons. The proliferation of terminology and the complexity of the different processes involved prevented the general public from engaging with this feature of sustainable practice. As a result of this lack of understanding, the purchase decision-making process had not been duly influenced by considerations of the lifecycle of products or packaging, and the possibilities of reuse or recycling. However, as the related terminology—degradable, biodegradable, and compostable—entered more general use, consumer awareness of the terms grew, though consumers still lacked a complete understanding. This has been stimulated by government campaigns encouraging more household recycling, coupled with local government recycling plans and curbside collections. Retailer and consumer lobbying actions, particularly those concerning packaging and the reduction of single-use plastic bags, together with increasing media attention have also contributed to greater consumer engagement.

Consumer confusion toward biodegradable material results from the use of terms without proper explanation or communication to inform and educate the public. *Recyclable*, *degradable*, *biodegradable*, *compostable*, and *home compostable* are terms all currently used by manufacturers, but studies in the United Kingdom indicated that just 52 percent of the population are aware of biodegradable plastic packaging and only 15 percent are aware of compostable packaging, which also implies that consumers would not know how to dispose of the different materials. Only 10 percent of those surveyed always looked for disposal information on packaging, while 55 percent never read disposal information. Consumer knowledge of the potential negative effects of biodegradable and compostable plastics is also lacking. For consumers, the terms imply that the material just disappears without them having to consider the fossil fuels and energy that are used in their production, or the consequences of their breakdown in the environment, which may result in carbon dioxide, methane or metal toxins being released into the ground, water or air.

When better informed of the consequences of biodegradability and the options for disposal or recycling, evidence suggests that consumers are willing to take action to avoid negative consequences, such as avoiding sending such material to landfills, particularly to avoid the production of methane, a more potent greenhouse gas than carbon dioxide. Biodegradable bags tend to be less strong than other alternatives, and can result in consumer dissatisfaction when they break or tear during use. When given the informed choice between biodegradable and home compostable, more consumers are in favor of composting at home, which may lead to establishing clearer standards for composting, such as the European standard EN 13432.

Consumer mobilization in different areas has led to usage reduction or elimination of all types of plastics: nonbiodegradable, biodegradable, and degradable plastics. For example, in 2006 the Women's Institute of the United Kingdom held a Packaging Day of Action, urging members to demand the removal of unnecessary and excessive packaging on food products, and to use only compostable and recyclable materials where packaging was required.

In 2007, after retailer Asda-Wal-Mart introduced Asda's Extra Special Tea Bags made from a nonbiodegradable, translucent, nylon mesh, there were online protests from composting enthusiasts. Although such bags were previously sold in specialty tea shops, it was the first time they became available in mainstream stores and had come to the attention of the wider general public.

In one instance, which gained worldwide attention, on May 1, 2007, the modest town of Modbury in Devon became the first town in Europe to stop using plastic bags. Rebecca Hosking is a wildlife filmmaker who worked for the BBC Natural History unit on productions such as David Attenborough's *Life of Mammals*. Hosking was born and raised in the area and was so distraught by how plastic was destroying marine life around the planet that she decided to act locally on her return to the United Kingdom from filming in Hawaii. She managed to persuade all 1,500 residents and traders of her town of Modbury to stop using plastic bags in favor of more sustainable, longer lasting alternatives.

Biodegradable material does not just refer to plastic and packaging, though much attention has been given to these aspects of biodegradability. Food waste, human waste, and animal waste are also termed as *biodegradable*. The consumer perception of this type of biodegradable waste is that it is not toxic or noxious to the environment and can be thrown away with the rest of nonrecyclable garbage in landfills, and consumers are unaware of the problems of leachate and methane. National and local governments are working to redress this misperception. For example, the EU Landfill Directive has set clear restrictions on the amount of biodegradable waste that will be sent to landfill by 2020. As a result, local councils are starting awareness campaigns promoting the use of cloth diapers instead of disposable diapers, which are responsible for tons of untreated sewage entering landfills each year.

Information available on the subject of biodegradability and its implications have been criticized by the media and professional associations. The U.K. Advertising Standards Authority (ASA) and the U.K. Department for Environment, Food and Rural Affairs (DEFRA) have produced guidelines for retailers for more effective communication with consumers to avoid misleading marketing, erroneous claims, and greenwashing, in particular for the terms *biodegradable* and *compostable*. For example, a retailer falsely claiming its bags as biodegradable would be objectionable not only on the grounds that it might mislead environmentally concerned consumers into choosing to shop there, but also on the grounds that it might cause consumers to throw the bags away in the countryside, which would constitute social harm. In addition to government initiatives, local authorities have also engaged consumers with informative campaigns to reduce confusion, including the standardization of labeling, logos such as the "Recycling Now" icon, color coding, and instructions for disposal and collection.

**See Also:** Packaging and Product Containers; Recyclable Products; Recycling.

**Further Readings**

Albertsson, A. C. and S. Karlsson. "Chemistry and Biochemistry of Polymer Biodegradation." In *Chemistry and Technology of Biodegradable Polymers*, G. J. L. Griffin, ed. London: Blackie Academic and Professional, 1994.

Arshady, R. "Biodegradable Polymers: Concepts, Criteria, and Definitions." In *Biodegradable Polymers*, R. Arshady, ed. PBM Series, Vol. 2. London: Citus Books, 2003.

Business for Social Responsibility (BSR). "Eco-Promising: Communicating the Environmental Credentials of Your Products and Services." San Francisco, CA: BSR, 2008.

Davis, G. and J. H. Song. "Biodegradable Packaging Based on Raw Materials from Crops and Their Impact on Waste Management." *Industrial Crops and Products*, 23 (2006).

Department for Environment, Food and Rural Affairs (DEFRA). "Green Claims—Practical Guidance: How To Make A Good Environmental Claim." London: DEFRA 2003. http://www.defra.gov.uk/environment/consumerprod/pdf/genericguide.pdf (Accessed September 2009).

Morissey, A. J. and P. S. Phillips. "Biodegradable Municipal Waste (BMW) Management Strategy in Ireland: A Comparison with Some Key Issues in the BMW Strategy Being Adopted in England." *Resources, Conservation and Recycling*, 49 (2007).

Nair, L. S. and C. T. Laurencin. "Biodegradable Polymers as Biomaterials." *Progress in Polymer Science*, 32 (2007).

Narayan, R. "Biobased and Biodegradable Polymer Materials: Rationale, Drivers, and Technology Exemplars." In *Degradable Polymers and Materials: Principles and Practice*, K. Khemani and C. Scholz, eds. Washington, D.C.: American Chemical Society, 2006.

Seal, K. J. "Test Methods and Standards for Biodegradable Plastics." In *Chemistry and Technology of Biodegradable Polymers*, G. J. L. Griffin, ed. London: Blackie Academic and Professional, 1994.

Stevens, E. S. *Green Plastics: An Introduction to the New Science of Biodegradable Plastics*. Princeton, NJ: Princeton University Press, 2002.

*Barry Emery*
*University of Northampton*

## Books

The destruction of forests that release oxygen and absorb hazardous carbon dioxide emissions, contributing to global warming, has been a major environmental concern. With the threat of global warming and the recognition that countries like the United States have enough agricultural waste to produce tree-free paper, there is now an emerging emphasis on producing environmentally friendly books. Therefore, some of the green movement principles of environmentally friendly book publishers include being socially responsible by minimizing the printing of books made from paper; using recycled paper, which will help save trees and lessen the amount thrown in landfills; using tree-free papers that are produced from either crops grown for this purpose, such as the kenaf plant, or from the remains of agricultural crops, such as corn and banana stalks; and using vegetable inks instead of petroleum inks in the printing process to reduce toxic chemicals.

In addition to slowing down human-caused global warming by not cutting down trees, another reason for book publishers to use tree-free papers is the length of time it takes for tree-free plants to grow in comparison to trees. Some tree-free plants, for instance, can grow in approximately five months, compared with 7–20 years for trees. An example is kenaf, a fiber crop of African origin, which the U.S. Department of Agriculture has recognized as being one of the best nonwood alternatives to replace trees. Growing quickly, kenaf reaches approximately 12–18 feet in height in about five months.

Clearly, in spite of some farmers using herbicides and fertilizers during the production process, kenaf is also more environmentally friendly than wood-based processing because of the nonuse of pesticides and the use of less energy and chemicals. Interestingly, though, in spite of kenaf being more environmentally friendly and the realization that deforestation is linked to global warming, the majority of papermaking facilities are still producing tree-based papers. The reason is that economically, it is more cost-effective to produce tree-based papers. Moreover, the machines are specifically designed for wood-based fiber instead of tree-free fiber; therefore, to process nontree paper sources effectively, the configuration of the machines would have to be altered. To do this would increase the cost of book production.

However, some book publishers have found alternative ways to become more environmentally friendly by changing part or all of their production process from tree-based books to tree-free books. One such company, Natural Source Printing, produces tree-free, recyclable papers made from discarded limestone. The production process also includes no water and chemicals.

Because of the nonuse of paper and fuel for transporting them, digitally downloaded books are also being embraced by many book publishers. Many university presses are selling their books this way. One such company helping university presses convert books to digital content is Tizra, a digital-content manager and packager. After a book is converted to PDF, it can be redesigned and resold according to the publisher's specification.

Likewise, ebrary is a subscription-based leading supplier of digital books for the academic library market and corporations. Book publishers can license or sell and market their digital books to libraries and corporations worldwide. Using search terms, end users have access to digital content from all over the world.

The U.S. government is also being more proactive by recognizing the importance of preserving forests to reduce global warming. Recognizing the importance of forest conservation, the Environmental Protection Agency recommends sustainable forestry instead of reckless deforestation without any regard to maintaining the ecological balance. The planting of trees to help maintain the ecological balance, in fact, was included in the plan.

Equally important, the U.S. government also recognizes that inks used by book publishers can have a damaging effect on the environment. Taking steps to help the environment, in 1994, Public Law 103-348 took effect, "requiring that all Federal lithographic printing be performed using ink made from vegetable oil and materials derived from other renewable resources, and for other purposes." Consequently, private-sector book publishers were also made more aware of the dangers of the nonenvironmentally friendly, petroleum-based inks used in the production process. In comparison with vegetable oil, many standard petroleum inks contain toxic heavy metals that are not biodegradable; therefore, when they are disposed of, they can emit hazardous fumes and leak into the soil, possibly contaminating groundwater and ending up in the water used in our daily activities, such as drinking and bathing.

Without doubt, book publishers are embracing more environmentally friendly methods and technologies, especially with the increasing popularity of digital books. Cheaper and more environmentally friendly ways to produce paper are also being explored. However, because of the economic aspect, only time will tell whether tree-free papers and vegetable-based inks will be fully embraced by the publishing industry.

**See Also:** Biodegradability; Carbon Emissions; Ecolabeling; Environmentally Friendly; Ethically Produced Products; Paper Products; Recyclable Products; Recycling; Sustainable Consumption; Waste Disposal.

### Further Readings

Hilts, Paul. "Ink From Soybeans: Lighter, Cheaper, Safer. Replacing Petroleum With Vegetable-Oil Base Marks a Printing Innovation." *Publishers Weekly,* 29 (1991).
Motavalli, Jim. "Beyond Wood." *E—The Environmental Magazine,* 9 (1998).
Van Putten, Mark. "Turning the Page on Saving Forests." *International Wildlife,* 29 (1999).
Witt, Clyde. "Stone-Age Solution." *Material Handling Management,* 63 (2008).

*Martha J. Ross-Rodgers*
*University of Phoenix*

## Bottled Beverages (Water)

British Bottled Water Producers Ltd. boasts that the U.K. bottled water market, including water coolers, has grown from 580 million liters in 1993 to almost 2.2 billion liters in 2007, or approximately 40 liters per person per year. In the United States, total annual sales of bottled water in 2008 were over 30 billion liters, having more than doubled in only a decade. Averaged out, each man, woman, and child in the United States consumes approximately 100 liters of bottled water, costing more per unit volume than gasoline. Worldwide growth has been even more explosive, posting yearly growth of more than 7 percent per year, and more than a dozen countries, including Mexico, France, and Italy, consume even more per capita. This makes bottled water the fastest-growing bottled beverage in the world and a very hot marketing category.

Given this, it is not surprising that large corporations such as Nestlé, PepsiCo, Danone, and the Coca-Cola Company have entered the bottled water industry, acquiring various bottled water brands or launching their own with exotic sounding names like Aquafina (PepsiCo) and Dasani (Coca-Cola), as well as those listed, alongside their market share, in Table 1.

In addition to basic bottled water products, many companies are seeking to add value to water in the following ways:

- Through the introduction of additives intended to make water more nutritional. So-called vitamin waters and flavored waters have become the latest variant of bottled water products, primarily being developed by Coca-Cola and Nestlé. Aimed at the style- and health-conscious 18- to 35-year-old demographic, these "functional waters" claim to promote health and sports performance.
- Through increasingly complex (and expensive) packaging. The epitome of this is Bling water, which is contained in a rhinestone-encrusted bottle designed by Swarovski and retails for $40/liter.
- By the creation of a distinction between basic and premium water. In March 2007, the first industry conference, "Towards the Creation of a Premium Water Society," was held in Barcelona, Spain, with the specific objective of developing strategies for enlarging the gulf between tap water and bottled water.

The bottled water industry has also developed a market for so-called premium waters. Such brands include Berg from Canadian glaciers, sold for 33 euros per bottle in hotels; Peteroa 9500 from Chile, which is marketed on the basis that it has been "filtered through glaciers" over 9,500 years; and Fiji, which touts its sources as among the most pristine in the world. Yet, such waters are shipped thousands of miles around the world, incurring huge carbon footprints. These waters are now increasingly evaluated in much the same way as fine wines would be described at a wine tasting, leading to the publication of the first connoisseurs' guide to "fine waters" in 2006. However, blind taste tests with consumers have failed to produced any discernible differences between bottled and tap water. Notwithstanding this fact, it is a standard part of many restaurants' business model to push higher-margin bottled waters on customers, raking in between $200 and $350 million per annum from bottled water sales alone, but also triggering an "Ask for Tap" backlash in the United Kingdom.

## Bottled Water: Health, Safety, and the Environment

Many bottled water brands base their publicity on the advertised "purity" of their product. In this regard, public health messages about drinking at least two liters of water per day appear to have been capitalized on by bottled water companies seeking to portray their products as the only safe way to meet this, by now well-recognized, healthy living requirement. Some dieticians warn that this form of advertising could in fact spur rates of obesity and hypoglycemia, as so many bottled water products—including "vitamin water"—have added sweeteners.

Bottled water companies have also been involved with safety scares and product

**Table 1** U.K. Bottled Water Brand Market Share (by value)

| | |
|---|---|
| Evian (Danone) | 9.7% |
| Volvic (Danone) | 7.3% |
| Powwow (Nestlé in bottled water coolers) | 7.2% |
| Highland Spring | 6.9% |
| Buxton (Nestlé) | 3% |
| Aqua Pura (Princes) | 2% |
| Strathmore | 2% |

*Source:* Zenith International (March 2005), cited in British Bottled Water Producers Ltd., 2008.

Worldwide growth of bottled water has posted year-on-year growth of more than 7 percent per year, making bottled water the fastest-growing bottled beverage in the world.

*Source:* iStockphoto

recalls, which have done nothing to slow the growth of the industry. The most recent recall was in 2004 in the United Kingdom when Coca-Cola's Dasani brand was found to contain up to 22 parts per billion of bromate, a known carcinogen. Another substance commonly found in many bottled water varieties is sodium, a recognized contributing factor to hypertension and coronary heart disease. Although sodium is usually only found in trace quantities in bottled water, it has been found to be as high as 18 mg/L in Fiji water—one of the world's best-selling waters—near to the maximum guide level in the European Union. Other bottled water samples have been found to be relatively high in sulfates, nitrates, phosphates, and other pollutants. Indeed, the French government advises people who drink bottled mineral water to change brands frequently, because the minerals in particular brands may be harmful in high doses or if consumed over a long period. Further, the National Resources Defense Council found that in a third of bottled water samples tested, there was at least one sample that violated state limits and guidelines.

More recently, a new menace has been uncovered, emanating not from the product but from the container in which it is packaged. The manufacture of polyethylene terephthalate (PET) bottles that are commonly used for soft drinks including bottled water is carried out using a heavy metal called antimony. This causes PET bottles to contain several hundred milligrams of antimony per kilogram, compared with less than 1 mg/kg in typical rocks and soils. In a test carried out at the University of Heidelberg in Germany on 63 Canadian and European bottled water brands, there was no sample that exceeded guidelines; however, the researcher commented that "in Japan, PET is manufactured using titanium, which is effectively insoluble and harmless—unlike antimony, which can be dissolved in water and is potentially toxic." Furthermore, other dangerous chemicals have been found to leach out of PET plastic bottles, including naphthalene (the chemical found in mothballs), provoking anaphylactic-type reactions in some consumers, and bisphenol A.

Despite being marketed as green or pure, the rapid growth of the global bottled water market may have negative environmental impacts. Many of the brands of bottled water currently available in the United Kingdom are transported hundreds of miles around the country by road, and some foreign brands are flown thousands of miles around the globe. Welsh Ty Nant water is available in British Columbia, Canada—hardly a water-stressed part of the globe—and Evian is an even more global phenomenon. Perhaps the most infamous of these, however, is Fiji, which is imported from the middle of the South Pacific Ocean to North American and Western European markets. Conversely, there are other bottled water brands that appear, from looking at the label, to have come from far-away pristine corners of the Earth that are in fact from far less exotic sources. A notorious example of such misleading marketing has been the case of Everest, sold in the United States. The bottle's label pictures snowcapped peaks, and the name suggests that the water has come from the sparsely populated region in Nepal, where the mountain is located, rather than from the municipal water supply of Corpus Christi, Texas, which is the actual source of the product.

## The Cost of Bottled Water

The figures regarding the comparative cost of bottled water show just how unacceptable it is for public money to be spent on bottled water. At 36 cents a liter, tap water is 141 times cheaper than the best-selling mineral water in Europe (Evian), which, even if you buy it in a supermarket, costs around $2 a liter. Arguably the costs of producing bottled water can be borne by Western industrialized countries, but from the perspective of middle-income countries, such as Mexico (ironically, one of the world's largest consumers of bottled water), such prices are disproportionately high, and it is worrying that the proliferation of bottled water could induce municipal authorities in developing nations to deemphasize public drinking water provision. A related issue involves the rapid proliferation of "dual water pipe" systems, especially in China and East Asia. On the basis of the assumption that standard municipal hook-ups are unfit for human consumption, developers have been quick to "add value" (and profit) to their developments by installing separate "pure water" supply lines, at great cost to the consumer. What is more, the greater the proliferation of such systems, the more municipal supply of drinking water (which depends on extensive cross-subsidies, from urban to rural, from rich to poor, etc.) is undermined.

Finally, bottled water companies are increasingly taking a page from the playbooks of their soda pop parent organizations, entering into exclusionary franchising deals with large institutions (such as prisons and universities) to secure monopoly access to "captive" populations. Bottled water companies have moved aggressively to monopolize access to groundwater resources throughout North America and Europe.

## Consumer Motivations and Perceptions

So what is persuading people to pay more per liter than they spend on gas for water, which can be obtained easily and, according to the Chartered Institute of Water and Environmental Management, is on average 500 times cheaper from the tap? Companies in the bottled water market have used a wide range of methods involving the psychology of perception, motivation, and buyer behavior to lead consumers to choosing their product to meet their "needs." A U.S. study into motivations behind drinking tap water alternatives found that among consumers who drink filtered water, 58 percent said safety was their primary reason for drinking filtered water, whereas 38 percent cited taste as their key criterion. The study found that across the United States, people who were concerned with the safety of their municipal tap water tended to drink filtered water as an alternative to water direct from the tap, whereas people who did not like the taste of tap water opted for bottled water, skipping the more economically logical option of filtration of tap water.

As previously mentioned, blind taste tests have shown that less than one quarter of consumers can tell the difference between bottled water and tap alternatives. In addition, researchers found that consumer preferences are further confused when the packaging of testing samples was altered, indicating that packaging has a great deal to do with the taste of water. Similarly, in a study conducted in December 2008, the authors found that taste testers did not appreciably prefer an expensive bottled water sample (Ty Nant) over other bottled waters (Evian and Highland Spring) or tap.

## A Question of Social Justice

The Mar del Plata agreement of 1977 declared that there should be a universal right to sufficient clean water to support the social, cultural, and economic development of every citizen on the planet. Subsequent international legal instruments have echoed and developed this call, although as yet it is unclear whether there is an enforceable right to water. If present trends continue, then more than a generation's worth of effort to widen access to clean drinking water could be undermined, as water authorities may use the consumer shift to bottled water to channel investment away from public water infrastructure development. The burgeoning bottled water industry represents a fundamental challenge to the ideal of clean water for all and needs to be controlled, through scholarship, communication, and activism.

**See Also:** Beverages; Consumer Behavior; Demographics; Needs and Wants; Water.

### Further Readings

British Bottled Water Producers. *Water's Vital Statistics: Industry Data.* http://www.britishbottledwater.org/vitalstats.html (Accessed September 2008).

British Broadcasting Corporation. "Coke Recalls Controversial Water." BBC News (2004). http://news.bbc.co.uk/1/hi/business/3550063.stm (Accessed September 2008).

British Broadcasting Corporation. "Volvic Investigates Water Contamination." http://news.bbc.co.uk (Accessed September 2008).

Chartered Institution of Water and Environmental Management. *Policies, Bottled Drinking Water* (2005). http://www.ciwem.org/policy/policies/bottled_water.asp (Accessed September 2008).

Cheung, K. and L. Hulbert. "The Taps Are Turning, Are We Ending Our Love Affair With Bottled Water?" In *Sustain: The Alliance for Better Food and Farming*. London: Sustain Publications, 2008.

European Commission. "EU COUNCIL DIRECTIVE of July 15, 1980 on the Approximation of the Laws of the Member States Relating to the Exploitation and Marketing of Natural Mineral Waters." (80/777/EEC)(OJL229,30.8. 1980, p.1) as amended.

Ferrier, C. "Bottled Water: Understanding a Social Phenomenon." Discussion Paper. Gland, Switzerland: World Wide Fund for Nature, 2001.

Landi, H. "Bottled Water Report." *Beverage World,* 127/4:12–14 (2008).

Mackey, E. *Consumer Perceptions of Tap Water, Bottled Water, and Filtration Devices: Project 90944F.* London: International Water Association Publishing, 2005.

Mascha, M. *Fine Waters: A Connoisseur's Guide to the World's Most Distinctive Bottled Waters*. Philadelphia, PA: Quirk Books, 2006.

Olson, Erik D. "Bottled Water: Pure Drink or Pure Hype?" Natural Resources Defense Council (1999). http://www.nrdc.org/water/drinking/bw/bwinx.asp (Accessed October 2008).

Royte, E. *Bottlemania: How Water Went on Sale and Why We Bought It.* New York: Bloomsbury, 2008.

Staddon, C. *Managing Europe's Water: 21st Century Challenges.* London: Ashgate, Forthcoming.

Vann, M. "Manager to Manager—Tap Water Beats Out Bottled Water in Birmingham Taste Test." *Journal of American Water Works Association,* 96/8:30–32 (2004). http://apps.awwa.org/WaterLibrary/showabstract.aspx?an=JAW_0060500 (Accessed January 2009).

Wanctin, L. *Have You Bottled It? How Drinking Tap Water Can Help Save You and the Planet.* London: Sustain Publications, 2006.

Which? Advice. "Switching From Bottled to Tap Water." http://www.which.co.uk/advice/switching-from-bottled-to-tap-water/tap-vs-bottled-water/index.jsp (Accessed March 2009).

*Andrew Fox*
*Chad Staddon*
*University of the West of England, Bristol*

## Carbon Credits

A carbon credit is an allowance or offset equal to one metric ton of carbon dioxide ($CO_2$) or other greenhouse gas (GHG) equivalent calculated in tons of $CO_2$. Carbon markets allow individuals, companies, or states to finance emissions reductions in other locations, for example, through projects that create or promote renewable energy, energy efficiency, or reforestation, as a means to lower their own carbon footprint. Trade of carbon credits within compliance and voluntary markets is a multibillion-dollar enterprise that makes up a cornerstone of international efforts to combat climate change. Buying and selling emissions allowances often means that mitigation can be achieved at lower costs.

Carbon trade was modeled after an experiment in the United States in the early 1990s involving the trade of sulfur dioxide and nitrogen dioxide credits to mitigate the effects of acid rain. The United States championed similar market-based mechanisms for inclusion in the Kyoto Protocol with the belief that placing a price tag on carbon emissions would create economic incentive to decrease the release of GHGs. The protocol adopted in 1997, and legally binding after 2005, included these market mechanisms in its climate governance regime, but it was never ratified by the United States, in contrast to nearly every industrialized country in the world.

Compliance markets involve transfers of carbon credits under a cap-and-trade system, meaning there are established goals for emission levels contained within set periods of time. If members cannot reduce internal emissions to the level of their established cap, they can purchase credits from other members who pollute less than their quota. Two of the best-known cap-and-trade systems are the Kyoto Protocol's Clean Development Mechanism (CDM) and the Emission Trading System of the European Union (EU ETS).

The EU ETS began operating in 2003 with the goal of limiting or reducing GHG emissions in a cost-effective way. About 12,000 energy-intensive plants in the EU, such as those in the power generation, iron and steel, glass, and cement industries, buy and sell carbon permits. Companies that exceed their individual limit are able to buy unused permits from firms that have cut emissions. Those that exceed their limit and are unable to buy spare permits are fined. Although the system was successful in that it rapidly expanded to cover a large portion of regional carbon emissions, the early caps established may have been too lenient, and a surplus of carbon permits contributed to market collapse by 2007. The EU ETS is currently under revision to reverse initial shortfalls.

Carbon credit transfers under the CDM must be validated to comply with the Kyoto Protocol, which differentiates responsibilities between industrial and developing countries. Industrialized countries must reduce emissions by 5.2 percent by 2012 compared with their 1990 levels. There are two flexible mechanisms established for trading carbon credits. Joint Implementation reduces emissions in projects created between industrialized countries. Because of the relatively high costs of reducing carbon in industrialized countries, there were only 22 Joint Implementation projects worldwide as of late 2008. In contrast, by the same time period, there were more than 1,000 CDM projects partnering industrialized and developing countries. Developing countries that host clean development projects receive payments for emission reductions, which are also intended to finance sustainable development. This stipulation is important so that offset purchasers do not merely pursue cheap offsets. CDMs are intended to provide wider social and ecological benefits and technological transfer—goals that are sometimes not actualized. The distribution of CDM credits is highly uneven: It is expected that by 2012, China will have received over 50 percent of the total credits. Energy efficiency and renewable energy projects have made up smaller portions of CDM credits than initially anticipated. The top project type, those that capture fugitive emissions, initially exposed the potential for corruption. Some plants started producing the potent GHG HFC-23 to receive carbon credits to capture it. This loophole has since been closed, but this example exemplifies the potential for perverse incentives and demonstrates why the rigorous requirements exist, as well as what can be perceived as an unwieldy CDM oversight bureaucracy.

Voluntary carbon markets allow individuals and businesses to elect to reduce their carbon footprint; for example, by offsetting air travel. They allow the purchaser to establish the level of their GHG emission abatement in the time frame that they select and with any of the numerous carbon credit providers supplying voluntary markets. These markets have been able to streamline implementation and sidestep some of the problems faced under compliance markets. However, validation requirements are less stringent, and oversight is inconsistent. Although lacking much of the bureaucracy and the high transaction costs of compliance markets, there is a concerning amount of variability and opaqueness in the voluntary sector. Increasingly, standards are being established, such as the Voluntary Gold Standard, to protect consumer rights, prohibit double-counting among providers or intermediaries, and ensure quantifiable emissions reductions. Use of the Climate, Community, and Biodiversity Standard promotes carbon credits from vegetation and soils, which are important carbon sinks, as they sequester and store carbon. To date, concerns about measurement, permanence, and leakage have largely kept biocarbon projects out of compliance carbon markets.

There need to be ongoing improvements to measurement, reporting, verification, and enforcement to avoid unscrupulous manipulation in both voluntary and compliance carbon markets. Carbon market critics suggest that actors will choose the easiest, least-expensive means to offset their emissions, thereby potentially constricting long-term planning and sustainable development. Carbon trade in well-regulated programs may serve to mitigate climate change but can also distract attention from wider systemic changes that are necessary, whether in personal consumption patterns or finance policy.

**See Also:** Carbon Emissions; Carbon Offsets; International Regulatory Frameworks; Kyoto Protocol.

**Further Readings**

Chafe, Zoe and Hilary French. "Improving Carbon Markets." In *State of the World 2008: Innovations for a Sustainable Economy*. Washington, D.C.: Worldwatch Institute, 2008.
European Commission. "Emissions Trading System." http://ec.europa.eu/environment/climat/emission/index_en.htm (Accessed April 2009).
Lohmann, Larry. "Carbon Trading: A Critical Conversation on Climate Change, Privatisation and Power." *Development Dialogue*, 48 (2006).
World Bank. "State and Trends of the Carbon Market 2008." http://carbonfinance.org/docs/State_Trends_FINAL.pdf (Accessed October 2008).

*Mary Finley-Brook*
*Curtis Thomas*
*University of Richmond*

# Carbon Emissions

Increasing concentrations of atmospheric carbon dioxide ($CO_2$) gas resulting from anthropogenic activities generates concern for possible changes in global climate. In 2005, global atmospheric $CO_2$ concentration reached nearly 380 parts per million (ppm), which is equivalent to 805 Pg C (1 Pg = 1 petagram = $10^{15}$ g) from a preindustrial (1850) concentration of about 280 ppm. An increase in atmospheric $CO_2$ concentrations to 450–600 ppm from current levels over the coming century will have an irreversible effect, including rainfall reductions in subtropical areas and an unprecedented rise in sea level, leading to the inundation of small islands and low-lying coastal areas. The United States is the largest single national source of fossil-fuel-related $CO_2$ emissions—1,577 million metric tons of carbon (MMTC)—followed by China (1,514 MMTC), the Russian Federation (410 MMTC), and India (382 MMTC) in 2005. Developed countries remain poised to be the main source of carbon emissions, whereas developing countries record high growth rates of carbon emissions.

The burning of fossil fuels and cement production are the major sources of carbon emissions internationally. The main forms of fossil fuels are petroleum, natural gas, and coal. Energy produced by burning of fossil fuels is used for electricity generation, industrial uses, transportation, and in private households and buildings. Globally, liquid and solid fuels accounted for 77 percent of the emissions from fossil fuel burning and cement production in 2005. In the United States, total $CO_2$ emission from fossil fuel combustion was 5,746 Tg ($10^{12}$ g) $CO_2$, which is a 22 percent increase during 1990–2007. Approximately 85 percent of the energy consumed in the United States was generated from the combustion of fossil fuels. Emissions from electricity generation contributed the largest portion (34 percent) of U.S. GHG emissions in 2007. Within the manufacturing industry sector, six industries—petroleum refining, chemical production, primary metal production, paper, food, and mineral production—constitute the majority of energy consumption. Emissions from these industries accounted for 20 percent of emissions. Transportation is the second-largest source of $CO_2$ emissions (28 percent) in the United States. Automobiles and light-duty trucks emit around two-thirds of emissions from transportation. Residential, agriculture, and commercial sectors contributed the remaining 18 percent of the emissions. In addition to farming operations, land use change and waste incineration also contributed to $CO_2$ emissions. The burning of crop residues and other farm operations also contribute to carbon emissions. Liming of agricultural soils and urea application release $CO_2$ to the atmosphere, accounting for 8 Tg $CO_2$ emissions in the United States during 2007. Clearing of forests for agriculture and logging for timber contribute to the reduction in carbon sink.

Mitigation of increasing carbon emissions to the atmosphere consists of two major options: reducing sources of carbon emissions to the atmosphere, such as reducing dependency on fossil fuel, and both encouraging the use of alternative renewable energy sources (wind, air, and solar) and increasing sinks of carbon in land and oceans. Strategies, including decreasing the rate of deforestation and increasing the carbon storage potential of existing forests, have the potential to reduce/maintain atmospheric $CO_2$ concentrations. Geologic sequestration of carbon is another promising method to reduce atmospheric carbon emissions. Factory-emitted $CO_2$ could be permanently stored in underground repositories, such as coal seams, abandoned natural gas reservoirs, and depleted and marginal oil fields. However, the success of these strategies depends on strict implications of governmental policies directed toward reducing global carbon emissions. At the household level, reduction in carbon emissions can be achieved by implementing a number of practices, including turning down the temperature of both heating and air conditioning, better insulation, using energy-saving appliances and lightbulbs, reducing dependence on private cars, minimizing energy waste, recycling (including water reuse), and using energy suppliers that use renewable energy sources. However, any policies directed toward creating change in households should be cognizant of the social, cultural, and material environments in which people live, the social networks and practices in which they are engaged, and the choices available to them.

**See Also:** Carbon Credits; Carbon Offsets; Overconsumption.

## Further Readings

Burney, Nelson E. *Carbon Tax and Cap-and-Trade Tools: Market-Based Approaches for Controlling Greenhouse Gases (Climate Change and Its Causes, Effects and Prediction).* Hauppauge, NY: Nova Science Publishing, 2010.

Houghton, R.A. “Balancing the Global Carbon Budget.” *Annual Review in Earth and Planetary Sciences*, 35:313–47 (2007).

Marland, G., et al. “Global, Regional and National $CO_2$ Emissions.” In *Trends: A Compendium of Data on Global Change*. Oak Ridge, TN: Carbon Dioxide Analysis Center, Oak Ridge National Laboratory, U.S. Department of Energy, 2008. http://cdiac.ornl.gov/trends/emis/tre_usa.html (Accessed August 2009).

Solomon S., et al. “Irreversible Climate Changes Due to Carbon Dioxide Emissions.” *Proceedings of National Academy of Sciences*, 106/6:1704–09 (2009).

U.S. Environmental Protection Agency. “Draft Inventory of U.S. Greenhouse Gas Emissions and Sinks: 1990–2007.” http://www.epa.gov/climatechange/emissions/usinventoryreport.html (Accessed August 2009).

*Amitava Chatterjee*
*University of California, Riverside*

# Carbon Offsets

A carbon offset is an emission reduction credit from another organization's project that results in less carbon dioxide or other greenhouse gases (GHGs) in the atmosphere than would otherwise occur. Carbon offsets are typically measured in tons of carbon dioxide ($CO_2$) equivalents and are bought and sold through a number of international brokers, online retailers, and trading platforms. Once it has been used, either for compliance or voluntary motives, an offset must be cancelled to avoid its reuse.

Voluntary carbon offsetting is particularly popular in Western countries that are concerned about the potentially negative environmental effects of climate change.

*Source:* iStockphoto

Before it is cancelled, a carbon offset can be considered a market commodity that represents a reduction in GHG emissions. Although most buyers of offsets use them for compliance purposes in a cap-and-trade GHG emission trading scheme, some entities purchase carbon offsets to become carbon neutral. This is called the voluntary market.

The compliance market involves companies, governments, or other entities that buy carbon offsets to comply with caps on the total amount of $CO_2$ they are allowed to emit. This market is currently much larger than the voluntary market, in which individuals, companies, or governments purchase carbon offsets to mitigate their own GHG emissions from transportation, electricity use, and other sources without being constrained to reduce or offset their emissions. Voluntary carbon offsetting is particularly popular in Western countries that have become aware of and concerned about the potentially negative environmental effects of climate change.

Offsets are typically achieved through the financial support of projects that reduce the emission of GHGs. The Kyoto Protocol established the Clean Development Mechanism (CDM) and Joint Implementation, which validate and measure projects to ensure they produce real benefits and are genuinely "additional" activities. Additionality is the fundamental criterion for the recognition of a project. Under this criterion, the project developers must, from a business-as-usual scenario, show that their project will result in GHG emissions reductions that would not occur otherwise. The difference between the level of emissions in the business-as-usual scenario and in the scenario with the CDM or Joint Implementation project determines rights to carbon credits (certified emissions reductions with the CDM, emission reduction units with the Joint Implementation). The additionality test is essentially composed of three elements: environmental additionality (does the project reduce emissions below the business-as-usual scenario?), investment additionality (does access to carbon credits make the project viable?), and technological additionality (does the project lead to a transfer of technologies in the host country?). The Marrakesh Accords implicitly recognized that investment additionality can be a means of proving the environmental additionality of the project but that it is not necessarily the only way (other barriers to investment such as technological barriers or availability of capital may also prove the environmental additionality of a project). Finally, economic additionality is also considered. This implies that the capital provided by the developed countries is not used as a substitute for traditional aid to developing countries.

Official compliance offsets (certified emissions reductions or emission reduction units) come mainly from the destruction of fluorinated gases, renewable energy, and waste management. Voluntary offsets come primarily from four types of projects: forestry, renewable energy, energy efficiency, and waste management. A feature of voluntary offsetting projects is that they are often small scale. For instance, a project in energy efficiency could be related to solar cooking in a school or the replacement of an inefficient diesel motor. More than one-third of the voluntary offset credits sold come from forestry projects or other sink projects. Forestry projects are considered more controversial than energy or waste projects (e.g., methane recovery in landfills), mainly because of the greater uncertainty in accounting for avoided emissions and the temporary aspect of carbon sinks (if the trees rot or burn, the carbon sink becomes an emission source, thus losing all offset benefits) and the potential for leakage resulting from displacement of the deforestation. Nevertheless, forestry projects may represent significant sources of income for people living in high-deforestation areas and protecting forests—a key element in the fight against global warming. There is clearly an urgent need for the development of financial mechanisms to encourage owners of primary forests not to exploit them. Here, too, only good quality controls can give the buyers the necessary guarantees.

Verification is a central element of project-based mechanisms generating carbon offsets. Avoided emissions need to be verified by an independent third party who can provide assurance on the calculations and scientific basis adopted. At this time, certified emissions reductions issued under the CDM and verified emissions reductions (also referred to as voluntary emissions reductions) assessed under the Gold Standard and the Voluntary Carbon Standard give the greatest assurance on the quality of the offset projects.

Multiple sales of the same $CO_2$ reduction credit are a risk in carbon offsetting. Once a credit has been sold to a customer, it should be cancelled to avoid the reuse of the same credit for offsetting purposes. Registries such as those of the CDM or the Gold Standard can guarantee the cancellation of sold credits and avoid the risk of fraud and double issuing.

**See Also:** Carbon Credits; Carbon Emissions; Green Politics; Kyoto Protocol; Overconsumption; Resource Consumption and Usage.

### Further Readings

Bayon, R., et al. *Voluntary Carbon Markets. An International Business Guide to What They Are and How They Work*. London: Earthscan, 2007.

Brohé, Arnaud, et al. *Carbon Markets. An International Business Guide.* London: Earthscan, 2009.

Hoffman, Andrew J. *Carbon Strategies: How Leading Companies Are Reducing Their Climate Change Footprint.* Ann Arbor: University of Michigan Press, 2007.

Miller, Debra A. *Carbon Offsets (Current Controversies).* Farmington Hills, MI: Greenhaven Press, 2009.
Tietenberg, T. H. *Emissions Trading: Principles and Practice.* Washington, D.C.: RFF Press, 2006.

*Arnaud Brohé*
*Independent Scholar*

# Car Washing

The three principal areas of environmental impact resulting from car washing are the contamination of surface water, as the substances used to wash the car are rinsed off and drain into local water sources such as rivers; contamination of soil and groundwater, as a result of surface runoff; and the use of water and energy. Washing one's car at home (or at a car-wash fund-raiser in a parking lot) is actually likely to have a greater effect than using a professional car wash, the industry of which has actually made significant strides in reducing its environmental impact. Professional car washes often use water reclamation systems—cleaning and recycling the water used to wash cars to reduce water waste. In some cases this is required by local regulations, but in many others it is simply practical and has been since long before the current push for greater environmental awareness. Car washes also use high-pressure devices that minimize the amount of water needed to wash a car compared with the traditional sponge, bucket, and hose method you may use in your driveway. Furthermore, surface water contamination is less common with professional car washes. In the United States, for example, the Clean Water Act requires that a car wash treat its wastewater before disposing of it, instead of letting it run off into the soil. Further reductions of environmental impact are found in green carwashes.

Surface water contamination is less common with professional car washes, because regulation requires treatment of such wastewater before it is released.

*Source:* iStockphoto

Although most detergents used for washing cars at home are harmful to surface water and groundwater, having unacceptably high levels of the toxic metals antimony, arsenic, beryllium, cadmium, lead, and thallium, and often also having high levels of chloride and naphthalene, there are a small number of brands of green biodegradable car-washing detergent. These are an especially beneficial alternative if the car is parked on a lawn or other dirt area, so that soil's natural capacity for neutralizing toxins will come into play, instead of letting the water run over pavement and concrete until finding a storm drain. Similar detergents are used by environmentally responsible, green car washes, which is particularly significant because many of the detergents that have typically been used by touchless car washes contain harsh corrosives. Even biodegradable detergents are toxic to surface water, and the nutrients used to make your car shine (typically nitrogen and phosphorous) can lead to algae blooms and spikes in bacterial growth in streams and lakes, whereas other substances poison the fish and other aquatic organisms. Detergents of all sorts can harm fish by damaging the natural oils of their gills, making it harder for them to breathe. Damage to the mucus membranes may occur, and damage to fish eggs begins at much lower concentrations than it takes to poison adult fish. Some surfactants also can break down in the water into hormone-mimicking compounds that disrupt the growth and development of fish and other aquatic species.

Green car washes use various methods to operate a more sustainable business. Better-than-standard filtering equipment can reclaim 100 percent of the water used, for instance, using fresh water only for the final rinse. This brings the amount of water used for each car to about eight gallons, down from the 15, 25, or even 70 gallons per car used by other methods, according to studies made by the International Carwash Association.

By some estimates, 800 million gallons of water are used every day to wash cars in the United States alone. A step better than water reclamation, especially in areas facing periodic or ongoing water shortages, is waterless car washing, sometimes called "dry wash." Dry wash uses a chemical blend to accomplish the work of cleaning without added water. Surfactants in the blend reduce the surface tension of the wetting agents used, allowing them to surround the dirt particles on the car the way water would. There are several green ecofriendly brands of dry wash for use at home, and in drought-prone areas there are more likely to be professional waterless car washes; this method has not, for the most part, caught on elsewhere. Because a waterless car wash lacks the dirt-dislodging power of high-pressure hoses, it is most effective for cars that are only mildly dirty or for keeping a car clean through regular waterless washings.

**See Also:** Automobiles; Cleaning Products; Green Consumer; Water.

### Further Readings

Brower, Michael and Warren Leon. *The Consumer's Guide to Effective Environmental Choices: Practical Advice from the Union of Concerned Scientists.* New York: Three Rivers, 1999.

Garlough, Donna, et al., eds. *The Green Guide: The Complete Reference for Consuming Wisely.* New York: National Geographic, 2008.

Leech, Eric. "Houston Car Wash Goes Back to the Basics." *Houston Business Journal.* http://www.treehugger.com/files/2009/03/houston-car-wash-goes-back-to-basics.php (Accessed July 2009).

*Bill Kte'pi*
*Independent Scholar*

# Certification Process

A certification process is a procedure whereby an accredited third party conducts an official on-site assessment of an organization's policy and procedures. If there is sound evidence that the organization's policy and procedures satisfy the standards set by the accrediting body and that they are being adhered to, then the organization is awarded a certificate demonstrating that it is following best practices. The Registration Process is reserved for the act of recording the organization's information in the register of the accrediting body. Founded in 1946, the International Organization for Standardization (ISO) is the world's largest developer of voluntary/nonlegal international standards. Its headquarters are based in Geneva, Switzerland, and in 2009 it had members from 161 countries and had produced in excess of 17,500 ISO standards. The ISO certification process is a formal procedure in which an organization registers and adopts an ISO template to design and document its policy, procedures, and working instructions. ISO 9001 (Quality Management), ISO 14001 (Environmental Management), and ISO 18001 (Health and Safety) are the most commonly utilized standards worldwide.

The certification processes for these standards are similar and are considered in detail. The ISO is the gatekeeper of the certification process, but it does not implement it. Each member country has established a national accreditation body that authorizes subscribed certification bodies. The United Kingdom Accreditation Service (UKAS) is the U.K. accreditation body, and examples of regulated certification bodies are the British Standards Institution (BSi) and Lloyd's Register Quality Assurance. The certification bodies also undergo a certification process in which UKAS assesses their technical competence via a company headquarters audit, an assessment of auditors in the field, and the collation of witness assessments from certified and registering organizations. In addition, the certification bodies periodically conduct an internal appraisal of their auditors. The outcome of these appraisals can contribute to an auditor training needs analysis, and in extreme cases can lead to disciplinary procedures.

Although the standards are voluntary, they are viewed as essential marketing attributes. The potential benefits of ISO certification can be summarized as providing a business model that enhances operational efficiency and effectiveness, consolidating internal auditing procedures, decreasing waste, stimulating a positive company culture, enhancing supply chain relationships, strengthening marketing and supporting the promotion of international trade, and promoting greater productivity and profitability. Critics challenge these benefits and state that the weaknesses of the certification process are that it is a costly and time consuming process, it promotes specification and conformance rather than understanding and improvement, managing directors underestimate the degree of personal commitment and resources required to implement the system, supply chain pressures may provide organizations with incentive to seek certification before quality, and expectations of sustained profitability fail to materialize.

## Steps to ISO Certification

The certification program in principle is a logical and comprehensive five-step model as follows:

*Step 1: Application.* Organizations are strongly advised to thoroughly research the core principles of ISO and carry out a detailed cost benefit analysis. Specific attention needs to address the level of company-wide commitment, in particular, strong leadership and a full appreciation of the work involved in successfully achieving and retaining certification status is required. An application needs to be requested and completed from a registered body such as the BSi. This will invariably involve a site visit to clarify and/or reassure the above expectations and registration costs. A formal contract is then signed and a provisional date is booked for a Document Review/Assessment.

*Step 2: ISO Assessment.* This involves a pre-assessment to benchmark the organization's systems against ISO requirements. In addition, the registering body will formalize the scope of certification with the organization. The assessment and advice will often not involve a complete rewrite of the organization's systems but is more akin to a redrafting of current systems in line with the recognized ISO format.

*Step 3: ISO Training.* Training is fundamental in the certification process and involves a holistic organization wide approach. All staff, including senior management, require training in two key aspects. Firstly, a thorough understanding of ISO, its terminology and vocabulary, procedural requirements, documentation, manuals, auditing procedures, and potential benefits is required. The second area of training involves a clear insight into the operational challenges of the certification process. Selected personnel will also need to be deemed competent in document design and/or completion and first party auditing skills. Effective monitoring procedures must be established to ensure that the support mechanisms are both adequate and realistic in order to achieve the set deadlines. It is not uncommon for organizations to seek the advice and expertise of "second-party" consultants to aid the process. Organizations need to assess the expenditure involved and the support systems provided. A key attribute of consultancy services is that they can provide an independent assessment of an organization's systems prior to the formal assessment. This process is commonly referred to as the second party audit.

*Step 4: ISO Documentation of Quality Procedures.* Trained personnel will systematically document the assessment scope, manuals, procedures, and work instructions in line with ISO standards. In addition, an audit plan and/or schedule is designed in which all of the procedures are audited and can be traced through an audit trail. This process is pivotal in the certification process. If the final third party certification/assessment process identifies nonconformances between documented systems and audited procedures, then the accrediting party may defer or suspend certification.

*Step 5: ISO Third-Party Assessment Visit.* This is the formal assessment to determine that the existing documented procedures meet the requirements of the standard. The third-party assessors will meet management and agree on an agenda for the assessment. Assessors then review documentation, observe the systems in operation, interview staff, review audit schedules and the results are then cross-checked. Noncompliance is recorded as an observation, or more significantly as an action request or noncompliance. The findings are then concluded on-site in the form of a report that is presented to management of the company. If nonconformances are identified, then the accrediting body may require the organization to amend these nonconformances in the form of an improvement plan within a specified time period, and in certain cases schedule an additional assessment visit. Ethical standards and a strict code of conduct prevent the third-party assessors from offering operational advice on how to correct noncompliances. If there are no noncompliances, then the accrediting body will recommend and award a "Certificate of Approval" detailing its scope. The certified organization is then entitled to use the logo symbolizing that the organization is currently operating to ISO standard, subject to usage limitations.

With specific reference to green consumerism, ISO 14021:2001 "Environmental Labels and Declarations—Self-Declaration Environmental Claims (Type II Environmental Labeling)" is a key guideline. It involves organizations seeking certification to submit a self-declared environmental policy, which assures the marketplace that the organization's declaration: is accurate and verifiable; encourages the organization to proactively seek environmental improvements; helps prevent or minimizes unwarranted claims; reduces marketplace confusion; facilitates international trade; and helps enable more informed choices in the supply chain. The process of certification is not solely guided by ISO, although their philosophy of standardization, verification, and continued improvement is similar to other certification bodies, such as Fairtrade Certification.

Fairtrade Certification enables consumers and organizations to identify and track products that meet set environmental, labor, and development standards. Similar to ISO, Fairtrade Certification is monitored by a standards setting body, namely the Fairtrade Labelling Organizations International (FLO) and FLO-CERT, the certification body, and producers are independently audited to ensure that specified standards are met and sustained. Upon submitting a "Business Overview Questionnaire," successful organizations that demonstrate compliance may subscribe for a license and market the Fairtrade Certification mark in conjunction with their company name and products.

The ideology of fair trade can be traced back to 1988 and the initiative of Mexican coffee farmers. This was acknowledged as the first Fairtrade labeling initiative and was held in the Netherlands. The model was replicated in various other geographical markets and was referred to as Transfair. In 1994, the need for an international standard was growing in momentum and this led to a partial unification between labeling organizations, also known as LIs (Labeling Initiatives). The outcome led to the TransMax working group, who strategically set up the FLO in 1997. The FLO sets fair trade standards, supports, inspects and certifies disadvantaged producers, and strives to clarify and concentrate on what fair trade stands for on a global scale. In 2002, an International Fairtrade Certification mark, which superseded the Transfair Certification mark, was introduced to ensure impartiality. In 2004, the FLO was separated into two autonomous groups. Fair trade standards (in compliance with the ISEAL Code of Good Practice) and producer support is now provided by the FLO. Additionally, FLO-CERT has the responsibility of auditing and certifying producer organizations. In order to do this effectively, FLO-CERT sources the expertise of a network of third-party auditors who report their findings to a certification committee.

**See Also:** Certified Products; Green Consumerism Organizations; Production and Commodity Chains.

### Further Readings

Hoyle, D. *ISO 9000 Quality Standards Handbook,* 4th Ed. Oxford: Elsevier Butterworth-Heinemann, 2004.

Hoyle, D. *Quality Management Essentials.* London: Butterworth-Heinemann, 2007.

Naveh, E. and A. Macus. "When Does ISO 9000 Quality Assurance Lead to Performance Improvement." *Engineering Management*, 51/3:352–63 (2004).
Randall, R. C. *Randall's Practical Guide to ISO 9000: Implementation, Registration and Beyond (Engineering Process Improvement)*. Upper Saddle River, NJ: Prentice Hall, 1995.

*Derek Watson*
*Mitchell Andrews*
*University of Sunderland*

# Certified Products (Fair Trade or Organic)

The trade in goods certified to achieve certain social or environmental goals or standards is the mirror image of the boycott. Whereas the boycott uses social ostracism to obtain its goals, certified schemes try to use the power of the consumer to bring into being products that embody a goal desired by those who are purchasing the good. These products are simultaneously a product that can be used by the consumer in the present and also a symbol of a possible future. Although recently brought into focus by organic foods, Fair Trade products, demands for No Sweat labels, and various ecolabels such as the Rainforest Alliance, the concept has its roots in the 19th century. The British antislavery movement promoted sugar that had not been produced using slaves to highlight the link between slavery and the sugar trade. It was coincidentally during this period that the boycott first became a form of protest in the British Empire, first in Ireland and later by those campaigning for Indian independence. As modern industrial society began to create products for mass consumption, consumers began to realize that they could highlight aspects of their production and the development of alternatives. Since that time, as the flows of trade and commerce have become more complex, so have the means of creating such alternative products, but the central logic of creating a collective protest through individual acts of consumption remains.

One of the founding groups of products for the contemporary labeling system is organic food and farming. Early experiments in Germany in the 1920s and 1930s developed a system of standards for organic products, but a very small market hampered this effort at first, and the Nazi dictatorship later quashed this movement. Organic food and farming reappeared in the early 1970s in a coordinated initiative by the Soil Association in the United Kingdom and the California Certified Organic Farmers, who introduced a scheme whereby farmers undertook to grow crops to a set of organic standards, were independently inspected, and in return could display a logo demonstrating to consumers that the farmer had complied with the standards. Fritz Schumacher pioneered this scheme when he worked for the British National Coal Board, and it was Schumacher, as president of the Soil Association, who brought it to organic food. Initially, this scheme was entirely voluntary and largely based on other members of the organic movement visiting the farmers once a year. By 1979, state law in California allowed those flouting the organic standards to be prosecuted for doing so, and later consumer law across the world began to recognize the organic standards. Globally, the most important factor in the development of these standards was the acceptance by the European Community in the 1980s of the necessity of common European regulations for products labeled as organic.

By the 1980s, the organic movement was no longer alone in labeling its products with various schemes. Chains such as the Body Shop were growing very quickly through what they described as "ethical trade," although at this time, this concept was often vague and open to accusations of marketing hype. Throughout the postwar period, church groups and development organizations such as Oxfam had traded a range of handicrafts, based on the concept of a fair—rather than market—price. These goods were sold exclusively through church meetings and networks, a range of specialist shops, and some charity shops. The "alternative trade organizations," which had pioneered the distribution of these goods realized during the 1980s that there was a limited market for these craft items, which led them to increasingly focus on commodities such as coffee and chocolate. The first of these initiatives was Max Havelaar coffee, which was not only Fair Trade but was independently certified as such. These products could be traded outside church and charitable networks, so the opportunity to benefit the trading partners in the global South was greatly increased.

The launch of Fair Trade products into the marketplace became very successful, as consumers were able to identify the products. Along with products that were organic, or those not tested on animals or manufactured in sweatshops, it was argued that ethical consumption could be conducted through large-scale retailers and on the main street. Retailers realized that this was part of a more general trend for consumers to be concerned about the provenance of their purchases and to trade up to more luxurious brands. Therefore, although seeking to change the terms of trade, Fair Trade and other ecological labels were viewed by retailers as just one niche in a general trend.

The organization of Fair Trade became increasingly more sophisticated with the launch in 1997 of Fairtrade Labelling Organizations International (FLO), which sought to coordinate the message of Fair Trade across the diversity of national initiatives. This led in 2002 to the launching of the International Fairtrade Certification Mark. The mark was supposed to simplify the cross-border certification and trade in Fair Trade goods, which by this time had grown to encompass not only agricultural commodities such as coffee and tea but also some fresh products such as bananas and mangoes and processed products such as honey, wine, spices, and increasingly textiles. In 2004, the World Trade Organization introduced a scheme to recognize Fair Trade organizations as part of their role in organizing global trade.

Some of these labeling initiatives are based on certifying the process of production, such as "organic," and other initiatives, such as Fair Trade, look to the conditions in which a product is processed and traded. With the success of this labeling initiative, a range of such labels have appeared, some tied to social movements and advocacy groups, and others that appear to be more commercially driven. Most of these labels command a premium price in the marketplace, which those in favor of them argue reflects the true cost of production, whereas for others it is an unnecessary charge on the gullible and the source of profiteering. With the popularity of these schemes, many commentators point to the multiplication of these schemes that compete with another and confuse the consumer with their conflicting claims.

All of these initiatives have attracted controversy as they have been called on to justify the claims that they make and the long-term consequences of their actions have been challenged. In the first instance, this is why the products go through a process of third-party certification—to ensure that they meet the standards claimed for them. This also provides a layer of legal protection in that trading standards will monitor and enforce these standards under consumer law. The controversy arises in the claims around difference, which are, in effect, claims for moral superiority, in that by not damaging the environment, exploiting children, or paying an unfair price, other products are morally less worthwhile. In agricultural production, these controversies arise around claims that the environmental protection is either marginal or gained through a drop in productive capacity, thus lessening the opportunity for others to enjoy cheap food. In the United States, this has recently been displaced by criticisms of organic farming, in particular, that it is not different enough from nonorganic food, in that big business has diluted the practice of organic farming. Claims that organic food is healthier than its nonorganic counterparts have been the subject of continued controversy, with bodies such as the U.K. Food Standards Agency publishing critical reports about these claims.

Fair Trade has similarly met with criticisms that revolve principally around discussions about what is meant by development and the correct relationship between nations. The advocates of Fair Trade argue that paying a reasonable sum for the goods can provide a decent income for the communities and individuals producing the commodity. This is a direct transaction between people based on mutual respect and solidarity, rather than a faceless one in which an unfair market exploits people. The market critique of this argument is that these schemes may prevent the market mechanism from working, confusing the signal that is being sent to these producers to either become more efficient or leave the market for ones in which they will be better rewarded. Criticisms from another part of the political part of the spectrum view these schemes as patronizing, in that they do not address the wider causes of inequality and perpetuate the image of people in the South as dependent. Others have argued that despite the intention of many of the founders of these initiatives, they do not resist the global market but are, rather, the exemplar of liberal theory, providing consumers with the information to behave virtuously in the market place.

These philosophical controversies are often matched by practical considerations that, as the schemes are driven by market success, they are not able to truly secure the changes that they are aiming at. An economic downturn may see these products as premium-priced ones dropped by consumers with a disproportionate effect on producers and processors locked into the system. Equally, a boom may see market prices rise above the agreed price and so disadvantage the producers, as it threatens the logic of the scheme.

These schemes are designed to allow people to engage in political consumerism, which some commentators such as Naomi Klein argue is a way of redressing the balance between consumerism and citizenship in capitalist societies. For Klein, commercial interests and consumerism have eroded the opportunity for people to act in traditionally political ways, and thus, by politicizing consumerism, the chance to put forward a political program is recreated. Social theorists such as Ulrich Beck and Daniel Miller point to how people use consumer products to project their social identity and, through this, create new ways to be politically active. For Beck, this has become the most powerful way of expressing a new form of citizenship in a globalized, consumer-driven society. Others such as Zygmunt Bauman argue that political consumerism cannot achieve these ends, as it is dependent on the inequalities of society, in that the wealthiest and most powerful get to set the agenda through their purchasing, which undermines democracy. Andrew Potter and Joseph Heath argue that some forms of political consumerism can further the exploitative aspects of capitalism, as the elements of revolt or resistance are repackaged as “cool.” That through convincing people they are expressing themselves, political consumerism can further embed the practices that it seeks to end. Potter and Heath are particularly scathing about organic food, arguing that it only serves to differentiate consumers and is based on scant evidence.

Labeling schemes such as Fair Trade and organic have certainly transformed many aspects of how consumerism operates, offering some consumers and producers the opportunity to relate to one another in a novel manner. The innovation that this represents has been subject to criticism from those who view it as unnecessary or not able to create the systemic changes that many would hope. How powerful this intervention has been is evident from the number of imitators it has attracted and the disproportionate amount of media coverage it gathers when compared with its share of sales. Institutionally, these interventions are now well founded and unlikely to be dislodged. Whether they are capable of spreading their influence beyond their existing limited but influential networks will mark the next stage of the development of the movement.

**See Also:** Ethically Produced Products; Fair Trade; Green Consumer; Organic.

### Further Readings

Bauman, Zygmunt. *Liquid Times: Living in an Age of Uncertainty.* Queensland: Polity, 2009.
Beck, Ulrich. *World at Risk*. Queensland, Australia: Polity Press, 2009.
Heath, J. and A. Potter. *The Rebel Sell. How the Counterculture Became Consumer Culture.* Chichester, UK: Capstone, 2006.
Klein, N. *No Logo*. London: Penguin, 2001.
Michelleti, M. and D. Stolle. "Mobilizing Consumers to Take Responsibility for Global Social Justice." *The Annals of the American Academy of Political and Social Science*, 611/1:157–75 (2007).
Miller, Daniel. *Capitalism: An Ethnographic Approach (Explorations in Anthropology).* Oxford: Berg, 1997.

*Matt Reed*
*Countryside and Community Research Institute*

## Cleaning Products

Detergents, household cleansers, soaps, and other products are used for cleansing in homes or businesses. Cleansing products numbering in the thousands are found in average homes, including products for specific tasks in kitchens, washing machines, bathrooms, tile surfaces, on carpets, on glass, floor strippers, floor cleansers, drain cleaners, stainless steel cleansers, and other uses such as deodorizers and disinfectants. The chemicals used range from natural agents to synthetics.

Cleansers that are made from chemicals have an environmental impact in three ways. The manufacture of these cleansers, which usually contain hazardous chemicals, affects the environment of their place of origin. When they are used, the hazardous chemicals are released into the atmosphere. They may affect breathing or may drift away into the atmospheres. The disposal of the containers for these products causes them to end up in landfills, where traces of the products add to the chemical mix of the landfill.

Studies of the harsh chemicals in industrially manufactured cleansers show that there are over 160 chemicals in these products. Examinations of people who use these products at home and never at work show that they had absorbed traces of many of these chemicals. Many of these chemicals are carcinogenic and can cause nerve or brain damage or contribute to birth defects. These chemicals are long lasting, though it may take considerable time and constant exposure for toxic effects to occur.

Other kitchen cleansers include soft soaps for washing dishes and utensils. These products can be quite mild and may be biodegradable. However, because of the toxic nature of so many chemicals contained within commercial cleaning products, consumers have turned toward "green compounds" for their cleaning materials.

Soaps available to green consumers are often available commercially in well-known products, such as Ivory Soap, or soaps made by organic crafters. In many countries the fatty acid used to make soap is a vegetable oil such as olive oil (castile soap). These soaps are viewed as green because they are made from renewable oils that are sustainable. In much of the world, beef tallow is rendered into soap. Glycerine can be included as a moisturizer; however, soap in past decades also has included triclosan, which acts as an antibacterial agent. Triclosan, however, breaks down in sunlight to form dioxin, which is a known carcinogen.

Green cleansing products are biodegradable and biorenewable products formulated to clean many types of dirt. In general, these products can be used on spills or dirt on surfaces that are not harmed if water is the only cleanser used. At times, "elbow grease" has to be applied to complete the cleansing task.

There are currently laundry products that are advertised as "all natural." These products do not use synthetic chemicals and are biodegradable cleansing agents without toxins or perfumes. Products of this type may be less toxic to allergy suffers with sensitive skin or for babies. Products of this type work as does all soap: Saponins are released when they come into contact with warm or hot water. Agitation increases the saponins released, which then act as a natural surfactant.

Laundry cleansers are available that clean, soften, bleach, and odorize laundry. Many of these products are not very green because they are derived from petrochemicals. The age-old method of washing using soap and air drying was replaced by the use of chemical-filled detergents and electric clothes dryers. Green consumers will find that reverting to the traditional method of washing and open-air drying uses the least amount of energy, is the least environmentally damaging, and is the least costly as well.

Many products are advertised as "green" cleansers that are used as antiseptics, glass cleaners, tile and tub cleansers, and for a hundred other uses. These products are made by well-known companies such as Procter & Gamble, Johnson & Johnson, and other smaller companies. The claims made for these products, which are often actually green in color, is that they are effective cleansing agents and are also less environmentally harmful than earlier products.

Products that are environmentally green, ecosafe cleansers are chlorine-free and biodegradable. They also usually include abrasives for scouring. All-purpose green cleansers include Bon Ami, Citra Solva, Seventh Generation, Ecover, Trader Joe's (dishwashing detergent that is phosphate and chlorine free), Bi-O-Kleen, and others.

Many people who strive to live a green lifestyle make their own cleansers, some using plant-based soaps. Other traditional products used for cleaning are baking soda (also used to remove odors from refrigerators) and vinegar, which is used as a cleanser to remove mold and bacteria. Vinegar is an acid, but it is also very biodegradable. Distilled white vinegar is the most popular type of vinegar used for cleaning.

**See Also:** Biodegradability; Electricity Usage; Environmentally Friendly; Green Consumer; Green Marketing.

### Further Readings

Hunter, Linda Mason and Mikki Halpin. *Green Clean: The Environmentally Sound Guide to Cleaning Your Home*. New York: DK Publishing, 2005.

Imus, Deidre. *Green This!: Greening Your Cleaning*. Vol. 1. New York: Simon & Schuster, 2007.

Lansky, Vicki. *Baking Soda: Over 500 Fabulous, Fun, and Frugal Uses You've Probably Never Thought Of*. Minnetonka, MN: Book Peddlers, 2004.

Lansky, Vicki. *Vinegar: Over 400 Various, Versatile, and Very Good Uses You've Probably Never Of*. Minnetonka, MN: Book Peddlers, 2004.

Loux, Renee. *Easy Green Living: The Ultimate Guide to Simple, Eco-Friendly Choices for You and Your Home*. Emmaus, PA: Rodale, 2008.

*Andrew Jackson Waskey*
*Dalton State College*

# Coffee

Coffee is a beverage made from roasted and ground beans grown in tropical climates. Coffee's origins can be traced back at least as far as the 9th century to areas of Africa, where it was initially consumed for religious rituals. Coffee consumption spread rapidly throughout Africa, from Egypt to Yemen, reaching the Middle East by the 15th century. Coffee trade and consumption spread to Europe, particularly Italy, from where it was introduced to the rest of the continent. The British East India and Dutch East India companies, as part of colonization, extracted coffee from many Asian and African countries. Coffee came to America during colonization but did not become widely popular until tea became unavailable during the War of 1812. The popularity of coffee grew considerably during the Civil War, when new and improved methods of production and consumption became available. Thereafter, coffee became an everyday American beverage. Worldwide, more than 500 million cups of coffee are served.

Nearly 125 million people around the globe depend on the cultivation and sale of coffee, including these fair trade coffee growers in El Salvador. After oil, coffee is the world's second most valuable commodity.

*Source:* Adam C. Baker/Wikipedia

The primary coffee-importing countries are the United States, Germany, Japan, Italy, and France. The U.S. coffee market is valued at $19 billion annually, serving approximately 161 million consumers by 150,000 full- and part-time workers. More than half of Americans aged 18 or older each drink on average 3.5 cups of coffee daily. Scandinavia has the highest coffee consumption, at more than four cups per day per person.

Nearly 125 million people around the globe depend on the cultivation and sale of coffee. After oil, coffee is the world's second-most-valuable commodity, worth approximately $60 billion annually to the world economy. However, less than 10 percent of that money ends up with the farmers who grow the beans. The world's coffee markets are dominated by four major corporations: Nestlé, Philip Morris, Procter & Gamble, and Sara Lee, accounting for 60 percent of the U.S. coffee trade and 40 percent of the global coffee trade. According to the World Bank, 25 million small producers of coffee depend on the trade for their livelihood. More than 500 million people are involved in coffee and coffee-related industries around the globe. Coffee is grown in more than 50 countries. The top coffee producing countries are, in order, Brazil, Vietnam, Colombia, Indonesia, India, Mexico, Ethiopia, Guatemala, Ivory Coast, and Uganda. Global production exceeds 16 billion pounds. This equals nearly 125 million bags in 2007, an increase of 3 percent from 2006.

The political economy of coffee is complex. There are significant inequities in compensation for those who do the labor of cultivation and destructive effects on the environment. The price that coffee brings does not keep up with the price of production, and it is small farmers who pay the price. At present, global revenue for coffee is in the range of $55 billion, of which only $7 billion (approximately 13 percent) actually is returned to the exporting countries. Since its high in 1997, the average price paid for coffee has fallen 80 percent. The poverty and political unrest resulting from inequities in profit and pay have shaken many coffee countries. In addition, the cultivation of coffee takes a tremendous toll on the environment, including degradation of water quality and soil through pesticide use and loss of biodiversity above and below ground. Of particular import is the effect that coffee cultivation has on loss of rainforest habitat.

There is a strong correlation between full-sun coffee production and deforestation. Traditionally, coffee was grown in the shade of trees, which also provided critical habitat for animals and insects and other vegetation. However, over time, larger yields are produced more quickly under direct sun, leading to deforestation. For example, of the 50 coffee-producing countries, the highest 37 were also the highest in deforestation. The primary environmental impacts include conversion of primary forest habitat, soil erosion and degradation, liquid waste and sewage from processing, and agrochemical use and runoff of pesticides.

In recent years, heightened public awareness about the effect of full-sun-grown coffee on people and place has resulted in increased demand for shade-grown coffee.

The Fair Trade–labeled coffee movement, started in the Netherlands in 1988, united 23 Fair Trade producer and labeling initiatives in North America, Asia, Africa, New Zealand, Europe, and Latin America. The Fairtrade Certification Mark designates coffee originating from circumstances in which farmers are paid a fair wage ($1.26/pound, regardless of market conditions), can participate in credit programs, and have decent working conditions. Fair Trade benefits individuals, families, communities, and the environment through improved working conditions and practices leading to environmental conservation.

**See Also:** Beverages; Certified Products (Fair Trade or Organic); Fair Trade; Markets (Organic/Farmers); Resource Consumption and Usage.

## Further Readings

Allen, Stewart Lee. *The Devil's Cup: A History of the World According to Coffee.* New York: Ballantine, 2003.

Pendergrast, Mark. *Uncommon Grounds: The History of Coffee and How It Transformed Our World.* New York: Basic Books, 2000.
Roseberry, William, et al., eds. *Coffee, Society, and Power in Latin America.* Baltimore, MD: Johns Hopkins University Press, 1995.

*Debra Merskin*
*University of Oregon*

# Commodity Fetishism

A characteristic of capitalist economies is that goods are bought and sold. A commodity is a marketable good or material that can satisfy human needs. It can be a raw good, such as gold, grain, or diamonds, or a finished good, such as shoes, bread, or heating oil. The term *fetish* arose in 19th-century anthropological discourse about religious cultures that attach intense feelings and desires to specific inanimate objects they believe have great power.

German economist and philosopher Karl Marx saw a relationship between these ideas. In the classic work *Capital* (1867), Marx analyzed capitalist society and the connection between forms of production and corresponding sociopolitical forms to understand the underpinnings of this economic system. Marx argued that a mystical, magical quality is derived from some commodities not because of what they can do for us (use-value) but, rather, from what they mean to us (exchange-value). He called this commodity fetishism.

Two forms of value are central to understanding how the process works: use-value (its inherent capability of satisfying human wants or needs) and exchange-value (its interchangeability with other products). In the first case, the value of a particular product is connected with what makes up the product—its physical characteristics. For example, bread is a commodity with physical properties that cannot be transferred. We cannot eat oil, for example, and we cannot heat our homes with bread. Use-value is life sustaining and has a specific concrete purpose. In traditional societies, in which agreements were often informal and social relations were intimate and close, products were often made and consumed within sight of one another. People often baked their own bread, for example, or knew the person who baked the bread they bought. But in capitalist societies, where workers are often invisible and the material production process hidden behind the veil of consumer culture and branding, people often have little sense of the effort that went into the making of a good.

Exchange-value only occurs under capitalism and is the expression of the value of one commodity as equivalent to a quantity of another commodity. So rather than the value based in the human effort and labor that creates the product, exchange-value is about what one might get by exchanging one thing for another. Thus, whereas wheat and coffee might have different values for different weights, a common value between the two is found in the process of the exchange that is unrelated to the creation of the product.

When a commodity is fetishized, people form a relationship with it—perhaps it is a car, a computer, a pair of shoes—and equate these things with qualities, such as the power to make us popular, or respected, or beautiful, and lose connection with another. Marx predicted that this relationship with things would come to define our relationships. Whereas use-value is directly related to what a product can or cannot do, exchange value is more abstract, and more about the meaning of a particular product. In commodity fetishism, what a product/object is originally intended for is less important than what it means to the culture. This stage goes beyond what people's needs for a product might be in a practical sense and reaches into what a product might say about the possessor of it.

In the religious use of the term, a fetish carries qualities and characteristics that believers likely do not themselves posses. By believing in the power of an object, the faithful thereby are able to come closer to experiencing whatever state of being is represented by the object. Similarly, a commodity, such as a cosmetic or face cream, for example, might not inherently have any special qualities, but the advertising and marketing that accompany it imbue that product with these powers. Thus, in some ways, the money spent on the product is exchanged for what people believe it can do. In this view, whether or not an expensive facial cream can actually make a woman look younger is less important in many ways than that she believes it will. What cosmetic maker Charles Revson said of his company (Revlon)—"we don't sell cosmetics, we sell hope"—captures the essence of commodity fetishism. Thus, the product, which, in the mind of the consumer, may have no connection to actually how, when, where, or by whom the product was made, is mystical. The mysterious quality contained in both concepts comes together in the idea of the transformative qualities of something created through labor. Instead, the magical qualities of what a product is thought to do for someone supersede what its fundamental function is.

Similarly, the person who makes the product might not have any idea where the product he or she has made will be sent, who will use it, or how much might be paid for the good. Thus, he or she is similarly alienated from the consumption of the product and might feel interchangeable as well, as who makes the product is less important in this phase of capitalism than the product is. Thus, at both ends, producer and consumer are alienated from one another and maintain a *false consciousness*, a term associated with Marx's colleague Friedrich Engels, because ideology of the society tells them that they are in relationship to each other when they are not. What feels like a compulsion to buy a product because of what it does for us is, in fact, informed by what we are told it can do. The maintenance of false consciousness is necessary to the continuation of the ideological system in which citizens participate in their own oppression by not peeling back the veil covering the working conditions of people who make products, for example, and continuing to consume what is produced. The argument is that industrial society has lost awareness of the labor necessary to create the product—that there is a distance between the production of material goods and their perceived value. Thus, currency and commodities come to have value because it is agreed that they do.

Therefore, the labor that produces the commodity has value, but it is not necessarily attached to the product itself. The product, containing properties often communicated through advertising, marketing, and branding processes, is thought to be able to do something for consumers, regardless of the conditions of its creation. The process of production through consumption remains largely invisible as well. Human relationships to and through products, regardless of the conditions of the production of the commodity, largely replace human relationships with each other. The intention of this Marxist theory is to reveal this underlying process, which, in modern times, is increasingly visible in the conditions of workers.

In this model, the market comes to seem to be a living, breathing entity that responds to human activity rather than a mechanism that is manipulated by human beings. Thus, the political, economic, historical, and cultural circumstances are obscured behind the façade of often geographically, as well as psychically, distant conditions of production. Consumer dissociation is the outcome of the loss of inner connections between the people who make goods and those who buy them. Neither side is fully aware of the circumstances of the other. The product becomes the unit of social relations. A person can thereby be defined by what he or she does in terms of the production of goods or services and by what he or she buys. Thus, commodity fetishism operates to naturalize and reify the believed value of something, and the power of imagination imbues goods with what we believe they can do for us.

**See Also:** Consumerism; Consumer Society; Materialism; Needs and Wants; Social Identity.

### Further Readings

de Graf, J., et al. *Affluenza: The All Consuming Epidemic.* San Francisco, CA: Berrett Koehler, 2001.

Lukács, Georg and Rodney Livingstone. *History and Class Consciousness: Studies in Marxist Dialectics.* Cambridge, MA: MIT Press, 1972.

Marx, Karl. *Capital: Volume 1: A Critique of Political Economy.* London: Penguin, 1992.

Miklitsch, Robert. *From Hegel to Madonna: Towards a General Economy of "Commodity Fetishisms."* Albany: State University of New York Press, 1998.

*Debra Merskin*
*University of Oregon*

## Commuting

For the past century, urban and suburban areas have increased exponentially in population. This growth, coupled with these communities' high cost of living and many families' concurrent desire to live within traveling distance of cultural and entertainment opportunities, has made commuting a ubiquitous part of contemporary life. Commuting alters how cities are constructed and marketed to potential community members. It also affects regulation of both personal and public transportation. Increased awareness of global warming and economic and planning concerns has caused commuting to become part of the public consciousness. Commuting is more complex, however, than carpooling, high-occupancy-vehicle lanes, mass transit, and light rail. Indeed, the necessity for and various forms of commuting may be viewed as the result of a number of subtle and intertwined events. Over the past generation, individual choices affected personal and professional lives that hinge on

The average worker in the United States spends over 100 hours commuting to and from work each year.

*Source:* iStockphoto

commuting options. These choices are, in part, a result of increased costs of owning and operating a car, population increases in the United States, the immigration bubble, and the decreased number of workers added to blue- and white-collar industries. Various patterns of commuting, from the traditional suburban to urban commute, reverse commuting, and cross commuting, have expanded conceptions of commuting from a way to save time and money to broader concerns of pollution reduction, population-resource efficiency, and job creation.

Since 1950, urban and suburban areas' populations have increased exponentially. For example, the 2000 U.S. Census found the assumption that the U.S. population would increase by 25 million people each decade from 1950 through 2050 to be too modest. Between 1990 through 2000, the United States increased by 33 million, largely as a result of immigration. In addition to immigration, fewer workers retired from their jobs than the historical model predicted. The increase in the number of immigrant workers, coupled with prolonged careers, overloaded local streets, city highways, and expressways. Municipalities emphasized the cost savings engendered by commuting practices such as carpooling and time-saving measures like the use of high-occupancy vehicle lanes on expressways as one way of alleviating the stress on streets and highways.

## Suburbanization and Commuting

Suburbanization also affected commuting practices because in recent decades more individuals have moved from cities to rural areas adjacent to metropolitan areas. In many instances, once the rural region reached a saturation point, these areas were, essentially, suburbanized. The 2000 U.S. census reported nine metropolitan areas with over 5 million inhabitants. According to the 2005 census, the number of metropolitan areas with over 5 million inhabitants had increased to 12. Thus, what changed was where people lived, but not necessarily where they worked. This increase in population in urban and suburban areas affected commuting practices.

Commuting can be considered in terms of time and pattern. The average worker in the United States spends over 100 hours commuting to and from work each year. Thus, population growth and distances that people will travel to work assume that the practice of commuting is necessary for citizens to earn a wage and to maintain or increase their quality of life.

Three patterns of commuting are traditional commuting, reverse commuting, and cross commuting. Traditional commuting is where an individual or group travels from the suburbs where they live to an urban area for work. Based on data from the 2000 U.S. census, traditional commuting constituted 19 percent of all commutes. Reverse commuting refers to individuals or groups traveling from an urban area where they live to a suburban area for work, an occurrence that accounts for 9 percent of all commutes. Cross commuting occurs when individuals or groups travel from a suburb where they live to a suburb where they work, and constitutes over 64 percent of all commutes. Each pattern, of course, is contingent on where workers reside and where their place of employment is located. As jobs have shifted increasingly to the suburbs, the existing infrastructure has been taxed.

In the United States, in cities with populations of greater than 250,000, travel times to and from work vary greatly, ranging from a high of 38 minutes in New York City to 16 minutes in Corpus Christi, Texas. Although individuals find it necessary to commute, the vast majority of commuting options fail to include the potential savings that alternative forms of commuting offer. As of 2007, over 75 percent of working-age individuals travel to and from work alone via automobile, 10 percent carpool, 5 percent use mass transit, 3 percent walk, and approximately 2 percent travel to and from work by other means. The remainder of the population works at home. Alternatives to driving such as walking, biking, carpooling, telecommuting, and mass transit are not currently seen as viable for many despite the realization that alternative means would save time and money.

## Attitude Adjustment and Alternative Modes

Programs that seek to change people's views regarding alternative modes of commuting have become increasingly common of late. In New York City, for example, "Bike Commuting 101," a course offered by Bike New York, teaches individuals to select appropriate gear and equipment, ride safely, and save money. In 28 communities across the United States, light rail systems transport individuals to and from work in an ecologically and economically sound manner. These communities have used complementary educational programming to support their task of altering people's commuting habits. In the United Kingdom and the United States, programs have been implemented to allow employees to commute during off-peak hours and to telecommute. In addition, many communities' bus and rail systems are integrated in a way that allows people to commute to work using multiple transportation systems.

Offering a variety of commuting options, and marketing these programs effectively, can entice individuals to use the transportation infrastructure provided. Despite these successes, individuals are often caught between the limitations of a commuting practice (e.g., the Bike New York organization suggests that their ideas work for commutes of less than 10 miles) and the desire to be a more environmentally responsible member of a community. Economic realities often constrain the ability to go green. The need to work in a particular area may proscribe walking or biking to work. So, too, individuals may lack options for mass transit because of a lack of train or bus service in their community. Interestingly, job satisfaction surveys often consider the amount of time spent and ease of access to commuting options.

Commuting can have a key role in an environmentally sound existence. However, until the infrastructure is designed to enable more cities to meet the needs of their inhabitants and there is a greater variety of jobs that make it possible for people to work from home or closer to home, commuting will remain necessary for many individuals and will make it difficult for groups of people to take advantage of the cost-saving and environmentally sound practices that other options provide such as carpooling, light rail, walking, cycling, and off-peak commuting. There are a variety of ongoing challenges that cities, states, and individuals will need to confront to make commuting options more accessible and attractive to the community members.

**See Also:** Automobiles; Carbon Emissions; Environmentalism; Lifestyle, Sustainable; Public Transportation.

### Further Readings

Cervero, R. and R. Gorham. "Commuting in Transit Versus Automobile Neighborhoods." *Journal of the American Planning Association,* 61/2:210–25 (1995).

Cullingworth, J. *Planning in the USA: Policies, Issues, and Processes.* New York: Routledge, 2003.

Noland, R. B. and K. A. Small. "Travel-Time Uncertainty, Departure Time Choice, and the Cost of Morning Commutes." *Transportation Research Record,* 1493:150–58 (1995).
Transportation Research Board. *Commuting in America III. The Third National Report of Commuting Patterns and Trends*. Washington, D.C.: Transportation Research Board, 2006.

*Stephen T. Schroth*
*Jason A. Helfer*
*Marie-Alix Powell*
*Knox College*

# Composting

Composting is an important agricultural practice of returning nutrients to agricultural soils. It is also a method for both maintaining and improving soil health and fertility in agricultural fields. Composting has specific goals and a relatively concise set of practices, and many international monographs and journal articles advocate its adoption for soil and farm health. By focusing on composting's goals and practices, this article outlines the role and rationale of composting in ecoagricultural regimes.

Turning consumer food waste into compost for organic agricultural soils is one way in which food consumption can go green.

*Source:* iStockphoto

There are two ways humans have added necessary nutrients to agricultural fields over human history. One is the method developed during World War II in which chemicals used in weaponry were applied to agricultural fields and were found to help boost agricultural productivity, speed up plant growth, increase yields, and reduce the need for various agricultural inputs, especially labor and compost. The result of this method of adding chemicals to agricultural fields, coupled with the development of high-yielding hybrid seeds, increased inputs of pesticides and herbicides, increasing the commercialization and consolidation of farming under national and international trade regimes that help subsidize such chemical-based farming, as well as the ascendant popularity of this farming method and the research and teaching of this method in agricultural land grant universities throughout the world, has come to be known as the "Green Revolution." The Green Revolution is coming under increasing criticism from eco-agriculture advocates because of this dependence on fossil fuels and the deleterious effect these chemicals have on soil health.

The other way humans have added necessary nutrients to agricultural fields is by composting. Composting is the process of combining organic items that are rich in carbon, such as animal manure or dried leaves, with organic items that are rich in nitrogen, such as grass and food waste, into humus. One practice of composting is to make sure there is a correct ratio between these "brown" and "green" components; the ratio that is traditionally sought is a chemical ratio of 30 parts carbon to 1 part nitrogen. These organic matters are transformed in a compost pile by the presence of microorganisms, especially bacteria, which use enzymes to chemically break down the organic matter. Invertebrates then help to physically break down the compost.

The method of creating compost entails gathering the organic matter into a large enough pile or bin and adding water; after this, heat from the sun and naturally occurring oxygen mix with the pile, and the process of decomposition begins. It is important for the compost pile to achieve a high enough temperature—over 100 degrees Fahrenheit—for seeds and harmful bacteria and blights to be successfully killed and broken down; otherwise, they are spread back onto agricultural fields. If the compost pile becomes warmer than 140 degrees Fahrenheit, then the beneficial bacteria and organisms in the compost can be killed and the compost loses its life and health-giving properties. Finding a successful mix and ratio of carbon/nitrogen and maintaining a proper temperature and moistness level must be learned. Thus, making compost is part art form and part agricultural practice.

Building soil health through composting is a many-year process, so ecoagricultural farmers usually analyze soil samples to see where they need to add compost, and especially other additives like lime and/or calcium to help balance soil pH levels. If done correctly, composting creates organic humus that is colloidal in nature. This humus is then added to agricultural soils and helps with all of the following: increasing nitrogen fixation in the soil, supporting the healthy presence of soil bacteria and organisms, continuing maintenance of soil humus, and providing a habitat for the proliferation of root rhizomes and worms

Properly composted humus is rich with trace minerals, bacteria, and organisms, so that the plants eating this soil via feeder roots become healthier. Many ecoagricultural farmers claim that this health is then passed up the food chain.

In addition to composting animal manures, food waste, and yard trimmings, various movements are underway to compost "night soil," or human excrement; sewage sludge; and greywater. Other composting methods exist in addition to the predominant method of placing organic matter into a compost heap. These methods include, but are not limited to, vermiculture and biodynamic composting methods.

Regarding green consumerism, studies suggest that approximately $14.5 billion of food is thrown away daily in the United Kingdom alone. This includes 4.4 million apples and 720,000 loaves of bread. Studies in the United States suggest that up to $100 billion worth of food, or more than 40 percent of what is grown, is annually lost or thrown away. Most of this "wasted" food ends up in landfills or is incinerated. Efforts to sort out residential food and yard waste could potentially capture this waste and, through local initiatives, turn it into nutrients to return to the soil through large-scale composting projects undertaken at local landfills or farms. Closing this loop and turning consumer food waste into compost for organic agricultural soils is one way that the consumption of food, in both rural and urban areas, can potentially be greened.

**See Also:** Green Food; Slow Food; Solid and Human Waste; Vege-Box Schemes.

### Further Readings

Bird, Christopher and Peter Tompkins. *Secrets of the Soil: New Solutions for Restoring Our Planet*. Anchorage, AK: Earthpulse, 1998, 2002.

Cotton, Matthew. "Nine Composting Trends." *Waste Age* (March 1, 2003). http://wasteage.com/mag/waste_composting_trends (Accessed December 2009).

Howard, Albert. *An Agricultural Testament*. Emmaus, PA: Rodale, 1972 [1943].

Ward, Megan. *Composting: A Beginner's Guide*. http://academics.sru.edu/MacoskeyCenter/Publications/Composting_Booklet.pdf (Accessed April 2009).

*Todd J. LeVasseur*
*University of Florida*

# Computers and Printers

The ubiquity of computers has had a considerable effect on energy consumption. Although generally making life and work more efficient, computers' electricity consumption is considerable, with their energy consumption being greater than anything they have replaced, burning energy even when not in use. Furthermore, rapid turnover of computers and peripherals—with many consumers and corporate customers replacing their computers every few years—leads to a great deal of material waste, little of which is recycled or biodegradable. The environmental impact of computer use was not fully understood or foreseen at the time personal computers and office computers were introduced; computing caught on with such rapid popularity that, as with automobiles before them and factory smokestacks before that, the problem was widespread by the time it was recognized. Green computing seeks to make computers and computer peripherals environmentally sustainable, reducing waste and energy consumption both in their use and in their manufacture.

The term *green computing* originated around 1992, when the U.S. Environmental Protection Agency (EPA) launched its voluntary Energy Star program, later adopted by the European Union, Canada, Australia, New Zealand, Taiwan, and Japan. A voluntary labeling program representing an international standard for energy-efficient consumer technology, Energy Star was one of many early 1990s initiatives designed to combat greenhouse gas emissions and energy inefficiency. The logo is permitted to be displayed on computers, home electronics, and other devices that meet a product-type specific standard for energy efficiency, typically representing a 20–30 percent efficiency gain compared with other products of that type. Arguably one of the EPA's most successful initiatives, Energy Star caught on, and within 15 years of its introduction, more than 40,000 individual products bore the logo, accounting for billions of dollars in energy savings every year. The standards for each product type are set by the EPA or the Department of Energy, depending on the product, and are updated periodically; as of summer 2009, Energy Star 5.0 is the applicable standard for computers, and Energy Star 1.0 was implemented for computer servers a few months earlier.

The motive for increasing the energy efficiency of computers incorporates more than concern for the environment and more than saving money on energy consumption: Inefficient electronic devices waste energy in the form of heat, which can be problematic for offices in which computers generate waste heat and run up the cost of air conditioning, but it is even more critical at server farms, where vast numbers of computers process data continuously in rooms that need to be cooled to keep components from overheating. Energy efficiency thus leads to computers that run better and longer, not just more cheaply.

## Computers and Energy Star

Some of the actions taken to meet Energy Star standards were simple, but were no less an improvement for that. Screensavers, for instance, originated because cathode ray-tube displays (as all early computer monitors were) were vulnerable to phosphor burn-in—the prolonged display of the same pixels over a long period of time (such as an unchanging computer screen) would result in permanently "burning" that image into the monitor, where it would remain as a ghost image, degrading display quality. Screensavers, by constantly updating the display, helped to prevent this from happening, but the same software could be used to blank or turn off the display when not in use, dispatching two birds with one stone. (In fact, the first screensaver did only that, blank the display—the fanciful graphics of later screen savers were selling points of commercial products.)

In 2007, the Energy Star 4.0 standards for computers incorporated the 80 PLUS standard, and independently developed energy-efficiency standards for computer power supply units. The 80 PLUS standard specifies energy efficiency levels at 20 percent, 50 percent, and 100 percent of rated load of at least 80 percent. Eighty percent efficiency at those load levels entitles the unit to 80 PLUS certification, 80 PLUS Bronze requires efficiency of 82 percent/85 percent/82 percent, 80 PLUS Silver requires 85 percent/88 percent/85 percent, and 80 PLUS Gold requires 87 percent/90 percent/87 percent. The Energy Star 5.0 standards incorporate the 80 PLUS Bronze requirements.

Various power management methods can reduce unnecessary energy consumption by computers and peripherals by turning off components that are not in use, including not only monitors and printers but also hard drives and (in "hibernation mode") the CPU. Physically smaller hard drives also consume less power, per unit of memory, than larger drives. Although flash drives—such as those used for low-capacity MP3 players or keychain backup drives—consume less power per unit of memory at low capacities, they exceed the power needs of equivalent-capacity hard drives at larger capacities and read/write at significantly slower speeds. The graphics processing unit is a heavy energy consumer in many computers, particularly these days, when the popularity of processing-intensive video games results in powerful GPUs being installed even in computers for work and home offices, where such games are not the need being addressed.

Just as commuting shorter distances reduces the emissions of your workday commute, regardless of changes to one's car, so too can computers be made more efficient by focusing on areas other than the hardware itself.

Algorithmic efficiency is the measure of an algorithm's consumption of resources; algorithms are sets of instructions, such as those in computer programs, and even the best programmers rarely compose a program of maximum efficiency in their first draft. Because early revisions and tweaks are aimed at functionality and effectiveness—the quality of the final product—rather than efficiency, subsequent versions and improvements can actually result in less-efficient software. With the price of memory and processing power increasing at substantial rates for the last few decades, software is often accused of being inefficient simply because it has the luxury of being so, whereas early programmers were forced to be efficient with their use of resources because they had so few resources to work with. Microsoft in particular is regularly accused of producing operating systems that are terribly inefficient in their use of resources, including energy and memory.

Algorithmic efficiency is typically considered in terms of a program's use of memory while running and the time it takes the algorithm to complete, but with regard to energy consumption, the measurement called "performance per watt" becomes relevant as well. Performance per watt measures the rate of computation produced for every watt of power consumed. A seemingly efficient program that runs quickly and seamlessly may do so because of the way it consumes power resources, and it could potentially be streamlined. There are disputes as to the energy-efficiency ramifications of algorithmic efficiency, particularly as computers consume power even when sitting idle. A Harvard study famously claimed that the average Google search was responsible for 7 grams of carbon emissions; Google contests the claim and has stated that the resulting emissions are less than 5 percent of the Harvard figures.

## Computers and Recycling

Electronic waste, including computers, is problematic even beyond the amount of plastic that could be reclaimed: the circuitry includes contaminants such as lead, cadmium, and beryllium, which are harmful in landfills and should be processed separately and reclaimed. Various journalistic exposés have revealed that extraordinary amounts of electronic waste are shipped from developed nations to developing countries, often under the guise of recycling. The fast obsolescence and frequent upgrades of computer technology make electronic waste a serious and growing concern. More than two-thirds of the heavy metals in U.S. landfills come from electronic waste, despite more than half the states requiring that electronics be kept out of such landfills. The amount of electronics discarded every year is measured in the millions of tons and greatly outnumbers the amount that is brought to recycling centers. Unwanted computers can also be donated to charities with low computing needs. Some manufacturers and vendors of "green computers" make a point to use recycled materials in as many of their components as possible. In recent years, major manufacturers have joined this effort, offering green options in addition to their existing product line. Dell, for instance, recently introduced the Dell Studio Hybrid, which is 80 percent smaller than the average desktop, made from (and packaged in) recyclable materials, and uses two-thirds less power than an average desktop, with an Energy Star 4.0 power supply.

Hewlett-Packard, similarly, has released a number of green printers, which consume half as much power as standard printers and use ink cartridges (a major source of waste) that last substantially longer. The fact that Dell and HP, two of the market leaders, are entering the green computing market indicates that this is more than a passing trend and is, rather, a new way of approaching computer use, design, and manufacture. Stricter European Union regulations have encouraged manufacturers who want to sell in that growing market, and some reports predict the green computer market will quadruple by 2013, despite the effect of the global financial crisis of 2008.

**See Also:** Audio Equipment; Carbon Emissions; Electricity Usage; Energy Efficiency of Products and Appliances; E-Waste.

### Further Readings

Lamb, John. *The Greening of IT: How Companies Can Make a Difference for the Environment.* Indianapolis, IN: IBM Press, 2009.

Poniatowski, Marty. *Foundations of Green IT*. New York: Prentice Hall, 2009.
Schulz, Greg. *The Green and Virtual Data Center*. Boca Raton, FL: CRC/Auerbach, 2009.
Velte, Tob, et al. *Green IT: Reduce Your Information System's Environmental Impact While Adding to the Bottom Line*. New York: McGraw-Hill, 2008.

*Bill Kte'pi*
*Independent Scholar*

# Confections

The greening of the confectionery sector has noticeably increased in pace toward the end of the first decade of the 21st century. This process has seen niche "ethical" confectionery products enter markets previously regarded as mainstream, with major global players such as Cadbury, Mars, and Nestlé adopting a variety of green strategies in order to keep in touch with changing consumer awareness of green issues and shifting demand. Green chocolate-based products continue to dominate the confectionery sector, although other products, such as green chewing gum, have also entered the market. A variety of green, ethical, and sustainable issues have gained ground in the sector, notably Fair Trade, supply chain considerations, the use of natural ingredients, packaging, health considerations, and consumer lobbying.

The majority of cacao trees grow in the Caribbean, Central and South America, and Africa, making cacao the second most important agricultural export commodity in tropical regions. Here, a *Theobroma* cacao tree in the Dominican Republic.

*Source:* Christopher Cooper/Wikipedia

Green, ethical confectionery started as a niche product within the sector, and the chocolate producer Green and Black's soon became emblematic of the ethical market. In 1994, Green and Black's Maya Gold 70 percent cocoa solids dark organic chocolate was the first in the United Kingdom to obtain the Fairtrade mark, and signaled the beginning of business interest in green, sustainable confectionery. Green and Black's demonstrated that sustainable principles could be applied to all aspects of the business, including supporting local communities in Bélize where their cocoa was sourced, and repairing the environmental damage of conventional farming methods. As interest and demand for organic, high cocoa content chocolate increased, Green and Black's was able to make the transition into mainstream retail outlets, extending its distribution. In 2005, Green and Black's was sold to Cadbury, marking a deliberate move into the green market by an international brand.

Nonchocolate confectionery is also present in the green market category. For example, The Natural Confectionery Company market their jelly sweets in the marketplace as containing no artificial colors or sweeteners. The Mexico-based, sustainably-managed company Consorcio Chiclero has produced the world's first 100 percent natural, biodegradable, and certified organic chewing gum. Its Chicza gum is nonsticky, and decomposes to dust within weeks. Most mass-produced chewing gums use artificial, petrol-based polymers and are difficult to dispose of, as they bind easily with any surface they touch. Conventional gum accounts for 78 percent of discarded litter in the United Kingdom, and costs an estimated 150 million pounds a year to remove it from public spaces.

The use of artificial ingredients in confectionery products has gained a good deal of public attention because of their effect on human health and the damage they cause to the environment in their production. For example, high-fructose corn syrup is frequently used as a sweetener instead of sugar or honey in many products, but is criticized for rising obesity, increasing environmental degradation from the intense farming of corn (its principle ingredient), as well as for its high-energy footprint. Palm oil is used as a texture agent by chocolate manufacturers, but its production is associated with the destruction of rainforests and the loss of natural habitat for the orangutan. Countries such as Indonesia, Malaysia, and Papua New Guinea have been identified by Greenpeace as being at risk from producing palm oil. Greenpeace is lobbying companies like Ferrero, Nestlé, Kraft, and Unilever to ensure that their supplies come from sustainable sources.

Cadbury's and Mars were criticized in 2009 for not adhering to a voluntary ban on the use of artificial food colors in their confectionery products, despite assurances made in 2007 that they would do so. It was reported that well-known products such as Cadbury's Creme Eggs, Cadbury's Mini Eggs, Starburst Choozers, and Mars Revels still contained the banned colors.

Consumer lobbying power against the use of certain ingredients was demonstrated when intense pressure from the media, consumers, and the Vegetarian Society forced Mars to reverse the introduction of whey containing animal rennet in Mars Bars, Twix, and Snickers. Mars ran full-page national newspaper advertisements in the form of an apology letter, stating that it was reverting to vegetarian ingredients.

Confectioners are also under pressure to reduce packaging and use more recyclable packaging in their products, particularly as a result of more restrictions on waste destined for landfill. In 2009 Cadbury's, Nestlé, and Mars all launched reduced packaging versions of their Easter eggs in response to consumer perceptions that the products were overpackaged. For Christmas 2009, Cadbury gave its Roses and Heroes chocolates a trial green makeover using recyclable cardboard boxes in place of special edition metal tins, as part of its "Purple Goes Green" environmental strategy to reduce packaging of seasonal and gifting products by 25 percent by 2010.

As the pressure to demonstrate ethical and sustainable credentials has increased, more mainstream confectionery companies have begun green initiatives. Most notable was the announcement by Cadbury that its iconic Dairy Milk brand—approximately 300 million bars sold per year—had achieved Fairtrade Foundation certification, making it the first mainstream chocolate bar to be sold with a commitment to pay cocoa suppliers the Fair Trade premium.

Other mainstream brands have followed Cadbury's lead. In 2009, Kraft launched Rainforest Alliance-certified premium dark chocolate, and Nestlé's Kit Kat launched a print ad campaign, "Have a break, have a Fairtrade Kit Kat." This confirmed that their flagship Kit Kat brand would carry the Fairtrade logo as of January 2010, as part of their sustainability initiative "The Cocoa Plan." The plan promised to invest 65 million pounds over 10 years into its cocoa supply chain. Additionally, as part of a multi-year, multi-country collaboration with the Rainforest Alliance, Mars agreed to use certified cocoa in its Galaxy chocolate in 2010, culminating in the certification of its entire cocoa supply being produced in a sustainable manner by 2020.

**See Also:** Ethically Produced Products; Fair Trade; Food Additives; Green Food.

### Further Readings

Candy Timeline. http://www.candyusa.org/Classroom/timeline.asp (Accessed July 2009).

Fair Trade Chocolate. http://www.globalexchange.org/campaigns/fairtrade/cocoa/background.html (Accessed July 2009).

Food Timeline. http://www.foodtimeline.org/foodcandy.html#aboutsugar (Accessed July 2009).

Hartman, Eviana. "High-Fructose Corn Syrup: Not So Sweet for the Planet." *Washington Post* (March 9, 2008) http://www.washingtonpost.com/wp-dyn/content/article/2008/03/06/AR2008030603294.html (Accessed August 2009).

Mariani, John F. *Encyclopedia of American Food and Drink*. New York: Lebhar-Friedman, 1999.

Sams, Craig. *The Story of Green & Blacks: How Two Entrepreneurs Turned an Ethical Idea Into a Business Success*. New York: Random House, 2009.

Toussaint-Samat, Maguelonne. *History of Food*. New York: Barnes & Noble Books, 1992.

*Barry Emery*
*University of Northampton*

## Conspicuous Consumption

First published in 1899, *The Theory of the Leisure Class*, written by Norwegian American sociologist and economist Thorstein Bunde Veblen, introduced the concept of conspicuous consumption. *Conspicuous consumption* is the term that describes the tendency of individuals to purchase expensive products as an outward display of wealth and a means of enhancing their status in society. Veblen used the term to describe the phenomena of gaining and holding the esteem of others in society through the evidential display of wealth. In this way, an individual is attempting to prove that they have the financial means to afford a particular product. Conspicuous consumption is therefore closely linked to demonstrating status, success, and achievement.

It has long been considered that material possessions, capable of being observed in society, carry social meanings and are used as a communicator to signal a person's wealth, status, and identity. In Plato's *The Republic*, Book II, Adeimantus declares to Socrates: "[s]ince... appearance tyrannizes over truth and is lord of happiness, to appearance I must devote myself." Through consumption behavior and product choices, consumers can send signals to society. Products and brands displayed conspicuously (overtly) have the ability to indicate to others in a particular society one's image identity, as well as wealth. Consuming conspicuously cannot be achieved without the presence of others and the visual display of that consumption. Therefore, those who consume conspicuously rely on other people's understanding the "signaling by consuming" and evaluating the person on the basis of their choices, known as the spectator's view. Aron O'Cass and Hmily McEwen defined "conspicuous consumption" as the tendency for individuals to enhance their image through overt consumption of possessions, which communicates status to others. It is through consumption decisions that an individual can benefit not only from the direct effects of their choice but also from the indirect and social effects emanating from society's observation of their choice. Private or fundamental utility is the theoretical framework that refers to the individual's evaluation of their own satisfaction from consuming certain goods. In this way, product styles and cost, rather than utility, determine how consumers are perceived by others.

Not all individuals desire conspicuous goods. The level of conspicuous consumption prevalent depends on one of a number of factors. First, the prevailing norms, values, customs, beliefs, and laws in a society may all be part of sociocultural context that underlies consumption patterns. In this case, conspicuous consumption occurs where the visibility of such behavior can be understood by those within the society. It is not only Western industrialized countries that can be characterized by conspicuous consumption. Russell Belk (1988) argues that even in Third World countries people are often attracted to and indulge in aspects of conspicuous consumption before they have adequate food, clothing, and shelter. Second, an individual's social network or reference group can influence their consumption patterns. Third, psychological variables, that is, the way in which an individual regulates their own behavior, otherwise referred to as "self-monitoring," plays a role in conspicuous consumption. According to O'Cass and McEwen, high self-monitors tend to place more importance on the overt self and be concerned with maintaining their appearance and overall image as a means of compensating for a lack of security in their own identity. Ottmar Braun and Robert Wicklund argue that people who are committed to an identity and who evidence incompleteness with respect to that identity will be more prone to exaggerate the prestige value of whatever symbols they have at hand. Last, gender has been found to also increase susceptibility to conspicuous consumption. S. Auty and R. Elliot observe that females use clothing and apparel more than males to tell others who they are and how much status they have.

In these ways, Veblen's theory implies a positive relationship between wealth and conspicuous consumption, in which the more costly the item, the greater the demand, although the utility remains the same as a similar item at a lower price. The rationale for this has been explained by S. Brehm in his theory of psychological reactance. A higher-priced product is more attractive because the affordability of the item decreases, which precludes the majority from obtaining the product. Intrinsically linked to the higher price of a product is the prestige value and status that intensify a product's attractiveness to consumers on the basis of exclusivity. It follows, then, that certain brand dimensions and associations can lead to increased marketplace recognition and economic success, although researchers have found that with greater utility and uptake of more expensive products, the prestige and symbolism of wealth can dissipate, as can be observed with the Burberry label.

Today, conspicuous consumption not only refers to the wealthy obtaining expensive and relatively exclusive goods not for their utility but for the prestige value, as Veblen described, but it also has come to be regarded as a broader term to explain the phenomenon of expenditures made by an individual from any socioeconomic background for the purpose of ostentatiously displaying wealth, status, image, or a certain identity that will be perceived by their particular social networks and reference groups. However, with the global economic crisis, there has been a distinct reduction in support for conspicuous consumption. Instead, Western societies in particular are focusing increasingly on inconspicuous consumption. This theory, adopted by scholars in 1980, provides the antithesis of Veblen's conspicuous consumption. Inconspicuous consumption is characterized by consumers buying cheaper items than they need to avoid ostentation. The underlying motivation of the inconspicuous consumer is either to avoid embarrassing others by their wealth or to discourage them from asking for financial support. So, although higher-priced items continue to be bought, many individuals, described as "furtive shoppers," are now choosing discretion over demonstrable, conspicuous goods.

**See Also:** Consumer Behavior; Consumerism; Consumer Society; Downshifting; Fashion; Materialism; Simple Living.

### Further Readings

Auty, Susan and Richard Elliot. "Social Identity and the Meaning of Fashion Brands." In *European Advances in Consumer Research*, B. G. Englis and A. Olofsson, eds. Provo, UT: Association for Consumer Research, 1998.

Belk, Russell W. "Possessions and the Extended Self." *Journal of Consumer Research* (September 1988).

Braun, Ottmar L. and Robert A. Wicklund. "Psychological Antecedents of Conspicuous Consumption." *Journal of Economic Psychology*, 10:161–87 (1989).

O'Cass, Aron and Hmily McEwen. "Exploring Consumer Status and Conspicuous Consumption." *Journal of Consumer Behaviour*, 4/1:25–39 (2004).

*Hazel Nash*
*Cardiff University*

## Consumer Activism

Consumer activism has evolved over the years, from the early days of the cooperative movement to the contemporary era of political consumerism. Consumer activism can take a variety of forms, but the rights, consciousness, and interests of consumer segments lie at its heart. Consumer activists are thus cause-oriented, as they seek to bring about change within the marketplace—usually through protest, campaigning, boycotts (refusal to buy particular products, e.g., that are highly polluting), and "buycotts" (only buying brands that fulfill particular criteria based on human, social, and environmental capital)—and their attention can focus on all or combinations of stakeholders who influence the ethos and behavior of the market, including other consumers. In turn, companies are increasingly recognizing the value of consumer activists in aiding their innovation in product design and service delivery. It is increasingly being suggested that consumer activism, especially political consumerism, is the new politics—particularly in Europe—enabling individuals to come together to voice their concerns and to lobby business leaders and politicians. We can see this with the increasing protests at summits of world leaders debating trade, fiscal measures, climate change, and declining natural resources. As a consequence, consumer activism is a major force in helping to drive the sustainability agenda forward, and thus to build a more ethical and responsible marketplace that facilitates greener consumption, and arguably has become recognized as a positive, persuasive force for change in democratic societies.

## History of Consumer Activism

Consumer activism has had a long, colorful, and cosmopolitan history, particularly within the context of boycotts, which enable much of the activism rhetoric to evolve into action. Among others, its roots can be traced back to Irish peasants in the 1880s, 18th-century America, and Gandhi's 20th-century India. The cooperative movement was the first organized consumer movement; it started as a working-class protest against highly priced, poor-quality goods, particularly food, supplied by commercial monopolies seeking to exploit their labor force. The premise of this movement was "self-help by the people," in which businesses and consumers cooperated together for mutual benefit without manipulation. This movement was so successful that cooperatives were set up all over the world; it is probably the only time in history when consumers were in charge of the goods available to them. Yet, over time, the power of cooperatives declined as markets became less threatening and competition facilitated choice of shopping outlets. Contemporary society is a major challenge for the cooperative movement, as the notion of consumerism, which so characterizes modern-day consumption, whether actual or aspirational, is somewhat of an anathema to it. Yet its underlying principles of fairness suggest that it will have a role to play in the future in supporting consumer activists who are pursuing environmental, human, and social capital through their own consumption choices and their influence on the consumption choices of others. Indeed, we can see its influence within the political consumerism movement, particularly in relation to Fair Trade.

Following in the footsteps of the cooperative movement, the value-for-money consumerist movement has been hugely influential in shaping modern-day consumer activism. Originating from the United States in the late 1800s, the explosion of consumerism in 1920s America caused activists to become concerned about the exploitative power of capitalism. As a result, organizations were formed to test products to increase consumer knowledge so that they could take advantage of the market. This is the foundation of consumer associations; for example, the Consumers Union in America, the U.K. Consumer Association and their well-known publication *Which?*, the *Konsumenten Bond* in the Netherlands, and the *Test Achats* in Belgium. In all cases, the aim of these organizations is not to change society but, rather, to make the market more efficient, to educate consumers, and to ensure that they get the very best products and services at the very best prices. From a sustainability perspective, this movement has created problems. It has never encouraged individuals to think about the consequences of consumption, essentially because it has encouraged consumer sovereignty—consumption rights without responsibility.

## Naderism

By the mid-1960s—in the United States predominantly—another form of activism emerged that challenged the dominance of large corporations. Ralph Nader reinvigorated consumer politics by feeding on the deep-seated anxiety of the American public about the power of large corporations. Nader also believed in arming the American public with information, and for them to become aware of the conspiracy of misinformation deliberately targeted at them by corporations. Thus freedom of information lies at the heart of Naderism, in which lobbies and rallies work in union to establish citizen actions. The core message here is for consumers to be frugal, to be wise, and to protect themselves in their interactions with the marketplace. Although the core principles of Naderism have been admired, it does not sit easily alongside consumer cultures nationally or internationally. Yet, interestingly, Naderism has been used by leaders of developing economies to help them to balance the rights of consumption with their responsibilities, akin to sustainable consumption and civic society. From a sustainability perspective, this is encouraging, as it enables these countries to consider the societal and environmental consequences of adopting consumerism as their dominant social paradigm, and it helps consumer activists to lobby corporations on their behalf.

It is likely that Naderism helped to trigger a more recent expression of consumer activism—political consumerism—particularly in more affluent societies. First appearing in the 1970s, political consumerism has taken 20 years to consolidate into its current form—as progressive activism that aims to integrate ecological, ethical, and social concerns into the mindset of the marketplace. Environmental concerns have been a key influence in shaping political consumerism, in which green consumer activists have attempted to balance their consumer rights with their environmental responsibilities by actively choosing products and services with green credentials from cradle to grave and/or actively reducing their consumption levels. Shades of the cooperative movement also reemerged here as green consumers challenged corporations and their environmentally damaging products and services and their greenwashing claims, using boycotts, buycotts, lobbyins, and protests to reform these nongreen business practices. As a result, from the mid-1980s onward, in many Western economies the power of activists was witnessed as they pressured companies to become greener. For example, removing chlorofluorocarbons (CFCs) from aerosols, reducing phosphate levels in household cleaning products, withdrawing the use of chlorine-bleached cotton in the manufacture of babies' diapers, using recyclable and recycled materials, and adopting cleaner and more resource efficient product design. In many ways, these activists had ceased to become oppositional—they were working within the dominant social paradigm to make the market more ethical. However, as a result, although their influence in greening individual and organizational behavior was considerable, the evolution of this movement was stifled because these activists were not questioning the dominant social paradigm of consumerism itself. As a result, this movement fractured, with more radical activists positioning themselves to challenge the dominance of consumerism. Alongside this green movement, and in many ways synonymous with it, the ethical consumer movement flourished. Its aim was to reintroduce morality into the marketplace, particularly the global implications of consumerism, by creating awareness and cultural change among individuals, societies, and organizations. By focusing on the global political issues surrounding the natural environment, and the welfare of people and animals, these activists politicized themselves, and thus their consumption choices and shopping behavior have become political acts. The Fair Trade cause illustrates this well. According to the Fair Trade Foundation, in 2004, Fair Trade linked 17 affluent food countries and 350 producers and represented 4.5 million farmers. As a consequence, in buying Fair Trade, and in encouraging others to do so, ethical activists have not only helped to increase demand for these products, thus helping to green business, but also have encouraged consumers to think about and take a more active stance in increasing human, social, and environmental capital through their consumption choices. Importantly, these acts possess a sense of immediacy, unlike traditional voting and the protracted discussions of governments.

## Who Are Consumer Activists?

This synopsis of the history of consumer activism leads us to consider activists themselves. Consumer activists tend to be very cognitively competent, media savvy and technologically smart, and confident and proactive. They possess high levels of personal efficacy, and they strongly believe their activism actions will be successful in changing the character and/or behavior of organizations and individuals—indeed, their past actions reinforce this belief. Their personal values may well be more civic in their orientation, embracing self-transcendence and benevolence, particularly for those pursuing a sustainability cause. Hence, their actions are strongly influenced by judgments derived from personal morality and ethics, fairness, equity, and responsibility, revolving around environmental, human, social, and economic capital. They are also likely to be highly politicized, particularly with the advent of single-issue politics, which is shifting the locus of power from government and political parties to political issues driven by small, organized groups of activists with the drive and commitment to pursue their cause without compromise. Accordingly, consumer activists have a high profile in democratic, prosperous economies, with their free access to the media, the World Wide Web, and online and offline communities, which they use as a resource to expose and attack business leaders, organizations, and governments in the pursuit of their cause.

Consumer activists frequently work for what they perceive to be the best interests of consumers who do not possess the skills or confidence to protest themselves; in an environmental context, they also pursue causes to promote human, social, and environmental capital within the marketscape. As a consequence, it is important to recognize the complex questions that emerge from the actions of activists. These arise particularly in relation to imperialism, elitism, and representation. For example:

- Boycotts of fashion clothing and leisure footwear in Europe and America, because of the manufacturers' use of child labor, have caused further economic deprivation to these families as their children are withdrawn from the workforce.
- The pursuit of more affordable and wide-ranging organic food has resulted in a significant increase in air miles, as this food is imported into the United Kingdom from growers in developing economies. Concerns over these air miles has sparked complaints from the governments of these developing economies, which believe that the economic prosperity of their country should outweigh any concerns campaigners have over air miles and their contribution to climate change.
- Protests surrounding battery farming in the United Kingdom have caused some supermarkets to withdraw battery eggs from sale and to insist on Royal Society for the Prevention of Cruelty to Animals animal welfare standards as a minimum in the rearing of all livestock for meat. This has caused concern that many British families will no longer be able to afford the food that is available to them, and thus their health and welfare will be compromised.

In these examples, it can be seen how the demands of Western economies can suddenly elevate the prosperity of developing countries, and just as quickly abandon them as agendas change. How attempts to protect one group of individuals, using Westernized values, can result in an even more serious crisis, regardless whether we believe the governments of these countries should be responsible for ensuring the welfare of its people. How the actions of unelected individuals can dictate the choices and behaviors of the rest of society, for a higher goal, even when harm might result. Thus, in judging consumer activism, readers are encouraged to consider a variety of perspectives in reaching their conclusions that reflect the complex and contradictory nature of this topic, the diversity of activists, their differing foci, and forms of practice.

How, then, can the success of consumer activism be judged? It has certainly had a sustained effect on the marketplace, denting market share and profits and causing shareholders to withdraw from particular product/service sectors, businesses to rethink their offerings, and affluent consumers to evaluate their consumption. However, it has not resulted in a shift in consumer culture in which consumerism has become less dominant. Essentially, consumer activists have realized they must work with the market, not against it, if they are to have any success in influencing values, attitudes, and behavior. Furthermore, although this may seem somewhat paradoxical, in reality, responsible consumption offers a pragmatic solution to the urgent environmental, human, and societal problems. How consumer activism will evolve in the future is unclear. Yet, as we are increasingly witnessing the consequences of our consumption on a changing climate and its consequences for the habitats of people and animals, and the tensions arising from rapidly diminishing resources like oil and natural gas, it is clear that consumer activists will still have a vital role to play.

**See Also:** Commodity Fetishism; Consumer Boycotts; Consumer Ethics; Consumerism; Ethically Produced Products; Fair Trade; Green Consumer.

### Further Readings

Friedman, Monroe. *Consumer Boycotts. Effecting Change Through the Marketplace and the Media.* London: Routledge, 1999.

Harrison, Rob, et al., eds. *The Ethical Consumer.* London: Sage, 2005.

Micheletti, Michele, et al., eds. *Politics, Products and Markets. Exploring Political Consumerism Past and Present.* London: Transaction, 2006.

*Janine Dermody*
*University of Gloucestershire*

# Consumer Behavior

This article introduces the three major actors involved in shaping consumer behavior: businesses, states, and consumers themselves. It provides a conspectus of the debates that have arisen over the effectiveness of green consumption, and of whether consumption can be green at all. A third section summarizes the effect on consumer behavior of misinformation (greenwash) and discusses the problem of "information overload," as well as the role of ecolabeling.

## Consumers, Firms, and States

The major actors involved in shaping consumer behavior are consumers themselves, alongside states and businesses (together with advertisers and marketing consultants). The first actor, the consumer, is an individual who professes or exhibits a concern for the environment in his or her consumption activities. Although states, businesses, and other organizations are major consumers in their own right, scholarly research into consumer behavior normally focuses on individuals. Relevant consumption decisions include the purchase of personal and household goods and services that are environmentally significant in their effect (notably automobiles, energy for the home, and travel), and the use and maintenance of environmentally significant goods (such as heating and cooling systems). Green consumer behavior includes the purchase of items that are produced or cultivated in ecologically sensitive ways. Examples include energy-saving lightbulbs, degradable freezer bags, locally farmed agricultural products, and a range of products such as timber or shade-grown coffee beans that are certified as "sustainable." Influences on consumer behavior can broadly be divided into two sets. One comprises personal factors, including norms and beliefs, capabilities, customs, and habits. The other, contextual factors, include interpersonal influences and community expectations, broader cultural influences (including advertising), and the material capabilities and constraints that encourage certain forms of action and discourage others—for example, the availability of cycle paths, or of a supply of mains electricity from renewable sources.

The second actor is the state. States shape the context in which consumers make choices that directly or indirectly affect the environment. By illustration, consider the transport industry, the infrastructure of which is to a large extent constructed by states such that consumers are pushed to choose certain forms of transport over others. One of the forms of transport with the most adverse environmental effects is the automobile. Compared with rail, bus, or tram, it is associated with high rates of air and noise pollution and extremely low energy efficiency per person kilometer. Its dominance was achieved through state action:

In the early 20th century, the advanced industrial states opted for private road transport over rail, setting a precedent that the rest of the world has followed. State policies, above all in road construction, favored the interests of the automobile–oil corporate bloc, serving to consolidate its strength, such that today, of the 10 largest global corporations, nine are in the oil and automobile sectors. Along with the airlines, these corporations form a commanding alliance that fuels the world's hunger for oil and leads the resistance to "regime change" in the transport sector. The result is that consumers around the world are obliged to buy and maintain automobiles, on pain of significant sacrifice to their mobility.

If in major respects states constrain consumers to make environmentally unfriendly choices, they can also favor responsible behavior. For example, in 1956, the British government introduced the Clean Air Act, which forced consumers in certain areas to use only smokeless fuel, thus reducing pollution from smoke and sulfur dioxide. More recently, the government of Bangladesh has banned the use of polyethylene bags, which had observably been clogging drainage systems, and several European governments have imposed levies on plastic bags to discourage their use. In Australia and the European Union, legislation has been passed that effectively phases out incandescent bulbs by setting higher efficiency standards than these bulbs can achieve. In such ways, states can and do take action to enforce mass changes toward greener consumer behavior.

The third actor is the firm. Given a competitive, profit-determined framework, businesses are obliged to seek to convince consumers to consume ever more goods, and advertising and marketing budgets are correspondingly large. This has contributed to the advent of a consumption ethic, with value invested in the latest fashion, the fastest automobile, or the designer label. With the arrival of the green consumer, however, the possibility arises of businesses persuading consumers to "buy green" without undermining their own prosperity. Surveys suggest that for a sizable proportion of consumers, the environmental attributes of a product or service play a significant role in determining which products they buy and which they avoid. Many are sufficiently committed to environmental goals that they are willing to pay a premium of at least 10 percent for green products. In company boardrooms, it is well known that trends toward green shopping provide opportunities.

## The Effectiveness of Green Consumption

Green consumer behavior is an area of concern for businesses and marketing consultants, but it has also come to function as a prism for disputes among environmentalists and social scientists. One debate, as simple as it is straightforward, pits "liberal consumerists" against "counter-consumerists." For the former, the roots of environmental crisis lie in the nature of mass consumption. They emphasize the ability of consumers to make a difference. As the awareness of environmental crisis spreads, and as individual consumers respond by opting for environmentally friendly products and services, the purchasing power of the mass market will come to force businesses to green their products and their manufacturing and distribution processes on pain of being shunned in the marketplace by green-leaning consumers. In contrast, for ascetically inclined "deep greens," the crisis results less from the quality than from the quantity of consumption. For them, the primary aim should not be consuming discerningly but, rather, consuming less.

A more complex and protracted debate concerns the effectiveness of ecoconscious consumption. An early proponent was the campaign group Friends of the Earth. In a 1989 pamphlet, it expressed the hope (albeit with reservations) that green consumer issues would inspire a growing number of people to critically examine their own lifestyles and the social system. Green consumerism would provide an incentive to businesses to clean up their act, empower individuals to accept personal responsibility for their own choices, and add an important dimension to the work of campaigning organizations. Both the condition of the environment and our quality of life, the pamphlet concluded, could be significantly improved through the cumulative effect of discerning consumer choices.

This position has since become the mainstream among environmentalists. Al Gore, for example, enjoined the audience at the 2007 Live Earth concerts to pledge that they would "buy from businesses who share my commitment to solving the climate crisis." In Britain, climate change activist Mark Lynas avers that although the "green shift" of corporations can appear to be, in the first instance, solely about improving their image, "in doing so they have to achieve a real transformation," as well as make themselves "ever more open to consumer pressure." Moreover, corporations are extremely powerful customers in their own right: when Wal-Mart in the United States opted to switch to green electricity, it sent a strong signal to energy generators that investment in renewables should be ramped up. Environmentally concerned consumers buy green, and because marginal differences in image translate into large differences in turnover, rendering companies highly accountable to their customers, "going green" becomes a vital brand asset. What customers have to do, Lynas concluded, "is ensure that big companies are global leaders in climate change."

Activists such as Gore and Lynas see responsible consumption as a catalyst, not a substitute, for environmental campaigning. The act of consumer choice establishes an entry point into the green arena, awakening individuals to the opportunities and imperatives of political engagement. For example, after a consumer buys energy-saving fluorescent bulbs, he or she may then move on to more radical goals, such as organizing with others to shut down a coal-fired power plant. There is evidence to suggest the existence of synergies between ethical consumption and green activism. One market researcher, for example, reports that focus group participants spoke of their purchases at "Eco Options" aisles as a beginning, not an end point. "We didn't find that people felt that their consumption gave them a pass, so to speak," he said. They knew that their purchases were not significant in themselves, but they performed them as a practice of mindfulness. "They didn't see it as antithetical to political action. Folks who were engaged in these green practices were actually becoming more committed to more transformative political action on global warming."

An alternative thesis holds that, with the exception of the mega-rich, individual consumers are not "empowered"—their market power and influence are negligible, and they know it. A comparatively low proportion of environmental degradation results from what individuals put in their shopping basket compared with decisions taken by states and corporations in areas such as transport and energy-supply systems, construction projects and building regulations, and war. As a consequence, even genuine and committed attempts by individuals to reduce their carbon emissions only succeed in achieving reductions on the order of 20–30 percent. (An example is the family of BBC Newsnight's "Ethical Man," who went so far as to renounce car and air travel and meat and dairy products.) Individual consumers may purchase comparatively fuel-efficient cars, but a vastly greater effect would be achieved were fuel-efficiency standards raised for the industry as a whole.

Some scholars have taken this argument further and propose that the focus on consumer behavior is a highly politicized gambit: a deliberate tactic to shift culpability for ecological crisis from corporations and states to individuals. Timothy Luke has described how the rhetoric of ecological responsibility slowly shifted from slogans of "Big business is dirty business," to "Factories don't pollute. Consumers do." The message is that voracious consumers carry the burden of guilt for the environmental crisis. They may ameliorate this by altering their consumption habits, so long as they do not thereby challenge the consumption ethic per se. There is in this, Toby Smith has suggested in *The Myth of Green Marketing*, a twofold confidence trick. First, the consumer is led to believe that his or her individual use of unleaded gasoline or a benign soap powder will actually affect the ecological crisis. Consumers are lulled into complacency by the mistaken belief that they are doing something significant. Second, the solution to the environmental crisis is defined as individual action, deflecting attention from power elites and structural issues.

A more substantial attempt to address these questions was outlined by Heather Rogers in her study, *Hidden History of Garbage*, which traces the history of business-led campaigns that explicitly set out to propagate the idea that individuals are the primary source of pollution. In the 1980s, manufacturers learned to exploit the rise of recycling as a bulwark against the adoption of more rigorous measures mandating reduction and reuse. Recycling also worked to further ingrain a sense of personal culpability for increasing levels of trash. The rhetoric of recycling designated individual behavior as the key to the garbage problem, helping to steer public debate away from the regulation of production.

It is probably no coincidence that the 1980s was a pivotal conjuncture in which green political activity came to be framed in terms of individual choice and morally responsible decision making, for it was the breakthrough decade of both the green movement and neoliberal ideology. The latter contrives to link liberty with market deregulation and individual consumption, yielding the tenets that identity and subjectivity are defined in terms of what we consume, that free markets empower the consumer, and that corporate behavior is driven by consumer choice. Increasingly, people came to be referred to as consumers, where previously they had been coach passengers, gallery visitors, university students, or hospital patients. A turn toward conspicuous consumption, most pronounced among the middle and upper strata, could be observed, entailing an obsession with the display of good taste and the "stylization" of existence. It linked to broader trends, notably a narcissistic preoccupation with the cultivation of looks in general and the body in particular. It was only fitting that the Body Shop came to symbolize green consumerism in that pioneering decade.

## "Seeing Green": Ecolabeling, Greenwash, and Information Constraints

The obstacles to green consumption are several. An obvious one is that consumers experience a clash between the greenness of a product and its other characteristics, such as price or effectiveness. Another is situational constraints, such as time pressure. (The time that consumers devote to selecting an everyday product is typically between 5 and 10 seconds.) These are compounded by informational problems. To establish the greenness of a product, consumers must rely on information supplied by the corporation—through labels and advertising—and the media. They must be able to recognize green behavior, yet there are difficulties associated with this. One concerns the criteria by which "greenness" is defined. Does one choose the factory-farmed local chicken or its free-range organically fed cousin, air-freighted from abroad? Is the recycling of glass bottles an environmental act even where, as in Britain, they are ground down and used as aggregate in road building? The second, related to but distinct from the first, concerns the sourcing of reliable information. Commercial secrecy prevents much necessary information from being known, and although corporations do issue corporate social responsibility reviews, there is little consistency between them as to the level of detail they report. This, according to one marketing consultancy, "makes it difficult to assess who has the most impact on the environment and who is doing more to reduce their impact on the environment." A major proportion of the information available to consumers comes from biased sources such as advertisements or press releases by marketing companies.

The outcome is that consumers find it difficult to assess the environmental friendliness of a product. Some companies put forward claims regarding the environmental qualities of their products without backing them up with real greening of production process. Where a firm "cheats" in this way, consumers may punish it by choosing the products of other firms, but if they feel bamboozled or discouraged by misleading advertising, the result can also be cynicism toward all green marketing. According to a report by WPP, although four of every five consumers believe that it is important that companies act in an ecologically sustainable way, their understanding of what this entails is "shallow, confused and easily swayed by company messages." The same survey revealed that the top 20 green brands included names such as Smart, Tesco, BP, Toyota, Virgin Atlantic, and Shell—testament indeed to the sway of corporate greenwash. Consider Shell and BP: Together they account for an astounding 40 percent of the carbon dioxide emissions of all FTSE100 companies; recently, their operations have become markedly more carbon intensive, with investments in Albertan tar sands and Chinese coal. Smart is seen as green, even though its parent company, Daimler, manufactures gas-guzzling limousines and sports utility vehicles, as does Toyota. Even some of Toyota's allegedly "green" hybrids consume 22 miles to the gallon on the highway. British company Tesco has developed an exceedingly fuel-intensive business model and relies heavily on out-of-town outlets. The outcome of greenwashing is cynicism. Although surveys show that in a clear majority, consumers feel strongly enough to take green and ethical issues into consideration where possible, a similar proportion is cynical about the green credentials asserted by corporations.

One solution to the greenwash problem, as utopian as it is obvious, would be for consumers to engage in a cradle-to-grave analysis of the products that they purchase, with regard to the energy and materials involved in their production and distribution. However, this would be not only unrealistic but problematic, in that it creates an "information paradox." People wish to feel in control of their lives and to avoid feelings of impotence. One might think that additional information would be empowering them, but it can have precisely the opposite effect. According to one focus group–based study involving over 900 participants in 29 countries, participants with the keenest awareness of environmental issues also evinced the highest levels of anxiety based on their sense of impotence. As the director of the research institute that conducted the survey concluded, consumers "are looking for clarity of information that can lead people to make decisions," yet "the more information that is pumped out . . . the more contradictory it seems, and the less people are able to translate that information into knowledge." Other studies have suggested that observed declines in concern about environmental issues are best explained by a sense of futility and helplessness—a feeling that is intensified by information overload.

There is here an evident case of market failure. Consumers with preferences for a specific type of product lack the information to reward its manufacturer through their purchasing decisions, and one cannot reasonably expect them to research the company's environmental record. The remedy would appear to be to reduce the inefficiency involved in dispersed information gathering by way of labeling schemes. Ecolabeling authenticates a product's environmental attributes, enabling customers to act appropriately. In Europe, for example, where a "carbon count" is printed on the packaging of cut flowers, consumers discover that the carbon footprint of those sourced in Africa is, against expectations, lower than that of their locally sourced equivalents. Some will compare labels and buy the African product.

Ecolabels, it is widely believed, represent an effective method of altering consumer behavior. They highlight the degree of greenness of products, enabling consumers to shop according to environmental criteria and without the need for individual research. However, few empirical studies have documented a positive effect of ecolabels, and some have suggested that a proliferation of labels can actually exacerbate the information paradox. In Denmark alone, one analysis estimated that there existed three dozen public labeling campaigns, as well as several hundred private labels. Many of these were not recognized by consumers, severely affecting the effectiveness of the campaigns.

**See Also:** Advertising; Consumerism; Consumer Society; Ecolabeling; Green Consumer; Green Marketing; Greenwashing.

## Further Readings

Connolly, J. and A. Prothero. "Sustainable Consumption: Consumption, Consumers and the Commodity Discourse." *Consumption, Markets and Culture*, 6/4:275–91 (2003).

Crane, A. *Marketing, Morality and the Natural Environment.* London and New York: Routledge, 2000.

Lodziak, C. *The Myth of Consumerism.* London: Pluto, 2002.

Moisander, J. and S. Pesonen. "Narratives of Sustainable Ways of Living: Constructing the Self and the Other as a Green Consumer." *Management Decision*, 40/4:329–42 (2002).

Smith, T. *The Myth of Green Marketing: Tending Our Goats at the Edge of Apocalypse.* Toronto: University of Toronto Press, 1998.

Wagner, S. *Understanding Green Consumer Behaviour: A Qualitative Cognitive Approach.* London and New York: Routledge, 2003.

*Gareth Dale*
*Brunel University*

# Consumer Boycotts

A boycott is a form of consumer activism that involves voluntarily abstaining from using, buying, or dealing with someone or some other organization as an expression of protest, usually for political reasons. The word entered the English lexicon during the Irish Land War and is based on the name of Captain Charles Boycott, the estate agent of an absentee, who was subject to ostracism organized by the Irish Land League in 1880. That year, protesting tenants demanded a substantial reduction in rent from Boycott, who not only refused but also evicted them from the land. Rather than resorting to violence, everyone in the locality refused to deal with him—local businessmen stopped trading with him, and even the local postman refused to deliver his mail. The boycott continued for some time, and it was used by *The Times of London* in November 1880 as a term for organized isolation. Captain Boycott withdrew to England in December 1880.

Even though the term was coined in 1880, the practice dates back to much earlier times. Americans boycotted British goods during the American Revolution. In 1830, the National Negro Convention encouraged a boycott of slave-produced goods. Later examples of boycotts were the use by African Americans during the U.S. civil rights movement and the United Farm Workers union grape and lettuce boycotts; Indians, organized by Mohandas Gandhi, boycotted British goods in the early 20th century; and the Jewish boycott against Henry Ford in the 1920s was also a success story.

Around 40 percent of Fortune 50 companies are shown to be boycotted at any given time, and research has demonstrated that their sales can plummet by 10 to 40 percent in the next six to 12 months after a boycott against them has been called. Organizations such as Nestlé (criticized for marketing breast milk substitutes), Nike (criticized for unfair labor practices in Asia), KFC (for allegedly mistreating chickens), Levi-Strauss (for closing production facilities), and Target (targeted for not using the words "Merry Christmas" in their advertising) have all been the victims of boycotts in recent years.

## Research on the Effect of Boycotts

Research on boycotts has demonstrated that providing negative information about one product from a multiproduct firm negatively affects perceptions of other brands from the same company. Expectation of success also affects the success of boycotts—studies have demonstrated that "full supporters" protesting rising prices have been more successful than "partial" supporters of such initiatives. They have been shown to be most effective when there is greater economic pressure, more "image pressure" (publicity), and "less commitment" by the target to the policies that prompted the boycott. There is also empirical evidence to illustrate that boycotted companies experienced significant decreases in their stock prices over the 60-day period after a boycott was announced. Concerns have nevertheless been raised about the potential effects of changes in production practices arising from boycotts; for example, where firms shift production practices elsewhere, or where boycotts against child labor result in diminished incomes for already impoverished families. As a consequence, "buycotts," where consumers are encouraged to purchase from particular (green) companies, can also be an option for promoting change.

Other studies have shown that the expectation of success with commitment to the boycott, and that fervor itself can lead to success. Boycotts are deterred by prior firm commitments to green production practices. In 1997, Greenpeace, the Rainforest Action Network, and the Natural Resources Defense Council formed the Coastal Rainforest Coalition to encourage Home Depot to stop buying timber from ancient forests. The International Fund for Welfare successfully boycotted Mitsubishi's plan to build a salt plant in gray whale calving grounds, with the support of over a dozen organizations. In the United Kingdom, Scott Paper abandoned plans to build a eucalyptus plantation and paper mill in Indonesia, which threatened the survival of indigenous people, in response to a threat of a boycott from Survival International and the Sierra Club. INFACT, an environmental organization, succeeded in pressuring General Electric to sell its last nuclear weapons–related division. Today, one of the prime targets of boycotts is consumerism itself, for example, "International Buy Nothing Day," celebrated globally on the Friday after Thanksgiving in the United States.

Boycotts are usually legal in most countries, especially because they are voluntary and nonviolent. There are, however, certain legal impediments, such as the "refusal to deal" laws, that prohibit concerted efforts to eliminate competition by refusal to buy from or sell to a party. The Internet has enabled greater successes of boycotts. Several countries have also instituted antidiscrimination laws that prohibit any place that offers goods, services, and facilities to the general public, such as a restaurant, from denying or withholding services to an individual because of his or her race, gender, social class, or any other characteristics.

Boycotts are important to companies from at least two perspectives. First, they help understand what role boycotters play as stakeholders and the powers that they wield: How should their claims to power be treated compared with those of the other traditional stakeholder groups, such as shareholders, customers, and employees? Second, they also help influence the changes in the marketing mix, that is, product, price, distribution, and promotional strategies before, during, or after the boycott. Accordingly, companies may use boycotts as an opportunity to build stronger relationships with their customers.

**See Also:** Consumer Activism; Fair Trade; Green Consumerism Organizations; Morality (Consumer Ethics).

### Further Readings

Davidson, D. Kirk. "Ten Tips for Boycott Targets: Consumer Boycott Targets." *Business Horizons* (March–April 1995).

Friedman, M. "Consumer Boycotts: A Conceptual Framework and Research Agenda." *Journal of Social Issues*, 1:149–68 (1991).

Friedman, M. *Consumer Boycotts: Effecting Change through the Marketplace and the Media.* New York: Routledge, 1999.

Friedman, M. "Consumer Boycotts in the United States, 1970–1980: Contemporary Events in Historical Perspective." *Journal of Consumer Affairs*, 19/1:96–117 (1985).

Garrett, D. "The Effectiveness of Marketing Policy Boycotts: Environmental Opposition to Marketing." *Journal of Marketing* (April 1987).

Gelb, B. D. "More Boycotts Ahead? Some Implications: Consumer Boycotts." *Business Horizons* (March–April 1995).

Hoffman, S. and S. Mueller. "Consumer Boycotts Due to Factory Relocation." *Journal of Business Research*, 62/2:237–49 (February 2009).

Miller, A. "Do Boycotts Work?" *Newsweek* (July 1992).

Pruitt, S. W. and M. Friedman. "Determining the Effectiveness of Consumer Boycotts: A Stock Price Analysis of their Impact on Corporate Targets." *Journal of Consumer Policy* (December 1986).
Smith, N. C. "Consumer Boycotts and Consumer Sovereignty." *European Journal of Marketing*, 5:7–19 (1987).

*Abhijit Roy*
*University of Scranton*

# Consumer Culture

Any culture that is based on the consumption of things, in which people use materials to identify themselves and to communicate, socialize, and relate to others, is called a "consumer culture." This particular phenomenon may be considered part of a "material culture"; that is, a culture in which the cultural values that shape our everyday practices and beliefs are driven by the desire to buy, consume, and show off purchases, goods, possessions, and ultimately, status. In contemporary society, consumer symbols of consumption are the cornerstones of the construction of self-identity.

## Consumer Culture: Why Do We Want Goods?

The concept of consumer culture is often discussed in the context of capitalism, industrialization, and modernity. Initially, discussion centered on recognition of the effect of consumption on the economic situation. John Maynard Keynes, who was both an economist and a proponent of this view, was in favor of promoting consumption, claiming that together with (industrial) investment, it would result in increasing the total income of a society. With the expansion of capitalism, which saw the focus shift from production to consumption, along with the increasing acquisition of goods and places of and for luxury and leisure, it has been recognized that the concept of consumption is not simply a derivate of production but also a behavior rooted in individual (and sometimes collective) wants and needs.

A further aspect of the question "Why do we want goods" is the relevance of the increase in both satisfaction and status gained from the display of goods and services. More and more emotional desires can become fulfilled by the experience of consumption itself; for example, the comforting familiarity, emotional stimulus, and material reward of shopping malls, the "temples of consumption," per se. This is where the act or experience of consumption is celebrated, for shopping malls serve as symbols and meanings that are consumed, rather than simply places in which actual tangible goods and services can be purchased.

Why do we want goods? Consumer culture is the representation of a perceived "good life"—the term *good* here being understood as a high-quality and achievable life, inasmuch as the consumer is able to consume products that not only fulfill her or his basic needs but also provide additional luxury and leisure—a life based on the consumption and exhibition of goods and possessions. It is what Thorstein Veblen so aptly termed the way of life or lifestyle of the "leisure classes" and the "show-offs." The buying pattern he is noted for labeling is "conspicuous consumption." Thus, consumer culture is a basic function of the need for fashion. Without fashion—the nigh-ostentatious owning and showing off of an acquisition—we would not consume on this grand scale: We would not consume items that do not fulfill any need other than to make one feel "somebody," that allow one to make one's mark both individually or within a group.

At the same time, the existence and degree of consumer culture is a perceptual entity rooted in the perceptions and motivations of the people investigating and talking about it. This is how advertising visualizes consumers' emotions, wants, and needs to sell the symbols, signs, and meanings of consumption. Although the basic assumption underlying consumption includes "free" (rational or irrational) choice of the consumer, consumer culture tends to imply the mass marketing of products and services accompanied by subsequent limited free choice. Concomitant with the merging of the cultural and commercial spheres, art too has come to be viewed as a commodity. Commercialized art such as advertising, for example, has now come to be regarded as goods.

"Consumer culture" has been a critical tenet of modernity. However, with the emergence of the postmodern era, and the understanding that everything is cultural, consumer culture has suddenly become celebrated and idolized, as it plays an increasingly meaningful role in the private lives and spheres of consumers. With this transition, however, the role of consuming has changed markedly. Originally, consuming carried the meaning "to destroy, to use up, and to waste." Today, consuming needs to be understood more as "constructing" (identity), "enjoying" (the product, social ranks, and signifier it provides), and "communicating" (subjectivity, meanings, and symbols, but also differences and delimitations).

## Cultural Imperialism

Since the 1980s, voices have been heard that have predicted the homogenization of cultures, resulting in a so-called world culture based on mass consumption. In an increasingly globalizing world, keywords such as "McDonaldization" and "Coca-Colonization" have signified the spread of a (fast food) consumption pattern in tandem with the spread of American cultural values to the rest of the world. As parts of a consumer culture that has changed societies both socially and economically worldwide, McDonaldization and Coca-Colonization are just two phenomena in the emergence order of consumption that underpins the trend toward valuing material abundance and consumption.

As a consequence of the aforesaid increasingly globalizing world, with its stronger and broader interconnectedness, an even more global consumer culture is emerging. However, the construction of globalization as increasing homogeneity is much debated, with consideration of local needs, products, and consumption patterns still widespread. The dream of marketers to sell one product around the world—to cater to a global consumer culture—has yet to become the reality for many products. In the final analysis, it may be that adaptation and customization are the way to go; that is, that our cultural values will work to determine the nature of our (often compulsive) attachment to goods and consumption per se.

**See Also:** Commodity Fetishism; Conspicuous Consumption; Consumer Behavior; Consumerism; Consumer Society; Social Identity.

### Further Readings

Douglas, Mary and Baron Isherwood. *The World of Goods. Towards an Anthropology of Consumption*. London: Routledge, 1979.

Featherstone, Mike. *Consumer Culture and Postmodernism*. London: Sage, 1991.

Hannerz, Ulf. *Cultural Complexity*. New York: Columbia University Press, 1992.

Lury, Celia. *Consumer Culture*. Brunswick, NJ: Rutgers University Press, 1996.
McCracken, Grant. "Culture and Consumption: A Theoretical Account of the Structure and Meaning of Consumer Goods." *Journal of Consumer Research*, 13:71–84 (1986).
Ritzer, George. *The McDonaldization of Society*. Thousand Oaks, CA: Pine Forge, 2004.
Slater, Don. *Consumer Culture and Modernity*. Queensland, Australia: Polity Press, 1999.
Veblen, Thorstein. *The Theory of the Leisure Class*. New York: Dover, [1899] 1994.
Williams, Raymond. *Keywords*. London: Fontana, 1976.

*Sabine H. Hoffmann*
*American University of the Middle East*

# Consumer Ethics

Consumer ethics deals with issues pertaining to moral behavior in consumer markets. Because consumers are an integral part of the business process, it is imperative to understand both the underlying motivations for their propensities for buying ethical products and the reasons why some engage in unethical practices, which may be helpful in curtailing many questionable practices. Consumer ethics deals with a variety of issues such as willingness to benefit from questionable actions, consumer reaction to ethical transgressions by sellers, the perception of company ethics and product purchase, willingness of consumers to pay for socially acceptable products, and the emergence of reasons for consumer boycotts of business organizations. In addition, research has also investigated responsible consumer behavior, attitudes, and intentions to purchase specific ethical products, such as environmentally safe products, to name a few. Consumer ethics has also been shown to be deeply intertwined with marketing ethics.

Although there are inconsistencies in research findings, there are a few generalities that can be safely assumed. First, there is evidence that there are more ethically oriented consumers than those who are not. However, in one study, as many as 39 percent of the respondents indicated no ethical concerns, and more interestingly, there was very little variance related to demographic variables. Second, studies examining consumer choice of products with socially embedded messages (e.g., avoiding child labor in producing goods) found that clusters of "responsible" consumers do exist, yet "ex ante," it was difficult to predict who was in these clusters based on just demographic factors such as age, gender, ethnicity, education, and so on. Third, the most significant predictor of future ethical behavior is one's past behavior relating to social causes, such as being involved in associations like Greenpeace or Amnesty International. Fourth, the actual behaviors of consumers are unrelated to the general attitude surveys or surveys measuring consumption ethics. Finally, ethical behavior can be affected by the nature of the product and the price that is charged. For low-priced and low-involvement products (e.g., bath soaps), more price-sensitive consumers were shown to be less ethically inclined, whereas for high-priced, high-involvement products (e.g., televisions), there was no relationship between price sensitivity and ethical attribute sensitivity.

## Greening and Consumer Ethics

Recent studies have shown that consumers today are more likely to be driven by commitment to ethical and sustainable methods of production than by the health attributes of food. Environmentally conscious individuals attach aesthetic quality to environmental goods. Green consumerism is on the rise, yet its environmental effects are contested. On one hand, green consumerism can contribute to the greening of consumer consciousness, yet it might also encourage corporate greenwashing. This tenuous ethical situation mandates that ecomarketers carefully frame their environmental products such that they appeal to consumers with environmental ethics as well as to buyers who consider both natural products and conventional items.

Some major companies like the Body Shop, Burt's Bees, and Tom's of Maine proactively promote their environmental consciousness. In an attempt to sell products to consumers based on a set of ideological values, these companies use two specific strategies to market their products: they fashion enhanced notions of beauty by accentuating the performance of their natural products, and hence, they infuse green consumerism with a unique environmental aesthetic. Moreover, they express ideas of health through community values, which in turn enhances notions of personal health to include ecological well-being.

## Cross-Cultural Consumer Ethics

With the rapid progress of globalization, there is also a need to understand how ethical interpretations and behavior may differ in various consumer markets in the world, and what rationales are used in justifying the ethical conflicts that arise in their consumption activities. In one particular study, it was shown that the ethical behaviors were not based on one's socioeconomic position in society, and that culture had relatively less effect on the perception of consumer ethics. The middle-class respondents from Germany and India had very different responses and understanding of ethical scenarios presented to them, yet their overall evaluations of whether or not their consumption behavior was ethical were remarkably similar.

In another study, researchers investigated the relationship between cultural values and marketing ethics in two diverse countries: India and the United States. The study showed that although the gaps in culture had narrowed considerably, there were significant differences in the interpretation of marketing ethical norms. The paradigm of "cultural relativism" (i.e., ethics vary from one culture to another on the basis of the business practices of the host culture) held true, and thus firms are advised that "when in Rome, do as the Romans." It would also be beneficial to organize training programs for managers from participating countries to develop and reinforce a formal or informal code of ethics. A common code of ethics should also be established from the perspective of different stakeholders with diverse cultural backgrounds to minimize the chances of subsequent misunderstandings.

**See Also:** Consumerism; Disparities in Consumption; Ethically Produced Products; Green Consumer; Materialism; Morality (Consumer Ethics); Poverty; Simple Living.

### Further Readings

Brady, Arlo. "Appealing to Consumer Ethics in a Price-Driven World." *Corporate Responsibility Management*, 1/3:3 (December 2004).

Chatzidakis, A. and D. Mitussis. "Computer Ethics and Consumer Ethics: The Impact of the Internet on Consumers' Ethical Decision-Making Process." *Journal of Consumer Behavior*, 6/5:305 (2007).

Ford, Charles W., et al. "A Cross-Cultural Comparison of Value Systems and Consumer Ethics." *Cross Cultural Management*, 12/4:36–50 (2005).

Kavak, B., et al. "Examining the Effects of Moral Development Level, Self-Concept, and Self-Monitoring on Consumers' Ethical Attitudes." *Journal of Business Ethics*, 88/1:115–35 (2009).

Maldonado, C. and E. Hume. "Attitudes Toward Counterfeit Products: An Ethical Perspective." *Journal of Legal, Ethical and Regulatory Issues*, 8/1/2:105–17 (2005).

Murphy, Patrick E., et al. *Ethical Marketing*. Upper Saddle River, NJ: Pearson/Prentice Hall, 2005.

Paul, Pallab, et al. "The Impact of Cultural Values on Marketing Ethical Norms: A Study in India and the United States." *Journal of International Marketing*, 14/4:28–56 (2006).

Rao, C. P. and Adel A. Al-Wugayan. "Gender and Culture Differences in Consumer Ethics in a Consumer-Retailer Interaction Context." *Journal of International Consumer Marketing*, 18/1,2:45–71 (2005).

Rawwas, Mohammed Y. A. "Culture, Personality and Morality A Typology of International Consumers' Ethical Beliefs." *International Marketing Review*, 18/2:188–211 (2001).

Rawwas, Mohammed Y. A., et al. "Consumer Ethics: A Cross-Cultural Study of the Ethical Beliefs of Turkish and American Consumers." *Journal of Business Ethics*, 57/2:183–95 (2005).

Steenhaut, Sarah and Patrick van Kenhove. "An Empirical Investigation of the Relationships Among a Consumer's Personal Values, Ethical Ideology and Ethical Beliefs." *Journal of Business Ethics*, 64/2:137–55 (2006).

Vitell, Scott J. "Consumer Ethics Research: Review, Synthesis and Suggestions for the Future." *Journal of Business Ethics*, 43/1/2:33–47 (2003).

Vitell, Scott J. and James Muncy. "The Muncy-Vitell Consumer Ethics Scale: A Modification and Application." *Journal of Business Ethics*, 62/3:267–75 (2005).

Vitell, Scott J., et al. "Consumer Ethics: An Application and Empirical Testing of the Hunt-Vitell Theory of Ethics." *Journal of Consumer Marketing*, 18/2:153–78 (2001).

Vitell, Scott J., et al. "Religiosity and Consumer Ethics." *Journal of Business Ethics*, 57/2:175–81 (2005).

Wagner-Tsukamoto, S. "Consumer Ethics in Japan: An Economic Reconstruction of Moral Agency of Japanese Firms—Qualitative Insights from Grocery/Retail Markets." *Journal of Business Ethics*, 84/1:29–44 (2009).

Ziad, Swaidan, et al. "Consumer Ethics: The Role of Acculturation in U.S. Immigrant Populations." *Journal of Business Ethics*, 64/1:1 (2006).

*Abhijit Roy*
*University of Scranton*

## Consumerism

The connections between consumerism and climate change are increasingly visible. Research by the Intergovernmental Panel on Climate Change indicates, with 90 percent certainty, that recent global warming is caused by human activity and, in particular, their patterns of consumption, and that this effect is accelerating at a much faster rate than previously thought. As a result, consumerism generates some profound issues for sustainability, particularly in relation to enhancing human, social, and environmental capital. The meaning of consumerism is multilayered. It can be regarded as a moral doctrine in developed economies, where, in offering consumers choice, it is perceived to bring freedom, power, and happiness, and has thus come to represent a better way of living. It can also be thought of as an ideology of conspicuous consumption, a means by which status can be created through the acquisition of material possessions. It is an economic ideology for global development, in which unfettered consumerism reinforces and increases capitalism—and its power—on a global scale. It is a political ideology that has facilitated the emergence of neoliberalism, where the market, market mechanisms, choice, and consumer dominate to provide an exciting and interesting blend of products and services that appeal to our functional and aspirational needs and desires, even in the provision of public services. It is therefore not surprising that consumerism is the dominant social paradigm in affluent societies and an aspirational goal for developing economies. However, it can also be viewed as a social movement of consumer advocacy, in which consumer activism is used to modify the types and patterns of consumption, as witnessed by the activism of environmental and ethical consumers. The challenges posed by climate change and diminishing resources, coupled with shifts in economic and political power, mean it is likely that consumerism will change in the future. Thus, particularly in prosperous societies, it is unlikely that the unrestrained consumerism of the 1980s and 1990s will progress too far into the 21st century, and it will be interesting to observe its advancement in developing economies.

## A Long History

Consumerism and the narcissistic, nihilistic consumer are not new phenomena—they form part of an ongoing societal evolutionary process. This behavior can be traced back centuries, including, for example, the lavish displays of wealth and status seen among the aristocracy of Elizabethan England in the 16th century. Although a historical perspective indicates that consumption enabled the "common man" to break the chains of their masters and their circumstances (e.g., hunger, ignorance, immobility) through the promise of liberty and prosperity, once it went beyond the satisfaction of utilitarian needs necessary for survival, consumption evolved into consumerism and became associated with hedonism, decadence, instant gratification, and social control. Consumerism thus became synonymous with unfettered freedom, and the more people grew to love their freedom and to view it as a distinct element of their lifestyle, the more they viewed themselves as having no obligation but to self-indulge, alongside the right to do so. Accordingly, some commentators argue that it is not hedonism or self-gratification that gives consumerism a bad name but, rather, the consumer sovereignty and liberty of choice that this perceived freedom engenders. Hence, consumerism is very much about the rights of the consumer to consume, with little regard for the consequences of this behavior. For sustainable consumption to flourish, these rights need to be balanced by consumers' acceptance of responsibility for the consequences of their consumption excesses, particularly given their increased spending on shopping and their use of shopping as a leisure pursuit. As a consequence, critics regard consumerism as a negative influence on the morals of society, encouraging "false values," materialism, unrestrained choice, and indulgence and the isolation of individuals from their traditional communities as they seek "never-to-be fulfilled" promises from their consumption choices, which in turn feed their anxiety and self-doubt. Furthermore, critics believe that our pursuit of material possessions undermines our sense of subjective well-being, making us more and more unhappy with our lives. Accordingly, consumerism can be charged not only with damaging the natural environment and ignoring human, social, and environmental capital but also with damaging the psyche of consumers themselves.

Perhaps paradoxically, then, for more and more people, the influence of consumerism on their lives is growing. As a consequence, around the world, mass consumer society has emerged as the major source of economic and social influence. This signals that the notion of the modern, self-disciplined individual, in which consumers are professed to be rational, self-maximizing economic individuals in control of their emotions, no longer fully explains this pursuit of consumerism, particularly given that the concerns about it are well known. A more contemporary approach to understanding the meanings and influence of consumerism on individuals' lives recognizes the importance of their "cultures of consumption."

## The Role of Consumers

Within this more culturally focused perspective, consumerism is regarded as a process of shared social learning, laden with emotion, symbolic meaning, and identity. Accordingly, consumers are no longer simply culture bearers but are also culture producers. As a consequence, the marketplace, in which the balance of power has, in many ways, shifted in favor of consumers, provides them with an assorted repertoire of mythic and symbolic resources enabling them to create their individual and collective identities through their (expanded) consumerism choices. Within this landscape of "consumption think, consumption behave," social relations are expressed through the language of marketing, and the identity and worth of individuals is understood via their material possessions. In this respect, consumers, through their consumerism choices, are actively engaging in this "meaning-process" (signification), embracing the experiential and tangible dimensions of consumerism, and the symbolic image conveyed. Yet this creative process is not equal for all consumers—their cultural capital is influenced by social class hierarchies, gender, ethnicity, and familial influences, which raises further concerns about the dominance of consumerism as a way of life in wealthy and developing economies. Thus, consumers are negotiating their way through their consumerism choices to seek and create their identity(s). This negotiation also is not culturally neutral: Although individuals are pursuing personal goals through their consumerism, the majority are enacting and personalizing cultural scripts that align their identities with the underlying parameters of consumerism embedded in capitalist ideology, and in so doing, they are reinforcing the dominance of the market.

What then emerges about contemporary consumerism is that consumers act as interpretative agents who, in creating meaning from their consumerism, play, individually and collectively, within a spectrum ranging from acceptance to (pseudo)rejection of the dominant identity and lifestyle images conveyed by advertising and mass media, and for many, are willingly seduced by these images. Consumerism has therefore become a powerful influence on both individual and collective behavior.

## Sustainable Consumerism

It is the power of this influence that raises a major challenge in the pursuit of more ethical, more sustainable consumerism and, perhaps paradoxically, offers a pragmatic solution that helps to address the excesses of consumerism itself. Accordingly, on the one hand, we see rampant consumerism as consumers pursue identity and status from the meanings ascribed to their consumption choices. This consumer sovereignty without responsibility is a major obstacle in engendering a more ethical, sustainable orientation in consumers' shopping behavior. On the other hand, however, the power that consumerism brings to individuals, coupled with a more citizenly orientation, enables them, as consumers, to become politicized—through boycotts and "buycotts"—as they choose greener brands and reject brands that damage the natural environment or compromise the welfare of others.

Overall, however, there continues to be apathy surrounding the fundamental changes in consumer behavior required to safeguard the future of the planet and its population. This is essentially because although politicians, business leaders, and individuals are beginning to witness the consequences of climate change, and can see the positive outcomes of Fair Trade practices, only members of the environmental movement are critiquing the ideology of consumerism itself, and they have no voice within the political corridors of power, the closed meeting rooms of business executives, or the private homes of individuals. Thus, for the majority of politicians, the challenge surrounding sustainability revolves around the efficient and effective use of environmental resources, and their policies reflect this. For business leaders, although their practices have become greener, the ethos of profit and loss—shareholder value and competitive advantage—still dominates their thinking. For individual consumers, Fair Trade consumption choices will suffice. For many, they are waiting for the latest techno-fix that allows this majority to sustain their lifestyle without compromise. In this respect, consumerism is sowing the seeds of its own destruction. As the planet warms up, resources become exhausted, extinction increases as habitats die, and human poverty and illness escalate, so too will consumerism, as we know it, cease to exist. Yet, ironically, at the time of writing, when the world is facing a global recession alongside environmental degradation, poverty, and inequality, the leaders of affluent economies are encouraging us to go shopping. Accordingly, the mindset of consumerism poses the greatest challenge to mankind and their longevity on this planet. Until we evaluate the dominance of this paradigm, the inherent ideals within sustainability can never be achieved.

**See Also:** Consumer Activism; Consumer Boycotts; Consumer Ethics; Green Consumer.

### Further Readings

Dermody, Janine, et al. "Shopping for Civic Values: Exploring the Emergence of Civic Consumer Culture in Contemporary Western Society." *Association for Consumer Research*, 36 (2009).

Featherstone, Mike. *Consumer Culture and Postmodernism*. London: Sage, 2007.

Gabriel, Yiannis and Tim Lang. *The Unmanageable Consumer*, 2nd Ed. London: Sage, 2006.

Micheletti, Michele, et al., eds. *Politics, Products and Markets. Exploring Political Consumerism Past and Present*. London: Transaction, 2006.

*Janine Dermody*
*University of Gloucestershire*

## Consumer Society

Alongside population growth, the development of a consumer society in the West and its worldwide diffusion (aided by globalization) is a significant contributor to the current environmental crisis—a main reason for concern about the chances of future generations to enjoy a minimally decent standard of living. The concept of a "consumer society" has been forged to describe the transformations that happened in Western industrialized societies, mostly in the aftermath of World War II, even though the rise of the consumer society has a long historical precedent. During this period, Western European and North American societies have undergone significant structural and cultural changes that can be described as a shift from production to consumption as the main structuring principle of daily life practices and social behavior, and as the ultimate criterion of efficacy for economy and the government. The shift of emphasis from production to consumption as a focal principle and the problem of affluent societies meant that people were more useful for the economy as consumers than as workers because, in a context of increasing globalization, productivity (thanks to technological innovation), and women's participation in the labor market, the main obstacle to economic expansion was less the risk of shortage in labor supply than an insufficient effective demand.

However, it is not so much consumption but the vital necessity of a steady and endless economic growth that characterizes the consumer society. If no society can survive without ensuring a sufficient level of consumption, only the consumer society depends for its very functioning and reproduction on permanent economic growth and, as a consequence, on expanding consumption to absorb the increasing supply of commercial goods and services. The crucial importance of consuming helps to explain the role of institutionalized waste, advertising, and consumption credit in consumer societies. They are necessary to overcome the two main barriers to a steady rise in consumption in a context of stagnant population: the lack of motivation to consume and the lack of money to do so. The lack of motivation can result either from actual satisfaction or from satiation—two similar but slightly different phenomena. In the first case, the consumer is simply satisfied by the goods and products already in their possession and sees no reason to replace them. Satiation refers to the fact that, for some needs such as food, for example, there is a threshold beyond which any additional unit of what satisfies the need brings no further satisfaction, if not sheer dissatisfaction. The satisfaction problem is addressed in consumer society by creating artificially the dissatisfaction of the customer, principally through the strategy of built-in obsolescence—a form of institutionalized waste organized by manufacturers to speed up the "buy-use-discard" cycle of their products. It consists in planning the short living or rapid outdating of their products (for material, functional, or stylistic reasons), so that the consumers are willing to replace them quicker than otherwise necessary. In other words, corporations regularly launch new versions of their products with minor functional or stylistic modifications sufficient for making the previous version comparatively less attractive and inducing customers to purchase the new one.

The share of the advertising industry in gross domestic product (1.5 percent, e.g., in Germany in 2006) and the pervasiveness of commercials in public spaces and mass media are probably the most reliable and significant indicators of the development of a society as a consumer society. Independent of its effects on the purchases of individual goods and services, advertising directly or indirectly (through its influence on TV programs, movies, etc.) contributes to the diffusion of an entire consumer culture—a system of values, beliefs, and representations that links happiness, as well as the formation, expression, and actualization of individual and group identity, to the acquisition and use of market goods and services (commodities). Notably, at the dawn of the consumer society, advertising helped in overcoming the Puritan fear of debt and the reluctance to consider borrowing for consumption purposes, which impaired the development of the third of a major consumer society's institutions: consumption credit. In particular, since its inception in 1950, the credit card has been highly effective in supporting the continuous expansion of consumption, even at the cost of worrying rates and levels of indebtedness among working and middle-class households in Western countries.

Because it is correlative of commoditization as "tendency to preferentially develop things most suited to functioning as commodities—things with qualities that facilitate buying and selling—as the answer to each and every type of human want and need," the consumer society sanctions the hegemony of the market as a system of provision for almost all needs and wants and undermines the role of other fundamental institutions such as the state, the community, and the household in the satisfaction of individuals' needs.

The necessity of ensuring the quickest circulation of ever-increasing amounts of goods and products that characterize the consumer society shapes many aspects of everyday life, and especially the way space and time is structured. As for space and the living environment, for instance, it leads to the replacement of small, local, independent retailers with large-scale (national or even multinational) corporate chains. The consequences of these changes on the vitality and diversity of local and community life are noteworthy. For instance, it means replacing small retail businesses (generally localized in the center of the city, where they contribute to the animation and the security of the streets) with Big Box stores located at the periphery of town, thereby increasing car dependency and traffic problems and encouraging the homogenization of neighborhoods and landscapes.

**See Also:** Affluenza; Consumer Culture; Needs and Wants; Overconsumption.

### Further Readings

Baudrillard, Jean. *The Consumer Society: Myths and Structures*. London: Sage, 1998.

Durning, Allan. *How Much Is Enough? The Consumer Society and the Future of the Earth*. New York and London: W. W. Norton, 1992.

Galbraith, John Kenneth. *The Affluent Society*. London: Hamish Hamilton, 1969.

Goodwin, Neva R., et al., eds. *The Consumer Society*. Washington, D.C.: Island Press, 1997.

Manno, Jack. "Commoditization: Consumption Efficiency and an Economy of Care and Connection." In *Confronting Consumption*, Tomas Princen, et al., eds. Cambridge, MA: MIT Press, 2002.

*Paul-Marie Boulanger*
*Institut pour un Développement Durable*

## Cosmetics

Cosmetics are defined as preparations—such as powders, oils, creams, and pills—used for healthcare and, most of the time, beauty reasons. They include products that people use daily, such as perfumes, creams, shampoo, and lipstick. The most obvious effect of cosmetics is related to the chemical components that they contain. The potential harmful effects of these components are, in the same way as chemicals, regulated by national and international authorities dedicated to their control.

A second effect of cosmetics that is increasingly attracting attention, and is particularly interesting with regard to the objective of sustainable development, is related to the exploitation of biological components by the cosmetic industry. Cosmetic recipes are often drawn from the properties of seeds, plants, and trees, as well as animals, to propose effective healthcare solutions. For example, coconut milk has been used since early times as a skin moisturizer, and rose petals have been used as a perfume for many years.

The consequences of the use of biological components by the cosmetic industry on the conservation of the world's biodiversity are the subject of several debates at the national and the international levels of policy making. In contrast to the first environment impact already mentioned—chemical pollution—which is mostly the subject of restrictive regulations (such as the European Union's REACH legislation), these discussions aim at creating a complementary relationship between the uses of the world's biodiversity in cosmetology and the conservation of natural resources. This positive equation would consist of the sharing of the benefits made by the users of the biological resources—mostly the pharmaceutical and cosmetic industry—with the actors responsible for the conservation of these resources—mostly local authorities and indigenous people.

At the international level, several international organizations have been, since the beginning of the 1990s, discussing the development of international principles to encourage the collaboration between the users and the providers of biological material. The main organizations concerned with this topic are the United Nations Environment Programme, the World Intellectual Property Organization, and the World Trade Organization.

Under the United Nations Environment Programme, the Convention on Biological Diversity, adopted in 1992 and ratified by 168 member states, recognizes in its third objective the need to share the benefits arising from the commercial use of biodiversity products. The convention therefore asks the users of biological components—cosmetic companies included—to redistribute part of their benefits to the local communities and indigenous people that have been conserving these components for centuries.

The World Intellectual Property Organization and the World Trade Organization are also currently discussing a special disclosure requirement for patent applications using biological material. Such a disclosure aims at enhancing the transparency related to the use of natural genetic resources to improve the possible financial contributions of firms for biodiversity conservation. Nevertheless, all these new regulatory elements are under negotiation and are not expected to be adopted until 2010.

Several initiatives at the national and the corporate levels are already in force. Several cosmetic firms have started to develop new products and ranges based on the responsible use of biodiversity components, as well as on the sharing of the benefits with local communities. In Brazil, the enterprise Natura is one example of such firms. Natura has recently elaborated a so-called ecologic range that reconciles biodiversity conservation with cosmetics development. For instance, the firm commercializes a fragrance produced from the sap of a tree harvested in the forests near Belém in the Brazilian state of Amapa. This sap is harvested by indigenous people, who have, through a partnership with the firm, been able to maintain their activities while improving their everyday life with the building of schools, hospitals, and so on. These populations are also responsible for the conservation of the natural species. The Brazilian state is very active in encouraging such partnerships.

Countries rich in biodiversity are more inclined to adopt some incentives to foster the involvement of biological material users in biodiversity conservation. American and European states and companies are, however, also starting to participate actively in these schemes. For example, the transnational corporation the Body Shop was one of the first firms to engage in natural and Fair Trade cosmetics. The company is now present in 61 countries worldwide.

The current market trends in cosmetology show that consumers recently favor biological products obtained through transparent and fair supply chains. It is probable that the growing demand for these kinds of cosmetics will improve the mutual supportiveness of biodiversity conservation and human healthcare.

**See Also:** Ethically Produced Products; Green Consumer; Healthcare; Personal Products.

## Further Readings

Gavenas, Mary Lisa. *Color Stories: Behind the Scenes of America's Billion-Dollar Beauty Industry.* New York: Simon & Schuster, 2002.

Laird, Sarah and Rachel Wynberg. *Access and Benefit-Sharing in Practice: Trends in Partnerships Across Sectors.* Convention on Biological Diversity Technical Series, 2008. http://www.cbd.int/doc/publications/cbd-ts-38-en.pdf (Accessed April 2009).

Purvis, Debbie. *The Business of Beauty*, 2nd Ed. Toronto: Wall & Emerson, 1994.

Rosendal, Kristin, G. "Balancing Access and Benefit Sharing and Legal Protection of Innovations from Bioprospecting: Impacts on Conservation of Biodiversity." *Journal of Environment & Development,* 15/4 (2006).

*Amandine J. Bled*
*Université Libre de Bruxelles*

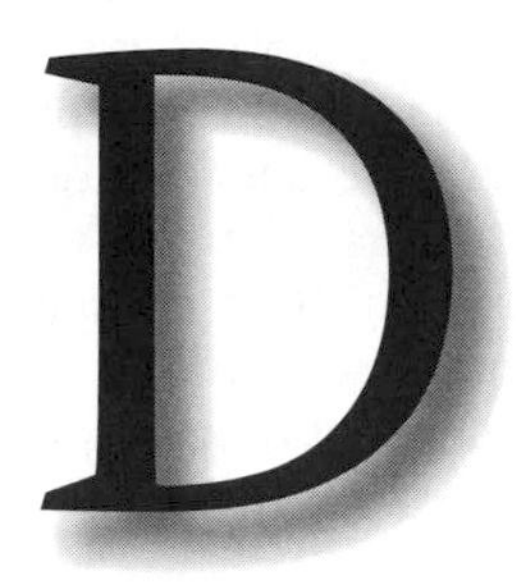

## Dairy Products

Green products are all-natural and nontoxic and are produced sustainably, and often locally, with renewable resources or postconsumer recyclable materials. Their production and use have a minimal negative effect on the environment and promote conservation of energy and natural resources. According to some definitions, green consumerism enhances quality of life now and for future generations. Current levels of production and consumption of dairy products cannot be sustained without compromising the environment, health, animal welfare, and economic justice. As consumers demand it, dairy businesses are beginning to offer safer foods produced with fewer negative environmental impacts.

Most dairy products in the United States are produced by cows, but worldwide more people consume milk from goats. Here, Holstein dairy cows in Louisiana are cooled during the summer to maintain milk production, on the premise that contented cows produce more milk.

*Source:* Bob Nichols/Natural Resources Conservation Service/USDA

In the United States, regulated dairy products are liquid milk, butter, cheese, whey products, and frozen products. Hundreds of billions of pounds of milk are now produced annually. In 1940, 76.4 percent of America's farms had dairy cows. By 2000, that percentage had decreased to 6.1. While herds grew, the number of farms with dairy cows and specialized dairy farms (those reporting milk sales as the primary source of income) decreased. From 1975 to 2000, milk production per cow jumped 76 percent. In 2000, farms with more than 500 head of milk cows accounted for 31.4 percent of the nation's cow inventory and 35.8 percent of milk production. The vast majority of dairy products in the United States are produced from cow milk, but production of market dairy products from goats and sheep has been on the rise. Worldwide, more people consume milk from goats than from cows.

Milk is mostly water. For every unit of milk she produces, a dairy cow must consume an equivalent unit of water. Commercial dairy operations convert millions of gallons of water into milk. They use thousands of gallons of water daily to keep their herds and megafacilities clean enough to pass inspections. This water becomes waste that contaminates surrounding aquifers. Production of methane gas is also problematic. Agribusiness dairies employ advanced technologies (animal genetics, food microbiology, biophysics, and biochemistry) and gigantic housing and production facilities stocked with specialized equipment (milking machines, piping, sprinklers, and sprayers) that are manufactured by other industries that also produce waste. Infrastructure must be disposed of as newer industry standards make it obsolete. Transporting milk long distances requires refrigeration and fuel. Milk used to be packaged in reusable glass containers, but it is now typically sold in plastic, most of which becomes solid waste, although much of it is recyclable.

Agribusinesses produce, process, market, and distribute their products, controlling the production chain from cow to consumer. This drives down the prices small producers can get for their products. Oversight of commercial dairy production is supported by powerful lobbyists and national and international industry groups. Government subsidy programs encourage large-scale production and distribution. The USDA's Dairy Export Incentive Program allows U.S. exporters to sell dairy products abroad for less than it costs to acquire them, which allows them to undercut subsidized prices for dairy products sold abroad.

When thousands of bovines are housed together in combined animal feedlot operations with little access to open space, and consume grain instead of grass, their health is compromised. Dairy products can threaten human health in a number of ways. Injecting dairy cows with recombinant bovine growth hormone (rBGH) boosts milk production, but there is evidence that rBGH also causes mastitis, breast cancer, and other diseases in humans. Overconsumption of dairy products can exacerbate asthma and respiratory problems. Dairy food chains can become infected with highly poisonous dioxins.

Lawmakers, entrepreneurs, educators, and advocacy organizations promoting sustainable development, healthcare, economic justice, and ethical treatment of animals see green consumerism as one solution. Green consumption movements in the United States, Europe, Australia, New Zealand, and elsewhere are changing industry practices and helping increase demand for sustainably produced food throughout the global system. Educational campaigns are preparing consumers to make better choices. A grassroots campaign to demand safe, sustainably produced milk in schools is ongoing. Green consumer activists believe that making small changes in their habits can have a cumulative effect and begin to change how dairy products are produced.

Availability of "green" dairy products is increasing as markets for rBGH-free, grass-fed, and organic dairy products expand. Certification and labeling standards help consumers identify sustainably produced food. Humane certification programs help consumers connect with producers who use humane animal husbandry and natural feeding practices. Green consumer buying guidelines are widely available, but consumers must beware greenwashing schemes, in which distributors and marketers promote their dairy products as sustainably produced when in fact producers are not taking necessary measures to conserve resources and protect the health and well-being of their animals.

More specialized dairy farms are implementing programs to recycle waste, minimize greenhouse gas emissions, and reduce their carbon footprints. More consumers are willing to pay more to support local, sustainable food production. Dairy farmer co-ops enable small-scale farms to market their products and compete more successfully with agribusiness producers in local markets. Artisan producers are successfully marketing high-quality, sustainably produced dairy products.

Sustainable consumption is possible for educated consumers who have access to and can afford "green" dairy products. However, producing and distributing the quantity of dairy products consumed today cannot be done without excessive resource use and practices harmful to the environment, local economies, people, and animals. Sustainable, local production of dairy foods is possible, but per capita consumption must decrease significantly.

**See Also:** Consumer Activism; Consumer Ethics; Greenwashing; Overconsumption; Sustainable Consumption.

### Further Readings

Blayney, Don P. "The Changing Landscape of U.S. Milk Production." U.S. Department of Agriculture, Economic Research Service Statistical Bulletin (No. 978, June 2002). http://usda.mannlib.cornell.edu/usda/nass/SB978/sb978.pdf (Accessed October 2009).

European Commission. "Reducing Water Consumption by 35% Through Reuse in the Dairy Products Production Process, Environmental Technologies Action Plan." http://ec.europa.eu/environment/etap/inaction/pdfs/feb07_dairy_no_water.pdf (Accessed October 2009).

Food and Agriculture Organization of the United Nations, Agriculture and Consumer Protection Department, Animal Production and Health Division. "Milk and Dairy Products." Updated December 2008. http://www.fao.org/Ag/againfo/themes/en/dairy/home.html (Accessed October 2009).

International Federation of Agricultural Producers. "A Global Dairy Agenda for Action—Climate Change." Doc.9_09. http://www.ifap.org/en/newsroom/documents/IDFDairyClimateChange.pdf (Accessed October 2009).

U.S. Department of Agriculture, Foreign Agricultural Service. "Dairy Export Incentive Program." http://www.fas.usda.gov/excredits/deip/deip-new.asp (Accessed October 2009).

U.S. Department of Agriculture, National Agricultural Statistic Service Reports. http://www.nass.usda.gov/Publications/Reports_By_Date/2009/October_2009.asp (Accessed October 2009).

*Stefanie Wickstrom*
*Independent Scholar*

# Demographics

Demographics are characteristics of human populations. Statistics on demographics as they pertain to consumer markets are important factors to be considered when attempting to understand consumer behavior. Groupings such as age, gender, ethnicity, income, mobility, education, and social class are considered to be predictors of consumer behavior and patterns. When market researchers conduct demographic analysis to identify purchase patterns or habits particular to specific consumer demographic segments, the output can aid businesses in matching individual consumer needs with products and services.

Consumer demographic analysis employed as a research instrument is part of a portfolio of tools used in marketing to improve understanding about consumer attitudes, characteristics, consumption tendencies, and lifestyle activities. The gathering of demographic data can occur through a variety of methods from online surveys to point-of-purchase tracking. In recent years, technology has enabled an extraordinary increase in the sheer volume of consumer demographic information available to market researchers. Parsing the data into statistically valid groups or segments based on relevant demographic criteria provides useful input into the marketing processes of a firm. For example, consumer demographic data can improve returns on investment in new product research and development, aid in the selection of the optimal advertising mediums to reach targeted consumers, and help set appropriate product/service pricing to incent consumer purchases from select outlets.

Consumer demographics can also help identify entirely new market segments resulting from changes in society, such as those consumers who are motivated to protect the environment. With concern growing about the deterioration of our natural resources in recent years, there has been a general trend indicating that certain consumers want products and services that can be considered environmentally friendly. This "eco-consciousness" among select consumers is important for businesses to understand, particularly as an input into their product development and marketing processes.

Translating a general market trend, such as protecting the environment, into specific buying patterns has to do with the nature of consumer demographics themselves. In general, consumer demographics are characterized as either antecedent or nonantecedent. Antecedent demographics, such as gender and nationality, refer to sociodevelopmental processes occurring early in life that affect a person's cognitive and emotional reaction to choices of consumption. Nonantecedent demographics are those that occur later in the life cycle, such as income level and education.

Researchers generally consider consumers' values as an antecedent demographic, as they represent enduring beliefs that a particular behavior is good. Valuing the environment is a prerequisite for behaving in environmentally friendly ways, such as purchasing green products. However, studies have shown that concern for the environment does not always translate into proenvironmental behavior. That is, a person who values protecting the environment might not actually purchase green products or even recycle. This value–action gap suggests there are a variety of conflicting factors that can affect a consumer's decisions. Nonantecedent consumer demographics such as income and education play a role. For example, a consumer might not be able to pay a premium to purchase a green product that has a higher price resulting from added supply chain and manufacturing costs. Consumers with little formal education on the sometimes nuanced cause and effect relationship between consumption and environmental deterioration might not feel compelled to bother recycling, particularly if there is an absence of peer group pressure.

There are multiple ways in which a consumer can act on their proenvironment values. For example, a green consumer might pay more for certain products that they perceive to have a greater positive effect on the environment. A hybrid car might fit this category. Environmentally conscious consumers might recycle paper, but not plastic.

In general, research studies consider demographics as important when the prerequisite values favoring environmental protection exist. In recent years, studies have offered further insights into consumer demographic segments that act on their proenvironment values through a willingness to pay more for green products. These studies have shown that consumers tend to be younger than middle age, female, at least secondary school educated, of above-average socioeconomic status, and married, with at least one child at home.

As described above, the motivation behind an individual consumer's commitment to the environment, as well as the depth of that commitment, can be difficult to ascertain. Analysis of consumer demographics and extrapolating patterns of consumer behavior as they relate to the environment is continuing to improve. The continued exposure of environmental issues within mass media will likely spur specific demographic segments into action in the coming years.

**See Also:** Advertising; Consumer Behavior; Green Consumer; Green Marketing.

### Further Readings

D'Souza, C., et al. "Green Decisions: Demographics and Consumer Understanding of Environmental Labels." *International Journal of Consumer Studies* (2007).

Finisterra do Paço, Arminda M. "Identifying the Green Consumer: A Segmentation Study." *Journal of Targeting, Measurement and Analysis for Marketing,* 17/1 (2009).

Flynn, Simone. "Consumer Demographics." *EBSCO Research Starters* (2008).

Kollmuss, A. and J. Agyeman. "Mind the Gap: Why Do People Act Environmentally and What Are the Barriers to Proenvironmental Behavior?" *Environmental Educational Research,* 8/3:239–60 (2002).

Straughan, Robert and James Roberts. "Environmental Segmentation Alternatives: A Look at Green Consumer Behavior in the New Millennium." *Journal of Consumer Marketing,* 16/6 (1999).

U.S. Census Bureau. "Population Profile of the United States: 2000." http://www.census.gov/population/www/pop-profile/files/2000/profile2000.pdf (Accessed November 2009).

*Timothy P. Keane*
*Saint Louis University*

# Diderot Effect

The Diderot effect addresses the relationship between consumption and culture, the complementarity of consumer goods and a key mechanism underpinning the escalation of consumer demand. It is named after Denis Diderot (1713–84) and his reflections in an essay titled "Regrets on Parting With My Old Dressing Gown." Here, Diderot details a process set in motion as a result of receiving a new dressing gown as a gift from a friend. He was initially happy to replace his old dressing gown, but in doing so, the other items in his study suddenly looked unsatisfactory. Piece by piece, he was compelled to replace everything until the whole room was transformed to match his "imperious scarlet robe." Diderot finds himself lamenting the loss of the original dressing gown, the objects in his old study, and above all, the unity that existed between these things. Grant McCracken has, through his engagement with this story, identified (and named) two important phenomena: Diderot unities—the consistency that links complements of consumer goods, and the Diderot effect—the force that maintains these unities.

Diderot unities are an example of how culture controls consumption; in this case, to govern consistency across the whole range of individual purchasing behaviors. To see this more clearly, it is worth noting that some goods somehow "go together," whereas others do not. For example, Rolex watches go well with BMW cars, but not with old Volkswagen vans. The reason for this complementarity lies in the meaning of consumer goods, which stems from their place in a wider system of goods, which in turn stems from a wider system of cultural categories. For example, the BMW corresponds to certain cultural categories of social class (professional), gender (male), and age (35–65 years old) in a way that other motor vehicles do not. Similarly, the Rolex corresponds to certain cultural categories in a way that other watches do not. Moreover, the BMW and the Rolex occupy the same relative position within their respective product categories, such that they carry a similar meaning and therefore "go" together. The cultural meaning of wearing a Rolex would be somewhat diminished if it were to be worn while driving that old Volkswagen van, suggesting that the meaning of a consumer good is best communicated when it is surrounded by goods that carry the same meaning. Diderot unities can be seen to apply to much more than cars and watches (or the contents of Diderot's study), insofar as symbolic consistency might be expected across the consumer behaviors, attitudes, and objects that make entirety of an individual's lifestyle.

## How the Diderot Effect Operates

Grant McCracken has defined the Diderot effect formally as the "force that encourages the individual to maintain a cultural consistency in his/her complement of consumer goods." Essentially, Diderot unities are a consequence of the Diderot effect, and there are three main ways in which the Diderot effect can operate.

1. *Continuity and Conservatism:* First, the effect works to ensure the cultural consistency of an individual's lifestyle and to prevent the entry of a new object that might disrupt an existing Diderot unity. According to Grant McCracken, we are able to maintain a consistent sense of self by maintaining the cultural consistencies in our things. In guarding against the intrusion of new objects and the meanings that they carry, we are protected from the new ideas and activities that might serve to destabilize our sense of who we are. In this framing, the Diderot effect ensures continuity in our material world and the purest of signals from our possessions, such that we are able to experience continuity and a clear set of meanings in our personal lives.

2. *Transformation and Innovation:* In another framing, as in the case of Diderot's dressing gown, it works to bring about a radical transformation of an individual's set of consumer goods. The introduction of a new consumer good can create a demand for other goods in the bundle to be replaced, such that they match the meaning of the object that initiated the change. Here, the Diderot effect is innovative and not conservative, raising the question of how an effect that preserves unity can become one that brings about a complete overhaul of one's existence. The answer lies in those purchases that have no precedent in the existing complement of goods and whose introduction causes the Diderot effect to work toward a new unity. These "departure purchases" can come about for any number of reasons. For example, the force of marketing has the potential to persuade individuals to make a purchase that sits outside their usual bundle, just as a new job with a higher income might provide the means to purchase a Rolex, which in turn will compel the individual to

purchase the BMW, and so on. Similarly, changes in personal circumstances such as the arrival of a new child or a move to a different city might prompt departure purchases that lead to a new unified bundle of consumer goods. Finally, as in the case of Diderot's dressing gown, gifts can be the point of departure from which the Diderot effect operates toward a new Diderot unity. Indeed, gifts very often carry new meanings into the existence of the receiver, which in turn brings about a new standard of consumption. It is not difficult to imagine a badly dressed young man being given a new shirt by his girlfriend, nor is it difficult to imagine the shirt acting as a catalyst for the young man to transform his entire wardrobe, haircut, and grooming regime. It might, however, be beyond the scope of this article to suggest that this was the girlfriend's intention all along!

3. *Personal Experimentation:* Finally, the potential exists for individuals to exploit the Diderot effect by deliberately purchasing consumer goods whose meanings disrupt the notion of unity. In doing so, these individuals exhibit the ability (or hope) to displace any existing sense of self in favor of creating a whole new lifestyle. Although the emphasis here is on choice and identity formation, it is interesting to note how the Diderot effect still comes into play and compels the individual to purchase further consumer goods in a potentially endless chain of new product complements.

## The "Ratchet" Effect

The important thing about the Diderot effect is that it operates a "ratchet" effect on consumer expenditure. In its "rolling form," the departure purchase is made at the top end of an individual's product complement and then operates a gravitational pull on the rest of the bundle, such that the next purchase matches the departure purchase and so on, until the entire bundle matches the departure purchase. In its "spiral" form, the effect operates on every purchase such that each purchase is drawn from a higher, more-expensive product complement than the last, and there is never a point at which the entire bundle catches up with and matches the departure purchase. In either case, it locks the consumer into an ever-ascending spiral of consumption and expenditure such that no product complement is ever complete and the consumer is never satisfied. Clearly, the Diderot effect can be understood as a key mechanism through which high and environmentally damaging levels of consumption are perpetuated. Nevertheless, Elizabeth Shove and Alan Warde have offered a speculative nod to the idea that the Diderot effect might be appropriated in favor of "greener" patterns of consumption. For example, it is feasible that a "green" departure purchase might compel the individual consumer to match this with further green purchases and environmentally less damaging practices across their whole lifestyle.

**See Also:** Consumer Culture; Lifestyle, Sustainable; Overconsumption; Social Identity; Sustainable Consumption.

### Further Readings

Gregory, Mary Efrosini. *Diderot and the Metamorphosis of Species (Studies in Philosophy).* London: Routledge, 2006.

McCracken, Grant. *Culture and Consumption: New Approaches to the Symbolic Nature of Goods and Activities.* Bloomington: Indiana University Press, 1988.

Schor, Juliet. *The Overspent American: Upscaling, Downshifting and the New Consumer.* New York: Basic Books, 1998.
Shove, Elizabeth and Alan Warde. "Inconspicuous Consumption: The Sociology of Consumption, Lifestyles and the Environment." In *Sociological Theory and the Environment: Classical Foundations, Contemporary Insights*, R. Dunlap, et al., eds. New York and Oxford: Rowman and Littlefield, 2002.

*David Evans*
*University of Manchester*

# Disparities in Consumption

The term *disparities in consumption* refers to the conditions of unequal use of economic goods to satisfy the needs of consumers. These disparities are stark—both within countries as well as across them. With world population at 6.8 billion and rising, the richest 20 percent of people consume 86 percent of all goods and services used, and the poorest 20 percent consumes just 1.3 percent. The world's middle 60 percent consume 22 percent of all goods and services used. Furthermore, the wealthiest 10 percent account for 59 percent, and the poorest 10 percent account for just 0.5 percent of all consumption. Here are some facts that illustrate this phenomenon:

The wealthiest fifth of the world:

- Consume 45 percent of all meat and fish; the poorest fifth consume 5 percent
- Consume 58 percent of the total energy; the poorest fifth consume less than 4 percent
- Have 74 percent of all telephone lines; the poorest fifth have 1.5 percent
- Use 84 percent of all paper; the poorest fifth use 1.1 percent
- Own 87 percent of the world's vehicles; the poorest fifth own less than 1 percent

Over 2.6 billion people lack basic sanitation, 1.3 billion have no access to clean water, 1.1 billion lack adequate housing, and nearly 900 million have no access to modern health services of any kind. Americans each consume 260 pounds of meat per year on average, most of it hamburger; the average in Bangladesh is 6.5 lbs. Meanwhile, according to the United Nations Development Programme, every year:

- Americans and Europeans together spend $17 billion on pet food
- Business entertainment in Japan amounts to $35 billion
- Cigarettes in Europe cost $50 billion
- Alcoholic drinks in Europe total $105 billion
- The business of narcotic drugs in the world is worth $400 billion
- Military spending in the world is over $800 billion

Compared with the above figures, the estimated additional costs required to achieve universal access to basic social services in all developing countries were as follows: basic education for all ($6 billion per year), water and sanitation for all ($9 billion a year), and basic health and nutrition ($13 billion a year).

## History of the Growth of Inequality

History plays a significant role in understanding the differential levels of consumption throughout the world. A few key factors explain why some parts of the world have grown rich and others have lagged behind. There are several specific historical issues that are important in the study of poverty for the global marketing student of the 21st century—for example, how traditionally, poverty has been concentrated in the toughest places, the inadequacy of market forces to overcome it, the relationship between social spending and reducing poverty over time in different regions of the world, the causes and consequences of the growth in the gap between rich and poor individuals and nations, and so on.

The gross domestic product per capita during 1500 to 1820 remained constant in Asia and Africa and grew at 0.1 percent per year in Latin America and Eastern Europe and 0.2 percent per year in Western Europe. The population growth during the 18th and 19th centuries made the leaders of European societies concerned about the future availability of natural resources. The famous demographer and political economist during that period, Thomas Malthus, discussed the fate of humankind in his "Essay on the Principle of Population." He famously predicted that because of the limited land on Earth, food production in the world would be hard to keep up with the geometric growth rate of human population. He did not, however, foresee the coming of the Industrial Revolution, which dramatically changed everything—from food production to standards of living—in the next two centuries.

The Industrial Revolution started slowly in England but quickly gained momentum because of its social openness, political liberty, scientific revolutions, and crucial geographical advantages. By the late 19th century, the people in Western Europe, North America, Australia, and Japan became prosperous because of the new technologies. Most of the innovations and industrialization processes started in the European countries and later were exported to the United States, Canada, Australia, and Japan.

In this period of modern economic growth, the world's average per capita income rose as fast as the population growth. However, gross domestic product per capita grew much faster in Western Europe, the United States, Canada, Oceania, and Japan compared with Asia, Africa, Latin America, Russia, and Eastern Europe. For example, the per capita gross national income of the United States grew at an annual rate of around 1.7 percent per year during the period 1820 to 1998 compared with 0.7 percent in Africa. This 1 percent difference in growth rate over 180 years leads to approximately a 20-fold gap in income between the United States and Africa. The gross world product rose nearly 50-fold during this period, with only some of these regions experiencing most of the growth. At this time, at the beginning of 21st century, 10 percent of world population collects 70 percent of the world income—an average of $30,000 per person.

## Disparities, Greening, and Quality of Life

Runaway growth in consumption in the past half a century is putting strains on the environment never before seen. If emerging nations follow the same path as the rich countries, their consumption patterns will also be damaging to the environment. World consumption patterns are undermining the environmental resource base. During the 1990s, one U.S. citizen was consuming 30 times what one citizen of India did; developed nations comprised 20 percent of the world's population, yet used two-thirds of all resources and generated

75 percent of the world's pollution and waste. Over 1 billion people living in absolute poverty require increased consumption so as to alleviate their malnutrition, disease, and illiteracy.

Global spending on advertising, which stimulates consumption, multiplied nearly sevenfold from 1950 to 1990, when the total was $257 billion, or $48 for each person on the planet. It has nearly doubled again since then, to about half a trillion dollars, and is increasing faster than incomes or population, especially in developing nations. Surveys of U.S. households found that the income desired to fulfill consumption aspirations doubled between 1986 and 1994. The definition of "necessity" is changing worldwide.

Increased consumption is, however, not producing a parallel increase in happiness. The National Opinion Research Center of the University of Chicago has found that the proportion of Americans who say they are "very happy" has remained at about one-third since 1957, although personal consumption has more than doubled.

**See Also:** Affluenza; Conspicuous Consumption; Consumerism; Final Consumption; Leisure and Recreation; Materialism; Needs and Wants; Overconsumption; Resource Consumption and Usage; Super Rich.

## Further Readings

Bennett, M. K. "International Disparities in Consumption Levels." *American Economic Review,* 41/4:632–49 (September 1951).

Bhattacharya, N. and B. Mahalanobis. "Regional Disparities in Household Consumption in India." *Journal of the American Statistical Association*, 62/317:143–61 (March 1967).

Dunning, Alan. *How Much Is Enough? The Consumer Society and the Fate of the Earth.* New York: W. W. Norton, 1992.

Edward, Peter. "Examining Inequality: Who Really Benefits From Global Growth?" *World Development*, 34/1 (2006).

Ferguson, Niall. *The Ascent of Money: A Financial History of the World.* New York: Penguin, 2008.

Lyons, Thomas P. "Interprovincial Disparities in China: Output and Consumption, 1952–1987." *Economic Development and Cultural Change*, 39/3:471–506 (April 1991).

Molini, V. and G. Wan. "Discovering Sources of Inequality in Transition Economies: A Case Study of Rural Vietnam." *Economic Change and Restructuring*, 41/1:75–96 (2008).

Sachs, Jeffrey. *Common Wealth: Economics for a Crowded Planet.* New York: Penguin 2008.

Seshadri, Ananth and Kazuhiro Yuki. "Equity and Efficiency Effects of Redistributive Policies." *Journal of Monetary Economics*, 51/7:1415–47 (2004).

World Development Group. *World Development 2009: Reshaping Economic Geography.* Washington, D.C.: World Bank, 2009.

Yao, Shujie, et al. "Rural-Urban and Regional Inequality in Output, Income and Consumption in China Under Economic Reforms." *Journal of Economic Studies*, 32/1:4-24 (2005).

*Abhijit Roy*
*University of Scranton*

# Disposable Plates and Plastic Implements

Disposable plates are probably as old as the human desire to be rid of used dinnerware. Contemporary consumers are faced with many choices of disposable ware that range from paper to plastic, but the distinction between green disposable implements and other products is not always straightforward.

Paper plates are convenient and do not need washing after use. In general, they can become wet and weakened from the food that is served on them, so they are not reusable. Because they are disposable, they can simply be thrown away. They are somewhat green because they are biodegradable. However, many consumers believe that using a paper plate wastes trees and, therefore, do not view paper plates as green consumer goods. It is also the case that paper plates require energy to manufacture, which reduces their green character as biodegradable. If paper plates are burned as garbage, then they add to the carbon dioxide in the atmosphere.

Chinet is a brand name for heavy-duty paper plates. The plates are thicker and do not leak easily. The manufacturer of Chinet, Huhtamaki Americas, is a consciously green company that seeks to recycle its wastes. Chinet is biodegradable and can be made from recycled paper, which reduces waste and energy. Gardeners can place their used Chinet plates into compost piles for composting. The disposal of heavy paper products including paper cups can be accomplished with environmental integrity and safe sanitation if the products are rinsed before disposal in the compost pile. If placed into compost while coated with food, vermin may be attracted.

Bamboo-veneered plates, bowls, and serving dishes are currently marketed as an alternative to paper plates. They are biodegradable like paper plates. Disposable plates are also made of sugarcane and other biodegradable materials.

Styrofoam plates, made from polystyrene, are disposable. They are much less absorbent than paper plates. Some brands could stand up to hand washing and reuse. However, they are usually used once and then discarded into the garbage. Because they degrade over a long time in landfills, they are not very green. In addition, they are made from petroleum-based chemicals, which is not a green renewable resource. They also require a considerable amount of energy to manufacture.

There is a variety of disposable plastic tableware. Plastic plates, cup, bowls, flatware, and other utensils are available in a variety of thicknesses for strength. They come available in a variety of colors. This type of disposable tableware is common at weddings and other social events where there is a large gathering that is catered. They are cheap and easy to use. They do not require washing, which saves on water and energy. They could be reused, but many people would see this as being "cheap."

Green consumers would not find disposable plastic tableware as green as paper plates because they are not readily biodegradable. However, if cleaned, they could be recycled if there were a recycling collection point that took plastic with their recycle number. Another negative aspect of using disposable plastic tableware is that they are made of plastic derived from petroleum. In addition, they require a significant amount of energy to manufacture. However, disposable tableware is not washed, and therefore water is not used, nor is dishwasher detergent or dishwashing soap. These would enter the wastewater stream as stresses on the waste disposal system and potentially become an environmental pollutant.

Disposable drinking cups are commonly used where coffees, teas, juices, or soft drinks are sold by the cup. They have the advantages that they do not have to be washed and that they can be used as "carry-out" drinks. Drinking cups for cold drinks are often paper with a light wax coating. Most wax is derived from petroleum tars. The cups are ultimately biodegradable. Paper cups are also used for hot drinks such as hot coffee.

Both cold and hot drinks can be served in Styrofoam cups. The Styrofoam has insulating properties that keep the drink cooler or hotter for longer than a paper cup does. As with plates, Styrofoam cups degrade very slowly, and are generally thought to be a poor "green" choice.

Plastic implements are utensils that have a number of uses, and there are a wide variety of styles available. Plastic forks, spoons, and knives can be used on picnics or in outdoor eating of virtually any kind. If the grade of plastic is heavy enough, they can be washed and used again. However, the cheaper, lightweight ones are designed for one-time use. Oddly enough, they have replaced many other types of biodegradable utensils such as flat wooden spoons used for eating frozen cups of ice cream; there is little difference in price between them, but the wooden ice cream spoons are readily biodegradable.

**See Also:** Biodegradability; Green Consumer; Paper Products; Recyclable Products.

### Further Readings

Andrady, Anthony L. *Plastics and the Environment.* Hoboken, NJ: Wiley-Interscience, 2003.

Clarke, Alison J. *Tupperware: The Promise of Plastic in 1950s America.* New York: HarperCollins, 2001.

Hoff, Al. Thrift Score: *The Stuff, the Method, the Madness!* New York: HarperCollins, 1997.

*Andrew Jackson Waskey*
*Dalton State College*

## Downshifting

Downshifting an automobile signals a shift to a lower gear. The driver wishes to slow down. Much like the automobile, downshifting (often synonymous with "voluntary simplicity" or "simple living") is a conscious decision to slow down one's life and adopt the values of simplicity. It is a voluntary decision to reduce income and consumption in pursuit of a less stressful, more meaningful life. Downshifting signals a return to a sense of balance lost in the frenzied pace of everyday life. Downshifting is a rejection of the "rat race" of consumerism, long work hours, and the frantic modern life. Predominantly an individual act, the decision to downshift is made for a variety of reasons:

- A lack of balance in one's life (i.e., working too much, not enough time with family)
- The unfulfilling "work-and-spend" lifestyle
- A rejection of consumer and material values
- The desire to live a greener, more environmentally and socially responsible life

Downshifting is a reevaluation of the relationship to time, work, and money. Downshifters find they have to work too much to earn the money to retain their materialistic wants. They find themselves worn out by the "rat race" of modern society that involves a work- and money-focused lifestyle. Downshifting is about taking control of one's life and emphasizes time rather than work or money. Many downshifters cite the high pressure and unfulfilling nature of their work as a primary reason for making the change. They report psychological and physical burnout from their stressful lives.

Even leisure time is influenced by today's frenzied pace. Although those in the United States and other Western countries have more free time than they did 30 years ago, they are busier than ever. Free time is filled with leisure activities such as travel and watching television. Downshifting highlights the paradox that a Caribbean holiday or the purchase of a new flat-screen plasma television is dependent on working enough hours to earn the income to afford these leisure items. Although travel and television can be enjoyable, downshifting emphasizes the lack of meaning that often coincides with contemporary leisure activities. Practicing religion or spirituality or participating in the community is supplanted by socializing while shopping at the mall. Downshifting recaptures time with family, friends, volunteering, or for meaningful self-fulfillment. It indicates a desire to return a sense of balance to one's life. Downshifters seek a life filled with purpose, passion, fulfillment, and happiness.

Downshifting is also a rejection of consumer and materialist values. Consumerism is predicated on the continuous acquisition of material goods, most often in an attempt to achieve happiness and competitively "keep up with the Joneses." Social competition and comparison demands physical indicators of status. Material goods are constantly acquired to keep with ever-increasing social standards. As a result, an expanding definition of "necessity" has occurred over the past 50 years. Items such as air conditioning, cell phones, and iPods are deemed necessary for living a good and socially acceptable lifestyle. The consumer treadmill is always on and running, yet with no purpose or end in sight. Many downshifters admit to being on the consumer treadmill and ultimately report burnout. Keeping up materially is both physically and emotionally exhausting. Downshifting rejects these unhealthy values in favor of values that cultivate contentment.

Living with fewer material goods is related to the rejection of consumerism and is another core tenet of downshifting. Goods take on a new meaning. Rather than reflect the social symbols of status and identity, material goods should reflect quality, durability, and reliability. The use-value of a good outweighs its symbolic value. As a result, downshifters buy less, but often buy better. More centrally, downshifting reflects a suspicion or uneasiness with the consumer culture in general.

Not all individuals downshift out of environmental concerns. In fact, downshifting most often is done primarily for personal reasons such as stress and a lack of time. However, consuming less, reusing goods, and recycling items make downshifting a natural ally with the green movement. Greens emphasize slowing down and consuming less for the benefit of the natural environment. They emphasize that the environmental cannot sustain our consumer culture.

Recently, downshifting has increased in popularity because of its environmental connection. The growing attention to environmental problems indicates that the public is realizing that human activities like automobile driving and the pursuit of materialistic wants are environmentally unsustainable. Organic gardening is a popular downshifting activity that brings happiness and a sense of self-sufficiency. It benefits the environment by reducing harmful chemical use and the effects of food transport. Not only does downshifting give the gift of time but it also lessens one's ecological or carbon footprint. So although downshifting may be done for personal reasons, there are large environmental benefits that enhance its attractiveness.

Downshifting is quickly becoming part of the mainstream. For instance, 25 percent of Australian adults are downshifters. Contrary to some public perceptions, most downshifters do not give up everything, move to the country, and become wholly self-sufficient. Downshifters are blue- and white-collar workers who represent the middle-class as well as the upper middle-class. They are mostly middle-aged. Fear and anxiety about future economic security often leads individuals to adopt downshifting values. For example, the global economic recession of 2009 has led many to make the conscious decision to consume less out of fear over their future employment, home values, and savings.

Downshifting has a large Internet presence. Multiple Websites educate the reader on what downshifting is, as well as extol the virtues of becoming a downshifter. Many of these Websites have "how to" manuals that direct individuals, and even businesses, on how they can simplify their homes and places of work. The Internet has served as a support group of sorts for downshifting. Blogs and other Internet "communities" connect downshifters, who share their personal experiences and strategies with the lifestyle. For example, "InterNational Downshifting Week" uses the Internet to mobilize individuals from around the world to take part in a variety of activities such as cutting up a credit card, eliminating three nonessential purchases, turning off the television, and volunteering in the community.

In a highly materialistic society, downshifting is quite subversive. Downshifting simultaneously challenges and redefines what society thinks of as "success." It challenges notions of social status, competitive consumption, materialism, and the "more is better" mantra. Success is subsequently redefined as contentment and meaningful work, as is having the time to participate in fulfilling activities with loved ones. In short, downshifting turns off the go-nowhere treadmill of an out-of-balance life.

**See Also:** Frugality; Green Consumer; Materialism; Quality of Life; Simple Living.

### Further Readings

Breaksphere, Christie and Clive Hamilton. "Getting a Life: Understanding the Downshifting Phenomenon in Australia." Manuka, Australia: The Australia Institute, 2004. http://www.tai.org.au/documents/dp_fulltext/DP62.pdf (Accessed April 2009).

Drake, John D. *Downshifting: How to Work Less and Enjoy Life More.* San Francisco, CA: Berrett-Koehler, 2000.

Schor, Juliet B. *The Overspent American.* New York: Basic Books, 1998.

Sevier, Laura. "Shifting Gear." *Ecologist* (November 28, 2008). http://www.theecologist.org/pages/archive_detail.asp?content_id=1186 (Accessed April 2009).

Smith, Tracey. "International Downshifting Week." http://www.downshiftingweek.com (Accessed March 2009).

*Austin Elizabeth Scott*
*University of Florida*

## Dumpster Diving

Dumpster diving involves picking through discarded trash for usable material goods and edible food. Although many engage in this activity solely out of economic necessity, Dumpster diving, as it is used here, is a critical or frugal reaction to the wastefulness of the consumer society. Dumpster diving is the modern-day variant of scavenging but is often done for a variety of political, environmental, and ethical reasons. Dumpster divers, also known as "freegans," live off the enormous amount of waste produced by the modern world. ("Freegan" combines "free" and "vegan," although not all freegans are vegans). Some Dumpster dive to protest capitalism and the hyper-consumptive society it produces. Others scavenge because they want to lessen their impact on the environment.

Dumpster diving highlights an interesting paradox of the consumer society: the majority of items taken from the trash or garbage are not trash at all. They are functional, usable, desirable, and often unused. The functionality and newness of so much trash generates much of the protest from Dumpster divers, even though they benefit from the refuse. They point to the excessively wasteful nature of an overconsuming society that often has never used the discarded goods. Usable "waste" finds its way to landfills and harms the environment. Critics claim that Dumpster divers are hypocritical because they benefit from the very capitalist, waste-generating society they protest.

Dumpster diving challenges mainstream environmentalism. The traditional response to the mounting ecological problems society faces is to buy greener (e.g., a hybrid vehicle rather than a sport utility truck). Dumpster diving lessens the human impact on the natural world by not buying anything and by keeping discarded items from entering landfills.

Although Dumpster divers search trash for discarded goods such as furniture, books, or clothing, part of the appeal is finding unexpected items. Dumpster diving can uncover seemingly treasured family photographs or personal diaries, as well as morally dubious items such as unused bullets. Discarded food is also a sought-after and treasured find for Dumpster divers and freegans. Most of the food is slightly bruised or wilted produce, day-old bakery items, dented canned goods, or just-past-expired items. Dumpster divers pride themselves on making gourmet communal meals using everything they scavenged from the trash—including the plates and flatware. They insist there are no health risks as long as common sense is employed.

Successful Dumpster diving is all about location. Urban settings provide the greatest diversity and amount of refuse. Cities have more wealth, and wealthy people throw away more than the poor. The more affluent one is, the more likely he or she is to follow fashion and trend cycles. Thus, goods become obsolescent more quickly and are discarded.

A college dormitory is one of the most popular places to dive. Every year at the same time, hundreds or thousands of students all move out of their residences. The more affluent the university, the greater the chance Dumpster divers will score big finds. Divers can find anything from half-used bottles of liquor to fully functioning electronics. For example, at New York University (NYU), Dumpster divers participate in an annual "NYU Dorm Dive," as they sift through the refuse of students moving out of NYU's residences. It is a cooperative group effort in which Dumpster divers share what is found. Although the Dorm Dive is a communal event, most Dumpster diving is done alone or in pairs to avoid unwanted attention from security or the police.

Dumpster diving has legal consequences of varying degree. Although the law varies from city to city, more locales are making it illegal to scavenge through trash left on the street and in Dumpsters. Enforcement of such laws is usually lax, but because Dumpsters are usually located on private property, divers are technically trespassing and/or stealing. Many stores and businesses, especially bakeries and restaurants, lock their Dumpsters. This is to prevent divers from accumulating in numbers and to deter theft or vandalism to the business. Locking a Dumpster is also done to avoid legal action that may be taken against the business if an individual was injured on the property while Dumpster diving.

There are numerous "how-to" Websites on the Internet that give instructions and warnings for anyone interested in Dumpster diving. The tips range from what to wear, to what to bring, to how best to avoid (or get out of) an encounter with the police. These Websites also include specific locations of Dumpsters that offer appealing items. The sites give practical advice for inspecting and eating food found in Dumpsters. The Web has become an integral part of Dumpster diving; divers network, exchange location information, and arrange Dumpster diving meet-ups.

Dumpster diving seems to be gaining momentum in today's society. The consumer society appears to be on the rise, as do the massive amounts of waste it generates. Unless socially and environmentally responsible solutions are crafted (e.g., donating "trash" or buying less), Dumpster diving will continue to be a frugal and political reaction to the consumer society.

**See Also:** Frugality; Lifestyle, Urban; Overconsumption; Secondhand Consumption; Waste Disposal.

### Further Readings

Eighner, Lars. "Dumpster Diving." http://www1.broward.edu/~nplakcy/docs/dumpster_diving.htm (Accessed March 2009).

Ferrell, Jeff. *Empire of Scrounge: Inside the Urban Underground of Dumpster Diving, Trash Picking, and Street Scavenging*. New York: New York University Press, 2006.

Grifter. "Dumpster Diving: One Man's Trash." http://web.textfiles.com/hacking/dumpster_diving.txt (Accessed April 2009).

Hoffman, John. *The Art & Science of Dumpster Diving*. Boulder, CO: Paladin, 1993.

Kurutz, Steven. "Not Buying It." *New York Times* (June 21, 2007).

*Austin Elizabeth Scott*
*University of Florida*

## Durability

The issue of durability concerns the material qualities of classes of particular consumer goods, the durability of the relationships that consumers have with those goods, and cultures of repair, reuse, and recycling. Durability is complex and not simply a matter of condemning an apparent rise in the disposability and disposal of frivolous and less-durable consumer goods.

In relation to classes of commonplace consumer goods, somewhat contradictory dynamics affecting durability can be identified. A general improvement in the quality of manufacture, reliability of components, and precision of engineering has increased the reliability and longevity of many consumer goods. However, durability is not reducible to longevity or reliability: Goods may last a long time, but with impaired and diminishing functional capacity, whereas reliability may be ensured in function, while overall durability diminishes over time. Durability requires reliability and functional longevity, and it is this complexity that may encourage practices that work against the effects of higher quality design and manufacture working their way through into always more durable products.

As a consequence, various forms of built-in obsolescence can be seen to work against this general increase in product durability. These include the following:

- Built-in technical obsolescence, such that products are designed to fail as a result of the inclusion of a particularly vulnerable, short-lived, or irreplaceable component. This has been more assumed than proven, but it remains a strong suspicion and was certainly widely advocated as making economic sense for manufacturers in the middle decades of the 20th century. There is, however, an evident form of this obsolescence in contemporary product design. Many objects increasingly tell the consumer when they have apparently worn out. For example, dye-pigmented toothbrush bristles that advertise the end of the brushes' functional efficacy. The discovery of surprisingly prolonged durability is often the outcome of ignoring these built-in suggestions to discard, dispose, and replace.
- Stylistic obsolescence denotes an object's loss of semiotic durability. Previously associated with a small range of goods, especially clothing, fashion, or product styling, it is now a strongly pronounced feature of practically all consumer goods. Functionally durable goods, it is suggested, stylistically wear out, necessitating replacement.
- Objects are increasingly nested in networks of supporting products. The proliferating range of misnamed peripherals associated with the purchase of a computer provides a good example of systems of objects in which there are multiple opportunities for a lack of durability to compromise the whole nested arrangement. Often, all forms of built-in obsolescence are seen to work in concert.

Manufacturers, keen to protect revenues, have also turned to other techniques to counter the general increase of product durability, much of which is abetted by environmental legislation. The increasing promotion of services as necessary adjuncts to products is commonplace, particularly the often-aggressive selling of extended warranties offering information services, upgrade options, and service contracts. These can be regarded as an attempt to recoup income previously accruing from the regular repair of physically less durable goods. They have proven to be highly controversial, as their cost is often a large proportion of the purchase cost of the goods. Moreover, against a background of increasing product reliability and consumer protection, it is argued that such warranties are unnecessary.

The offer of such services and guarantees resonates strongly with many green critiques that have explicitly argued for the move toward a product–service economy, in which objects are embedded in supportive networks of services. However, the underlying motivation is different; the various projected green service and guarantee schemes would mean manufacturers would have no reason to plan obsolescence. Instead, it would be in their interests to ensure durability.

The durability of consumer goods can also be assessed in relation to the qualities of individual objects. Concerns include the following:

- The appropriate use of materials: Is the object constructed from materials and components that promote durability of use? Allied with this is the question of the inappropriate use of materials. For example, plastics have a number of particular material qualities that render their use as disposable packaging inappropriate.
- The relationship between product styling and engineering design: Many products are increasingly designed such that there is an inability to get to internal components because of sealed carcasses, the need for specialized tools, and the prospect of warranty invalidation. "No user-serviceable parts" stickers work against vernacular repair that may often extend durability.
- The aesthetic durability of the object: Fashion represents a lack of semiotic durability and suggests the premature disposal and replacement of functionally useful goods. Critics have

articulated the need for new product aesthetics consonant with extended ownership and environmental benevolence.
- The durability of the relationship between object and user: The emotional durability of objects has become central to green thought, and the investigation of both aesthetic and emotional durability has resulted in various product "strategies." These include promoting objects that require ongoing attention from the user, such that their usefulness remains in mind. In addition, the material qualities of objects are considered, with commentators often looking to previous incarnations of durable products for suggestive clues. How materials age, how they disclose their various properties, and the nature of their sensual affordances have all been considered.

The less durable objects are, the more material impacts they tend to have. One ameliorating force is provided by practices of repair, reuse, and recycling. For example, throughout the 19th century, packaging use grew dramatically, but most of this was specifically intended for reuse. However, from the late 19th century forward, the growth of packaging specifically made for disposal increased and now is ubiquitous. Moreover, during the 19th century, nondurable household material was routinely recycled and reused, and there existed an economy attuned to this, including neighborhood collection trades (rag pickers, rag and bone men, swill girls) that passed on these nondurables as raw materials for processes of industrial revalorization. On a domestic scale, many product containers, especially glass bottles, were returned and reused—a practice that, apart from milk bottles, has been replaced by the breaking up of glass containers and their energy-inefficient recycling into new forms of glass.

Overall, vernacular cultures of repair, reuse, and recycling have given way to state-provided, professional services often backed by strong legislative adjuncts. Indeed, it has been argued that the rise of official recycling itself acts to mediate concerns about durability to the advantage of the producers of nondurable goods. Recycling legitimates a disposal culture while apparently mitigating its effects. Many argue that it would be far more rational to consolidate a material culture of durability in which such official recycling is unnecessary and in which self-organized practices of repair, reuse, and recycling address issues of durability. They point to a growing subculture of vernacular reuse, along with the persistence of many unacknowledged product "disposition" practices, as evidence of this potential.

**See Also:** Conspicuous Consumption; Product Sharing; Recycling; Services.

### Further Readings

Moles, A. "The Comprehensive Guarantee: A New Consumer Value." In *Design Discourse: History, Theory, Criticism*, Victor Margolin, ed. Chicago: University of Chicago Press, 1985.

Packard, Vance. *The Waste Makers*. London: Penguin Books, 1960.

Slade, Giles. *Made to Break: Technology and Obsolescence in America*. Cambridge, MA: Harvard University Press, 2006.

*Neil Maycroft*
*University of Lincoln*

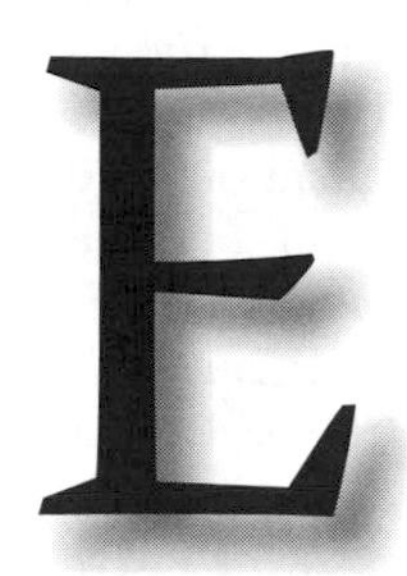

# ECOLABELING

Ecolabeling is a voluntary initiative to enhance environmental performance during the production or consumption of products and services. Ecolabels have emerged as a main tool for green marketing. Ecolabeling is proposed as a market incentive to create incentives and to force compliance for business to operate in an environmentally conscious and ecologically friendly way. An ecolabel is a seal awarded by a third party (ecolabeler) that provides information for consumers regarding environmental commitments of a specific good or service. Through this, ecolabeling is expected to inform consumer choice, stimulate spontaneous environmental improvement, and diminish aggregated pollution, enabling a healthy environmental competition among firms.

Since the 1990s, there has been a proliferation of ecolabels worldwide in the business community and at the policy level. Ecolabels have emerged as a result of consumers' willingness to pay a premium for goods that are environmentally friendly. According to the Organisation for Economic Co-operation and Development (1997), the process of development of different ecolabels is similar. Analyses based on life cycle approaches are usually used to identify the consequences and the complexity of the environmental impacts caused by a product in its life cycle. These analyses allow the ecolabeler to draw the first ecolabel process, which contains two phases: a phase of negotiation and a market phase. The phase of negotiation consists of a negotiation between the ecolabeler and the firm to set up the environmental criteria to be met. In this phase, the firm suggests the modifications to the initial proposal drawn by the ecolabeler. The market phase consists of the labeling of those firms that met the criteria and that therefore can compete in the marketplace.

Ecolabels are designed on the basis of the assumptions that consumers have a choice in the marketplace, that they will prefer products and services that were produced under environmentally friendly conditions, and that they are willing to pay a premium to satisfy this preference. According to Eva Eiderström (1998), there are elements that have been considered successful in ecolabeling: clearly defined objectives, scope, and strategy; independence with respect to funding; schemes that are accessible to all producers, regardless of their size; formulation of criteria taking into account the current market situation; consumers being represented and having a strong influence in ecolabeling schemes; transparent ecolabels, to avoid criticisms; cost-efficiency; and a focus on national, rather than international, action—"think global, act local."

The main difference between ecofriendly pictograms/green logos and ecolabels attached to certain products and services is that ecofriendly/green logos are environmental statements claimed by producers and do not require auditing, monitoring, or inspections, whereas ecolabels are generally dependent on third-party evaluation. Ecolabelers establish the terms of trade between the seller and the buyer by defining environmental protection standards to be met by the producer or the trader.

According to the International Standards Organization, there are three types of ecolabels—type 1, type 2, and type 3. Type 1 ecolabels are multiattribute labels developed by a third party. It is known as third-party practitioner scheme. Within type 1 ecolabels, the labeling criteria are elaborated under the International Standards Organization system guidelines. Type 2 labels are single-attribute labels developed by a producer. These ecolabels are described as self-declarations, as the producer uses a logo to represent an environmental attribute, such as "recyclable," without a third-party auditing. Type 3 labels are ecolabels based on life cycle assessment, and they are also evaluated by a third party.

The best-known ecolabels are attached to official governmental schemes in terms of financing and governance. In these ecolabels, the ecolabelers are the national (or regional, in the case of the European Union) standardization institutions that deal with the definition of the criteria, evaluate the processes, and grant the labels.

The E.U.'s ecolabel is a voluntary scheme intended to promote businesses to market services and goods that are friendly to the environment, making it easier for consumers in Europe to identify them. This scheme is part of the Integrated Product Policy led by the European Commission. This initiative was one of the European Union's responses to Agenda 21, agreed on in the 1992 Rio Summit toward sustainability, and it was developed in conjunction with European customers, retailers, manufacturers, and services providers.

## Criticisms

There are different criticisms of ecolabeling. First, it has been questioned whether voluntary ecolabeling might cause producers' profits in a competitive industry to decline and whether ecolabeling will inevitably create different prices for labeled and unlabeled goods. This could be improved by providing information to consumers and producers on how and why ecolabeling schemes work.

Another criticism is that ecolabeling normalizes existing power relationships that are a feature of trade between the north and the south. As a consequence, firms and the industry should be educated about the environmental effects of practices that occur through commodity networks, encompassing production, trade, and consumption. Where firms' interests dominate over the protection of the environment, this also decreases the credibility of ecolabeling campaigns. Finally, although ecolabeling campaigns provide consumers with the ability to identify and choose "green products," ecolabels are not legally binding, and favorable social and environmental outcomes cannot be guaranteed.

**See Also:** Certification Process; Ecological Footprint; Environmentally Friendly; Green Consumer.

### Further Readings

Amacher, Gregory S., et al. "Environmental Quality Competition and Ecolabeling." *Journal of Environmental Economics and Management*, 47/2:284–306 (2004).

Clift, Roland. "Life Cycle Assessment and Ecolabelling." *Journal of Cleaner Production,* 1/3–4:155–59 (1993).
Dosi, Cesare and Michele Moretto. "Is Ecolabeling a Reliable Environmental Policy Measure?" *Environment and Resource Economic,* 18/1:113–27 (2001).
D'Souza, Clare. "Ecolabel Programmes: A Stakeholder (Consumer) Perspective." *Corporate Communications: An International Journal,* 9/3:179–88 (2004).
Du Toit, Andries. "Globalizing Ethics: Social Technologies of Private Regulation and the South African Wine Industry." *Journal of Agrarian Change,* 2:356–80 (2002).
Eiderström, Eva. "Ecolabels in the EU Environmental Policy." In *New Instruments for Environmental Policy in the EU,* Jonathan Golub, ed. New York: Routledge, 1998.
Hale, Monica. "Ecolabeling and Cleaner Production: Principles, Problems, Education and Training in Relation to the Adoption of Environmentally Sound Production Processes." *Journal of Cleaner Production,* 4/2:85–95 (1996).
Mason, Charles F. "An Economic Model of Ecolabelling." *Environmental Modeling and Assessment,* 11/2:131–43 (2006).
Nadaï, Alain. "Conditions of Development of a Product Ecolabel." *European Environment,* 9/5:202–11 (1999).
Organisation for Economic Co-operation and Development (OECD). "Eco-Labelling: Actual Effects of Selected Programs." OECD Working Paper, 5:44 (1997).
Rex, Emma and Henrikke Baumann. "Beyond Ecolabels: What Green Marketing Can Learn from Conventional Marketing." *Journal of Cleaner Production,* 15/6:567–76 (2007).
Sedjo, Roger A. and Stephen K. Swallow. "Voluntary Ecolabeling and the Price Premium." *Land Economics,* 78/2:272–84 (2002).
van Ravenswaay, Eileen O. and Jeffrey R. Blend. "Using Ecolabeling to Encourage the Adoption of Innovative Environmental Technologies in Agriculture." In *Flexible Incentives for the Adoption of Environmental Technologies in Agriculture,* Frank Casey, ed. New York: Springer, 1999.
West, K. "Ecolabels: The Industrialization of Environmental Standards." *Ecologist,* 25/1: 16–20 (1995).

*Maria-Alejandra Gonzalez-Perez*
*Universidad EAFIT*

# Ecological Footprint

The term *ecological footprint* is used to indicate the impact of a single individual on the Earth's environment. It can also be used to indicate the impact of a single society on the Earth's environment. The image of a footprint is visual and universally understood: Footprints can vary in size from small to large, and they can be shallow or deep. The goal of green living for consumers is to have the least footprint possible, with no trace of a footprint as ideal.

At this time, the world's best, or at least leading, environmental indicator is the ecological footprint calculation. Using currently prevailing technology, a given group of people needs a certain amount of land and water to sustain itself and to dispose of its wastes. Using an ecological footprint as a measurement gives the clearest answer to the question, "Is the lifestyle of a certain group of people sustainable at current rates of consumption?" The ecological footprint measurement is similar to the carbon footprint measurement.

*Carbon footprint* is a term to describe the effect human activity has in terms of the carbon dioxide emitted by the activity. The amount of jet fuel burned by different types of jet planes flying across the Atlantic Ocean can be measured specifically. An engine can be mounted in a laboratory and operated using jet fuel, and its carbon dioxide exhaust captured and measured. If a jet plane carries 200 people, then the carbon footprint of a single individual would be one two-hundredth of the weight of the carbon dioxide. If this amounts to 100 pounds of carbon dioxide, then that is the carbon footprint for the trans-Atlantic traveler. A similar footprint would be derived if the trans-Atlantic traveler sails on an ocean liner instead.

Climate change is an important ecological concern today. Rising global temperatures are believed to be to the result of increases of carbon dioxide in the atmosphere that began with the Industrial Revolution. Scientifically based studies of changes in the chemistry of the atmosphere and the biosphere have raised concerns about the apparent trends. The trends predict dire consequences if carbon dioxide continues to increase in the atmosphere because it is a greenhouse gas that contributes to global warming.

## Carbon Footprint "Wars"

Serious disputes have emerged over the subject of carbon footprints. These carbon footprint "wars," or more accurately power struggles, are part of the politics of environmentalism. Efforts to legislate limits to carbon dioxide emissions inside countries involve caps on industrial production, energy usage, transportation, and other limits. Because the limits interfere with businesses, industry, agriculture, and a great many people, they have been the subject of controversy.

The same political conflicts are at play internationally. Third World countries, especially those with developing economies such as India and China, have been reluctant to restrain their drive toward industrialization and accuse the richer industrial nations of Europe, North America, and Japan of seeking to solve environmental issues at their expense.

Carbon footprints are a part of the large, more comprehensive subject of ecological footprints. Many scholars, scientists, politicians, administrators, engineers, industrialists, and others are adopting ecological footprinting as a practical yardstick for measuring human impacts, whether large or small, on nature. It is a measure of the effect of humans on the ecosystems of the Earth. As a measure, it represents the quantity of land, water, and air that is needed to provide food, shelter, and all the other things that the people of the Earth need and, in addition, the ability of the Earth to absorb their waste products.

Recent estimates of the current population of the Earth are that it is overpopulated: The current population of the Earth should be living on 1.5 Earths. In other words, the current population is using the Earth's resources at a rated of 1.5 times the capacity of the Earth to supply them and absorb their waste.

The methods for measuring ecological footprints vary widely. The concept was first presented by Mathis Wackernagel while he was a doctoral student at the University of British Columbia in Vancouver, Canada. As part of his requirements under the guidance of Professor William E. Rees, Wackernagel presented a discussion of the idea of an ecological footprint in his dissertation. The term *ecological footprint* was substituted for the term *appropriate carrying capacity*. The latter term has connotations that ecological footprint does not have and is easier to understand. In 1996, Wackernagel and Rees coauthored *Our Ecological Footprint: Reducing Human Impact on the Earth*.

Ecological footprinting measures the effect of humans on the biosphere and its power to regenerate resources. As previously stated, comparison is between the land and marine areas and human resource extraction as they involve carbon, water, minerals, food, goods, and services. However, the comparison is not just for a locality—it compares how much is required for everyone in the world to live at the highest level. If the mode of transportation were something else, such as the trans-Atlantic airplane described above, then more than the carbon footprint of a single journey is the total impact (ecological footprint) for all the flights the plane makes; the supplies it uses to provide food, water, and other beverages for the passengers and crew; all of the materials of which it is made; and the servicing costs generated when it is grounded for service.

The measures developed to measure ecological footprints differ between nations. It obviously requires more heating for people in the polar regions or the temperate zones than it does for people living in the tropics. For example, people in the Philippines or the oceanic islands can often supply themselves with fresh fruit or coconuts with little effort, but these foods are available in temperate and polar climates only when imported. However, tropical areas have impacts that are different.

## Why Measure Ecological Footprints?

The goal of ecological footprint measurements is to have a comparative index that can be used to see which nations are having the largest impact and which are having the least. The footprint measure shows which lifestyles are sustainable and which are not. It also provides policy makers with a way to educate people so that there is a public understanding of overconsumption and its relationship to carrying capacity. The measure can be brought into the political arena to argue that a lifestyle is unjust because its ecological footprint is much heavier than that of other people in the poor countries of the world.

The ecological footprint of a country is measured in terms of global hectares per capita. The measurement of the different resources used to support a particular lifestyle is similar to using life cycle assessment (LCA), which is also called "life cycle analysis," "ecobalance," and "cradle-to-grave analysis." The LCA calculates the value or benefit at each step in a production process from beginning to end. The benefits are, however, matched by the losses or negative effect that each step has on the environment. For example, the gathering of sea salt through solar evaporation in salt pans is a benefit with little if any negative environmental impact other than the loss to nature of the salt pan. The salt is gathered using rakes or other tools, which are losses to nature because they are made from material (probably wood) drawn from nature. The salt is packaged, transported, and distributed. If the salt is used in animal feed, it is a plus, but it is a negative impact if the domestic animal is raised as a beast of burden or for slaughter. The successive steps include the transportation cost(s) to deliver the salt, the disposal cost of the packaging, the utilization cost for the animal, and other associated costs. In general, the interconnectedness of any one part of human production processes often involves hundreds of resources. So whatever may be the benefits, the costs are also quite significant. LCA seeks to measure all of these negative costs.

If ecological footprints are measured in terms of per capita, the goal is to compare consumption and lifestyles. The real goal is to understand Earth's carrying capacity for any kind of lifestyle. To put it another way, the Earth is finite, but the wants of people, if not infinite, certainly are measured as a very large number. Therefore, whether measured as environmental footprint or LCA, the question is, "Can the finite resources of the Earth sustain consumption of a particular lifestyle at current levels of consumption?" For many who are green scientists or activists, the answer is that the industrial-urban lifestyle of the leading industrialized nations, and especially of the United States, cannot be sustained at current levels.

## Calculating Ecological Footprints

Because of its ability to compare economies and lifestyles, the measure of the ecological footprint is now used widely. There are a number of Websites that calculate environmental footprints. Some of the Websites on which ecological footprints can be calculated are those of nongovernmental organizations. The Website http://www.footprintstandards.org is one such network, hosted by the Global Footprint Network. Other allied organizations are linked to this Website.

The Global Footprint Network issues a "Living Planet Report" that presents both the accounting methods for measuring ecological footprints and the results of their analysis. Other organizations have used different methods to calculate ecological footprints. Factors such as sea area, energy consumptions, and the power of technology vary, as do estimates of population, biodiversity, and other factors.

At this time the environmental footprint of the United States is estimated at 9.4 global hectors per capita. China, in comparison, is well over two global hectors per capita, but still less than three. National footprints include calculations that add and subtract imports and exports. The figure yielded can be used to show whether a nation is sustainable or whether it is exceeding its carrying capacity. These figures would very likely vary some, depending on which methodology is used. Efforts have been made to standardize the ecological footprint calculation approach. Critical examinations of ecological footprint calculations have been offered by a number of critics, and several countries have commissioned studies of the validity of the method. At stake is the political use that can be made of the ecological footprint calculation. If it is not accurate, but is accepted as such, it could do harm to the economy of a country without ecological benefit.

An example of the negative conclusions that can be drawn from an ecological footprint calculation is the negative number that a densely populated area such as Tokyo or New York might receive. The conclusion could be made that such a dense population is really parasitic, when in fact these regions are vibrant and very productive. They draw from vast hinterlands, so that their ecological footprint is really much smaller than if the only area considered is their densely populated urban area.

**See Also:** Consumer Culture; Disparities in Consumption; Quality of Life; Sustainable Consumption.

### Further Readings

Casper, Julie Kerr. *Trends: Ecological Footprints*. New York: Facts on File, 2009.

Chambers, Nicky, et al. *Sharing Nature's Interest: Ecological Footprints as an Indicator of Sustainability*. London: Earthscan, 2001.

Crocombe, Angela. *A Lighter Footprint: A Practical Guide to Minimizing Your Impact on the Planet*. Philadelphia, PA: Scribe, 2009.

Goleman, Daniel. *Ecological Intelligence: How Knowing the Hidden Impacts of What We Buy Can Change Everything*. New York: Broadway Books, 2009.
Mayhew, Wil. *SuN LIVING: Developing Neighborhoods With a One Planet Footprint*. Gabriola Island, British Columbia, Canada: New Society, 2008.
Merkel, Jim. *Radical Simplicity: Small Footprints on a Finite Earth*. Gabriola Island, British Columbia, Canada: New Society, 2003.
Rees, William. *Our Ecological Footprint: Reducing Human Impact on the Earth*. Gabriola Island, British Columbia, Canada: New Society, 1998.
Sim, Stuart. *Carbon Footprint Wars: What Might Happen If We Retreat From Globalization?* Edinburgh, UK: Edinburgh University Press, 2010.

*Andrew Jackson Waskey*
*Dalton State College*

# Ecotourism

Ecotourism is a form of nature-oriented tourism based on more sustainable practices that suggest a lower environmental impact than conventional mass tourism. It offers nature-based adventure and environmental education and interest at destinations that are considered ecologically intact. At the same time, ecotourism emphasizes a more responsible form of travel that is more sustainable, operates under a higher ecological and cultural awareness, and offers a higher benefit to local income and employment opportunities.

Some aspects of ecotourism compared with conventional tourism have an actual positive effect on the destination resulting from local ecotourism operations. Many ecological arguments, however, refer to a mere reduction of a nevertheless negative effect, while accepting current consumption levels in society. The concept of ecotourism is subject to some critiques, and the industry faces several unsolved problems in practice.

Ecotourism emerged as an alternative tourism concept in the 1980s, when negative environmental and social impacts of conventional mass tourism were starting to be felt on a larger scale. Ecotourism has increased its market segment continually to an estimated 35 percent, with suggested further growth in the future. In Costa Rica, Kenya, Nepal, Ecuador, Madagascar, Antarctica, and many other places, ecotourism has become a significant economic activity.

## Ecological Sustainability and Conservation

Ecotourism has been connected to various benefits in comparison with conventional mass tourism, such as a lower environmental impact on local ecosystems. Ecological low-impact efforts range from technical measures (e.g., increased energy efficiency, use of renewable energy sources, recycling, water conservation, use of more sustainable building materials, environmentally less harmful pesticides, or infrastructural planning leaving more natural habitat for local flora and fauna) to preventative approaches that focus on an overall lower consumption of the facilities and the individual tourist (e.g., low-impact activities, accommodation, and transport) and/or compensation efforts that offset the ecological impact (e.g., carbon credits, reforestation projects).

Ecotourism may offer educational programs and a rich implicit environmental learning experience that includes a heightened overall environmental awareness, green consciousness about individual consumption, ecological and cultural knowledge, and a stronger emotional connection to "nature" and sense of place that may support conservation efforts. Ecotourism projects can also encourage environmental conservation efforts toward the protection of biodiversity and ecosystemic integrity as a major economic agent for change. However, ecotourism projects have been related to indigenous dislocations when endorsing the creation of protected areas, suggesting negative influences, where selected economic interests are put above the cultural survival and basic livelihood needs of the local population.

## Cultural Effects and Concepts

Ecotourism has a potentially less adverse effect on local culture than conventional mass tourism, and even a positive effect on some communities. Cultural appropriateness in interactions with the local population and a higher cultural awareness are easier to learn in smaller-scale ecotourism operations. Overall, ecotourism represents a significant opportunity for economic growth in rural communities because of its preference for remote destinations with little prior industrial activity. The small-scale nature of ecotourism makes local start-up businesses a viable option. Locally owned ecotourism operations can contribute to the income opportunities of the local population more effectively and make the local community socioeconomically stronger, thus contributing to local sustainability and poverty reduction. This may also have positive indirect effects such as capacity building or local empowerment. However, environmentally competitive tourism businesses may in some cases require investment capital that is locally unavailable, thus causing the need to attract investors to the area and diverting the tourism revenue into anything but local income. The suggested solutions range from Fair Trade elements to community-based management and local benefit sharing to a diversification of local income opportunities for poverty alleviation, but the economic needs and pressures of external investments remain problematic.

A related critical point is the frontier-seeking nature of ecotourism. Ecotourism is fairly limited to a number of exquisite destinations with pristine nature, exotic cultural traditions, and overall unexplored locations that are perceived as "untouched." Most areas in the world are either too close to industrial activity and human settlements, affecting the aesthetic view and nature-seeking experience, and other areas are too remote and lack transportation access to make ecotourism economically viable. Places that show a successful ecotourism development may also be considered potential candidates for further tourism development on a larger scale, attracting more external investors and tourists. Ecotourism is therefore sometimes considered a pioneer stage of tourism, as it may lead to new infrastructure and diversified mass tourism. When this occurs, ecotourism will tend to move on to new destinations, repeating the process.

Another important conceptual critique is the link of ecotourism to green consumerism. The idea of a sustainable and nature-emphasizing form of tourism not only resonated with growing public concerns about the environment but also offered a diversification of consumption opportunities to the tourist who was searching for guilt-free alternatives, while taking the range of travel experiences to a new level. Where the only true low-impact activity may be to refrain from consumption and stay home, ecotourism offers an environmentally marketed excuse to not reduce consumption but, instead, to modify and possibly reduce the environmental impact while not lowering our lifestyle expectations.

## Consumer Participation and Standards

One of the main implementation problems is lack of market regulation across the ecotourism industry. Effective regulation, control, and enforcement mechanisms are needed to guarantee a sustainable system of local economic benefits, sound environmental standards, and overall responsible tourism operations in practice. These elements are often missing, especially in, but not limited to, the global South, where most ecotourism destinations are located. Very few state-controlled or market-based regulations have been created beyond voluntary self-regulation, and even fewer standards are made known to the public. National and international ecotourism certification initiatives are currently underway worldwide and have partly been implemented, for example, in Australia, Sweden, Costa Rica, and Kenya. The difficulties in moving toward stronger regulations highlight persistent controversies between environmental conservation goals, local and indigenous interests, and the tourism industry and show that the current problems of the industry need to be addressed more effectively globally, as well as locally.

A related problem is lack of transparent consumer information, especially the number of government-regulated information requirements, as opposed to voluntary market information. There are very few control mechanisms so far, and the transparency of ecotourism businesses varies widely, with little legal requirement regarding public disclosure of information. Alternative tourism has popularized a variety of terms such as "green," "bio-," "eco-," "responsible," "appropriate," "small-scale," "nature-," "soft" tourism, and related forms range from adventure tourism to cultural tourism, geotourism, ecotourism "lite," true ecotourism, overall "green" tourism, and "voluntourism." All of them suggest a less conventional form of tourism with a lower impact on the social, economic, and ecological fabric of the destination, but no commonly agreed definitions of these terms exist. Given the positive marketing potential, the imagery of "ecotourism" and its related terms has become used for an extremely broad spectrum of travel activities that range from dedicated low-impact projects to fairly conventional tourist destinations with only a few references to environmental education and/or outdoor activities, which may lead to false impressions about the sustainability of the tourism business. This can cause adverse competition forces in the ecotourism industry that reward lower environmental and social standards and create economic disadvantages for those ecotourism businesses that aim to operate on high standards. Dedicated ecotourism requires detailed information, encouraging responsible tourism through consumer awareness and informed choices.

**See Also:** Environmentalism; Green Consumer; Greenwashing; Leisure and Recreation.

### Further Readings

Honey, Martha. *Ecotourism and Sustainable Development: Who Owns Paradise?* Washington, D.C.: Island Press, 1998.

International Ecotourism Society. http://www.ecotourism.org (Accessed November 2009).

Nature Conservancy. "Ecotourism." http://www.nature.org (Accessed November 2009).

*Conny Davidsen*
*University of Calgary*

# Electricity Usage

The impact of using electrical devices is in part determined by the mode of generation used to produce the electricity they consume. One measure of the environmental impact of using electricity is to look at its "carbon content." The carbon content depends on a number of factors: the blend of generators—coal, oil, gas, nuclear, or renewable energy—used to supply the power and the efficiency of those generators and any losses in the system. Increasing the grid penetration of renewable energy will reduce the carbon intensity of grid electricity.

Many electrical devices have a "soft off," which enables a device to appear to be off while it is actually powered down to a "standby" mode. Although this provides some marginal benefit to the consumer—the ability to switch the item on again using a remote control—it also consumes power. Although for each device the amount of power consumed is small, with households typically possessing a large number of electrical devices, the cumulative total of these "phantom loads" can be considerable—around 5 percent of domestic electricity consumption, with each device drawing between 10 and 15 watts of power in standby mode.

## Reconciling Supply and Demand

Electricity supply and demand must be matched instantaneously in an alternating current grid, as there is no provision for storage. When power quality deteriorates as the result of a lack of supply, it is referred to as a brownout; if there is a total failure of the grid, this is called a blackout.

An electricity grid operates as a single synchronous alternating current machine; that is to say, where synchronous generators and motors are used, all parts of a single electricity grid will be rotating at a synchronized speed. Power is conventionally generated by a device, whether that is a steam turbine or wind turbine blade, that turns a generator that produces power that is synchronized to the grid's frequency. Some renewables, such as photovoltaic cells, produce direct current, which must be converted to alternating current to be fed back to the grid.

Some advocate the move from large, centralized generators of power to a more "distributed" arrangement of local sources of renewable energy providing power for the local area, supplemented by a more lightweight grid.

In a grid in which a greater percentage of power is generated by renewable energy sources, users adapting their usage habits to the power that is available through market incentives could help significantly toward reducing the need for additional power storage to be added to the grid.

## Monitoring Electricity Usage

Many consumers know very little about their electricity consumption habits. Electricity that is consumed from a public utility is often metered by a sealed meter in a low-profile location at the consumer's home. The meter is read quarterly, and then the customer is either billed or charged by direct debit from their account. This provides little or very weak feedback to consumers.

Strengthening the feedback loop, ensuring that consumers are acutely aware of their energy consumption, is a necessary precondition for helping consumers of electricity reduce their demand. It is impossible to control what you are not monitoring. To that end, there is an increasing array of metering devices that can help consumers to monitor, evaluate, and control their electricity usage, from simple plug-in meters that allow an assessment of the instantaneous energy use and cumulative energy use of a device to more complex meters that transmit information about energy consumption to a central display that can be displayed in a prominent location. Some designers have even used this feedback to creative effect by creating objects whose color changes or by providing some other visual stimuli in response to household energy usage.

Traditional "dumb" energy meters provide a simple count of the cumulative amount of electricity consumed over a given period; in contrast, "smart" meters collect a much richer set of information, data logging energy consumption habits in a memory that can be interrogated by software or more sophisticated display and output systems.

In a conventional metering arrangement, electricity usage is billed at a fixed price per unit, with a set rate for a unit of energy. In some areas, there is also an additional "off-peak" rate, encouraging the consumption of electricity overnight when there is low demand for power, and inflexible generators such as nuclear power stations must be kept spinning to operate efficiently.

In the United Kingdom, such an arrangement is referred to as "economy 7," and a dual-metering system is installed that counts "standard units" and "economy 7 units," available during seven hours of low demand overnight. The switch between the two metering regimes is done remotely, by a synchronized radio signal transmitted on the 198 kHz BBC Radio 4 LW carrier.

There is the potential for smart metering systems to connect to information systems networks, allowing homeowners to download data about their household energy consumption.

## Dynamic Pricing and Smart Devices

Selling electricity at a fixed cost per unit transfers the burden of matching supply and demand to the electricity supplier. Often this is accomplished on the supply side through maintaining "spinning reserve" on the network—power stations in which the turbines are kept spinning and ready to react to changes in demand for power, but that do not generate power. Managing demand is more complicated; often, large industrial consumers of power may be on flexible supply contracts, which entitle them to cheap energy that is subject to interruption. Many domestic and commercial consumers of energy are on a fixed-unit cost supply arrangement.

To help match supply and demand more closely, especially in the context of future electricity grids with increasing penetration of renewables, one method that has been advanced is to adopt a "dynamic pricing" regime, which would incentivize users to consume more power at times of oversupply by making power cheap and would deter users from consuming excess power at times of high demand by increasing the cost of power. Such a system would require feedback to users as to the spot price of electricity.

Smart devices take the concept of smart metering a step further by regulating their power consumption according to the availability of power. It is possible, via home automation systems, for smart devices to connect with smart metering solutions, through wireless protocols such as ZigBee and Bluetooth to exchange information about power availability. These devices then have the potential to enter power-saving modes or adapt the amount of power that they use, depending on either the price or availability of electricity.

### Off-Grid Living

Off-grid living refers to the practice of meeting the household's energy requirements without a connection to the public electricity grid. A combination of solar, wind, microhydro, and/or combined heat and power can be used to produce electricity, which must then be stored on-site. At present, a battery bank usually fulfills this role; however, there is potential for the growth of hydrogen and fuel cell technologies to provide energy storage in remote regions.

Off-grid living requires building occupants to carefully regulate their electricity consumption to ensure that it is matched with the availability of electricity from ambient, renewable sources. This may entail saving energy-intensive activities for days when there is high-wind speed, rainfall, or insolation.

**See Also:** Energy Efficiency of Products and Appliances; Green Homes; Needs and Wants; Resource Consumption and Usage.

#### Further Readings

Ayres, Robert U., et al. "On the Efficiency of U.S. Electricity Usage Since 1900." *Energy*, 30/7:1092–145 (June 2005).

Kirschen, D. S. "Demand Side View of Electricity Markets." *Power Systems IEEE Transactions*, 18/2:520–27 (2003).

Patterson, Walt. *Keeping the Lights On: Towards Sustainable Electricity.* London: Earthscan, 2007.

Patterson, Walt. *Transforming Electricity: The Coming Generation of Change.* London: Earthscan, 1999.

*Gavin D. J. Harper*
*Cardiff University*

## Energy Efficiency of Products and Appliances

Appliances and, more broadly, energy-using products have totally changed our lives in less than a century. However, these modern marvels convey too often an image of lightness and of lack of economic and environmental costs, as their consumption of energy (electricity) is not directly related to practices. Public policies are increasingly encouraging the production of more efficient energy-using products. The energy efficiency of appliances is seen as a step toward reducing the energy consumption of households. There are good arguments for the increase of energy efficiency of products and appliances: energy independence, energy cost, and climate change. According to different models, improvements in energy efficiency since the 1970s have contributed more to our economic prosperity than traditional sources of energy supply. Energy efficiency is for this reason sometimes called "negawatt," the biggest energy source. The potential of energy efficiency improvements is still huge, but there are also doubts that it will be enough to face the major problems linked to energy consumption, as the case of appliances shows.

Consumption of energy in households can be divided into the following sectors: space heating, water heating, lighting, cooking, and appliances. In terms of energy consumption, space heating uses the most (53 percent in 2005), followed by appliances (21 percent) in Organisation for Economic Co-operation and Development (OECD) countries. In terms of carbon dioxide emissions, however, appliances will soon catch up with residential heating. This is because of the low conversion factor from fossil energy to electricity and the steady increase of appliances in households. In OECD countries, the electricity use in appliances grew by 57 percent between 1990 and 2004, despite energy savings from improvements in energy efficiency. The energy share of larger appliances (refrigerators, freezers, washing machines, dishwashers, and televisions) is currently about 50 percent. However, this share is declining, as the most rapid increase in appliance energy consumption comes from increased ownership of a wide range of mostly small appliances such as computers, mobile phones, personal audio equipment, and other home electronics. Standby power accounts for around 10 percent of residential electricity demand. In some countries, air conditioning is also a key factor. Despite the decrease of the average unit energy consumption of big appliances put on the market (apart from televisions), their total energy consumption has increased since 1990 as households possess and use more of these appliances. For televisions, energy efficiency gains have been undermined by the consumer trend toward wide screens, which use more energy. In OECD countries, the demand for big appliances is almost saturated. However, this is not the case in other countries, where increase of energy consumption for appliances and products is expected.

According to life cycle analysis, energy-using products consume much more energy when used than when manufactured—even in the case of computers, which require many resources during the production phase. It is then important that households be aware that use of appliances is energy consuming. Most of the countries have developed energy-labeling schemes to educate consumers about the most efficient products. Energy labels are progressively improving the appliances market because producers are encouraged to manufacture more efficient products. When market mechanisms are not sufficient, some countries develop mandatory performance standards, for example, on lamps and on standby.

Energy efficiency has not reduced the global energy consumption of appliances but has merely slowed it down. Some technologies (e.g., washing machines and dishwashers) seem to have reached their energy efficiency limit. Unless there are technological breakthroughs, energy consumption of appliances will not decrease significantly. Furthermore, different effects and associated consumption should also be considered. Although the direct rebound effect for appliances seems to be small (less than 10 percent), indirect rebound effects are important. The direct rebound effect is the increase in demand for an energy service (washing, heating, refrigeration, lighting, etc.) when the cost of the service decreases as a result of technical improvements in energy efficiency. Although researchers have reported households keeping their old refrigerator when buying a new one or putting efficient lamps in places previously not lit, this effect seems to represent less than 10 percent of the energy gained through the increase of efficiency. However, the indirect rebound effect is high: the saved energy through improvement of energy efficiency is offset by the increase of the equipment rate and more frequent use. In addition, associated energy consumption of appliances can be considerable, as is the case with computers (energy consumption of Internet servers), with washing machines (the detergent can take up to half of the total energy of a wash cycle), or with printers (paper).

Energy efficiency improvements are thus necessary, but they are not enough if energy consumption has to be reduced. Efficiency is defined as the rate between output and input or between benefits and costs. Energy efficiency is using less energy to provide the same level of service. This equation is therefore very helpful in comparing devices that provide the same service. It is, however, much more difficult to compare when the service changes. This is the case when technology takes over an old practice—washing machine versus hand washing, or computer versus typewriter—to save time: the new appliance provides new services that have to be taken into account. The notion of service itself depends on the description: if the service is entertainment, high- or low-energy-intensity devices can provide it. Energy efficiency is supposed to be purely technological and discards any "behavioral interference." An improvement in efficiency corresponds to a reduction in the amount of energy consumed for a given result without changing human behavior. In contrast, when there is reduction of energy consumption by changing human behavior, we are talking about conservation. Energy conservation is broader than energy efficiency, in that it encompasses energy efficiency and using less energy to achieve a lesser energy service; for example, through behavioral change.

Energy efficiency focuses on the equipment and leads to policies dealing with the acquisition of new appliances. However, these measures will increase their benefits during a whole cycle of stock replacement, which will take between 10 and 20 years for large appliances, and during this time, the demand for more consuming products will grow, as illustrated by the increase of wide-screen televisions and the ever-growing capacity of computers. From the point of view of green consumption, it should be considered whether limits should be set on the growing demand for energy. Many want stronger energy policies and programs to implement the necessary mix of market- and regulatory-based instruments, including stringent norms and standards. Appealing to the global warming threat and energy security concerns, policy makers are being asked to consider devoting more attention to influencing lifestyles and behaviors.

Studies from different disciplines (psychology, sociology, economics) have shown that increased demand for energy from households depends on a wide range of mechanisms. As Elizabeth Shove has substantiated, expectations of comfort, cleanliness, and convenience have changed radically over the past few generations. Social norms have evolved quickly, leading to an increase of energy consumption. Homes, offices, domestic appliances, and clothes play a crucial role in our lives, but not many of us question exactly how and why we perform so many daily rituals associated with them. There is clear evidence supporting the view that routine consumption is controlled by conceptions of normality and profoundly shaped by cultural and economic forces. Comfort is a need, but it also is a social trend that can be adapted. When considering finite energy resources, the case of appliances shows the need to make efficiency and sufficiency strategies complementary. Sufficiency is not abstinence or lack of the necessities; as with efficiency, it is the intelligent use of limited resources.

**See Also:** Audio Equipment; Automobiles; Computers and Printers; Consumer Behavior; Diderot Effect; Downshifting; Ecolabeling; Electricity Usage; Frugality; Garden Tools and Appliances; Green Homes; Heating and Cooling; Home Appliances; Lighting; Mobile Phones; Resource Consumption and Usage; Television and DVD Equipment.

### Further Readings

Geller, Howard and Sophie Attali. "The Experience With Energy Efficiency Policies and Programmes in IEA Countries: Learning From the Critics." International Energy Agency Information Paper. Paris: International Energy Agency, 2005.

International Energy Agency. *Worldwide Trends in Energy Use and Efficiency: Key Insights From IEA Indicator Analysis*. Organisation for Economic Co-operation and Development/ International Energy Agency, 2008.
Nye, David E. *Consuming Power: A Social History of American Energies*. Cambridge, MA: MIT Press, 1998.
Rüdiger, Mogens, ed. *The Culture of Energy*. Cambridge, MA: Cambridge Scholars, 2008.
Shove, Elizabeth. *Comfort, Cleanliness and Convenience: The Social Organization of Normality*. Oxford and New York: Berg, 2003.

*Grégoire Wallenborn*
*Université Libre de Bruxelles*

# Environmentalism

Environmentalism refers to a collective set of perspectives and actions that together represent an approach (rather than a single movement) to the environment. This article explores environmentalism by examining its origins, varying definitions, development, and manifestations and demonstrates how environmentalism is developing in society today.

## Origins of Environmentalism

Unlike many other terms with the ending "-ism," environmentalism is neither conceptually coherent nor manifested in a discrete movement. Indeed, there is no definitive time at which we can determine its origin. This imprecise nature of the term reflects the subject matter of environmentalism, which is concerned with our relation to and interpretation of the natural environment. Accordingly, the first major issue that scholars have tackled is how society–nature relations have come to be represented in something termed *environmentalism*. To adequately address this point, we need to explore the varying social interpretations of nature and how these have evolved over time. For example, scholars have argued that in preindustrial societies, nature was and is often set in contrast and opposition to civilized society, being feared, worshipped, or marginalized. The notion of nature as the "other" is important and helps us to appreciate that the resources postindustrial societies regard as aesthetically valuable, such as wilderness, have often been portrayed as valueless or only of limited value by other societies.

Accordingly, society's interpretation of the natural environment has and is constantly evolving, and there is a significant body of scholarship that has explored both the history of society–nature relations and contemporary understandings of these relationships. Indeed, this scholarship illustrates the varying ways in which former societies and cultures have both used and abused natural resources for their development. Overall, this scholarship points to one critical conclusion, which posits that the environment is socially constructed and mobilized by societies at particular points in their development within a wider social context. However, to explicitly define the term *environmentalism*, we need to look to a proenvironmental or environment-centered perspective; it is generally accepted that the modern forms of environmentalism began to emerge in the middle of the 20th century in developed nations, although we should be cautious in asserting that this is when individuals or communities started to "care" for their environments, as it is likely that preindustrial societies were and are considerably more environmentally sustainable than industrial societies.

## The Environmental Crisis

If we take the development of modern environmentalism in developed nations as a convenient starting point, then what we know today as environmentalism emerged from several events that occurred during the 1960s in a period termed the *environmental crisis*. Although these events are unconnected, they had a major effect on Western society's perspective on the environment. Indeed, it is interesting to note that early works considered to be indicative of environmentalism are in fact focused on wider issues of public health and population growth, but were later reinterpreted as classic examples of environmentalism.

The major events and publications in early environmentalism include the publication of Rachel Carson's *Silent Spring*, which portrayed the devastating effect of the chemical dichlorodiphenyltrichloroethane (DDT) on natural environments; several pollution incidents, including oil tanker spillages and the notorious problems of smog (smoke from coal fires mixed with fog, causing visibility and respiratory problems); and the publication of Garrett Hardin's *Tragedy of the Commons*, in which the depletion of natural resources on common land is used as an illustration of global resource depletion.

These and other events during the 1960s and early 1970s were placed within a political context driven mainly by population concerns; the scientific understanding of the time was framed by the writings of Thomas Malthus, an 18th-century demographer who had predicted that exponential population growth would outstrip the ability of society to feed itself within three generations. On the basis of these assumptions, a group of concerned scientists, known as the Club of Rome, sought to test the extent to which population growth trends could be supported by food production. Their model, published in 1972, predicted that the combination of increasing population growth and related pollution would lead to a collapse in population by around 2060.

The publication of Donella Meadows et al.'s *The Limits to Growth* report in 1972 was the start of political action that we now come to recognize as environmentalism. Through the 1970s and 1980s, governments began to implement programs with an explicitly environmental focus, which was reflected in actions by both the United Nations at the global scale and the local authorities at the municipal level.

## Defining Environmentalism

Accordingly, modern environmentalism has its origins in a series of critical events that concerned population growth, health, pollution, and more broadly, an emerging concern among Western societies about the natural environment. Yet at the start of this article, the complexity of environmentalism was emphasized—this is because although the events of the 20th century enabled environmentalism to flourish, it did so in many different ways, and scholars have therefore attempted to use a number of different methods for defining environmentalism.

The most basic definition of environmentalism relates to the understanding individuals have of their relation to nature and the environment. A range of scholars have used a universal spectrum of values as a way of interpreting individual, organizational, and political values toward the environment. This spectrum ranges from biocentric to anthropocentric: the former emphasizes unity with nature and a belief that humans are equal to nature, whereas the latter places humans above, superior to, and distinct from nature. This basic spectrum has been added to in various different ways; one of the most well-known attempts to define and codify environmentalism in practical terms was Timothy O'Riordan's spectrum of values, from ecocentric to technocentric. This spectrum plotted individuals, governments, policies, and organizations along a line, from those who argue that working within natural limits and capacities is preferable (ecocentrists) to those who posit that humans are able to use technology and development to overcome environmental limits (technocentrists).

Such definitions have also been applied by scholars to a range of both generic and specific situations and scenarios, and perhaps the most comprehensive definition was supplied by David Pearce, who derived his Sustainability Spectrum on the basis of four dominant approaches to developing a "green economy": cornucopian, accommodating, communalist, and deep ecology. These approaches in turn provide a progressively more biocentric/ecocentric position.

## Environmentalism Today

These and other definitional approaches set out the ways in which scholars have attempted to explore environmentalism in theoretical terms and to explain the various approaches adopted by individuals, policy makers, and organizations toward the environment. However, environmentalism is more than a theoretical exploration of relationality and the environment, it is also a way of being or doing for many people. Accordingly, this second part of the article explores some of the manifestations of environmentalism.

Scholars have explored environmentalism in a range of contexts and at different scales, and the following provides a brief overview. Four dominant manifestations of environmentalism can be identified: political, activist, organizational, an individual.

On the political scale, scholars have focused on the ways in which governments and other political networks have interpreted their role as agents for change in both regulating and managing environmental resources. Considerable focus on the international scale has focused on the development of United Nations (UN) policy, and specifically the moves the UN has made in promoting sustainable development. Indeed, scholars have explored the way in which the UN acts as a broker between developed and developing nations and the conflicts that major environmental decisions generate. On the regional scale, researchers have explored the ways in which groups of nations (such as the European Union) have sought to regulate interstate environmental reform. Indeed, regulators such as the European Union have been studied by researchers who have sought to understand the decision-making processes and underlying values that make certain reforms (such as the regulation of waste) more prominent than other potential reforms (such as changes in transport regulation).

However, in studying political aspects of environmentalism, it is at the national scale at which most researchers have focused their attention. Within this context, the study of political parties and their environmental credentials has become a critical focus of study, affording insights into the environmental philosophies of decision makers. More broadly, it is generally accepted that in many Western societies, environmental issues emerge as important factors in national politics when social and economic conditions are good; in times of economic hardship, the environment generally reduces in importance. Nonetheless, in most nations political parties do have significant differences in either philosophy or approach toward the environment, and in general, those parties that are to the right of the political spectrum tend to have more conservative and technocentric environmental policies, whereas those on the left tend to have more ecocentric policies.

Environmentalism has also been explored in terms of the activist movements that have emerged since the 1960s. These movements represent a wide range of both political as well as environmental perspectives: some are entirely passive, and some have resorted to what is known as "direct action," which can be anything from hijacking a whaling vessel to a sit-down protest at a new road development. Such groups have complex sets of values and are often driven forward by a small group of very committed individuals. In many cases, such groups also have relatively short life spans and are issue-specific, although other groups have developed into global entities, such as the World Wildlife Fund or Greenpeace. Overall, such groups reflect specific interpretations of environmentalism and are highly focused, meaning that they can have a powerful influence within the political system on occasion.

The role of environmentalism in organizations is perhaps the field of study in which research has faced the biggest challenges. Particularly in the private sector, researchers face problems in gaining access to key documents or policies, and the complexity of large organizations means that exploratory work is difficult. However, in recent years many organizations have attempted to engage with environmental issues in a number of ways; the best illustration of this is the development of corporate social responsibility policies, which set out an organization's role and responsibilities within its community and the wider environment. Research on corporate social responsibility policies illustrates that in many cases, organizational commitments to the environment are basic, and that considerably more commitment is likely to be needed to address global environmental issues such as climate change.

Finally, research on environmentalism has also focused attention on the values, attitudes, and behaviors of citizens toward the environment. Since the 1960s, researchers have been exploring the commitments individuals make toward the environment and have identified a host of barriers and motivations for engaging in environmentally responsible behavior. However, the growing interest and concern about the state of global climate change has meant that such research has received greater prominence, and researchers are now focusing their attention on understanding how individuals can be encouraged to make behavioral changes throughout their lifestyle, from recycling waste to saving energy. Indeed, scholars are now arguing that an approach is needed that explores the environmental commitment of citizens across both contexts (e.g., home, work, and leisure) and behaviors.

Environmentalism is a challenging concept—it is definitionally complex and is manifested in numerous ways at different scales. However, modern environmentalism is without doubt a reflection of society's values and its current understanding of the natural environment, both of which are constantly evolving.

See Also: Consumer Activism; Demographics; Green Consumer; Green Consumerism Organizations; Lifestyle, Sustainable; Sustainable Consumption.

### Further Readings

Barr, Stewart. *Environment and Society: Sustainability, Policy and the Citizen*. Aldershot: Ashgate, 2008.

Carson, Rachel. *Silent Spring*. Boston, MA: Houghton Mifflin, 1962.

Castree, Noel and Bruce Braun, eds. *Social Nature: Theory, Practice and Politics*. Oxford: Blackwell, 2001.

Connelly, James and Graham Smith. *Politics and the Environment.* London: Routledge, 2003.
Hardin, Garrett. "The Tragedy of the Commons." *Science*, 162 (1968).
Jordan, Andrew. *Environmental Policy in the European Union.* London: Earthscan, 2001.
Meadows, Donella, et al. *The Limits to Growth.* New York: Universe Books, 1972.
O'Riordan, Timothy. *Environmentalism.* London: Pion, 1976.
Pearce, David. *Blueprint 3: Measuring Sustainable Development.* London: Earthscan, 1993.
Pepper, David. *Modern Environmentalism.* London: Routledge, 2003.
Roberts, Jane. *Environmental Policy.* London: Routledge, 2004.
Young, John. *Post-Environmentalism.* London: Belhaven, 2000.

*Stewart Barr*
*University of Exeter*

# Environmentally Friendly

Environment change is an issue that negatively affects both developed and developing countries. Although many governments are challenged with protecting the environment, companies have an interest in protecting the environment and the well-being of future generations, as it provides them with opportunities to develop environmentally friendly products for environmentally friendly consumers. Further, globalization has given consumers access to information they did not previously have, which has made consumers more informed of the manufacturing processes of products and proactive. As a result, consumers seek environmentally friendly products. The term *environmentally friendly* can encompass numerous aspects but generally refers to products, services, or practices designed to minimize harm to the natural world. Clearly, the expectation is that firms will be candid about how they conduct their businesses and about how they manufacture their products. For example, consumers may be interested in the green attributes of products (e.g., whether they are biodegradable, recyclable, etc.), as well as in the manufacturing process (e.g., they may look for assurance that environments were not polluted or that rivers were not contaminated by the processes used to manufacture the products).

## Opportunities to Alter Consumer Behavior

Environmental concerns have led people to adopt new behaviors worldwide; for example, recycling, pollution-free transportation, and environmentally friendly products, among others. The challenge for companies is to determine the extent to which consumers' new behavior toward protecting the environment has resulted in purchasing products from companies that are perceived to be environmentally friendly. A study by S. Singh and D. Carvalho (2005) reveals that consumers prefer purchasing gasoline from an oil company they perceive to be the most environmentally friendly and that they are willing to drive farther to locate their preferred gas stations. The implication is that companies need to understand consumers' purchasing behavior, as most do assess the effect on the environment of using a nonenvironmentally friendly product. Companies should seize the opportunity to alter consumer behavior by educating consumers about the value of purchasing environmentally friendly products. Governments and insurance companies may want to subsidize environmentally friendly products to create further value to the consumers and to induce behavior that encourages this. As a result, individuals may even choose to seek employment with environmentally friendly companies.

## Opportunities to Develop Environmentally Friendly Products

Market economies encourage companies to invest in developing countries, giving marketing managers opportunities to globalize their brands. Although this activity gives rise to economic development and improvement in quality of life, it is imperative that companies develop environmentally friendly products (e.g., energy-saving lightbulbs and fuel-efficient cars, refrigerators, and houses, etc.) to satisfy the needs of consumers by keeping the natural ecological system intact. Further, environmental protection measures have stimulated the development of innovative technologies that conserve energy, reduce greenhouse gases (GHGs), and thus contribute to quality of life. Some companies are making energy efficiency and GHG emission reduction a key component of their corporate strategy. For example, Daimler Chrysler has reduced the GHG emissions per vehicle manufactured by 42 percent over the past decade. Syncrude Canada has reduced GHG emissions per barrel of production by 26 percent since 1988. Given this awareness, automakers are poised to introduce innovative green gasoline-based cars; for example, Volkswagen's desire to use hydrogen as a fuel and General Motors' invention, using a hybrid-drive full-size sports utility vehicle (the Cadillac Escalade), among others. Indeed, consumers may demand environmentally products from globally branded companies such as these, as they associate these companies with superior brands.

## Positioning a Company as Environmentally Friendly

It is wise to position a company as an environmentally friendly company. P. Shamdasani, et al., found in their 1993 research that customers were willing to expend effort to search for environmentally friendly alternatives—they are aware of where such products are sold and are willing to visit these stores more frequently. The implication is that environmentally friendly companies have high brand recognition. For example, Shell has a commitment to environment that runs throughout the company, which is evident from their raw material acquisition to processing, from waste management to saleable products. Fact-supporting energy improvements have earned the company very favorable responses from environmental groups compared with Esso, which has only taken a small step toward sustainable environment management. As a result, consumers may boycott non–environmentally friendly companies through institutions such as Greenpeace or Internet efforts such as stopesso.com. Thus, the apparent growth of green consumerism should make marketing managers aware of the possible opportunities to attract consumers through green processes, green packaging, and green products. Past research indicates that firms take advantage of image-enhancing efforts by positioning brands as global in their communications and by using messages such as brand name, logo, and visual themes.

## Improving Business Performance

Although initiating an organization-wide environmentally friendly system may be costly in the short term, savings in waste management, reduction in waste, improvement in processing, and gains in efficiency should result in superior business performance in the long term.

In fact, some of the largest multinational companies, such as British Petroleum, DuPont, IBM, Toyota, and Royal Dutch Shell, among others, are reducing their emissions and profiting by doing so. For example, British Petroleum has reduced its worldwide emissions to 10 percent below 1990 levels seven years ahead of schedule at a net savings to the company. Shell management has led to a 10 million metric ton reduction in carbon dioxide emissions while increasing revenue by about 50 percent over a four-year term. As a result, companies may opt to pass on the savings to consumers either by decreasing the price of the products or by increasing dividends to shareholders. In fact, at the Clinton Global Initiative conference in New York in September 2006, British businessman Richard Branson and former U.S. president Bill Clinton pledged $3 billion to battle global warming and $1 billion to an investment fund for renewable energy, respectively.

Indeed, companies and consumers need a healthy environment, and thus should work together to protect the environment. Protecting the environment is not only good for the planet but is good for business as well.

**See Also:** Consumer Behavior; Green Consumer; Green Consumerism Organizations; Green Marketing.

### Further Readings

Alden, D. L., et al. "Brand Positioning Through Advertising in Asia, North America, and Europe: The Role of Global Consumer Culture." *Journal of Marketing,* 63/1:75–87 (1999).

Clinton Global Initiative. http://www.ClintonGlobalInitiative.org (Accessed March 2009).

Green Fuels Forecast. http://www.greenfuelsforecast.com (Accessed March 2009).

Hailes, J. *The New Green Consumer Guide.* New York: Simon and Schuster, 2007.

Shamdasani, P., et al. "Exploring Green Consumers in an Oriental Culture: Role of Personal and Marketing Mix." *Advances in Consumer Research,* 20/1:488–93 (1993).

Shell Canada Ltd. "Showcase Sustainable Development in National Campaign." Press Release. Calgary, Canada, January 9, 2002.

Shocker, A. D., et al. "Challenges and Opportunities Facing Brand Management: An Introduction to the Special Issue." *Journal of Marketing Research,* 31:149–58 (1994).

Singh, S. *Business Practices in Emerging and Re-Emerging Markets.* New York: Macmillan, 2008.

Singh, S. and D. Tornavoi de Carvalho. "Green Orientation and Oil Shopping." European Marketing Academy Conference, Milan, Italy, May 24–27, 2005.

*Satyendra Singh*
*University of Winnipeg*

## Ethically Produced Products

Ethically produced products are typically manufactured under three conditions: First, the firm should attempt to have positive stakeholder relations; for example, a commitment to diversity in hiring and installing safe production practices. Second, it must use sound environmental practices, such as using ecofriendly products and technology. Finally, it should

In an experiment that tested consumer willingness to buy ethically produced products, results showed that premium consumers would pay an additional $1.40 per pound for fair trade coffee.

*Source:* iStockphoto

also demonstrate concern about human rights by not employing child labor or forced labor in factories located in the developing world.

The market for ethically produced consumer products has soared over the last decade, and sales are expected to reach $57 billion by 2011. Research has shown that U.S. consumers want goods from ethical manufacturers but are having a hard time finding products at reasonable prices. During recessionary times, there is less of an attraction to such goods because of higher prices.

A recent study by Remi Trudel and June Cotte involved three experiments—two on coffee purchasers and the other on T-shirt purchasers. In all cases, they demonstrated that not only does it pay to be good but firms may be punished if they are bad.

In the coffee experiment, consumers of the product were divided into three groups with different ethical manipulations. The results showed that although premium consumers would pay an additional $1.40 per pound as a reward for Fair Trade practices, the punishment/discount for unfair practices was almost twice the effect of positive information (i.e., $2.40 per pound) on the coffee consumers' willingness to pay.

In the T-shirt experiment, the goal was to look at degrees of ethical behavior. Were consumers ready to pay for 100 percent ethically produced shirts versus one that is 50 percent or 25 percent ethically produced? To find out, cotton T-shirt users were divided randomly into five groups, and information was provided on a fictitious manufacturer. The first four groups were told that the shirts had 100 percent, 50 percent, 25 percent, and 0 percent cotton, respectively. The fifth (control) group received no information. The groups were also shown a short paragraph detailing the detrimental effects of nonorganic cotton. The subjects were willing to pay a premium for all levels of ethical production, and they would discount an unethical product more deeply than they would reward an ethical one. Yet, there was not a linear relationship between increasing levels of ethical production and increasing price premiums. The 25 percent organic cotton received a mean price of $20.72, which was not significantly different from 50 percent ($20.44) and 100 percent ($21.21). It appears that once the ethical threshold is reached, consumers agree to pay a premium for the products.

The final study tested whether, if consumers expected companies to behave ethically, that changed how much consumers rewarded and punished behavior, especially if they felt that the companies are in it for the money and for maximizing profits and are not concerned about ethics. The respondents' attitudes toward the firms were measured and labeled as high expectation or low expectation, and they were split into three groups that received positive, negative, or no information about the manufacturer and their methods. Yet again, it was found that irrespective of their expectations, consumers were willing to pay more for ethical goods than unethical ones, or ones for which they had no information, indicating reward for socially responsible behavior. Similarly, negative information had a much greater effect on consumer response than positive information. Subjects in the study punished unethical goods with a bigger discount (approximately $2 below the control group) than they rewarded ethical ones with premiums (approximately $1 above the control group). As far as consumer attitudes are concerned, individuals with high expectations handed out larger rewards and punishments than those with low expectations.

The managerial implications of the findings of the study are that companies can charge a premium for their products if they act in a socially responsible manner. Moreover, consumers will punish firms making unethical goods. The negative effects of acting unethically have a much greater effect on consumer willingness to pay than the positive effects of ethical behavior.

Because of consumers' willingness to pay a premium for "natural" products from value-oriented companies that make them feel good, big marketers are trying to capitalize on this profitable trend. Colgate bought 84 percent of Tom's of Maine, the all-natural personal care brand based in Kennebunk, Maine, that sells products such as lemongrass soap and calendula shaving cream made with natural and biodegradable ingredients. L'Oreal, the French cosmetics giant, recently bought the Body Shop, a personal care chain best known for its avoidance of animal testing and its support for human and animal rights causes. Unilever, the U.K. conglomerate, bought Ben and Jerry's, the socially conscious, flavorful ice cream brand, in 2000, and has tried to maintain its social platform.

The challenge of thc corporate buyer in all cases is to maintain the smaller brand's roots and cachet and sense of belonging to the local community. Corporations should respect the unique approach of creating efficacy with natural products and the overall ethical values approach to doing business to continue the success of the acquired brands.

**See Also:** Advertising; Affluenza; Baby Products; Biodegradability; Carbon Credits; Consumer Boycotts; Consumer Ethics; Genetically Modified Products; Green Consumer; Morality (Consumer Ethics); Recyclable Products; Regulation; Resource Consumption and Usage; Waste Disposal.

### Further Readings

Crane, Andrew. "Unpacking the Ethical Product." *Journal of Business Ethics*, 30/4:361–73 (2001).

Giaretta, Elena. "Ethical Product Innovation: In Praise of Slowness." *The TQM Magazine*, 17/2:161–81 (2005).

Hartlieb, S. and B. Jones. "Humanizing Business Through Ethical Labelling: Progress and Paradoxes in the UK." *Journal of Business Ethics*, 88/3:583–600 (2009).

Howard, Theresa. "Big Companies Buy Small Brands With Big Values." *USA TODAY*, March 21, 2006.

Irwin, J. and R. Naylor. "Ethical Decisions and Response Mode Compatibility: Weighting of Ethical Attributes in Consideration Sets Formed by Excluding Versus Including Product Alternatives." *Journal of Marketing Research*, 46/2:234–45 (2009).

Low, W. and E. Davenport. "Organizational Leadership, Ethics and the Challenges of Marketing Fair and Ethical Trade." *Journal of Business Ethics*, 86:97–108 (2009).

Luchs, M., et al. "Is There an Expected Trade-Off Between a Product's Ethical Value and Its Effectiveness? Exposing Latent Intuitions About Ethical Products." *Advances in Consumer Research*, 34:357–58 (2006).
McEachern, Morven G., et al. "Exploring Ethical Brand Extensions and Consumer Buying Behavior: The RSPCA and the 'Freedom Food' Brand." *Journal of Product and Brand Management*, 16/3:168–77 (2007).
Pelsmacker, Patrick de, et al. "Do Consumers Care About Ethics? Willingness to Pay for Fair-Trade Coffee." *Journal of Consumer Affairs*, 39/2:363–85 (2005).
Trudel, Remi and June Cotte. "Does Being Ethical Pay?" *Wall Street Journal* (May 12, 2008).
Warrell, Helen. "Fifteen Point Rise in Number of Ethical Retail Funds." *Third Sector*, 458 (January 11, 2007).

*Abhijit Roy*
*University of Scranton*

# E-Waste

The term *e-waste* refers to old end-of-life or discarded appliances, such as computers, consumer electronics, refrigerators, and so on, using electricity, that have been disposed of by their original users. It is also known as waste electrical and electronic equipment. These products are a relatively recent addition to the waste stream and are getting increasing attention from policy makers, as the quantity being generated is rising rapidly. E-waste contains valuable materials as well as hazardous materials that require special handling and recycling methods. Some specific examples of hazardous materials include televisions and computer monitors that contain cathode ray tubes, liquid crystal display (LCD) desktop monitors, laptop computers with LCD displays, LCD and plasma televisions, and portable DVD players with LCD screens.

## E-Waste Statistics

The use of electronic products has skyrocketed in the past couple of decades, transforming the means of communication, information, and entertainment. According to the Consumers Electronics Association, Americans own approximately 24 electronic products per household. Currently, over 50 percent of U.S. households own a computer, the average life of which is between three and four years. Mobile phones have been discarded at a rate of 130 million per year in 2008, resulting in 65,000 tons of waste.

Electronic appliances are composed of several materials that can be both toxic and of high value. According to some estimates, as much as 75 percent of old used equipment is in storage, where it takes up space and becomes more obsolete and less valuable. Although a majority of the materials such as iron, aluminum, plastics, and glass account for over 80 percent of the weight, toxic materials, albeit in smaller quantities, are very harmful for the environment. An average personal computer (not including the monitor) is typically 40 percent steel, 30 to 40 percent plastic, 10 percent aluminum, and 10 percent other metals, including gold, copper, silver, cadmium, and platinum. It uses 530 pounds of fossil fuel, 49 gallons of chemicals, and 410 gallons of water to produce.

In 2001, only 11 percent of the personal computers in the United States were recycled. Consumers in the United States now dispose of between 300 million and 400 million electronic items per year, and less than 20 percent of this e-waste is recycled. This is equivalent to 2 percent of the trash in landfills, yet equals 70 percent of overall toxic waste. E-waste legislation in the United States rests currently at the state level, and to date, only 24 states have passed or proposed take-back laws. The European Union has banned e-waste from landfills since the 1990s, and current laws hold manufacturers responsible for e-waste disposal.

Less than 20 percent of cell phones are recycled each year—recycling just a million cell phones is equivalent to reducing greenhouse gas emissions by 1,368 cars in a year.

As a consequence of consumer processing power doubling approximately every two years, many older computers also are being discarded. In 2008, over 70 million computers were abandoned, up from 20 million in1998.

Even though it is energy efficient to rebuild old computers, only about 2 percent of personal computers are handed down to another user. Flat-panel computer monitors and notebooks often contain small amounts of mercury in the bulbs used to light them, and cathode ray tubes in older televisions and computers typically contain between four and seven pounds of lead.

Much of the e-waste from the developed world is sent to developing countries like China, India, and Kenya, where lower environmental standards and working conditions make processing e-waste more profitable. Approximately 80 percent of the e-waste in the United States is exported to Asia.

## Why Is E-Waste Harmful?

Some metals like gold, silver, copper, platinum, and so on, are valuable substances that turn recycling of e-waste into a lucrative business opportunity. For example, precious metals are used in computer circuit boards and electronic components, and glass and plastic are used for television monitors. Recycling these products reduces the need to mine the Earth for raw materials. In contrast, the recycling of hazardous substances, for example, carcinogens such as lead and arsenic, is critical and poses serious health risks and environmental dangers if not properly handled.

Computers or television displays contain an average of six pounds of lead each. The lead from these components is released into the environment when they are illegally disposed of and crushed in landfills. Lead can cause damage to the central and peripheral nervous systems, blood systems, and kidneys in humans. It also accumulates in the environment and has highly acute and chronic toxic effects on plants, animals, and microorganisms. Other hazardous materials used in computers and other electronic devices include cadmium, mercury, chromium, polyvinyl chloride plastic, and brominated flame retardant. Computers, LCD/cathode ray tube screens, cooling appliances, mobile phones, and so on contain precious metals, flame-retardant plastics, chlorofluorocarbon foams, and many other substances. The presence of these harmful chemicals also makes waste collection hazardous to workers.

## Managing E-Waste Through Source Reduction

Source reduction is the most proactive and least expensive strategy to manage e-waste. Reduce, reuse, and recycle are three good ways of addressing this problem.

- *Reduce:* Consuming less of everything will prolong the shelf life of electronic equipment and contain the growth of e-waste.
- *Reuse:* Not-for-profit organizations, lower-income families, and schools can benefit from equipment that is not in good working order.

- *Donate:* Several charitable organizations and training programs repair equipment for reuse.
- *Recycle:* A growing number of electronics manufacturers offer fee-based recycling services. Many municipalities offer electronic collection as part of household hazardous waste collections or special events.
- *Buy Green Products:* Increasingly, more products are addressing the environmental issue and are being made with fewer toxins and more recycled content. They are more energy efficient, are designed for easy upgrade or disassembly, and use less packaging. Some firms also offer lease and take-back options to help consumers dispose of their electronics.

## Recommendations for Governments and Firms

Every government should enforce proper disposal of electronic waste (e-waste) because of the potential implications for the sustainability of the global environment. Such tough rules should ensure that e-waste from developed economies is not arbitrarily dumped in developing countries. Today, exports of electronic waste from developed countries are largely unregulated. Many of these objects are recycled using rudimentary and unsafe methods. Governments should also encourage recycling e-waste rather than mining virgin raw materials for production. Switzerland, for example, has had a decade-old law holding the producers responsible for their products beyond sale, until end of product life, through the concept of extended producer responsibility. Recycling electronics by selling them through online exchange marketplaces such as eBay should also be encouraged. Individual firms' inconsistent claims about environmental responsibility and corrupt practices in the e-cycling industry should be investigated.

Moreover, everyone in the electronics industry should also consider e-waste a serious issue. Firms can play a proactive role in disposing e-waste and educating consumers about their harmful effects on the environment. For example, the Eco-Mentors program by Sharp Electronics of Canada Ltd. on Earth Day Canada involved high school students touring Sims Recycling Solutions, an electronics recycling facility used by Sharp, to teach them how to adopt a more ecofriendly lifestyle. The Remade phone by Nokia is made from recycled materials. Terra Cycle Urban Arts Pots and Crafts are made from 100 percent e-waste from old computers and fax machines, which allows the company to address the e-waste problem while also allowing local graffiti artists to execute their work in a legal manner.

**See Also:** Biodegradability; Computers and Printers; Ecological Footprint; Mobile Phones; Recyclable Products; Television and DVD Equipment.

### Further Readings

Betts, Kellyn. "Producing Usable Materials From E-Waste." *Environmental Science and Technology,* 42/18 (September 15, 2008).

Consumer Electronics Association. Press Release, January 2009. http://www.ce.org/Press/CurrentNews/press_release_detail.asp?id=11679 (Accessed July 2009).

Desposito, Joe. "Parting With Old Electronics Can Be Hazardous to Someone's Health." *Electronic Design,* 53/23 (November 7, 2008).

"Eco Mentors Get a Lesson on E-Cycling." *Maclean's*, 122/14 (April 2009).

Elgin, Ben, et al. "The Dirty Secret of Recycling Electronics." *BusinessWeek* (October 27, 2008).

Hong-Gang, Ni and Eddy Y. Zeng. "Law Enforcement and Global Collaboration Are Keys to Containing E-Waste Tsunami in China." *Environmental Science and Technology*, 43/11 (June 1, 2009).

Khetriwal, Deepali, et al. "Producer Responsibility for E-waste Management: Key Issues for Consideration—Learning From the Swiss Experience." *Journal of Environmental Management*, 90/1:153–65 (January 2009).

Manchen, Ken. "Do You Know Where Your E-Waste Is Going? *Design News*, 64/3 (March 2009).

McFedries, Paul. "E-cycling E-Waste." *IEEE Spectrum*, 45/10 (October 2008).

Nixon, Hilary, et al. "Understanding Preferences for Recycling Electronic Waste in California: The Influence of Attitudes and Beliefs on Willingness to Pay." *Environment and Behavior*, 41/1:101–24 (January 2009).

"Report Reveals Flawed U.S. E-Waste Policies." *World Watch*, 22/1 (January/February 2009).

Schaffhauser, Dian. "The Dirt on E-Waste," *T H E Journal*, 36/3:20–25 (March 2009).

U.S. Environmental Protection Agency. "eCycling." http://www.epa.gov/ecycling (Accessed July 2009).

U.S. Environmental Protection Agency. "Electronic Waste Management in the United States, Approach 1, Table 3.1 EPA 530-R-08-009." July 2008. http://www.epa.gov/osw/conserve/materials/ecycling/docs/app-1.pdf (Accessed July 2009).

Walsh, Bryan. "E-Waste Not." *TIME* (January 19, 2009).

*Abhijit Roy*
*University of Scranton*

# Fair Trade

Fair Trade is a system of private governance that aims to improve working conditions and compensation for producers of certain goods. The primary intention is to create a more socially just and environmentally sustainable trade arrangement. Having said this, Fair Trade is subject to various interpretations and, as such, the concept, its practice, and subsequent results are not consistent. Although there is considerable support for Fair Trade, there has also been much criticism of the approach. In particular, many have questioned the power and spread of Fair Trade discourse on the grounds that empirical understanding remains only modestly (if not increasingly) developed.

Although Fair Trade is usually considered to have emerged at the end of World War II, more recent scholarship has suggested that in some countries, such as the United Kingdom, the movement only really began in the 1970s. The literature shows that Fair Trade developed as a multitude of institutions sought to channel greater benefit from the international economy to producer communities seen to be otherwise disadvantaged. These activities became known as the Alternative Trade movement, which had two primary aims: to link materially poor producers in the developing world with markets in richer countries as a means to facilitate resource transfers, and to organize this interaction around the aim of maximizing the benefit to these initial producers. Operationally, the concept of alternative trade was based on the idea of a partnership between socially orientated northern alternative trade organizations (ATOs), responsible for the purchase and import of goods, and southern producer organizations, which provided services to their members in marketing, product development, financing, and distribution services. To facilitate equality, producer and trade organizations structured their operations and interaction on predefined socially orientated norms. These principles were then used as part of the marketing of the final product, the "alternative" identity of which was legitimized by the social reputation of retailers. It was also considered that by embedding products with information about conditions of production, consumers and producers were brought closer together in links of solidarity. Although alternative trade mainly dealt with handicraft items, ATOs also moved into a range of basic commodity crops, including coffee.

However, the scope and impact of alternative trade proved to be limited by its nature as an "alternative" niche market, worsening financial conditions, and increasing competition from mainstream commerce. As a response, the Mexican cooperative Union de Comunidades Indigenas del Region del Istmo and a Dutch nongovernmental organization, Solidaridad, developed a system of external third-party governance and certification. The Max Havelaar mark, launched in 1988, guaranteed that certified coffee had been bought by importers direct from cooperatives for a price of up to 10 percent higher than world market prices, that importers had underwritten additional costs and provided prefinancing of up to 60 percent of the final price, and that producers were maintaining long-term relationships with producer communities. For the first time, alternatively traded coffee could be commercialized through mainstream channels and sold by conventional retailers.

This model proved tremendously popular, and gradually other national certification systems began to emerge; for example, the Fairtrade Foundation in the United Kingdom (founded in 1992) and Transfair in the United States (founded in 1998). While the number of products eligible for certification broadened, national organizations slowly assimilated under a unitary set of regulations, represented by a single certification "brand" and administered by the Fairtrade Labelling Organizations International (FLO). Although certification requires producer organizations to operate in a socially just and sustainable way, it also requires initial buyers to support the process by providing up to 60 percent up-front credit on request, engaging in long-term and predictable relationships, paying minimum guaranteed and above market prices, and paying an additional social premium. Certification is available for 16 different product categories, most of which are food commodities.

Another system of Fair Trade certification was developed by the International Federation for Alternative Trade, founded in 1989 and renamed the World Fair Trade Organization in 2009. In contrast to the FLO governance, this system certifies whole organizations and guarantees that 10 Principles of Fair Trade are all upheld. As such, the World Fair Trade Organization mark is closer to the original ATO principles in that certification refers to a type of relationship, rather than the fulfillment of minimum requirements, and certification is usually applied to handicraft products for which the FLO mark is unavailable.

Although Fair Trade practice and certification have been associated with many positive outcomes for producer communities (such as higher income, more stable livelihoods, and better health and infrastructure), the benefits of Fair Trade are far from agreed on. Many question the scope and efficiency of the mechanism for reducing poverty and facilitating development, and it is often argued that the producers most in need remain isolated by the requirements of the system. In contrast, supporters point to the complexity of the benefits brought by Fair Trade, as well as those supportive interactions that go above and beyond minimum standards. A further—and perhaps the most powerful—critique of Fair Trade suggests that guaranteed minimum and above-market prices will discourage the diversification necessary for producers to escape reliance on low-income and unstable commodity markets. However, this argument is countered by questioning the assumptions made by highly theoretical analysis, and more nuanced approaches reveal much more positive appraisals.

Another criticism has been the decreasing producer control of governance, and one particular problem has been relatively static minimum prices despite rapid inflation in producer living costs. Although producer representation and some prices have been increased by the FLO, some say that the response has been far from timely or adequate in its adjustment. This is particularly problematic when significant price premiums at the retail level are still lost to intermediaries, with price increases at the farm gate remaining modest. However, supporters argue that this is not caused by Fair Trade but, rather, by conventional supply system, and experience shows that as the Fair Trade market has become more competitive, retail margins have fallen to reduce the cost to customers. A final area of discussion has been the degree to which Fair Trade defetishizes commodity products and reconnects producers and consumers in solidarity. Although some have seen Fair Trade as contributing to great understanding, others highlight the commodification of poverty, the construction of a stereotypically romanticized poor and honest producer, and that the views of certified farmers do not always reflect the ideas of Fair Trade projected in consumer countries. One factor that constantly underlies the discussion of Fair Trade is the necessity of better understanding both the processes and the outcomes of the system, and for this reason, debate remains very much open on the merits and shortcomings of the Fair Trade system.

**See Also:** Certified Products (Fair Trade or Organic); Commodity Fetishism; Consumer Ethics; Production and Commodity Chains.

### Further Readings

Hayes, Mark. "Fighting the Tide: Alternative Trade Organizations in the Era of Global Free Trade—A Comment." *World Development*, 36/12 (2008).

Le Mare, A. "The Impact of Fair Trade on Social and Economic Development: A Review of the Literature." *Geography Compass*, 2/6 (2008).

Raynolds, Laura T., et al. *Fair Trade: The Challenges of Transforming Globalization.* London: Routledge, 2007.

*Alastair M. Smith*
*ESRC Centre for Business Relationships, Accountability, Sustainability & Society*

## Fashion

The greening of the fashion industry presents a number of challenges. The industry is immense, worth over $1 trillion worldwide. Its global influence on sustainability results from its long extended chain of activities, including production and supply of raw and man-made materials (e.g., cotton, polyester), clothing manufacture, distribution and sale of finished goods, fashion goods purchase and usage by the public, and final disposal after use. As a result, the fashion industry has multiple, significant sustainability effects on society at every stage of the clothing life cycle—environmentally, economically, and socially. In recent times, High Street fashion retailers have exacerbated the unsustainable aspects of the sector. The advent of fast fashion—the shortening of the lead times between the appearance of a garment on the catwalk and its availability in the High Street store—has resulted in encouraging more intense consumption of fashion. Shortening lead times has effectively shortened the life cycle of fashion, increasing the frequency of purchase, and the accompanying reduction of quality of garments has made fashion disposable in the mind of the consumer. Many consumers, particularly in younger age groups with less disposable income, purchase fast fashion from a growing number of budget fashion retailers, despite knowing that the product will not last long in good condition. The fashion industry has

More than any other crop, the mass cultivation of cotton is associated with the overuse of pesticides and results in the misuse of water supplies in some countries.

*Source:* David Nance/Agricultural Research Service/ USDA

received constant criticism for its environmental performance, as well as for its societal impacts.

Environmental impacts of fashion occur throughout the life cycle. For example, the mass cultivation of cotton is associated with the overuse of pesticides, more than any other crop worldwide, and also results in the misuse of water supplies in some countries. The treatment of material in preparation for dyeing and the dyeing process itself uses water and energy, but it is the toxicity of the chemicals used that causes serious problems for the workers who are exposed to them; the environment also suffers, as untreated effluent is dumped in the local water supply. Fierce High Street competition has led to brands seeking to reduce costs by moving production to India and China, adding to the fashion miles and carbon footprint of fashion, as material and clothes are transported around the world. However, the energy use and carbon footprint associated with washing, drying, and ironing garments once bought is far greater than that of transport. At the end of the life cycle, synthetic materials can take decades to biodegrade, and not enough clothes are reused or recycled, so many end up in landfills.

Economic and social effects of fashion are particularly evident in the materials sourcing and manufacturing stages of the fashion garment life cycle. Poor, unsafe working conditions; low wages; and sweatshop and child labor practices have all been present in the fashion industry as brands have sought to reduce costs through outsourcing production to developing countries.

The nature of consumer behavior and consumption with regard to fashion is associated with the building of the consumer's self-identity and the consumer's pursuit of pleasure. Both of these motivations are likely to encourage consumption for the sake of consumption and are not compatible with a green consumerism that would be better served by less consumption and more restraint to reduce carbon footprints, reduce pollution, and protect natural resources. The alternative option in green fashion, the ecochic niche, is growing but has not fully entered the mainstream. The extent to which something is fashionable and its price dominate consumer decision making in most segments, and even proenvironmental segments consider fashion, self-identity, and price ahead of sustainability impacts.

However, there are signs that the fashion industry is learning from other sectors and is moving toward greener and more sustainable practices. Fair Trade and organic are not out of place in the food sector, but these terms also are now used in the fashion industry. People Tree was one of the world's first Fair Trade fashion labels; it now supports some 50 Fair Trade producer groups in 15 countries. As a business, it has been growing at a rate of more than 30 percent a year. People Tree was the first to build a Fair Trade and organic supply chain for cotton and cotton manufactured products in the developing world in

India; their clothes appear on the catwalk at London Fashion Week, and the company sees the fashion business as a vehicle for environmental and social change.

The high end of the fashion market appears to be reacting more quickly to the opportunities presented by being able to demonstrate environmentally friendly credentials. For example, the Fur Council of Canada is now seeking to reposition its fur products as a natural, renewable, and recyclable resource for the creation of fashion garments in a clear attempt to move away from the ethical arguments against the use of fur.

In general, the luxury designer labels that are beginning to exploit the growing market for environmentally friendly, ethical fashion are the antithesis of the disposable mainstream fashion market. Sharkah Chakra describes itself as following slow fashion and produces its up-market denim jeans from Fair Trade, organic cotton, handwoven following traditional craftsmanship in southern India and dyed using only natural dyes. Wastewater is recycled, electricity used in the processes is green, and packaging is 50 percent recycled waste, 50 percent from managed, certified forest. Even the labels and brochures are printed with vegetable dyes. Greencaste jeans, created by the luxury fashion label Earnest Sewn, also aims to tackle the bad reputation of the denim industry, which is known for using vast amounts of water and contaminating waterways with the chemicals used to treat the material. Earnest Sewn takes dirty river water, cleans it for use in its wash processes, and then re-treats it to leave it cleaner than before.

**See Also:** Apparel; Ethically Produced Products; Linen and Bedding; Water.

### Further Readings

Black, Sandy. *Eco-Chic: The Fashion Paradox*. London: Black Dog, 2008.
Fletcher, Kate. *Sustainable Fashion and Textiles: Design Journeys*. London: Earthscan, 2008.
Hethorn, Janet and Connie Ulasewicz. *Sustainable Fashion: Why Now? A Conversation Exploring Issues, Practices, and Possibilities*. New York: Fairchild Books, 2008.

*Barry Emery*
*University of Northampton*

# Final Consumption

According to the Organisation for Economic Co-operation and Development, final consumption consists of goods and services used up by individual households or the community to satisfy their individual or collective needs or wants. The cumulative effects of ordinary product consumption and everyday activities become relevant to understanding current sustainability challenges. Moreover, countries and organizations have developed systems perspectives to address current and potential consumer behavior effects on sustainability.

Each and every product, from home furniture to casual clothing for work, required a specific set of resources to achieve its final form. These products are considered human-made capital and provide specific features to support functionality and lifestyle at the household and community level. Mass customization practices made possible the development of product solutions available at stores and retail centers. Current products and services offerings are intended to satisfy—and ideally exceed—customer expectations. Simultaneously, design and development processes for human-made capital significantly rely on natural capital and ecosystem services. Natural capital comprises all living organisms and nature available on the planet, such as the air we breathe, lakes, oceans, vegetation, and animal species.

The human population is approximately 6.7 billion and could grow to 8.9 billion by 2050. Therefore, the amount of resources needed to support households and communities is gradually going beyond Earth's carrying capacity. Historically, developed countries have invested in infrastructure for consumer goods production and witnessed the environmental impacts associated with their final consumption at the household level. Those impacts were initially acknowledged by regions in which products were locally manufactured and used, but as international trade increased, their environmental impacts became widespread across regions, following the path of product availability.

Products such as washing machines and laundry detergents have been widely studied to identify critical stages in their life cycles that cause important environmental impacts. Main life cycle stages generally defined by life cycle assessments are material sourcing, manufacturing, distribution, retailing, use, and end-of-life. From them, use-phase has been consistently identified as the stage in which additional resources are required to actually use the product and in which an increased potential for greenhouse gas emissions exists. Compared with production or transportation stages, use-phase contributes 75 percent or more of total product environmental impacts over its life cycle. For example, the amount of detergent that an average American family uses for their 468 laundry loads per year could be multiplied by the 84.1 million households in the country that own a washing machine. In addition, if an average washing machine requires 30 gallons of water for each laundry load, the amount of water and detergent that a nation such as the United States requires each year just to take care of dirty laundry becomes relevant. All without considering the amount of energy required to mechanically dry all those laundry loads.

## The Impact of Point of Sale

On a cumulative basis, simple everyday activities such as doing laundry or cooking dinner generate a significant impact on the environment. For this to happen, however, all products in the form of clothes, food, electronics, and appliances must arrive home first. In this regard, the role of the "point-of-sale" is critical. This location is represented by all regional and global retailers that consolidate and present all the products that fill our living and work environments in an appealing fashion. Current markets provide different retail formats, from brick-and-mortar value discounters and specialty stores to complete online businesses.

Modern society associates consumption with satisfaction, identity formation, status, and distinction; this social practice and routine denotes the importance that material goods have achieved in contemporary living. In addition, consumption fosters social cohesion and allows group norms development. Most of these attributes are related to traditional retail formats, which determine the way customers approach products, how often they visit these establishments, the specific experience one retailer provides versus another, or whether some kind of membership is required. All these elements provide a platform for social interaction around consumption.

Other emerging business models conduct transactions with customers over the Internet. Part of their initial value proposition was to do business directly with end customers and gradually phase out "the middle-man" and associated fees resulting from infrastructure and overhead. This notion was instrumental to account for the environmental effects from resource and product consumption at the retail phase. Moreover, logistics and transportation professionals became increasingly aware of their environmental impact when handling and distributing products, from factories to warehouses and retailers. Their insights considered the amount and nature of extra packaging materials for products and the amount of fuel associated with product delivery, take-back routing, or deadhead trips.

The food industry provided a clear example of these implications: eventually, the concept of food miles became relevant to customers. Food miles represent the distance food travels from where it is grown to where it is ultimately purchased or consumed by the end user. From a study conducted by the Leopold Center for Sustainable Agriculture at Iowa State University, local produce travels on average 56 miles to reach institutional markets, whereas conventional produce travels 1,494 miles, nearly 27 times farther to reach the same points of sale. Overall resource consumption is amplified when processed food requires specific refrigeration support during transport and when being displayed at stores. Local harvesting and consumption emerged as an option to reduce food miles.

## Sustainable Consumption: A Balancing Act

The previous examples illustrate some elements that contribute to the increasing complexity of promoting sustainable consumption. According to the National Consumer Council, sustainable consumption is a balancing act. It is about consuming in such a way as to protect the environment, using natural resources wisely and promoting quality of life now, while not spoiling the lives of future generations. Final consumption is affected by economic constraints, institutional barriers, access inequalities, and choice restriction. In addition, customer expectations, habits, and social norms resulting from specific cultures and regions shape the type and magnitude of consumption. Consumer behavior plays a critical role within final consumption.

A survey conducted by the Taylor Nelson Sofres Group found that on average, 59 percent of customers from Australia, Japan, Korea, Singapore, Thailand, Malaysia, Hong Kong, the United Kingdom, Germany, Spain, Italy, France, Russia, Brazil, Argentina, Mexico, and the United States are willing to pay higher prices for environmentally friendly products. In addition, 61 percent of the respondents consumed locally grown food to avoid unnecessary transportation, 71 percent considered that product choice editing made by retailers improves customer experience, and 51 percent considered the environmental reputation of the organizations from which they consume the most.

According to the Grocery Manufacturing Association, 54 percent of the U.S. market is leaning toward green products, with grocery stores experiencing the most visits (35 percent of the market over value discounters and specialty stores). Regarding product categories with an environmentally friendly choices available for customers, everyday grocery, produce, and household cleaning products are the leaders, with 33 percent, 31 percent, and 28 percent, respectively. Other product categories moving in this direction are health and beauty, pet products, and apparel.

These green shoppers usually hold some type of graduate degree, and their annual earnings are above the national average of other shoppers. Most of them usually have one or two kids, and they belong to the mid-forties age group. This customer description was not originally expected by researchers or organizations looking for price-sensitive and frugal customers interested in reducing their impact on the environment. Most recently, the

Natural Marketing Institute coined the term LOHAS, which stands for Lifestyles of Health and Sustainability, to identify the portion of the population for whom environmental, social, and healthy lifestyle values play an important role in their purchasing decisions.

Final consumption of resources, products, and services at the household and community levels has definitely acquired a new meaning under the lens of sustainability. Economic, environmental, and social considerations are now strategically considered by global organizations, as they will affect their supply chain stakeholders, from material suppliers to actual customers from secondary supply chains after initial service life of products. As a consequence, several value propositions are emerging from the production and consumption sides of the market, both converging at household and community final consumption.

The pressing issue of natural resources availability to support contemporary lifestyles is clearly related to consumer behavior, from product selection to actual use and to decisions on how to dispose of the used products. Manufacturing and retailing organizations have a clear opportunity to scientifically inform the market and drive changes toward sustainable consumption. Customers around the globe are gradually incorporating the notion of sustainability and becoming increasingly aware of their impact on the environment and future consequences.

**See Also:** Consumer Culture; Green Consumer; Lifestyle, Urban; Sustainable Consumption.

### Further Readings


Cullen, Jonathan. "The Role of Washing Machines in Life Cycle Assessment Studies." *Journal of Industrial Ecology,* 13 (2009).

Delloite Development LLC. "Finding the Green in Today's Shoppers: Sustainability Trends and New Shopper Insights." http://www.deloitte.com/dtt/cda/doc/content/US_CP_GMADeloitteGreenShopperStudy_2009.pdf (Accessed August 2009).

Jackson, Tim. "Motivating Sustainable Consumption." http://www.ces-surrey.org.uk/people/staff/tjackson.shtml (Accessed August 2009).

Taylor Nelson Sofres Group. "Our Green World." http://www.tnsglobal.com/_assets/files/TNS_Market_Research_Our_Green_World.pdf (Accessed August 2009).

Van Hoof, Gert, et al. "Comparative Life-Cycle Assessment of Laundry Detergent Formulations in the UK." *Tensile Surfactants Detergents,* 40 (2003).


*Marco Ugarte*
*Arizona State University*

## Finance and Economics

One of the central dilemmas of "going green" is that although it means lower costs in the long run—at least on the societal level—on the short-term, individual level, it can be a significant increase in expense, depending on how much one is spending already.

The gains of being more environmentally friendly may not be experienced right away by the individual consumer. Using a professional car wash that reduces its water usage, reclaims it, and treats it, rather than releasing harmful chemicals into the environment the way washing one's car at home does, achieves a definite good. However, the effect on the actor is minimal. The real effect is felt when everyone switches to green professional car washes, and even then some of the good achieved is distributed to entities never involved in the action—the fish and other wildlife who are spared the effects of chemicals in their ecosystem. Although green philosophy perceives a good shared by humankind when the environment is spared from harm, it is difficult for the individual actor to experience and appreciate that good.

The effect of one's actions on collective consequences, and one's participation in those consequences regardless of one's actions, is relevant because there is a dollar value associated with these actions. Furthermore, that dollar value can be complicated. When the cost of gasoline is high enough, the extra initial cost of a more fuel-efficient vehicle—or one that does not use gasoline at all—can seem less expensive, even if one does not quite believe the car will "pay for itself" through its fuel savings. However, if enough people were to buy those vehicles—in sufficient numbers to reduce the demand for gasoline—the cost of gasoline would come down. Then again, so would the cost of the vehicles, thanks to economies of scale. There is a push and pull of cost and relative savings that is difficult for consumers to predict, as it is affected so greatly by their collective actions. It brings to mind economist John Maynard Keynes' description of stock-market investing: a beauty contest in which the winner is not any of the contestants, but the judge whose scores come closest to the average scores of the judges collectively.

## The Cost of Going Green

The green movement is sometimes criticized for downplaying the cost of going green, and it is perhaps a valid concern. Just as food movements have sometimes acted unaware of the difficulty of a working-class family negotiating not only its food budget but also the effect on that budget of the time constraints of working parents and the need to appease children whose tastes are affected by what they eat at school and at friends' homes, so too did the early environmentalism movement sometimes pay too little attention to the financial effect of environmentally responsible behavior. Recycling was an easy win because wartime rationing had conditioned people for it, and the gains were obvious and short-term. Switching to energy-saving lightbulbs is fairly easy to convince people to do, because of their longevity and the obvious long-term savings. Reducing water usage is more difficult, particularly given how low water bills tend to be relative to other utilities—a change in behavior may amount to only a few dollars a month, and those few dollars may seem like a more-than-reasonable price for longer showers and a well-watered lawn. Even water conservation is only asking for a behavioral change, however, and perhaps the expense of a low-pressure showerhead. Asking people to spend money on going green has historically been difficult.

The Environmental Defense Fund claimed in 2009 that the climate change problem could be solved at a cost of 10 cents per day per person. They arrived at this figure based on the U.S. Environmental Protection Agency's estimate of the cost of reducing carbon emissions by 42 percent from 2005 levels by 2030. There are a great many factors informing that figure, however—for one thing, it assumes a healthy economy, with steady growth and low unemployment It also assumes that between 2007 and 2030, the population will increase by 25 percent, the economy will grow by 70 percent, solar power usage will increase by 1,800 percent, wind power will increase by 600 percent, and the gas mileage of new cars will be 50 percent better. It is clear that all of these things are possible, but it is not clear that they are guaranteed, and a 20-year period is a difficult one to make accurate predictions over, particularly when it comes to technology and energy usage.

The 20 years preceding 2007 saw the ubiquity of personal computers—often more than one per household—and the subsequent ubiquity of household Internet access, along with the rise and fall of the dot-com boom and the invention of MP3 file and electronic music piracy. Cellular phones went from oddities to multiple-per-household commonplace devices. At the same time, the Soviet Union and Eastern European communism collapsed, China adopted free-market initiatives but remained a thriving communist nation, multiple wars in the Middle East affected oil prices, electric cars were introduced and recalled, gas-guzzling sports utility vehicles became popular, hybrid cars were introduced, and financial crises struck the housing market, pension funds, and consumer credit. All of those things have an effect on the accuracy of any economic or energy-usage forecasts that could have been made in the late 1980s; many of them were either completely unforeseen or were foreknowable only in the broadest of broad strokes.

## What Happens to Cost on a Larger Scale?

In many cases, the costs of an action or a technology are known at a current small scale, but it is not clear whether we can accurately extrapolate them to a larger scale. In the recent past, this has proven to be a problematic trait of biofuel. When there are a small number of biofuel users, it is—for those users—a cleaner and more efficient fuel. When demand for biofuel increases—such as when the price of oil rises high enough to make biofuel attractive—meeting that demand requires repurposing land or crops previously devoted to some other task. Corn syrup experienced a spike in price when corn crops were diverted for biofuel in response to an oil crisis, for instance. Other biofuel sources face the same problem—there simply is not such a surplus of arable land that it would be a feasible task to replace oil with biofuel. Although algae can be farmed in ways that make this less of a concern, few commercial biofuel blends use it to a significant degree—a recently developed biofuel for aviation, for instance, uses only 5 percent algae.

Energy Star, the Environmental Protection Agency's energy efficiency standards program, maintains a fairly realistic list of "payback periods" for various green products. An Energy Star–compliant washing machine, for instance, will earn back the cost margin over a standard washing machine in 3.5 years. An Energy Star refrigerator takes nearly as long—3.1 years—even though the difference in price compared with a standard refrigerator is quite low. The great success story—energy-saving compact fluorescent lightbulbs—have a payback period of only four months.

It is harder to estimate the cost of going green at a community level. The renewables research firm New Energy Finance has estimated the cost per kilowatt-hour of different sorts of power plants: 4 cents for gas and 25 cents for solar, with nuclear (subject to a number of factors) somewhere between the two. The dilemma is evident. Gas is volatile in price, nonrenewable, and harmful to the environment, but solar costs seven times as much when gas prices are reasonable, and nuclear power—in some senses a compromise between the two, if the waste problem is solved—has a severe public image problem in the United States.

The post-2008 economy will provide a petri dish in which to examine consumer behavior with regard to green products on both an individual and collective basis. Before the financial meltdown of 2008, demand for green products was increasing. There has been much speculation that the subsequent recession will hurt the sales of green products, but it may depend on public perception of the costs and benefits. Further, it may depend on the choices of manufacturers: when marketing a green washing machine, although it will be more expensive than a standard washing machine one way or the other, it need not be placed in competition with the premium appliances on the market. Doing so sends the message that going green is the appliance equivalent of going gourmet—something that is not for everyone, and that not everyone can afford.

**See Also:** Advertising; Biodegradability; Carbon Credits; Carbon Emissions; Carbon Offsets; Composting; Conspicuous Consumption; Consumer Behavior; Disparities in Consumption; Energy Efficiency of Products and Appliances; Green Politics.

### Further Readings

Brower, Michael and Warren Leon. *The Consumer's Guide to Effective Environmental Choices: Practical Advice from the Union of Concerned Scientists.* New York: Three Rivers, 1999.

Carter, Alan. *A Radical Green Political Theory.* London: Routledge, 1999.

Dobson, Andrew. *Green Political Thought.* London: T&F, 2009.

Garlough, Donna, et al., eds. *The Green Guide: The Complete Reference for Consuming Wisely.* New York: National Geographic, 2008.

Radcliffe, James. *Green Politics: Dictatorship or Democracy?* New York: Palgrave Macmillan, 2002.

Rogers, Elizabeth and Thomas Kostigen. *The Green Book.* New York: Three Rivers, 2007.

Soderlind, Steven D. *Consumer Economics.* New York: M. E. Sharpe, 2001.

Torgerson, Douglas. *The Promise of Green Politics: Environmentalism and the Public Sphere.* Durham, NC: Duke University Press, 1999.

Wagner, Sigmund. *Understanding Consumer Behavior.* New York: Routledge, 2003.

*Bill Kte'pi*
*Independent Scholar*

## Fish

The harvesting of and the use of fish as a consumer product have several environmental effects that are increasingly playing a role in consumer choices regarding the consumption of fish. The environmental effects of fishing can be divided into issues that involve the availability of fish to be caught, such as overfishing, sustainable fisheries, and fisheries management, and issues that involve the effect of fishing on the environment, such as bycatch. As world populations grow, the gap between how many fish are available to be caught and humanity's desire to catch them continually increases, raising questions about the sustainability of modern fishing practices. Governments and consumer groups have responded to these questions with the implementation of increasingly rigorous fisheries management and by demanding sustainably produced seafood.

One of the longest-standing environmental concerns of fishing is overfishing. Overfishing occurs when fishing activities reduce fish stocks below an acceptable level. This can occur in any body of water of any size and is a concern in both freshwater and saltwater fishing. Ultimately, overfishing may lead to resource depletion, low biological growth rates, and critically low biomass levels. In particular, overfishing of sharks has

The shrimp-trawl fisheries produce over one-third of the world's total by-catch (the portion of the catch that is not the target species). Here, shrimpers in North Carolina separate shrimp from by-catch.

*Source:* National Oceanic and Atmospheric Administration

disrupted food chains and led to the upset of entire marine ecosystems.

Similarly, fishing may disrupt food webs by targeting specific, in-demand species. There might be too much fishing of prey species such as sardines and anchovies, thus reducing the food supply for the predators. It may also cause the increase of prey species when the target fishes are predator species, such as salmon and tuna. Fisheries can reduce fish stocks that cetaceans rely on for food.

Direct impacts on the environment are another significant aspect of fishing practices. Of particular significance is by-catch, or the portion of the catch that is not the target species. Often, this catch is all discarded, leading to higher levels of decomposition-related pollution. The highest incidence of by-catch occurs with shrimp trawling. Shrimp trawl fisheries catch 2 percent of the world's total catch of all fish by weight but produce over one-third of the world's total by-catch. Sea turtles, already critically endangered, have been killed in large numbers in shrimp trawl nets. Estimates indicate that thousands of Kemp's Ridley, loggerhead, green, and leatherback sea turtles are caught in shrimp trawl fisheries in the Gulf of Mexico and the U.S. Atlantic annually.

Consumer concern over by-catch has led fishermen and scientists to develop devices they can attach to their nets to reduce unwanted catch. The by-catch reduction device, for example, is a net modification that helps fish escape from shrimp nets. By-catch reduction devices allow many commercial finfish species to escape trawling nets. To be federally approved in the United States, by-catch reduction devices must reduce the by-catch of finfish by a minimum of 30 percent. In addition to efforts to reduce the amount of by-catch caught in nets, some fisheries are starting to implement programs to effectively use by-catch species, rather than throwing the fish back into the ocean. One such use of by-catch is the formulation of fish hydrolysate that can be used as a soil amendment in organic agriculture.

Many governments have implemented fisheries management policies designed to curb the combined environmental impacts of fishing. Fishing conservation aims to control the human activities that may completely decrease a fish stock or wash out an entire aquatic environment. These laws include quotas on the total catch of particular species in a fishery, limits on the number of vessels allowed in specific areas, and the imposition of seasonal restrictions on fishing.

Likewise, consumer groups have taken action to attempt to alleviate the negative environmental effects associated with fishing. Most notable is the development of the sustainable seafood movement. Sustainable seafood is a movement that has gained momentum as more people have become aware of overfishing and environmentally destructive fishing methods.

Sustainable seafood is harvested from either fished or farmed sources that can maintain or increase production in the future without jeopardizing the ecosystems from which it was acquired. Slow-growing fish that reproduce late in life, such as orange roughy, are often vulnerable to overfishing. Seafood species that grow quickly and breed young, such as anchovies and sardines, are much more resistant to overfishing and thus can be fished in larger numbers while maintaining sustainability.

Several organizations certify seafood fisheries as sustainable, most notably the Marine Stewardship Council. These organizations have developed an environmental standard for sustainable and well-managed fisheries. Three common factors used to identify sustainable fisheries are as follows:

- The condition of the fish stock(s) of the fishery
- The effect of the fishery on the marine ecosystem
- The fishery management system

Environmentally responsible fisheries management and practices are rewarded with the use of ecolabeling. Consumers concerned about overfishing and its consequences are increasingly able to choose seafood products that have been independently assessed against standardized environmental standards and labeled as such. This enables consumers to identify fish products that have been harvested sustainably and, in turn, to play a part in reversing the decline of fish stocks.

Following from the increasing consumer demand for sustainable seafood, restaurants are increasingly offering more sustainable seafood options, with some restaurants beginning to specialize in sustainable seafood. Although there is no official certifying body like the Marine Stewardship Council in the United States, the National Oceanic and Atmospheric Administration has created the FishWatch guide to help advise concerned consumers about sustainable seafood choices.

The move toward sustainability and increasing regulation of fisheries management are changing the way that fish is harvested, as well as the way it is consumed. These processes are driven by increased demands for green seafood and are likely to follow larger patterns in green consumption of food products.

**See Also:** Certified Products; Ecolabeling; Green Food; Sustainable Consumption.

### Further Readings

Castro, Peter. *Marine Biology*, 7th Ed. Boston: McGraw-Hill, 2007.

Clover, Charles. *End of the Line: How Overfishing Is Changing the World and What We Eat*. London: Ebury, 2004.

Ellis, Richard. *The Empty Ocean*. Washington, D.C.: Island Press, 2004.

*Ryan Zachary Good*
*University of Florida*

## Floor and Wall Coverings

The walls of human dwellings have been decorated since the first cave paintings and petroglyphs. Purely functional walls are not very interesting, so decorating them has been a continual challenge. In addition to decoration, wall and floor coverings can add to the

Because it is a renewable resource, bamboo is a green insulation material. Fashioned into sheets it can be used to cover walls, or it can be rendered into a pulp that can be molded into sound-absorbing material.

*Source:* iStockphoto

insulation of buildings, making them more energy efficient

In the Middle Ages, tapestries were used for decoration and for the warmth they provided as insulation against the cold in castle walls. Victorian-era wallpaper was used extensively for decoration, but it also had the effect of blocking drafts of air coming from between the cracks in the boards of the wall. With the development of drywall, wallpaper declined as a decorative material in part because of the additional cost of wallpaper (including hanging it) compared with that of painting drywall. However, the use of decorative wallpaper, which has been viewed by art critics as a kind of inferior form of wall decoration, has experienced renewed interest since 2000. Wallpaper is now viewed as an environmental medium that is both aesthetically pleasing and environmentally helpful because it can aid with the insulation of the rooms of a building.

A useful green insulation material that is also very decorative is bamboo. Fashioned into sheets, it can be used to cover walls, and it can also be rendered into a pulp that can be molded into decorative and sound-absorbing material. It is also a renewal resource. Other types of wood are used as paneling or as planks or laminated sheets for wall coverings that are decorative and that also contribute to reducing energy cost for heating and cooling. Cork wall coverings absorb sound, creating a green sound space.

There are several types of plaster that can be used for wall covering, and there are mixtures of clay that can be applied to add beauty and warmth. The Venetian style of wall plaster creates a covering that resembles marble or natural stone. Plaster also contributes to insulation. Avoided by green-conscious consumers are products that use energy-intensive processes such as cement or that use petroleum products in the wall-covering product or in the product's manufacture.

Leather as a wall covering was used in Italy in the 19th century and has since been brought back as a decorative medium. It also is renewable and a good insulation material.

To be avoided by green-conscious consumers are wall-covering products—including wallpaper—that include polyvinyl chloride (PVC)/vinyl. In addition, glues that are used should contain either low– or no–volatile organic compound glues.

Among the wallpaper-type products are silk-screened paper that is PVC-free. In addition, many products are capable of being cleaned with a wet sponge. Some wallpaper products are made of recycled paper.

Green textile wall coverings are made of recycled spun glass or mixtures of materials such as wood pulp, bark, and straw in a matrix. Recycled glass or ceramic wall-covering materials can be used in kitchens and bathrooms.

Floor-covering materials throughout the centuries have included packed earth, leaves, stone, wood, animal skins such as bearskin or sheepskin, linoleum, ceramic tiles, carpets, and other materials. Handwoven carpets have been made for centuries in central Asia and in China. They were usually made of some kind of sheep or goat wool.

The modern tufted carpet industry is centered in Dalton, Georgia, where over half the tufted carpet in the world has been manufactured since the 1950s. The industry has increasingly turned to green technologies to make new types of ecologically friendly carpets available for green consumers.

Many new types of green carpets and green backing are being developed. The new types of carpets include organic or renewable-source backing for the carpet and renewable materials for the carpet itself.

Carpet backing was historically made of jute fiber, but the advent of cheap petroleum-based fibers allowed for the creation of a synthetic backing. The new green backings are part of a holistic design that is part of a sustainable approach to the manufacture of carpeting and other forms of flooring. Among the inventions is the development of a non-PVC carpet tile that can be recycled back into itself as a renewable resource. Old automobile windshields are being melted into a form of spun glass fiber for carpet backings. Other forms of renewable resources are backings made from soybean oil.

Green carpeting is also being kept down in price with a dematerialization approach that uses a smaller amount of materials in the production process; however, increasing the performance of backing materials increases their longevity and reduces total costs for the consumer. The dematerialization is aided by the total recycling of industry waste streams. Everything from old carpet that is replaced is recycled. This green recycling reduces the use of nonrenewable resources. Recycling includes using the waste from other companies (e.g., wastes from coal-fired electrical generating plants) in the manufacturing of both backing and carpeting.

In several earlier civilizations, floors or beds were made of brick or some other kind of heat-conducting material that was heated by a fire that passed hot air through a system to warm the floor or beds. Variations on this kind of radiant-heated flooring have been in existence for a long time, but recent years have seen drastic improvements. In addition, heated flooring systems are now greener because they use less energy, have much longer life cycles, and are composed of materials that are in abundant supply.

**See Also:** Adhesives; Green Consumer; Green Homes; Insulation.

### Further Readings

Stanley Complete. *Complete Flooring*. New York: John Wiley & Sons, 2008.

Tannenbaum, Judith. *On the Wall: Contemporary Wallpaper*. Providence, RI: Museum of Art, 2004.

Torre, Francesca. *Materials: A Sourcebook for Walls and Floors*. New York: Harry N. Abrams, 2008.

Wilson, Alex, et al., eds. *Green Building Products: The Greenspec Guide to Residential Building Materials,* 3rd Ed. Gabriola Island, British Columbia, Canada: New Society, 2008.

Woodon, R. *Radiant Floor Heating*, 2nd Ed. New York: McGraw-Hill, 2010.

*Andrew Jackson Waskey*
*Dalton State College*

# Food Additives

The use of food additives—introduced at the food-processing stage—is the largest single difference between organic and nonorganic foods. Common additives include food coloring, bulking agents, preservatives, sweeteners, and acids. Each is useful, but there is a long-standing and lively debate over whether they do more harm than good. The increasingly widespread industrial processing of food in advanced societies has brought with it a similar increase in the use of additives, in turn resulting in closer monitoring of food safety on the part of consumers and food-regulatory agencies. The U.S. Food and Drug Administration (FDA) is charged with monitoring and inspecting most foods in the United States.

The increasingly widespread industrial processing of food in advanced societies has brought with it a similar increase in the use of additives, including food coloring, bulking agents, preservatives, sweeteners, and acids.

*Source:* iStockphoto

There are two types of food additives: direct and indirect. A direct additive is a substance that is purposefully added to a food product to achieve a desired effect, such as a sweetener. These are generally listed on the food's ingredients label. An indirect additive is one that enters the food product by way of its packaging, storage, or transportation. Organic food has been available for many years and has become ever easier to find in stores, no matter what your income or preferences. Whether such organic food products are in some way "superior" to processed foods or not, they continue to gain popularity. One of the biggest advantages organic foods offer the consumer is that they contain no additives. Although a majority of grocery stores carry at least some types of organic foods on their shelves, production of organic foods represents only 2 percent of the total volume of foods produced for sale in the United States. Significant research has been conducted on the healthiness and growing methods, and the wealth of scientific evidence available shows conclusively that organic foods sold in the United States are healthier, are grown with more careful attention, and are far less adulterated than processed foods.

Conventional and organic farming are distinctly different. Agribusiness—conventional industrial agricultural farms in the United States and elsewhere—uses large quantities of additives in their food processing activities. Among these additives are chemical fertilizers, pesticides, and herbicides used to grow vegetables and grain and hormones, prophylactic antibiotics, amphetamines, and other medications to promote growth and prevent disease in market animals, including beef cows, fryer chickens, and pigs. These additives are also given to producing animals like laying hens and milk cows, for example, and many of these animals are kept in tightly packed pens. Organic farmers apply natural "biofertilizers"; use insects and birds to reduce pests and disease; feed animals with organic, nonbonemeal feed; and provide wider, less-restricted access to outdoor movement to promote "happy and healthy" animals.

## Labeling

Food labeling varies widely globally, but in the United States, the U.S. Department of Agriculture (USDA) promulgates a number of labeling regulations for organic foods. Consumers are expected to examine the small print on all packaged-food product labels and compare them; understanding these labels is the buyer's responsibility, so caveat emptor. Agricultural products labeled "100 percent organic" are allowed to contain only organically produced ingredients. A product labeled as "organic" must be at least 95 percent organic. To apply the "USDA Seal" to an organic food package, the percentage of other ingredients must have been approved and placed on the USDA's National List—the official government compilation of substances that may be used in crop production and food additives. Foods labeled "made with organic ingredients" must contain at least 70 percent organic ingredients and list three of those included organic foods or food groups on the label. The USDA Seal may not be used on packages with the label "made with organic ingredients." Beyond labeling, the FDA mandates that no organic foods be produced by any method that incorporates radiation or sewage.

Nutritionally, there is much debate over the health advantages of additive-free food. One opinion is that organic foods may inadvertently lead to more food poisoning by *Escherichia coli* and other pathogens. Another argument is that chemicals and other additives used in food processing might be linked to cancer, birth defects, or hormonal disturbances. The nutritional values of organic and nonorganic products were measured in 41 studies. Organic foods were found to have 27 percent more vitamin C, 21.1 percent more iron, 29.3 percent more magnesium, and 13.6 percent more phosphorus. Organics had 15.1 percent fewer nitrates than nonorganic foods. Children are more susceptible than adults to toxins in food, so by eating food that has no preservatives or hormones added, children get a more balanced, nutritional diet.

## Processing

Food processing is the set of methods that uses additives and other methods to transform ingredients into food for humans or animals. Examples of common processing methods include baking, pasteurizing, and the addition of preservatives. Although these methods have many benefits, such as lowering production costs resulting from mass production and increasing storage capabilities, processed foods are also convenient for families that may live some distance from shopping centers or who do not have the time to prepare fresh meals. The most obvious benefits of packaging and preservatives are retardation of spoilage and ease of transportation. Drawbacks to food processing include a substantial reduction of nutritional value and the consumption of a variety of nonfood additives that may entail heath concerns. To ensure that processed foods do not spoil, additives like food dyes, flavoring, and nitrates are added with no nutritional value. "Fast food" is the prime example of mostly prepackaged, highly processed food, overloaded with fats, starches, and sugars that have negative health effects.

The generally higher price of organic foods is one reason people may decline to buy them. The average shopper must spend roughly 20 percent more on organic food than on nonorganic products. Much of the basis for these high costs is the intensive farm labor required. It is far less labor-intensive for a conventional farmer to plant, harvest, and grow crops; organic farmers often harvest manually and use other, more costly methods. Organic farms are often maintained better and cleaner than conventional farms. These expenses work together to limit the market for organic foods; as demand for more natural foods continues to grow, prices should decline.

## Regulating

The U.S. Food Additives Amendment of 1958 requires that additives must be tested and approved by the FDA. The additives must also pose a reasonable expectation or certainty that they are not harmful, and the agency bans any chemicals that have been found to cause cancer in animals. Although this law helps to establish a food additive's safety, legal loopholes that diminish its effectiveness exist. For example, part of the amendment allows certain chemicals widely used in commonly used foods or ingredients to bypass FDA testing if recognized as GRAS, or "generally recognized as safe."

This loophole allows companies to declare certain chemicals GRAS without notifying the FDA. The processed food industry may allow chemicals that could be known to cause health risks to be used in their products. Two examples are salt and partially hydrogenated vegetable oils. Both ingredients are commonly found in many foods in our daily diet, yet they are known to contribute to the incidence of high human-system cholesterol, heart attacks, and high blood pressure. Chemicals declared GRAS have often not been tested fully or were tested by the very companies that use them, leading to possible conflicts of interest. Trying to prove that a chemical is harmful to the consumer's health is also difficult. Food-quality activists have criticized the GRAS concept and how it is manipulated in practice, sometimes allowing cancer-causing chemicals to be passed through to the consumer, and has called on the FDA to work harder to keep harmful additives off grocery shelves.

A more recent law is the Food Quality Protection Act (FQPA) of 1996. The FQPA included amendments like the Federal Insecticide, Fungicide, and Rodenticide Act and the Federal Food, Drug, and Cosmetic Act. These amendments allow the Environmental Protection Agency and the FDA to ensure that all pesticides used on food meet certain safety standards. The law specifically addressed pesticides in children's diets at both the pre- and postnatal stages of child development. The law also required that, within 10 years, tests be reviewed to ensure they still meet the requirements. A brochure was required to be placed in every grocery store, listing the health effects of pesticides and how to avoid associated risks. The FQPA focused on public participation in the process, educating the public to assess risks along with federal agencies.

Government regulation and scientific studies on food additive safety should be continually monitored and reviewed by consumers and public-interest advocates. By keeping up with additives that go into our foods and teaching others about possible dangers, consumers can make informed choices. As more food products become more highly processed, more consumers seek alternatives. By understanding the additives that go into the food we consume every day, we can be better equipped to make the decision of what substances to consume.

**See Also:** Dairy Products; Green Food; Markets (Organic/Farmers); Meat; Organic; Pesticides and Fertilizers; Poultry and Eggs; Vegetables and Fruits.

### Further Readings

Burke, Cindy. *To Buy or Not to Buy Organic: What You Need to Know to Choose the Healthiest, Safest, Most Earth-Friendly Food*. Boston: Marlowe, 2007.

Farlow, Christine Hoza. *Food Additives: A Shopper's Guide to What's Safe and What's Not*. Grand Rapids, MI: KISS for Health, 2007.

Shan, Yaso. "Going Organic—Is It Nutritionally Better?" *Primary Health Care*, 16/3 (April 2006).

U.S. Department of Agriculture. http://www.usda.gov/wps/portal (Accessed April 2009).

U.S. Environmental Protection Agency. http://www.epa.gov (Accessed April 2009).

*Anthony R. S. Chiaviello*
*Kelli Bristol*
*University of Houston-Downtown*

# Food Miles

The "food miles" concept has arguably captured the public imagination more than any other term when it comes to debates about sustainable consumption. This popularity is largely a result of the apparent simplicity of its application—basically, it calculates the distance food travels from primary producer to end consumer. Transport is one of the main sources of pollution in the food chain in terms of getting produce from farmer to depot to processor to retailer. Consumers also increasingly use transport to buy their food. The food miles concept exposes this transport dependency and the often bizarre routes food takes as it travels from farm to fork. Academics, policy makers, and campaigners have used it as a powerful tool to expose the negative environmental impacts of global agrifood systems and to call for more sustainable agricultural practices and systems of food supply. This article explains the origins of the term and summarizes some key studies, including recent critiques

According to some estimates, milk from these cows at a small dairy farm in western Maryland might travel 1,500 to 2,500 miles to reach a consumer's table.

*Source:* Scott Bauer/Agricultural Research Service/USDA

that argue that the concept fails to recognize wider global trade benefits or to capture the carbon footprint of a product.

When it was first developed in the 1990s, the food miles concept was broadly conceptualized in terms of environmental impact. Reducing food miles was therefore about making a more explicit link between particular foods and where they came from. It was about "respatalizing" food production to reverse the aspatialities of globalized food provisioning, embedding food systems in local ecologies, and responding to consumer demands for quality foods. Today, however, food miles have been linked to carbon accounting and the climate change debate. As a consequence, the food miles argument has moved away from debates about sustainable agricultural production systems per se to food distribution and retailing, and particularly to the use of carbon in transport.

There is a wealth of evidence in the academic and public press about food miles. One of the earliest studies that revealed the bizarre distance food travels was of a West German yogurt bought "locally," but in fact made using ingredients (food and packaging) assembled hundreds of kilometers away. Adding the ingredients up to make a meal also tells its own story. If one were to buy a supermarket chicken casserole ready meal, for example, containing chicken, tomatoes, and peas, it may travel as much as 15,000 miles. Other academic studies are equally staggering:

- It has been calculated that the average food item in the United States travels between 1,500 and 2,500 miles from farm to fork.
- In the United Kingdom, a recent estimate suggests that food items travel on average 5,000 miles from field to table.
- British consumers are also traveling farther to get their food, with most using cars to do so. The distance traveled for shopping rose by 60 percent between 1975–76 and 1989–91.

Measuring distance alone, whether the product itself or the distance the shopper goes to buy the product, can be hugely misleading. The mode of transport (e.g., air freight vs. sea freight) is a critically important, if sometimes forgotten, variable. A recent application of food miles to two contrasting food distribution systems in the United Kingdom showed this very well. The comparison was between the carbon emissions resultant from operating a large-scale vegetable box scheme and those from a supply system in which the customer travels to a local farm shop. On the basis of fuel and energy-use data, the study suggests that if a shopper drives a round-trip distance of more than 6.7 kilometers to purchase their organic vegetables, their carbon emissions are likely to be greater than the emissions from the large-scale vegetable box scheme, which involves cold storage, packing, transport to a regional hub, and final transport to the customer's doorstep. This study challenges some of the ideas behind localism in the food sector, where it is assumed that local food is a solution to the problem of food miles.

So, although the food miles concept has been hugely successful because of its simplicity of message and ability to resonate with the public, a growing body of work is beginning to reveal its shortcomings in terms of scientific rigor. Here are four examples:

- Some argue that product life cycles are the only accurate way to determine carbon labeling. There is a public misconception that food miles measure the carbon footprint of a product.
- Fair Trade organizations argue that food miles and the close association with localism amount to a form of "green parochialism" that simplifies debates about local and global trade. The main problem they suggest is that the concept undermines the social and

economic benefits of global trade, particularly for developing countries. U.K. consumers, for example, spend more than £1 million every day on fresh fruit and vegetables from sub-Saharan African countries. This air-freighted trade accounts for less than 0.1 percent of the United Kingdom's total carbon emissions—minimal when compared to domestic U.K. food miles. However, this revenue contributes enormously to poor rural communities in Africa.

- Others argue that it may be more environmentally desirable to transport food rather than degrade local resources. Growing rice in Texas, for example, would produce negative environmental costs associated with water pumping that may exceed the costs of importing rice.
- New research by the Worldwatch Institute also suggests that as consumers seek out food with a reduced carbon footprint, a dietary shift from red meat and dairy consumption may be more effective than buying locally produced products.

**See Also:** Carbon Emissions; Consumer Ethics; Ecological Footprint; Fair Trade; Markets (Organic/Farmers); Organic; Sustainable Consumption.

### Further Readings

Born, Branden and Mark Purcell. "Avoiding the Local Trap: Scale and Food Systems in Planning Research." *Journal of Planning Education and Research,* 26 (2006).

Coley, David, et al. "Local Food, Food Miles and Carbon Emissions: A Comparison of Farm Shop and Mass Distribution Approaches." *Food Policy,* 34 (2009).

Edwards-Jones, Gareth. "Food Miles Don't Go the Distance." http://news.bbc.co.uk/1/hi/sci/tech/4807026.stm (Accessed May 2009).

Lang, Tim and Michael Heasman. *Food Wars*. London: Earthscan, 2004.

MacGregor, James and Bill Vorley. "*Fair Miles? The Concept of 'Food Miles' Through a Sustainable Development Lens.*" London: International Institute for Environment and Development, 2006.

Pretty, Jules, et al. "Farm Costs and Food Miles: An Assessment of the Full Cost of the UK Weekly Food Basket." *Food Policy,* 30 (2005).

*Damian Maye*
*Countryside and Community Research Institute*

## Frugality

Frugality is perhaps best thought of as the ghost at the banquet of modern consumerism. If consumption in the consumer society is characterized by excess and hedonism, frugality represents the antithesis. Frugality values work over leisure, saving over spending; restraint over indulgence, deferred over immediate gratification, and the satisfaction of "needs" over the satisfaction of wants and desires. To be frugal is to be moderate or sparing in the use of money, goods, and resources, with a particular emphasis on careful expenditure and the avoidance of waste. As such, it has an obvious resonance with environmental issues, despite its origins as an approach to economic affairs. Most notably, frugality has been championed by the voluntary simplicity movement and forms an integral part of efforts to "downsize" and "downshift." For example, giving up the so-called trappings of consumerism and materialism is seen to go hand in hand with a shift to a way of being premised on frugality.

This frugality can include a range of practices such as moving to a smaller home and growing one's own vegetables, giving up a BMW in favor of a smaller (or no) car, buying secondhand goods, going without luxury consumer goods that one does not "need," and switching from branded goods to supermarket "own brands" while grocery shopping. Similarly, there exists a tacit assumption that living frugally is an essential component of living sustainably, insofar as spending less money is seen to somehow translate into a smaller environmental impact. For example, commuting to work on a bicycle instead of using a car or taking public transport can be understood both as frugal and environmentally friendly. Furthermore, the Sierra Club (U.S. environmental movement) has put forward the idea that the survival of the planet depends on the emergence of a modern frugality, just as Raymond de Young has identified psychological links between undertaking proenvironmental behavior and deriving satisfaction from resourcefulness. These links are no doubt important; however, a better understanding of what frugality is and is not can paint a fuller picture that is perhaps more useful for the study of consumption and green consumerism.

Frugality has been something of an undertheorized and elusive topic in the study of consumer behavior and consumer culture. The reason for this, at least in part, lies in the fact that frugality has historically been defined in its opposition to consumption. The legacy of Puritanism and asceticism has constructed frugality as a virtue and, in doing so, has positioned consumption as trivial, frivolous, and having nothing to do with that which is virtuous. Nevertheless, frugality has not been entirely absent from the study of consumption. For example, Max Weber's classic account of how the emergence of industrial capitalism was grounded in the spirit of Calvinism and a distinctly Protestant ethic illustrates how a firmly moral restraint on economic activity came to dominate the conduct of economic affairs in the 19th century. The gist of this moral restraint was the avoidance of spending and self-indulgence, discipline in work and the acquisition of money, and control over waste and taking good care of monies acquired. Of course, the ethic of material and economic restraint is not unique to the Protestant thought—it is common to all Judeo–Christian traditions and virtually all major religions.

Similarly, as John Lastovicka and his colleagues have noted, the ethic of frugality can be found in economic theories, self-help books, and the moral virtue accorded to those who exhibit care and restraint in their economic activities. Of course, frugality can just as easily be understood in terms of being miserly, tight-fisted, and avoiding the spending of money for no other reason but to accumulate money for its own sake. So although frugality is not necessarily a moral restraint on economic activity, it certainly operates to restrict levels of spending and the acquisition of goods. As such, from the point of view of consumption, frugality can be understood—quite unambiguously—in terms of consuming less.

## Frugality and Thrift

In contrast, the same cannot be said of thrift. It is important here to note that—quite possibly as a consequence of frugality lying outside the consumer research radar—the concepts of frugality and thrift are often used interchangeably. This is entirely understandable, but deeply problematic when it comes to thinking about the idea of green consumerism. True enough, it is reasonable to think about thrift in terms of resourcefulness, saving, and certain restrictions on expenditure; however, unlike frugality, it does not necessitate less economic activity or less consumption. In fact, the practice of thrift is one of saving money through spending.

For example, it is thrifty to purchase an unbranded shirt that looks exactly the same as a branded one but costs half the price, just as it is thrifty to buy a particular item of food when it is on special offer in the grocery store. However, this does not bring into question whether these items needed to be purchased. In fact, it may well have been the case that a packet of cookies was only purchased because they were on special offer. Where frugality would ask whether the item actually needs to be purchased, and save money by not purchasing the item if the answer were "no," thrift would save money by asking whether the same item (or similar) could be purchased for less. By definition, thrift is the practice of getting more for less and saving by spending. To illustrate: With thrift, if one purchases a CD player that has been reduced from $500 to $300, it is more likely to be understood as a savings of $200 as opposed to a spending of $300. In contrast, frugality would understand the purchase as expenditure (regardless of the savings), and the CD player would only be purchased if it was worth the price paid for it.

Crucially, thrift does not necessitate less consumption in the same way that frugality does. In fact, it can actually result in higher levels of consumption insofar as the savings made through thrifty conduct can be used to purchase other goods and services. For example, the $200 saved on the CD player may well be used to purchase a new selection of CDs to go with it. Similarly, thrifty purchases are very often "throwaway" purchases that are easily disposed of and replaced. For example, thrifty conduct may well result in the purchase of five poor-quality T-shirts in the place of one good-quality, albeit more expensive, garment. From an environmental perspective, of course, this is far from desirable. This has led some commentators to suggest that, perhaps counterintuitively, lifestyles based on the purchase of luxury goods have a lower environmental impact than the thrifty lifestyles that we might automatically associate with being environmentally friendly. For example, the purchase of an expensive, high-quality item of furniture might be preferable—from an environmental point of view (although please note that there are certain distributional issues here, insofar as not everybody can afford luxury goods and are thrifty out of necessity)—to a mass-produced item of self-assembled furniture from Ikea. The reasons for this are twofold: it is likely to last longer and not require replacement following a change in fashion or simply falling into disrepair, and it means that less money is available to spend on other items that will inevitably have an environmental impact. Interestingly enough, the purchase of luxury goods is not incompatible with the idea of frugality, as it implies careful investment, resourcefulness in the use of goods, and the avoidance of waste.

Finally, even though frugality entails less consumption at the individual level, it is by no means a guarantee of a reduction in consumption and environmental impact at the global level. For a start, the positive environmental effects of consuming less need to be matched by the consumption of goods that have a lower environmental impact. For example, less consumption of environmentally damaging goods and services is not necessarily any better than high consumption of environmentally friendly ones (and vice versa). Similarly, a shift to frugality by rich consumers is likely to lead to a situation in which poorer consumers increase their consumption, insofar as these reductions in expenditure will create a surplus of supply, which will in turn lower their price and make them available for consumption by consumers who would have previously been unable to consume them.

**See Also:** Downshifting; Lifestyle, Sustainable; Secondhand Consumption; Simple Living; Sustainable Consumption.

### Further Readings

Alcott, Blake. "The Sufficiency Strategy: Would Rich-World Frugality Lower Environmental Impact?" *Ecological Economics*, 64 (2008).

Johnson, Warren. *Muddling Towards Frugality.* San Francisco, CA: Sierra Club, 1978.

Lastovicka, John, et al. "Lifestyle of the Tight and Frugal: Theory and Measurement." *Journal of Consumer Research*, 26 (1999).

*David Evans*
*University of Manchester*

# Fuel

Air pollution from vehicle exhaust emissions is the end result of burning fuels to power the car. This has caused much environmental concern. Gasoline has been the dominant fuel since the invention of the automobile, although other fuels have been tried—some, such as diesel oil and liquefied petroleum gas (LPG) or propane, with considerable success. Others have been experimented with primarily because of their lower or less-harmful emissions. On the whole, consumers have limited choice in what fuels to buy, yet lack of consumer demand is often used by suppliers to justify not offering alternatives. Growing environmental pressure is increasingly leading regulators to mandate new and improved fuel types. Although conventional gasoline and diesel fuels will be with us for several more years, their dominant position is increasingly being challenged by alternative fuels and energy sources, such as electric power.

Compressed natural gas (CNG) is readily available in many countries from the domestic distribution infrastructure that supplies natural gas for heating and cooking. Here, a bus powered by compressed natural gas.

*Source:* U.S. Environmental Protection Agency

Gasoline and diesel oil are produced from mineral oils formed many millions of years ago from decaying organic matter. Petroleum is a mixture of a wide range of hydrocarbons, varying from the volatile, such as butane, to heavier materials, such as lubricating oil, paraffin, and tar. Refining is the process used to separate out these different fractions and render them suitable for their individual tasks. From the early 19th century, oil was mainly used for lighting, with some fractions used for medicinal purposes. Both these uses led to early oil exploration in the United States from about the middle of the 19th century.

Carl Benz, who built the first automobile, used gasoline to power the engine of his Patentmotorwagen of 1885. The motor spirit used in early cars was relatively pure, as low demand made it possible to use only the purest fractions from the distilling process. However, oil companies felt that higher volumes at lower cost were essential if the car—and thus the potential market for gasoline—were to take off. As demand rose from the growing band of motorists, the proportion of crude oil used for gasoline increased, with a resulting decline in quality and gradual reduction in cost as less and less suitable—but much cheaper—fractions were used. This led to less complete combustion, and hence an increase in harmful emissions.

To make them suitable as a fuel, these fractions had to be chemically transformed to turn them into gasoline. This process also led to the discovery of new ways to purify and "engineer" gasoline for automotive use. The need for this arose with the increasing sophistication of engine designs. Both engine speeds and compression ratios increased steadily between 1920 and 1960, which was possible not only because of advances in metallurgy but also because of the simultaneous changes in fuel. Advances in engine design and performance could lead to spontaneous combustion of the gasoline–air mixture in the cylinder. This preignition "knocking" or "pinging" problem did not only lead to loss of power but could also damage the engine. For this reason, racing engines were often run on alternative—usually high-octane alcohol-based—fuels.

Researchers at General Motors discovered during the 1920s that the addition of tetraethyl lead could prevent knocking by increasing the octane rating of the gasoline. The sale of this so-called leaded gasoline started in 1923. By the late 1960s, supplies of the higher grades of low-sulfur, low-nitrate crude were becoming scarce. The oil industry favored these higher-quality oils, as the removal of the sulfur and nitrates is an expensive and complex process. The oil producers were reluctant to introduce the dramatic increase in gasoline prices that would be required to move to cleaner gasoline, and thus it was that the motor industry came up with a partial solution to this problem in the shape of the catalytic converter. The regulated three-way catalytic converter successfully removes most of the oxides of nitrogen ($NO_x$), carbon monoxide (CO), and unburned hydrocarbons (HC), whereas simpler, unregulated two-way catalysts remove more than half of the CO and HC.

Diesel is a commonly occurring fraction in oil and is quite readily available. However, its popularity for powering car, as opposed to truck or bus, engines is today largely confined to Europe. Rudolf Diesel carried out his early experiments with coal dust, though a change to a light oil was made as experiments progressed. Diesel's 1892 patent was awarded for the compression-ignition principle without specifying any fuel type. The first trucks powered by diesel engines were launched in Germany in 1923 by Benz and MAN, and over the next 30 years diesel gradually became the dominant fuel type for heavy commercial vehicles. The first diesel cars appeared in the late 1930s. Both Citroen and Mercedes introduced cars with diesel engines at this time, and several others experimented with diesel, among them even luxury carmakers such as Lagonda in England. Although the visible smoke emissions of diesel engines were also noted, this was outweighed by the advantage of much-improved fuel consumption, greater reliability, and longer service life.

The lower-fuel-consumption characteristics of the diesel engine, resulting from greater efficiency and the lean-burn nature of the combustion, made diesel particularly attractive in Europe, where fuel prices were high because of high excise taxes. The greater efficiency of diesel prompted the French government's prodiesel policy in the early 1950s, which persisted until the 1990s. French leaders regarded these characteristics as a way of limiting reliance on imported oil, thus improving the balance of trade. Many European countries have adopted similar incentives for diesel, either through lower excise duty or tax advantages for diesel-powered cars and commercial vehicles.

Compared with gasoline, diesel also has a number of emissions advantages. Emissions of HC are much lower, as a result of the more complete combustion process, and CO levels are also lower, because more air is used in combustion. However, the latter also leads to higher levels of $NO_x$. The lower fuel consumption per driven mile also means that carbon dioxide ($CO_2$) emissions are lower, although the carbon content per liter is actually higher. For these reasons, diesel was initially regarded as more benign than gasoline and was largely bypassed by the earlier phases of emissions legislation. However, by the late 1980s, research in Germany suggested a potentially harmful effect from particulates of around 10 microns in size—the so-called PM10s. With this began a gradual opposition to diesel in many countries, further exacerbated by the fuel's high $NO_x$ emissions. With catalyzed gasoline engines now producing much lower levels of these substances, diesel suddenly looked less acceptable. With the focus on improving gasoline engines to meet emissions standards, diesel developments had been somewhat neglected, and during the 1990s the industry began to rectify this situation. The more efficient, but noisier, direct-injection diesel was made acceptable for use in cars, and new injection systems such as Fiat's common rail and unit injectors were being introduced together with dramatic improvements in refinement. By the late 1990s, cars with diesel engines began to be compared favorably, or at least on a par, with their gasoline engine equivalents. By this time, many diesel cars had oxidation catalysts, and experiments with particulate traps and de-$NO_x$ catalysts were looking promising. Further improvements in emissions throughout the existing diesel fleet would come from the gradual reduction of the sulfur content in diesel.

The threat of a form of energy tax from the European Commission to reduce $CO_2$ emissions made the diesel engine a logical area for further development. This prompted a further take-up of this fuel in Europe, including for more sporty cars—a move that has been made possible by further developments in diesel engine technology, using the high-torque characteristics of these engines. Audi even won the prestigious Le Mans 24-hour race with a diesel-powered racing car.

It became a vital technology for meeting the tough $CO_2$ limits set to come in the early years of the 21st century in the wake of the Kyoto agreement of 1997. In fact, the U.S. Partnership for a New Generation of Vehicles was moving toward favoring diesel as the most likely powertrain technology to meet their ambitious standards, thus bucking the antidiesel trend in the United States. A further move toward cleaner diesel fuels has come from gas-to-liquids technology, whereby diesel fuel is derived from natural gas. Similar technologies to derive diesel fuel from coal via coal-to-liquids technology are hampered by the excessive $CO_2$ generated by this process as a by-product.

## Alternative Fuels

During the later years of the 20th century, many pinned their hopes on gaseous hydrocarbon fuels as a solution to the air pollution caused by vehicle emissions. In addition, known reserves of natural gas are longer lasting than those for oil. LPG, sometimes known as propane, is a by-product of the refining process and is therefore widely available wherever oil refineries operate. It is a mixture of butane and propane. Since the late 1950s, LPG has enjoyed a tax advantage in several countries, such as the Netherlands, with its many oil refineries around Rotterdam. Italy also has had a sizable LPG fleet, and in countries like Japan and South Korea, as well as Hong Kong, LPG is popular for taxi use.

The emissions advantage of LPG resulting from its simple molecular structure has been known for many years, yet it was not until the late 1980s that it began to be actively promoted as a cleaner alternative for gasoline and diesel. In diesel-dominated France, LPG became the alternative fuel of choice, alongside electric vehicles, during the 1990s. LPG is considered particularly attractive for commercial vehicles operating in an urban environment. Thus, a growing number of buses and local-authority vehicles are being converted. Heavy diesel engines are relatively easy to convert to run on LPG or CNG (compressed natural gas), although they have to be turned into spark-ignition engines.

Disadvantages of LPG include the cost of conversion, the need for an additional fuel tank, and the remaining emissions. Natural gas has an even lower energy density, and therefore needs an even larger additional tank to do fewer miles, although it has a slight emissions advantage over LPG. It burns fairly cleanly and is readily available in many parts of the world, with a more equal distribution than oil. It has proven popular with fleet users in countries such as Canada, Russia, Sweden, Germany, and many others. It comes in two forms, including CNG, in which form it is still gaseous. The alternative, liquid natural gas (LNG), requires storage at low temperatures, although more can be carried in the same volume than with CNG. LNG availability is more limited, as this technique is used mainly for storing and shipping longer-term reserves, whereas CNG is readily available in many countries through the same domestic distribution infrastructure that supplies natural gas for heating and cooking. Its main flaw is that methane, its principal constituent, is a greenhouse gas and is believed to be several times more harmful than $CO_2$.

Alcohol-based fuels have been used in automotive applications for a long time, particularly as high-octane fuels for racing cars. For this reason, they burn more completely and thus produce lower emissions, although they are still hydrocarbon fuels. Two types of alcohol are distinguished—ethanol and methanol. Ethanol is the type of alcohol we drink and can easily be produced from the fermentation of a range of different crops. In the mid 1970s, the Brazilian government launched the “Proalcool” program as an import substitution project. In the wake of the oil crisis of 1973–74, Brazil felt it spent too much on importing oil to run its cars, and a means was devised to substitute this fuel with ethanol produced from sugarcane. Bioethanol is also popular in the United States, and 10 percent is routinely mixed with gasoline in many states. More recently, biodiesel produced from oil-rich crops has also grown in popularity in some European countries and parts of the United States. In practical terms, there are limitations to this approach, as vast areas of dedicated crop cultivation would be required to run a significant proportion of the world’s cars on this fuel, although where surpluses of crops rich in oil exist, it may be feasible locally. Waste cooking oil from the food industry also has proven an alternative source of biodiesel.

Methanol is a different product. It is more dangerous to handle than ethanol, or even gasoline, and requires a completely different fuel delivery system, as it corrodes most existing fuel system materials. Nevertheless, it enjoys some popularity in the United States as an alternative fuel. In practice it is usually mixed with gasoline to control its effects somewhat and to make cold starting easier. M85 (85 percent methanol, 15 percent gasoline) is produced this way. M85 became a popular alternative fuel in parts of the United States from the early 1990s onward. Volvo was among the manufacturers to offer test vehicles capable of running on M85 for evaluation by California's Air Resources Board and South Coast Air Quality Management District. The 1992 Volvo 940FFV used a modified version of the company's 2.3-liter, four-cylinder engine. When run on M85 rather than on pure gasoline, power output increased from 120 hp to 130 hp, although range was reduced from 285 miles to 200 miles. These figures are a result of the higher octane level, but lower energy density, of methanol. More recently, methanol has come to be regarded as a useful source of hydrogen for feeding fuel cells. In this application it may prove more useful than as a direct fuel for internal combustion engines.

Hydrogen appears to offer the ideal vehicle fuel solution. It is a good fuel for internal combustion engines, and its emissions are essentially water. Hydrogen can be burned in existing internal combustion engines, so a wholesale move away from existing engine technology and production facilities would not be required. BMW offers a hydrogen-powered 750 model in very small numbers, and Mazda has reported that its Wankel rotary engines are particularly suited to running on hydrogen. However, there are some problems, mainly centering on hydrogen production. Hydrogen does not occur naturally in its pure form and is usually produced from water or some hydrocarbon fuel such as methanol. This process can be quite energy intensive and therefore raises the question of what energy source to use to make hydrogen. This process can itself be polluting, and currently only around 4 percent of hydrogen is produced from sustainable energy sources such as hydro.

Hydrogen also presents storage problems. Existing storage solutions such as compressed hydrogen tanks or metal hydride are quite bulky and are unable to contain the fuel for more than a few weeks. By the late 1990s, thinking therefore moved more toward generating hydrogen onboard the vehicle from a hydrocarbon fuel such as methanol, or even gasoline. The latter method was promoted by Chrysler, among others, as it allows the retention of the existing gasoline station infrastructure. However, it does not solve the problem of our overreliance on scarce oil reserves. Where hydrogen may have a role to play is in fuel cells. This technology is expected to appear around 2015, although the hydrogen-generation problem has yet to be solved.

Many other alternatives have been tried in the search for cleaner and more sustainable automotive fuels, and for the foreseeable future, the monopoly of gasoline and diesel is likely to be further eroded around the edges by such experiments. Reliance on a wider range of fuels is probably more prudent in any case, although it will be some time before gasoline and diesel disappear from the scene.

**See Also:** Automobiles; Carbon Emissions; Consumer Behavior.

### Further Readings

Nerad, J. *Hybrid & Alternative Fuel Vehicles: The Complete Idiot's Guide*. New York: Alpha, 2007.

Sawyer, Peter. *Green Hoax Effect*. Wodonga: Group Acumen, 1990.

Zapata, C. and P. Nieuwenhuis. "Driving on Liquid Sunshine—The Brazilian Biofuel Experience: A Policy Driven Analysis." *Business Strategy and the Environment*, 17 (2008).

*Paul Nieuwenhuis*
*Cardiff University*

# Funerals

Funerals are the final acts of individual consumption. Environmental impacts include consumption of various resources (especially wood, metal, stone, concrete, and land), pollution (embalming fluids, mercury, and other heavy metals), and energy use (cremation, concrete manufacture, mowing of cemeteries, and travel to funeral and memorial services). Greener alternatives are increasingly offered, particularly by organizations outside the traditional funeral industry.

Several Websites, including Green Funeral Guide, state that U.S. funerals—more resource-intensive than those in other nations—annually use

- 30 million board feet (70,000 $m^3$) of hardwoods (caskets),
- 90,272 tons of steel (caskets),
- 14,000 tons of steel (vaults),
- 2,700 tons of copper and bronze (caskets),
- 1,636,000 tons of reinforced concrete (vaults),
- 827,060 gallons (3,130 $m^3$) of embalming fluid.

The director of the Green Burial Council (GBC) estimates that the typical 10-acre cemetery contains enough casket wood to build 40 homes and enough embalming fluid to fill a typical backyard swimming pool. He suggests that the amount of concrete and steel associated with U.S. burials each year is equivalent to that in the Golden Gate Bridge.

Caskets (coffins) are made of various materials and contain an outer shell and an interior cloth liner. Approximately 75 percent of U.S. caskets are steel, manufactured using automobile assembly line techniques and having similar environmental impacts. Burial vaults and liners are used—predominantly in the United States—to prevent crushing of the casket and subsidence. The concrete typically used is extremely energy intensive to produce. Land set aside for traditional cemeteries and the stone or metal in grave markers and monuments are additional concerns.

Pollution concerns center around the widespread use of embalming fluids, including formaldehyde and other materials designed to disinfect and sanitize the body and slow decomposition. Embalming is commonly done to allow time for friends and family to gather and to permit open-casket viewings. Formaldehyde, although naturally produced by the human body, is a widely used industrial chemical and a suspected nasal and lung carcinogen. Direct exposure can cause effects on the immune system. Risks are mainly to mortuary workers, with possible effects on soil bacteria in cemeteries from eventual leakage. Embalming fluids are released as air pollution when cremated bodies are embalmed to be preserved for a funeral service. Other sources of pollution include mercury from amalgam fillings and other heavy metals from biomedical and electronic devices, such as cell phones placed with the deceased; these pollutants are released more widely and immediately in the case of cremation. U.S. cremations are estimated to release between 1,000 and 7,500 pounds of mercury per year. Although cremation has a higher immediate carbon footprint than burials, long-term cemetery maintenance—water, pesticides, fertilizers, and mowing—and periodic graveside visits mean a larger footprint for burials.

Greener funeral options are increasing. Removal of teeth with fillings before cremation reduces mercury pollution. "Clean cremation" employs a small pine, cardboard, or wicker coffin with a natural fabric liner. Cremation remains can be kept in an urn, scattered, buried in a shallow grave, or incorporated in a permanent structure such as Eternal Reef's "reef balls"—hollow-form structures used in reef restoration programs, providing a footing for marine coral species. Reef ball locations can be documented and "visited" using global positioning system and underwater video camera technologies.

Green burials primarily target coffin materials, vaults, embalming, and cemetery land. Green consumers should opt for simple cloth burial shrouds or biodegradable pine, bamboo, willow, grass, paper, or cardboard coffins with natural fabric liners, avoiding metal or exotic woods. Certified sustainably harvested wood or nonwood Fair Trade–certified coffins are available. In the United States, the Funeral Rule prevents funeral homes charging extra for or refusing to use externally purchased caskets. Reusable coffins with an interior liner or container to prevent body fluids from leaking onto the coffin are available, but interiors may be made of synthetic materials.

Green burials do not employ vaults, allowing natural subsidence to occur and adding dirt over time to maintain a level surface, or simply allowing depressions to form and be filled in naturally. Refrigeration or dry ice can be substituted for embalming to retard decomposition.

As of July 2009, GBC has certified over 170 funeral services in 34 states. GBC also certifies four types of green cemeteries—15 in 11 states to date. Hybrid cemeteries permit avoiding embalming and burial vaults and allow use of ecofriendly caskets. Low-impact cemeteries require low-energy and nontoxic burial practices. Natural cemeteries meet low-impact standards in a naturalistic setting, employing a variety of plant and animal species. Conservation cemeteries meet natural standards and are legally designated as permanent conservation areas, often through land trusts. All but hybrid typically prohibit plastic flowers and limit the size and/or material of grave markers, using plants or engraved field stones or avoiding markers altogether. Video or global positioning system locations may be offered. In the United Kingdom, over 200 green cemeteries exist.

A final consideration concerns energy used to transport mourners to funerals, particularly at-sea burials or those in the small number of green cemeteries. Attendees can be encouraged to mitigate these impacts by purchasing carbon offsets.

**See Also:** Biodegradability; Carbon Offsets; Fair Trade; Lawns and Landscaping.

### Further Readings

Green Burial Council. http://www.greenburialcouncil.org/ (Accessed June 2009).

Harris, Mark. *Grave Matters: A Journey through the Modern Funeral Industry to a Natural Way of Burial.* New York: Scribner, 2007.

iMortuary. "Green Funeral Guide." http://www.imortuary.com/resources/green-funerals.php (Accessed June 2009).

*Gordon P. Rands*
*Pamela J. Rands*
*Western Illinois University*

## Furniture

Furniture's sustainability is difficult to gauge. First, there is no universally accepted means of calculating carbon footprints. In addition, it is estimated that manufacturers ignore

75 percent of a product's footprint by omitting supply chain emissions. Furniture is composed of various raw materials, each with its own environmental considerations. Different manufacturing venues produce unique environmental impacts. Perhaps most significantly, an increasing amount of the furniture sold in North America and Europe is now imported from Asia. "Domestic furniture manufacturers" have increasingly become retail outlets for furniture manufactured abroad. As a result of importation, suppliers and producers are less subject to stringent conservation and pollution standards, and environmental impact increases with the transportation of furniture. In addition, life cycle information for a piece of furniture becomes much less certain, or perhaps unavailable. However, options exist for the green consumer.

## Environmental Problems Associated With Furniture

Although a renewable resource, wood is often harvested at unsustainable rates. Rainforest woods are particularly affected—less than 0.1 percent of tropical forests are sustainably managed, despite strenuous efforts of groups such as the Forest Stewardship Council (FSC). Metal furniture requires the mining of nonrenewable ores. Plastics and foams used in various furniture parts are almost entirely derived from petroleum, and recycling is minimal. Textiles are made from natural and/or synthetic materials, contributing their own environmental problems.

The carbon footprint of every component in the supply chain of a piece of furniture needs to be considered. Even furniture that is advertised as sustainably manufactured, packed, stored, or shipped has a considerable footprint. Energy use has increased as advanced technology has replaced craft production, although new factories can be more energy efficient through use of technology. Some of the largest, most high-tech furniture manufacturing facilities are now in China, and Chinese furniture imports to the United States have increased by 565 percent in a decade. However, despite the possibility of greater energy efficiency, supply chains are longer and less certain, and transport-associated pollution increases.

Increased furniture shipping has resulted in dramatic increases in air and water pollution and in energy use, both at sea and in ports. Because ships are powered by "bunker fuel," emitting 5,000 times more sulfur dioxide than diesel, one ship causes more pollution than 2,000 diesel trucks. Imported "green furniture" is thus not very green. A possible partial solution was showcased in 2008 when the *MS Beluga* sailed from Germany to Venezuela, reducing its fuel consumption by 20 percent with the aid of a giant parasail.

Renewable textiles are often grown using pesticides, herbicides, and chemical fertilizers. Cotton uses approximately 25 percent of the world's insecticides and more than 10 percent of the pesticides. Glues and finishes used in furniture production have traditionally been solvent based, resulting in high volatile organic compound (VOC) emissions and outgassing. An average solvent-based finish contains approximately 6 pounds of VOCs/gallon, and 70 to 90 percent of these end up as air emissions. An additional concern is the use of polybrominated diphenyl ethers (cousins of banned polychlorinated biphenyls) as flame retardants in foam cushions and mattresses, again resulting in outgassing. Products that permit more efficient use of wood fiber such as plywood, pressboard, and medium-density fiberboard all require the extensive use of glues in their manufacture. The most common VOC in glues and finishes is formaldehyde, and the more glue used in a piece of furniture, the greater the formaldehyde outgassing. Some manufacturers have switched to water-based glues and finishes, finding that less water-based than solvent-based product is needed for the same project. Thus, flammability risk and disposal costs are reduced and employees are protected from potentially harmful chemicals.

Solid waste is generated by many sources during the life cycle of a piece of furniture. Shipping and packaging materials account for one-third of the contents of an average municipal dump. By replacing standard disposable packing material with blankets and straps, Perkins Logistics was able to pack 65 percent more furniture product in its trailers, resulting in a 20 percent reduction in carbon dioxide emissions. Disposal of broken furniture and of usable cabinetry from remodeling projects and building demolition also contributes to solid waste. Although 30 percent of metal furniture is recycled, plastic and less expensive wood composites are almost always disposed of in landfills. Solid wood and high-end wood veneer furniture can be repaired and refinished, with a carbon footprint 0.009 percent that of manufacturing new furniture. One company, The Refinishing Touch, has refinished over 1.5 million rooms of furniture for organizations ranging from hotels to the U.S. military, keeping the equivalent of approximately 52 million hardwood trees out of landfills. Furniture reuse is another significant option. The Furniture Recycling Network (U.K.) oversees a network of 400 centers for collecting and redistributing mostly household furniture. Office furniture is often replaced only because fabric and color schemes are outdated and/or it has been fully depreciated. Furniture manufacturer Steelcase and the Institution Recycling Network have joined together to find ways for organizations to recycle or reuse office furniture. Used office furniture accounts for $1.2 billion of the $13.6-billion office furniture industry and costs businesses 30 to 50 percent less. Despite this, 3 million tons of U.S. office furniture are put in landfills annually.

## Greener Alternatives

Green consumers should look for some or all of the following features in furniture:

- Easy repairability and/or ability to be refinished.
- Short supply chains, including components such as locally or regionally harvested wood and local manufacturing. Consumers should request material safety data sheets and "chain of custody" assurances.
- High postconsumer recycled content for plastic and metal furniture, or wood furniture made from reclaimed or salvaged wood. Such wood sources include trees downed by storms or construction, trees from nonproducing orchards, logs lost in transport reclaimed from lake and river bottoms (sometimes centuries ago), and wood salvaged from old buildings, bridges, ships, and so on, which can be remilled.
- Wool instead of foam cushions. Wool is renewable and naturally flame retardant.
- Low-VOC glues and finishes, which are typically water based.
- Certification. The U.S. Business and Industrial Furniture Manufacturer's Association has developed a "Level" certification system. Some companies offering Level products include Allsteel, Gunlocke, HON, Herman Miller, Kimball Office, National Office Furniture, and Steelcase. The American Home Furnishings Alliance has developed the Enhanced Furniture's Environmental Culture audit program. Wood furniture harvested using sustainable practices can be certified under the FSC's program. FSC-certified furniture can be found at IKEA and other diverse retailers such as Wal-Mart, Target, Lowe's, Costco, Home Depot, Crate and Barrel, Sam's Club, and K-Mart, among others.
- Sold, manufactured, or designed by companies listed on the Website of the Sustainable Furnishings Council, an industry-based organization that applies life cycle analysis to identifying sustainable furniture. The council works with firms in the industry to green their practices via a number of different initiatives.

Finally, green consumers can purchase used furniture from sources such as thrift stores, auctions, flea markets, or yard sales and have it repaired, reupholstered, and/or refinished. The next step is to donate unwanted furniture to thrift stores or directly to individuals using online sites such as Freecycle or Craigslist.

**See Also:** Adhesives; Certified Products; Fair Trade; Locally Made.

### Further Readings

Forest Sustainability Council. http://www.sustainablefurnishings.org/ (Accessed June 2009).

Great Lakes Regional Pollution Prevention Roundtable. http://www.glrppr.org/contacts/gltopichub.cfm?sectorid=25 (Accessed June 2009).

Heath, Oliver. *Urban Eco Chic*. London: Quadrille Publishing, 2008.

*Gordon P. Rands*
*Pamela J. Rands*
*Western Illinois University*

# GARDENING/GROWING

Gardening—the cultivation of flowers, fruits, and vegetables on enclosed pieces of ground—has a long and varied history. It ranges from being a practical activity that supplies people and households with fruits, vegetables, and herbs, to being a multifaceted and elaborate leisure pursuit characterized by ornamental and ostentatious landscape design techniques. Gardening also varies from being a personal and individual experience to being a social and community activity, although the boundaries are becoming blurred. In recent years, gardening has come to play a major role in environmental and sustainability issues.

Gardening is one of the most popular pastime activities in the United States and Europe. In 2007, according to research conducted by the National Gardening Association, a nonprofit organization that promotes gardening education, an estimated 82 million households, or 71 percent of all U.S. households, participated in at least one type of garden activity. The average spent was $428. The growth and expansion of gardening centers, which range from small businesses to big box retail stores like Home Depot, Lowe's, and Wal-Mart, as well as the proliferation of gardening clubs and associations, television and radio shows hosted by celebrity gardeners, newspaper columns, specialized magazines, seed catalogs, books, videos, Websites, and blogs devoted to gardening all attest to the popularity of gardening.

The growth in gardening as a leisure activity is bound up with changing consumption patterns and the commoditization of nature by the constantly expanding lawn and garden industry. Although gardens, garden design, and gardening have become ubiquitous lifestyle matters that have been branded as an aspect of modern consumer culture, there is an ongoing tension between gardening as a leisurely pursuit in which individuals seek personal fulfillment through acts of privatized green consumption and gardening as a social movement based on creating a sense of community and a new relationship between people and the natural environment.

## Gardening for Food

Over the last decade, concerns over rising food costs, oil dependency, climate change, and lack of access to locally grown, healthy food have led to a surge in community and home-based food gardening. Gardening for food is nothing new. During World War II, 20 million victory gardens contributed to the war effort by growing approximately 40 percent of the food produced in the United States. According to research conducted by the National Gardening Association in 2009, 31 percent of all U.S. households, or an estimated 36 million households, participated in food gardening, spending an annual average of $70 and a total of $2.5 billion in 2008, making it the number one gardening activity. Similar developments have taken place in Europe, especially in the United Kingdom, with the revitalization of allotment gardens—small plots of land provided to people at nominal rents by local government agencies. Although allotment gardening has a long history, it dropped out of fashion until the 1970s, when environmental and sustainability concerns made them popular once again.

Community gardeners, both those who grow in individual plots and those who cultivate as a collective, have long been involved with growing food. Several community gardens began as victory gardens during World War II and evolved into their contemporary form. Many of the approximately 18,000 community gardens in the United States and Canada have been at the forefront of the food justice movement, which is concerned with the unequal distribution of fresh nutritious food and its deleterious effect on the health of poor and minority communities. Community gardeners have become heavily involved in educational campaigns that teach people about health and sustainability issues surrounding food production and consumption practices. They view growing food as a practical solution to heath, safety, and economic and sustainability issues and feel that good nutrition and health should be affordable and available to all, and not an elitist affair.

A small but growing "de-lawning" movement, whose aim is to transform environmentally wasteful and monocultured suburban lawns into ecologically friendly food gardens, began to emerge in the 1990s. Since 2007, several suburban tract developers in the United States have begun to include some type of community garden or working farm in their planned subdivisions that serve as amenities that provide food, open space, and environmental benefits for local homeowners.

Throughout the world, approximately 800 million city dwellers grow their own food, and 200 million produce food for sale at local markets. Historically, U.S. "kitchen gardening" functioned as both a leisure activity that supplied food and as a source of extra income. Many community gardeners sell a portion of their produce at local farmers markets and/or community-supported agricultural ventures. Since the late 1990s, there has been a steady growth in intensive, high-yield, commercial gardening throughout the United States, which in turn has generated an interest in garden sharing, whereby people who have horticultural knowledge and skills but lack land work on a share basis for those who have land but have no interest in cultivating it.

## Sustainable Gardening

Environmental and sustainability concerns have changed the way people think about and work their gardens. According a 2008 National Gardening Association Environmental Lawn and Garden Survey, Americans have become increasingly concerned with maintaining lawns and gardens in an environmentally friendly way. The number of households using only all-natural fertilizers, insect, and weed controls increased to 12 million in 2008, roughly double the number doing so in 2004. In 2008, people spent approximately $460 million in retail sales for organic lawn and garden products. Similar trends are occurring in Europe.

Gardens are increasingly coming to be viewed as habitats that provide vital ecological functions. Along with producing desired crops, they provide shelter for beneficial birds and insects and help to conserve soil, water, and energy. Although sustainable gardeners use fewer or no pesticides and engage in activities such as waste recycling through composting, water conservation, and rainwater harvesting, they are not preservationists engaged in an elegiac exercise. Sustainable gardening based on scientific ecological principles is a very efficient, low-impact form of land use that produces high yields while protecting the environment

Gardening is one of the major ways that people connect with nature in the modern world. Gardens are sites where people develop an awareness of natural processes and ecological connections. They also illustrate the hybrid nature of that connection. Although the garden is a widespread symbol for nature, the ultimate referent being the Garden of Eden, there is nothing natural about gardens: They are sites where nature and culture meet. Gardens are products of human cultivation, the process by which raw and wild nature is manipulated and transformed into a cultural form. Amid rising concerns over global climate change and environmental decline, gardeners are helping to forge a new environmental stewardship ethic and changing the way people think and feel about their place in nature.

**See Also:** Consumer Culture; Lawns and Landscaping; Leisure and Recreation; Pesticides and Fertilizers; Plants; Vegetables and Fruits.

### Further Readings

Crouch, David and C. Ward. *The Allotment, Its Landscape and Culture*. Nottingham, UK: Five Leaves, 1997.

Haeg, Fritz. *Edible Estates: Attack on the Front Lawn*. New York: Metropolis Books, 2008.

Millsone, E and T. Lang. *The Atlas of Food: Who Eats What, Where and Why?* London: Earthscan, 2003.

National Gardening Association. "The Impact of Home and Community Gardening in America." http://www.gardenresearch.com/home?q=show&id=3126 (Accessed August 2009).

Saguaro, Shelly. *Garden Plots: The Politics and Poetics of Gardens*. London: Ashgate, 2006.

*Steven Lang*
*LaGuardia Community College*

## Garden Tools and Appliances

For the purposes of this article, garden tools are defined as those implements roughly equivalent to unpowered hand tools that are used to materially work the garden for functional or cosmetic purposes—spades, hoes, saws, and manual lawn mowers, for example. Appliances include both powered versions of tools and those powered devices that are employed in the garden but that are not used to materially work it—for example, gas-driven lawnmowers and patio heaters. Clearly, there will be some overlap between these categories.

The U.S. Environmental Protection Agency estimates that American homeowners annually spill 17 million gallons of gasoline while attempting to fill the tanks of power lawn tools and appliances.

*Source:* iStockphoto

One marked characteristic of garden tools is the historical longevity of their form, despite much historical variation in the purpose, form, and apprehension of gardens. For example, one 12th-century treatise on the subject advocated the acquisition of an essential garden toolkit that would include knives, a shovel, a billhook, and a wheelbarrow. The recognizable antecedents of many such forms are traceable back into human prehistory and across cultures. Appliances, however, are generally historically recent inventions. The ability to power traditional hand tools has only come about via the development of portable power: batteries, electric cables, small gasoline engines, gas canisters, and so on. The widespread consumption of devices that are not used to work the garden but have become clearly identifiable garden appliances is an even more recent innovation, and one that has gathered pace over the last 20 years or so.

As with household tools and devices, the proliferation of garden tools into multiple lines of increasingly differentiated forms from competing manufacturers gathered pace during the 19th century, a consequence of the technological possibilities of industrialization, the super-productivity of reorganized labor, and the economic imperatives of capitalist political economy. Despite this expansion, there were relatively few genuine innovations, with the appearance in the mid-1800s of the manual lawn mower being a notable exception. Throughout the 20th century, proliferation and differentiation of recognizable tools into product lines was intensified further. This included both the addition of power—the gas-driven lawn mower, and later the electric lawn mower—and the development of alternative tools to achieve recognizable ends; for example, the widespread availability of the hover lawn mower in the 1960s, whereby a cylinder of blades rotating around a horizontal axis was replaced by a flat, circular blade revolving around a vertical axis.

As such, garden tools have been prey to the same ecological issues as tools in general. First, we can identify several forms of built-in obsolescence. Technological obsolescence is discernible in relation to cheap or value ranges of garden tools—their relatively low price is a direct consequence of decisions to use less durable, poorer-quality materials, inferior design, and quite often weak points, especially in the joining of components or materials. Moreover, garden tools and appliances have become increasingly styled, often with function being compromised by superficial stylistic elements. Even where functional efficacy is not compromised, highly styled garden tools are deemed by various promotional industries to have worn-out aesthetically and to be in need of replacement by functionally equivalent but new and superficially differing alternatives. Second, the overelaboration of traditionally simple garden tools is apparent; for example, complex and functionally vulnerable garden twine dispensers. Invitations to replace simple hand tools with powered alternatives are also commonplace; for example, the electric hedge trimmer for the hand shears, the leaf blower for the rake. The assumed and unquestioned universal benefits of technology have played a key role in the promotion of such alternatives.

More general concerns are also apparent. The manufacturers and promoters of much contemporary garden paraphernalia rarely either advocate the vernacular adaptation of tools already owned and in use or offer solutions to garden problems that do not rely on the private purchase of new products. Repeat consumption is a necessary condition of an industry worth over 330 million euros annually in the United Kingdom alone. The consequences may be environmentally, and socially, disadvantageous. First, ownership of many duplicated or functionally very similar tools occurs, which in itself often requires further consumption of, for example, garden sheds for storage. Second, and consequently, enormous numbers of garden tools remain unused all year round. Third, the private mass ownership of garden tools leads to situations in which they spend much of their time in an irrationally redundant state. For example, the most expensive, technologically advanced lawn mower will only be used a small number of times a year, maybe once a week for 30 weeks. Communal ownership or simple lending between neighbors makes far more ecological and social sense.

One of the most ecologically damaging recent developments has been the increasing deployment of garden appliances that are intended to enhance the garden environment. The widespread rhetoric of the last 40 years of the garden as an outside room has been pushed to extreme limits: the routine installation of elaborate outside lighting and entertainment systems, ever-more-power-hungry barbecues, water-feature pumps, patio heaters, and even outdoor air-conditioning. Technologically transforming gardens into active leisure and entertainment spaces that efface natural daily and seasonal conditions is hardly an ecologically sustainable accommodation to nature.

**See Also:** Consumerism; Durability; Green Design; Overconsumption; Tools.

### Further Readings

Illich, Ivan. *Tools for Conviviality*. London: Marion Boyars, 1973.

Lamp'l, Joe. *The Green Gardener's Guide: Simple, Significant Actions to Protect & Preserve Our Planet*. Franklin, TN: Cool Springs Press, 2008.

Papanek, Victor. *Design for the Real World*. London: Thames and Hudson, 1984.

Sanecki, Kay N. *Old Garden Tools*. Oxford: Shire, 1987.

*Neil Maycroft*
*University of Lincoln*

## Genetically Modified Products

The development of the biological science of genetics began in the mid-1800s, when Gregor Mendel (1822–84) studied the characteristics of plants, and peas in particular, over several generations. His studies were followed by the development of hybrids by

Luther Burbank (1849–1926) through selective breeding. Between 1945 and 1965, the green revolution in agriculture greatly increased the world's food supply through the use of selective breeding of hybrid seeds, along with large quantities of pesticides and fertilizers.

The great advance in genetics came in 1953, with the discovery of the DNA molecule by James D. Crick (1928– ) and Francis Watson (1916–2004). The secret of life is in the DNA, and especially in the genes on the chromosomes. Genes are attached to the DNA double helix. Genes are codes that process the information to manufacture proteins. To synthesize a protein, a complementary RNA molecule is produced. Its sequence orders a specific amino acid sequence of the protein.

As understanding of the DNA molecule advanced, it became apparent that genes contain genetic information that directs the biochemistry of the human body and all other life. In the case of the human genome, there are 23,000 genes on the chromosomes of the human genetic code. Between 1990 and 2003, the human genome was mapped. Since then, many other life-forms—both animal and plant—have been mapped. This information has been used to create genetically modified (GM) foods or breeds of animals.

Knowledge of the genes in food products allows breeders and seed developers to manipulate the genes in a great many ways. In some cases, genes have been removed. In other experiments, genes have been spliced to form new genetic combinations. The ability to manipulate food products or flowers or other biological forms has given scientists powerful tools for creating insect- and disease-resistant strains of plants.

In the 1990s, laboratories began developing GM organisms (GMOs). Some of the GMOs were agricultural seeds that were produced by manipulating genetic material in a variety of seeds. It was discovered that genetic material could be rearranged, and it was also discovered that genes could be eliminated or replaced by genes from other plants or animals. The development of GM crops allowed strains of seeds or livestock to be bred that grew faster and larger and were resistant to insects and disease.

Between the 1990s and 2009, the acreage devoted to GM crops has increased greatly around the world. Over half of such production is in North America. The GM seeds have greatly increased food supplies; however, there have been numerous concerns raised by scientists over this use of GM products. The concerns are centered on the unknown effects of consuming GM foods. Questions were raised concerning the fact that GM foods experience changes in genes. Can these gene changes affect human biology? Can they lead to human genetic changes that would be undesirable? Can the changes in GM products induce cancer? Do they play a role in the great rise of obesity?

## Opposition to GM Foods

In Europe, opposition to GM foods grew rapidly. The anti-GM movement was not able to marshal a strong opposition in North America. Despite opposition, many agricultural and scientific organizations have sanctioned the use of GM seeds. Research is currently focused on putting drugs into foods that will be consumed in tropical areas to combat diseases and also offer other benefits. Testing the safety of GMs is also an ongoing project.

The biotechnology industry has developed from modern genetic research. Plants and animals have been the subject of extensive research, and now development from genetic engineering. A major goal is the production of biologically useful proteins. Recombinant DNA is used to produce natural proteins. In addition, cloned genes are used to produce proteins with modified amino acid sequences. Transgenic products are created by altering the genomes of plants and animals. Cloning may be the ultimate act of genetic development.

GM foods have been growing in number since their introduction in the 1990s. Genetic engineering changes the DNA of specific foods in ways that are more advanced than the long tradition of scientific breeding. The first GM foods on the market were commodities such as canola, corn, soybean, and cotton seeds, which are processed for vegetable oil. Other commodities that were modified included potatoes, rice, sugarcane, sweet corn, field corns, and other foods. These products are usually transgenic plants that possess at least one gene that has been transferred into it from different species. Transgenic seeds and plants are created in laboratories using recombinant DNA technology, in which plants gain different characteristics by the artificial insertion of a gene from one species into another.

Genetic engineering may be seeking more than just food. For example, pigs were modified to pass manure that was much lower in phosphorus because the modified gene they inherited increased their phosphorus dietary absorption. When genetic engineers identify a gene that has a useful trait, it is then inserted into a plant or an animal. The gene becomes part of the genome of the recipient plant or animal, which is then grown or bred, and the effect of the new gene is examined.

A tomato variety, Flavr Savr, was modified in California in 1994. It was modified to ripen and not turn soft. However, it was not competitive with other varieties, so this GM product was rendered into tomato paste. It was sold in Europe before the opposition to GM foods gained strength.

A GM variety of cotton has been introduced into India, and farmers are also having success cross-breeding it with local varieties. Research in this area of agriculture is continuing as cottonseed oil is important to the Indian economy and diet.

Some researchers have claimed that GM products do not increase yields. This has yet to be substantiated. If this were to prove true, then it is likely that genetic engineering would be curtailed.

## Labeling GM Foods

Labeling foods as GM is not mandatory in North America; however, it is in the European Union, Australia, Japan, and other countries. The goal of labeling is to give consumers a choice between GM products and non-GM products. The Organisation for Economic Co-operation and Development is an international organization that currently has 30 member countries, most of which are wealthier countries. The organization has developed a system for tracing GM products. When the product is first developed by bioengineers, it is given its own discrete identifier. The Organisation for Economic Co-operation and Development then approves the identifier, which must be included at every step of the production process. Critics claim that trying to trace the modified product through all stages of the production process will be difficult, if not impossible.

Some GM products have not been successful. For example, in the 1990s, Brazil nuts were modified to see whether they could develop a nut that did not cause allergic reactions in people sensitive to Brazil nuts. However, the modification did not work—people who were allergic to Brazil nuts were also allergic to the new GM product.

Opponents of GM products have argued that these products are not necessary to increase the world's food supply. They have argued that hunger problems have much more to do with food distribution problems and politics than with production. However, this claim from earlier decades has been challenged by international changes.

Other critics challenge the safety of GM products. Claims of digestive changes have been shown in laboratory rats. However, some of these studies have been rejected by proponents of GM products on the grounds that the research was flawed.

Critics have argued that GM products are a threat to nature, which they fear will change if GM products escape into the wild. Another issue is the question of intellectual property. There have been cases in which farmers growing corn next to an experimental GM plot have been sued because the wind blew pollen from the experimental plot to the farmers' fields. The claim is that the pollination by nature has violated their intellectual property rights in the GM corn experiments.

Given the vast amount of GM production in North America, green consumers probably do not have much choice unless they turn to organic farmers. Organic farms have to be certified and do not use chemicals to fertilize or as insecticides. Because some pesticides are organic, this is an unusual requirement. Many organic farmers use heirloom seeds, which are seeds of old varieties of plants that are in danger of disappearing because they are no longer planted by the agricultural industry.

Green consumers can purchase organic foods and other supplies from a variety of suppliers. Products include non-GM seeds as well as well as meats, breads, beans, vegetables, wines, and other foods.

**See Also:** Consumer Boycotts; Final Consumption; Green Consumer; Green Food; Organic; Organisation for Economic Co-operation and Development (OECD); Pesticides and Fertilizers; Vegetables and Fruits.

### Further Readings

Avise, John C. *Hope, Hype, and Reality of Genetic Engineering: Remarkable Stories From Agriculture, Industry, Medicine, and the Environment*. New York: Oxford University Press, 2004.

Bernauer, Thomas. *Genes, Trade, and Regulation: The Seeds of Conflict in Food Biotechnology*. Princeton, NJ: Princeton University Press, 2003.

Conford, Phillip. *The Origins of the Organic Movement*. Edinburgh, UK: Floris Books, 2001.

McHughen, Alan. *Consumer's Guide to GM Food: From Green Genes to Red Herrings*. Oxford: Oxford University Press, 2000.

Murray, David R. *Seeds of Concern: The Genetic Manipulation of Plants*. New York: CABI, 2003.

Noussair, Charles, Stephanie Robin and Bernard Ruffieux. "Do Consumers Really Refuse to Buy Genetically Modified Food?" *Economic Journal*, 114:102–20 (January 2004.)

Pringle, Peter. *Food, Inc.: Mendel to Monsanto—The Promises and Perils of the Biotech Harvest*. New York: Simon & Schuster, 2005.

Thomson, Jennifer A. *Seeds of the Future: The Impact of Genetically Modified Crops on the Environment*. Ithaca, NY: Comstock, 2007.

Weirich, Paul, ed. *Labeling Genetically Modified Food: The Philosophical and Legal Debate*. Oxford: Oxford University Press, 2007.

Wu, Felicia and William Butz. *The Future of Genetically Modified Crops: Lessons From the Green Revolution*. Arlington, VA: RAND Corporation, 2004.

*Andrew Jackson Waskey*
*Dalton State College*

# Gifting (Green Gifts)

Recently, the term *green gifting* has entered into mainstream sustainable parlance, though, arguably, the practices encompassed by the phrase have existed for millennia. Today, green gifting straddles a paradox within sustainability movements: it has risen to prominence both as a cutting-edge capitalist marketing tool and as the foundation for radical barter-based anarchism.

What unites most references and instances of green gifting is an abiding interest in replacing excessively wasteful conventions of buying presents with more ecologically sustainable alternatives. Green gifting has emerged as a response to the avid consumerism and commercialism that has come to be negatively associated with gift-oriented holidays, as well as birthdays, weddings, and baby showers. In particular, the Christmas season has become saturated with consumption and its expanding expenses and refuse. According to the Environmental Protection Agency, U.S. household trash increases by 25 percent between Thanksgiving (and the notoriously consumptive Black Friday) and New Year's Day. Accordingly, the proposed simplicity of green gifting has captured the imagination and interest of increasing numbers of would-be shoppers and garbage tossers.

The central premise of green gifting is the shift in what constitutes an acceptable and desirable present. Previous de rigueur elements of fine gifts are suddenly distasteful within the new green aesthetic: the shiny wrapping paper and big bow become as uncomely as a Hummer. Green givers oppose excessive packaging, both of the item being given and within the process of giving it. Eschewing new rolls of wrapping paper, green givers encase presents in newspapers, calendars, magazines, or reusable canvas gift bags.

Packaging is also avoided through buying secondhand from thrift and consignment stores or yard and garage sales. Regifting has gained new, green legitimacy—rewrapping and recirculating an unwanted present precludes the ecological impact both of buying something new and of throwing something away. Once a private decision, now online, postholiday barter sites flourish, with unsatisfied recipients swapping newly acquired, undesired merchandise.

Even greener gifts include handmade items. Canned foods and homemade breads constitute the edible and agrarian examples of exemplary—even ancient—green gifting. Meanwhile, quilted, knitted, sewn, crocheted, or hooked handmade textiles have enjoyed a revival as especially green alternatives to store-bought or "brought-on" products, whose production, processing, packaging, distribution, branding, and advertisement bear deleterious ecological and social consequences.

Green gifting has also come to entail gifts pertaining directly to the outdoor world of ecological services: givers can offer the gift of a few trees, lily bulbs, a window box herb garden, heirloom seeds, rain barrels, or even the associated labor itself (of planting trees or applying compost). Often, green gifts comprise services rather than objects, such as baby showers wherein guests offer postnatal household or babysitting help.

Things local and seasonal constitute green intentions in current parlance, but sustainably minded givers embrace the global as well through Fair Trade products, whose finer quality and higher prices often lend themselves particularly to gifts.

This costliness characterizes the more elite end of the green gift spectrum, in which green retailing is a lucrative marketing strategy. Here, green givers do not buy less, they just buy more organic, low-impact, biodegradable objects: high-end ecobeauty products, high-tech solar chargers, and low-flow, chlorine-removing showerheads. Though still grounded in new purchases, such green gifts do improve on conventional counterparts. The long-term worth of recyclable batteries as stocking stuffers outweighs their plastic packaging, as 40 percent of all batteries (which leach toxic metals into the ground at landfills) are bought during the Christmas season. Green jewelry gifts maintain the "bling" while shifting from ecologically and socially suspect mining practices and conflict diamonds to recycled metals and gems. Meanwhile, the World Wide Web abounds with ecological services gift certificates, wherein donors "adopt" an endangered species or save an acre of rainforest in the recipient's name.

On the other end of the spectrum, however, green gifting has come to herald a more radical restructuring of the capital-based economy at large. Recessions and depressions are catalysts for barter systems. In the 1930s, with money scarce, elaborate exchange networks and co-ops developed. As contemporary credit flows slow, tighten, and even cease, mutual trade systems again emerge to help make ends meet—serving necessity as well as ecological and social justice consciousness.

Community bartering networks and time-banks are flourishing in the United States and around the world. The Internet has facilitated impressively intricate and adaptive exchanges of a host of goods and services including trading yard work for car repair, household cleaning for haircuts, Website design for used textbooks, and guitar lessons for language lessons. Reciprocity is ensured through a per hour credit system—a profoundly egalitarian premise that all labor merits the same unwaged hourly wage. Modern-day green-gift economies encompass traditional food co-ops, in which participants purchase in bulk; vacation home swaps; cohousing communities; dinner co-ops; child care co-ops; couch "surfing"; exchange of farming labor with portions of harvest; heirloom seed swaps; vehicle trades; and bike shares. Meanwhile, waves of even more radical "really" free markets have arisen in cities and online, dedicated solely to simply giving away surplus goods and needed services.

Here, green gifting partakes of more ancient gift economies in general, affirming the goal of mutual benefit, but also of fundamental interdependence. Reciprocity remains a central social, ecological, and metaphysical premise for numerous indigenous and non-industrialized communities around the world, serving as altruism and informal insurance. Gift-based economies and now, to various extents, green gifting, inextricably re-embed the economy within social and ecological relations and consequences.

**See Also:** Local Exchange Trading Schemes; Secondhand Consumption.

## Further Readings

Environmental Defense Fund, "Green Gifts for the Holidays." http://www.edf.org/article.cfm?contentID=5616 (Accessed November 2009).

Hyde, Lewis. *The Gift*. New York: Random House, 2007 [1983].

Vaughan, Genevieve. *For-Giving: A Feminist Criticism of Exchange*. Austin, TX: Plain View, 1997.

*Garrett Graddy*
*University of Kentucky*

# Government Policy and Practice (Local and National)

Government action has proven necessary to the green revolution for the simple reason that market forces are not enough to effect change. For instance, it is cheaper to wash your car at home than to bring it to a professional car wash that will recycle the water and treat it to prevent the release of harmful chemicals into the soil and surface water. Similarly, it was cheaper to buy leaded gasoline than unleaded, when leaded was available. The behavior of the market would not change that: government intervention was necessary to gradually phase out leaded gasoline. Phasing out home car washing is less simple, but it has been done in areas with water shortages, and in many U.S. states, the U.S. Environmental Protection Agency (EPA) actively promotes the use of professional car washes by making the public aware of the environmental concerns of washing one's car at home.

Key to the insufficiency of market forces are externalities. An externality of an economic action is any effect on parties not involved in that action. These can be positive, negative, or a mixture. Environmental examples are textbook cases of negative externalities: the water pollution caused by industrial waste, the air pollution caused by a factory's operations, and so on. A specific type of negative externality is "the tragedy of the commons," a nearly poetic term coined in Garrett Hardin's 1968 *Science* article. In the tragedy of the commons, a common resource (such as natural resources) is damaged or depleted by multiple individuals with access to and use for that resource, acting rationally in their own self-interest, even though such depletion is clearly against their long-term interest. Overfishing is a common case of this externality: even when the stock of fish dwindles, rationally acting fishermen will continue to fish unless prevented by government or industry intervention—they need to do so to compete with one another. Negative externalities mean that the cost incurred in the transaction is not the full cost of the transaction—what remains is the external cost, such as the cost of cleaning up pollution or the combined cost of settling a class-action lawsuit and the unquantifiable losses of life and health as the result of contaminated natural resources. Precisely these externalities are invoked in arguments against nuclear power, which because of Three Mile Island and Chernobyl is vividly associated with environmental disaster in the imagination of Americans older than 30 years, despite nuclear power's general safety.

Positive externalities, furthermore, are an argument in favor of spending more on green consumption. The EPA lists "payback periods" for various green products on its Website: a water-saving, energy-efficient washing machine may take three years, for instance, to earn back its cost margin above the cost of a standard washing machine, in the form of lower water and electric bills. There are external benefits beyond this, however, in the form of the benefit to the environment.

Remedies to negative externalities include criminalization of the conditions that bring them about; government funding to prevent or make up for them (agricultural subsidies essentially combat the negative externality of agricultural productivity resulting in prices too low to keep farmers in business); Pigovian taxes, which are taxes imposed to cancel out the cost of the negative externality; and redresses available through tort law (such as class action lawsuits against polluters). All of these are used to varying extents as part of U.S. federal and state environmental policy, although Pigovian taxes (named for

Arthur Pigou, an economist noted for his work on externalities) are more of a European, especially Scandinavian, strategy. The American model tends to turn the Pigovian tax concept on its head: Rather than taxing polluters, companies can essentially purchase the ability to pollute. The effect is the same, and the amount is the same, but the American model leads to situations such as the emissions trading market, which Americans and industrialists are more comfortable with.

Government involvement is not necessary in every case of negative externalities. When the externality affects only one or two parties outside the transaction, they can be negotiated with by the transactors. The greater the number of parties affected by the externalities, or the lesser their negotiating power, the more likely it is that government involvement is the only source of resolution, and it is for these reasons that so many examples of negative externalities in the U.S. economy are environmental in nature.

The environmental policies of state, at various levels of government, encompass issues related to the protection of biodiversity, wildlife, and endangered species; forest management and conservation; natural resources; pollution, including air, water, and soil pollution; waste management; energy efficiency and clean energy use; and human actions that adversely affect the environment, particularly if there are human consequences (such as the contamination of drinking water or the release of carcinogens into the atmosphere or soil). The next section examines how these aspects emerged in American environmental policy.

## U.S. Environmental Policy

U.S. environmental policy, though pieces of it originated earlier (particularly pesticide regulation and the creation of the National Park Service in specific, and antitrust and consumer protection legislation in similar philosophies of governance), essentially grew out of the environmentalist movement of the 1960s and 1970s and the growing public awareness of environmental hazards. Natural history writer Rachel Carson's 1962 book *Silent Spring*, a *New York Times* bestseller and Book-of-the-Month Club selection, focused public attention on the effect of the chemicals industry on the health and welfare of birds, wildlife, and humans. Carson was criticized by the chemicals industry and dismissed as a "hysterical and unqualified woman," despite her master's degree from Johns Hopkins University and her 30 years of experience as a scientist and science writer. The critics picked a poor target, however, and public sentiment sided with Carson—and with her cause. For the next eight years in particular, environmentalism was often seen as subversive and radical, as it opposed the titans of industry—the Goliaths of American progress. As a movement, it attracted members from social movements of the 1960s—feminists and social justice activists who saw the disadvantaged state of victims of environmental hazards; radical leftists who opposed environmental abuses as simply one of a long list of crimes of capitalism; vegetarians; and members of New Age or Earth-based religions who had personal, ethical, or spiritual reasons for preventing harm to nature.

In 1970, however, Republican U.S. President Richard Nixon—by his own assessment, no friend to radicals—created the EPA, an action that could be seen as reflecting the essential and rational nonpartisan nature of environmentalism, as legitimizing the environmentalist movement, or as co-opting it. The EPA has remained the principal federal agency responsible for environmental policy, though its work overlaps with that of the Department of Energy and several other federal agencies, and it delegates some of its power to the states.

Even as the harmful Agent Orange was sprayed across war-torn Vietnam, the 1970s became the environmental decade—a natural sequel to Lyndon Johnson's Great Society programs and, as Vietnam wound down into an eventual American loss and the Cold War cooled off into detente, a concern that most Americans could get behind. Environmental legislation in the 1970s was the first such U.S. legislation to set mandatory standards for environmental cleanliness that existing technology could not meet, requiring the relevant industries to develop new technology that could meet the new standards. Had industry not rattled its saber against its critics so much in the past, such faith might not have been placed in its capacities.

Smog, particulate air pollution, polluted swimming holes and drinking water, and acid rain were particular issues for the average American because they affected everyday life. Public health outcries over first dichlorodiphenyltrichloroethane (DDT) and, later, asbestos insulation, lead, and mercury led to strengthened pesticide controls and consumer protection legislation. The broader mission of the EPA, however, and of the environmentalist movement—to preserve the health of the ecosystem, not just defend against the short-term effect on human welfare—led to characterizations of environmentalists as "tree huggers," as radicals who would blockade loggers to save a rare owl, and as impractical people with no sense of perspective or scale. The characterization was unfair, but in politics, cherry-picking and guilt by association make it easy to make one's opponent seem irrational, inconsistent, or ridiculous.

By the 1980s, the association of environmentalism with radicalism and impracticality was strong enough that President Ronald Reagan was able to openly oppose the movement and remain one of the most popular presidents in history. In the interest of business, Reagan reduced the EPA's budget, downsized the agency's staff, appointed conservatives to various positions, and reinterpreted environmental laws to be more favorable to business interests. This continued to some degree under President George H. W. Bush, whose administration redefined "wetlands" in a way that protected far less land from industrial development and pollution, contributing—among many other factors—to the damage done by Hurricanes Katrina and Rita when they struck the vastly receded wetlands of the Gulf Coast in 2005.

The Clinton administration brought a staunchly environmentalist vice president but also a president forced into compromise after compromise and determined to show he was a moderate willing to work with the conservatives, not a traditional liberal. The Green Party's criticism of Bill Clinton throughout his two terms in office led to their nomination of Ralph Nader and to his surprising and controversial showing in the 2000 election. Nevertheless, Clinton did stand up to the attempts of Newt Gingrich's House Republicans to scale back environmental legislation, and he supported and signed the Kyoto Protocol (a climate change environmental treaty), though he did not submit it to the Senate, knowing it would not be ratified.

As expected, the George W. Bush administration set the Kyoto Protocol aside. By the dawn of the 21st century, conservatives in the main did not deny the need to reduce greenhouse gas emissions but sought to limit the effect of these reductions on U.S. industry and argued for smaller or slower reductions, or different definitions of which gases would be included. Clean coal technology was funded, whereas carbon dioxide emissions were dismissed as a subject of concern because they were not named as a pollutant under the Clean Air Act, and limiting them would affect energy prices. Forest protection was decreased in favor of more development, and Alaska's North Slope was opened to oil and gas development, despite reports from the U.S. Geological Survey that less than a third of the oil there was "economically recoverable" (meaning for the other two-thirds, the cost of getting the oil out would exceed the money that could be made from it), yet tax credits were offered to businesses using renewable energy. Environmentally, it was an administration of mixed messages and few clear goals.

## Critical Areas of Concern

Acid rain was one of the earliest environmental policy concerns, if we discount the pesticide legislation that predates the EPA. Sulfur and nitrogen dioxide released into the air from factory and vehicle emissions affect the acidity of rain, causing damage not only to the ecosystem but also to property—a problem publicized by the reconstruction of the Statue of Liberty. Though the Reagan administration played up the idea of scientific uncertainty about causation, and delayed doing much about the problem, George H. W. Bush's administration specifically targeted acid rain as a problem to fix. It has been pointed out by many environmentalists since then that the original estimated cost to reduce emissions to the levels called for by Bush's 1990 Clean Air Act was $4.6 billion, plus a 40 percent spike in the cost of electricity. Almost 20 years later, we know that the real cost was only $1 billion, with a bump in electricity cost of about 3 percent.

The handling and disposal of hazardous waste has been regulated at the federal level since 1976, and many states and municipalities have additional legislation. The gist of the legislation is that the producer of the waste is responsible for seeing that it is dealt with in such a way that its environmental impact is minimized, and cleaning up spills and other accidents is the responsibility of those producers. Though there has been less disagreement over this area of legislation than many others under the environmentalism umbrella, implementation and enforcement have repeatedly proven difficult, and it is in some senses easier for companies to illegally dispose of their waste than it is for them to bypass emissions standards for factories or vehicles. As with many other sorts of consumer protection legislation, there are generally whistleblower protections in place, encouraging employees to report their employers for illegal behavior.

The depletion of the ozone layer began in the 1930s with the use of chlorofluorocarbons (CFCs) but was not recognized for another 40 years. This is one reason greens call for "anticipatory democracy"—making decisions now with an eye to what might happen, and what we might know, many years from now. One of the areas of disagreement in environmental legislation, varying from administration to administration and among officials of those administrations, is the attitude to take toward risk. Must we prove something does no harm to approve it, or must it be proven to be harmful to restrict it? If we find no experimental evidence of harm, do we give the substance or practice the benefit of the doubt, do we look again, or do we restrict it anyway just to be safe? It took us decades to come to the understanding we have of the environmental consequences of our technologies and our chemicals; we can assume that understanding is not complete, but no matter how we decide, it is that incomplete understanding we must act on. The proper response to ozone layer depletion is still debated, though production of CFCs has been phased out. However, because they take so long to dissipate from the atmosphere, promising signs of recovery are expected to be seen in the 2020s.

## Environmental Justice

Modern-day environmental concerns have been championed primarily by the greens and the Left, and so it makes sense that they are often coupled with a concern for the disadvantaged—the link is natural. When the wealthy become wealthier from a coal mine, it is the working class who work the mine and face its hazards; when environmental damage is done to a neighborhood, it is the working class who cannot afford to move away from it and are forced to live in someone else's mess. The consequences of environmental harm are real, but they are often inequitably distributed. The EPA defines *environmental justice* as

> The fair treatment and meaningful involvement of all people regardless of race, color, national origin, or income with respect to the development, implementation, and enforcement of environmental laws, regulations, and policies. EPA has this goal for all communities and persons across this Nation. It will be achieved when everyone enjoys the same degree of protection from environmental and health hazards and equal access to the decision-making process to have a healthy environment in which to live, learn, and work.

In the 21st century, it is generally agreed that minority communities are overburdened with environmental risks and enjoy fewer benefits of greening. It is sometimes further argued that the environmental advances of the last 30 years have, in the short term, primarily benefited the middle class. Air, water, and soil pollution still persist, after all. That they are less common means the middle class experiences them less. That polluted conditions will very often be found in poorer neighborhoods—where authorities do not rally to clean up the local body of water that children use to swim in, or where drinking water toxicity is not responded to as quickly as it would be if someone in the neighborhood had a friend in the governor's office, and where people have reduced means by which to seek redress—is not necessarily deliberate and can in fact be an indirect consequence, as property values of neighborhoods are driven down as a result of polluted living conditions, and only the poor remain in residence.

For instance, situated between New Orleans and Baton Rouge is 85 miles of the Mississippi River, along which 125 companies have factories producing 25 percent of the petrochemical products made in the United States. The predominantly black population of "Cancer Alley" has been disproportionately affected as a result. Similarly, on Native American reservations, indigenous people have historically been subject to the effects of uranium mining; until 1980, the two largest mining companies declared outright that they did not consider federal pollution laws to apply to reservation lands and so took no actions to prevent contaminating the water supply with uranium. Farm workers are also often exposed to harmful substances in the course of their work—not only toxic pesticides but also animal waste. Workers at poultry factory farms have been identified as being susceptible to a slew of infections and lung disease as a result of the bacteria- and virus-breeding conditions of their work environment, which also contribute to water and soil pollution.

The charge of "environmental racism" is sometimes made when we consider the victims of these disparities in terms of their race; further, nonwhites have generally poor representation in environmental associations.

Legislation that seeks to remedy environmental injustice falls under the umbrella of *environmental compensation:* resource management that remedies environmental harm.

On the international stage, the victims of environmental injustice are not simply the working class of the United States but also the developing nations of the world, which suffer the consequences of the industry of the global North but do not have the means to combat them. At the same time, environmentalists of the developed world have been accused of eco-imperialism—imposing the values of comparatively prosperous environmentalists on a developing world ill-equipped to uphold those values—by commentators like American lobbyist Paul Driessen, who coined the term.

**See Also:** Carbon Credits; Carbon Emissions; Certification Process; Ecolabeling; Electricity Usage; E-Waste; Food Additives; Fuel; Green Politics; International Regulatory Frameworks; Regulation.

### Further Readings

Carter, Alan. *A Radical Green Political Theory*. London: Routledge, 1999.
Dobson, Andrew. *Green Political Thought*. London: T & F Books, 2009.
Radcliffe, James. *Green Politics: Dictatorship or Democracy?* New York: Palgrave Macmillan, 2002.
Torgerson, Douglas. *The Promise of Green Politics: Environmentalism and the Public Sphere*. Durham, NC: Duke University Press, 1999.

*Bill Kte'pi*
*Independent Scholar*

# Grains

The term *grains* refers to grasses (members of the botanical family Gramineae Juss) that have been domesticated and selected by man to be used for food. The main characteristics of grains are having a low percentage of water in their fruit; an abundance of starch; and a relative scarcity of indigestible substances such as cellulose and lignin. After grinding, flour, when mixed with water and fermented, is used for bread making and other similar products, all of which are very important for nourishment.

Wheat, corn, and rice account for about 85 percent of global grain production. Here, an irrigated wheat field in Arizona.

*Source:* Jeff Vanuga/Natural Resources Conservation Service/USDA

Grain cultivation began in the Neolithic period, between 9000 and 8000 B.C.E. Since then, grains have played a crucial role in human nutrition. From a broad point of view, it is fair to say that the replacement of some grains that provide low nutritional value and that are difficult to digest (e.g., sorghum and millet) with others that are decidedly superior is a constant phenomenon in the progress of various civilizations. In fact, the evolution of civilizations has always been linked with the evolution and expansion of cereal cultivation. Among the oldest cultivated species are barley and wheat in the Mediterranean basin, rice in eastern Asia, millet and sorghum in southern Asia and Africa, and maize in the Americas.

The traditional cereal cultivation system underwent a revolution after the discovery of America; European colonists did, in fact, contribute to the diffusion throughout the

Americas of Euro-Asiatic cereals (wheat, barley, oats, rye, rice) and, at the same time, introduced corn into European agricultural cycles. During the colonial era, European expansion contributed to the global diffusion of monoculture practices for the three principal grains: wheat, corn, and rice.

In the course of the 19th century, in Europe, the first true revolution in grains cultivation techniques occurred: an increase in yields resulting from the use of natural fertilizers, such as guano, imported from South America.

Between the early 1930s and 1940s, a second revolution in the cultivation of grains occurred, thanks to the creation of the first artificial grain (triticale) and to the development of high-yielding varieties. The triticale is a hybrid derived from the cross-fertilization of wheat and rye, that, thanks to its high protein content, has proven very valuable both for human and for animal consumption. "High-yielding varieties" refers to new seeds that are created through genetic hybridization to obtain varieties of grains that have a higher resistance to disease and are capable of higher yields. Experiments began in Mexico around the mid-1940s by U.S. scientist Norman Ernest Borlaugh. After the first experiments, beginning in the 1960s, the Food and Agriculture Organization of the United Nations promoted the use of high-yielding varieties with the intention of promoting the Indicative World Plan for Agricultural Development, known as the Green Revolution.

The new high-yielding varieties have, in effect, been able to make available higher quantities of food (between 1950 and 2000, world production of grains has increased from 631 to 1,863 million tons), but the environmental cost has also been high. High-yielding varieties have, in fact, demanded intensive irrigation and the employment of phytopharmaceuticals; this has caused radical territorial transformations caused by the construction of elaborate hydraulic infrastructures indispensable to providing the necessary volume of irrigation water, and also the increase in soil and water pollution. Furthermore, the nutritional elements contained in plant protection products (between 1950 and 2000, the consumption of soil fertilizers has increased from 14 to 141 million tons, and per person from 5.5 to 27 kg) bear the highest responsibility for the eutrophication that has affected the internal seas such as the Mediterranean, Baltic, and North seas. The use of high-yielding varieties has also meant a considerable loss in biodiversity (wheat varieties alone have shrunk from some thousands to a few dozen) and has contributed to the salinization of agricultural soils and increased erosion phenomena.

Beginning from the second half of the 1990s, new genetically modified varieties of grains have been introduced. The most common transgenic varieties worldwide are maize and rice. The effects of the introduction into the ecosystems of such genetically modified grains are difficult to predict. According to scientific literature, the most likely environmental risks are the transmission of new genes from the modified plant to plants of the same species or similar through the diffusion of pollen; the transfer of the new gene to soil microorganisms; the loss of biodiversity caused by the diffusion of identical transgenic seeds throughout the world, with the resultant disappearance of existing species; and an increase in the use of pesticides following a new genetically induced degree of tolerance in plants.

As of the beginning of the 21st century, according to predictions by the Worldwatch Institute, grain production is concentrated in China, India, the United States, and Europe (together, including the former Soviet states, these countries account for 67 percent of global production). Wheat, corn, and rice account for about 85 percent of global production, and barley, oats, and other less-common grains account for the remaining 15 percent. Nowadays, 48 percent of world production is destined for human consumption, 35 percent becomes livestock feed, and 17 percent is employed in the production of ethanol and other fuels. The domination of large and complex agrocommodity food networks mean there may be limited choices available to consumers with respect to the consumption of grain and grain derivatives. For the consumer wanting to make green choices, there are difficulties surrounding how to source and identify appropriate products. Aside from issues of availability and cost (e.g., of organically grown grains), making informed purchases may be challenging. A consumer wanting to avoid products with high fructose corn syrup or genetically modified material must rely on the accuracy of food labeling and/or must trust the certification and ecolabeling schemes that are in operation.

**See Also:** Ecolabeling; Genetically Modified Products; Green Food; Pesticides and Fertilizers; Plants; Resource Consumption and Usage; Water.

### Further Readings

Boomgaard, Peter and David Henley, eds. *Smallholders and Stockbreeders: Histories of Foodcrop and Livestock Farming in Southeast Asia*. Leiden: KITLV, 2004.

Charles, Daniel. *Lords of the Harvest: Biotech, Big Money, and the Future of Food.* Cambridge, MA: Perseus, 2001.

Harlan, Jack R. *The Living Fields: Our Agricultural Heritage.* Cambridge, MA: Cambridge University Press, 1995.

Warman, Arturo. *Corn and Capitalism: How a Botanical Bastard Grew to Global Dominance*. Chapel Hill: University of North Carolina Press, 2003.

Worldwatch Institute. *Vital Signs 2009. The Trends That Are Shaping Our Future.* Washington, D.C.: Worldwatch Institute, 2009.

Zohary, Daniel and Maria Hopf. *Domestication of Plants in the Old World: The Origin and Spread of Cultivated Plants in West Asia, Europe and the Nile Valley*, 3rd Ed. Oxford and New York: Oxford University Press, 2000.

*Federico Paolini*
*University of Siena*

## Green Communities

Green communities represent groups of people who are embodied by the ethos of green consciousness. The key focus of such groups is founding a dynamically sustainable ecological system with renewable energy and nutrients through responsible civic activities by human communities.

Green communities can be represented by business groups, nonprofit groups, and consumers as a generic group of advocates and practitioners. They all seek safer products and greater rights of information so as to make better-informed choices about the future. Green communities and their activist concerns of representation and social justice can traverse various aspects of consumer groups, such as race, gender, ethnicity, and class. Their purpose is to ensure that different identities and associated life experiences as repositories of knowledge do not become marginalized in the neo-construction of an ecologically effective world. Furthermore, there is constant dialogue between the various green communities (such as researchers, industries, consumer groups, and ecolabeling organizations) that exist within various networks. Some examples are the European Union–financed research network SCORE! and the United Nations network that creates the organization's Human Development Report. These relationships help to generate new theories and practice for sustainable development.

Green communities, as grassroots movements, seek actively to raise people's consciousnesses. Their key tools are local and international educational campaigns about improving people's lives and health and protecting the environment through effective resource management. Such efforts are driven by an understanding that power circulation at various levels of governance effects environmental change, conflict, and management. Hence, such efforts are aptly aided through the generation of governmental standards, active philanthropic support from major foundations like the Carnegie Trust, and enactment of charters of corporate social responsibility. These efforts focus not on up-front cost projections or net present value of projects but on their long-term or life cycle cost-effectiveness and generation of ecological integrity.

Such efforts are increasingly ushering in an era of green rush through creation of a Green New Deal (as evident in the current U.S. fiscal stimulus focused on generating more green technology and jobs). In the current economic crisis, such actions are also generating a shift in consumer mentality. A large amount of consumer research in the social sciences suggests that few consumers are "mindless hedonists" and that most are, instead, engaged in skilled, knowledgeable, and socially reciprocal consumption practices. This subjective well-being is reflected in widening consciousness about broader public and ethical issues of production and consumption, including a desire to live within one's own means, both financially and ecologically.

In this change of attitudes, religion is playing an important part. Faith-based communities are increasingly protesting the desecration of God's creation and demanding responsible and compassionate stewardship of nature and human relations. They are encouraging people to not be in denial or give in to despair. Instead, they are urging proactive action and self-sacrifice to save the planet for future generations. These include efforts such as stopping overfishing, avoiding excessive abstraction of water from water-stressed catchments, and investing in soil care.

Riding on this global change in attitudes away from atomistic individualism and toward more shared values and endeavors, green communities try to engender community development. Their attempts are reflected in efforts to organize people around their built environment and issues such as greening building, developing culturally relevant ecotourism, fighting against pollution, creating sustainable local energy resources, ensuring proper recycling and composting of waste, and ensuring sustainable nurturing of water resources.

By organizing such activities and training people in the requisite skills, green communities seek to build a civic ethic and to generate green jobs that enhance people's material life chances. Their goal is to ensure sustainable growth through pragmatic and innovative solutions. Given the complexity of such processes, activists seek to ensure that creative ideas generated in consultation with the locals, as well as with architects, engineers, and various professionals, are evolutionarily blended with trials of new modalities of living and sustaining biodiversity. This helps to engender the necessary reconfiguration of projects on the basis of the availability of funding and requirements of budgets. However, most critically, changes are made on the basis of pragmatic compromises of effectiveness and scale; that is, of ideals and existential constraints.

In rural areas, green communities are often focused around Fair Trade and ecotourism. Fair Trade is a significant modality of removing human rights abuses such as child labor and slavery. This is ensured through the generation of prosperity and normative validation of ecolabels for products produced with respect for human dignity. Fair Trade seeks to generate prosperity through the elimination of middlemen and ensuring that the rightful producers get a fair and predictable price for their products in the international market. This price includes a "social dividend" that enables families to lift their lives out of abject poverty, develop their communities, and send their children to school. Companies such as Cadbury have made significant strides down the Fair Trade route.

Similar to Fair Trade, ecotourism also focuses on helping local communities. The locals benefit through green jobs generated by the construction of ecolodges from local materials, managing the local flora and fauna with profits generated from tourism, and conservation of their culture through funding generated by various educational, research, and development programs. Ecotourism has an added dimension of providing authentic, rich, memorable, interpretive experiences for tourists, who constitute a significant buzz-network for generating further interest in such projects.

Community development is not just limited to rural areas but also is increasingly being embraced by megacities around the world (e.g., the large-scale Leadership in Energy and Environmental Design [LEED]–certified Quay Valley community in California). In the case of cities, the attempt is to convert landscapes into "lifescapes" through various measures. First, there is an attempt to create econeighborhoods through affordable housing. This affordability is ensured through the sweat equity of potential owners/residents and through reduced initial land costs via projects that seek to transform abandoned sites, such as degenerating residential subdivisions or misused industrial sites in need of remediation, or older shopping malls with decreasing viability. Such housing focuses on a healthy living environment through creation of good internal air quality.

Second, there is an attempt to create rooftop "rain" gardens and small-scale urban farming that generates cultivation of appropriate organic plants as sources of food and medicines. This fresh source of food ensures a healthier quality of food as a source of both nourishment and pleasure. It also intends to reduce one of the largest sources of carbon footprints—refrigeration.

Third, there is an attempt to reduce dependence on petro-products and eliminate nonbiodegradable waste. This attempt can be seen in movements that seek to ensure litter-free zones and unclog sewage systems by promoting alternatives to plastic bags. It is also evident in pressures on the governments to pass legislations that levy the use of plastic bags, especially by supermarkets, to generate reduction in consumer demand and change consumer habits. In addition, houses in such neighborhoods are installed with green technologies such as compost toilets that convert human wastes and waste food leftovers into compost without using water.

Fourth, energy-conservation consciousness is also evident in neighborhood designs. Houses are equipped with solar panels or draw energy from central sources of wind- or solar-generated energy. Roads in such planned neighborhoods are not constructed from asphalts but, rather, from permeable surfaces. There is also a reduced emphasis on the use of private cars, especially gas guzzlers, and an increased emphasis on using low-cost public transit for commuting to work and accessing community amenities.

Fifth, there is creation of green parks tailored to diverse cultural meanings of space and creation of public wildlife space. On the basis of all of the above-mentioned scenarios, an effective econeighborhood with a smart green infrastructure can become a live laboratory that generates incentives for the relocation of green industries through the generation of network externalities. This can help create a value-added bioeconomy.

Green communities, by focusing on community planning, hence seek to empower communities. This bottom-up activism, by generating demands for effective democracy of voice or the right to debate on issues that affect broad swaths of human life, keeps the authorities on their toes. Authorities are nudged toward passing benign green legislations such as green audits based on carbon-reduction, green certification, transparency policies (e.g., effective nutritional labels that enable informed choices), and so on. They are also forced to be conscientious about implementing and monitoring said legislations. Green communities are thus a constructive and powerful resource in the movement for a more sustainable world.

**See Also:** Ecolabeling; Ecotourism; Fair Trade; Green Design.

### Further Readings

"Creative Solutions to Save the World." *Herald Express.* (April 14, 2009).
Field, K. "Green Acres." *Chain Store Age,* 85/3 (2009).
Figueroa, R. M. "Environmental Justice." In *Encyclopedia of Environmental Ethics and Philosophy,* Vol. 1, J. B. Callicott and R. Froderman, eds. Detroit, MI: Macmillan, 2009.
Marzluff, J. M., et al. *Urban Ecology: An International Perspective on the Interactions between Humans and Nature.* New York: Springer Science + Business Media, 2008.
O'Riordan, T. "On the Politics of Sustainability: A Long Way Ahead—Sustaining Europe." *Environment,* 51/2 (2009).
Petricevic, M. "Giving Priority to the Spiritual Crisis; Green Party Leader Elizabeth May Sees a Role for Faith Communities to Play in Responding to Global Environmental and Economic Woes." *Waterloo Region Record.* Ontario, Canada. (March 7, 2009).
Reed, M. G. and S. Christie. "Environmental Geography: We're Not Quite Home—Reviewing the Gender-Gap." *Progress in Human Geography,* 33/2 (2009).
Stock, R. "Chocolate's Dark Secret: Slavery Claims Leave Bitter Taste." *Sunday Star-Times.* Wellington, New Zealand. (April 12, 2009).
Weeks, K. "Eco-Chic." *Contract,* 50/3 (2009).
"Why Communities Will Shape a New Caring and Sharing Era." *Marketing Week.* London. (April 9, 2009).

*Pallab Paul*
*University of Denver*

## Green Consumer

A green consumer is someone who is aware of his or her obligation to protect the environment by selectively purchasing green products or services. A green consumer tries to maintain a healthy and safe lifestyle without endangering the sustainability of the planet and the future of mankind. According to a recent study, 50 percent of consumers buy green products today. The top three reasons for not buying green products are a lack of awareness, availability, and choices. Green consumers are highly motivated to change their buying behavior for the good of the planet and are willing to pay 10 to 30 percent more to save the planet from environmental damage. However, businesses sometimes find it difficult to predict consumers' reaction to green products. For example, consumers were not excited about the smaller packages of concentrated detergents that were introduced to reduce the packaging cost, and the products were withdrawn from the market. However, niche markets of environmentally friendly consumers sustain products such as green cars, solar energy, organic food, and so on.

These consumers often use different sources of information or decision-making criteria for different types of products and services. The mass media helps them grow awareness of the harms caused by the products that are not ecofriendly, the availability of various green products in the market, the policies and practices of companies, existing and new national/global regulations regarding environmental safety and sustainability, and other helpful information. For example, in 2009, *Newsweek* magazine published a list of top 500 U.S. companies based on their actual environmental performance, policies, and reputation. Hewlett-Packard, Dell, Johnson & Johnson, Intel, and IBM are the five greenest companies on that list. Widespread Internet access has become a critical tool for consumers to spread information and advice, demand transparency from the corporations and manufacturers about their environmental data and policies, and be vigilant about company activities. Green customers have become a very powerful force in coercing large and small companies to accept their environmental responsibilities.

The future generation is growing up as green consumers. They have already taken part in maintaining sustainability and saving the planet from harm. For example, Paul Pressler, the former chief executive officer of Gap, was asked by his daughter whether the company owned sweatshops before he joined the company. As a consequence of his sensitivity to this issue, Gap's 2004 corporate social responsibility report improved substantially the transparency about worker conditions. Similarly, Dave McLaughlin, Chiquita's chief environment officer, confessed to the key role played by the appeals made by thousands of kids to reform the company's social and environmental policies. Often green consumers work with nongovernmental organizations to voice their environmental concerns to the big companies. For example, Dell had to reform its social and environmental practices after huge publicity about the intense protest organized by nongovernmental organizations at the annual consumer electronics show.

The ever-growing segment of green consumers currently includes

- consumers who carefully choose the products and services that are safe for themselves, their children, and the environment;
- business-to-business customers who inquire about how their suppliers make their products, and what is in those products;
- employees who are bringing their personal values to their profession and share their companies' mission;
- bank personnel who factor environmental variables such as financial cost of greenhouse gas emissions, projects that have adverse effects on natural habitats, and so on, in making loan decisions;
- insurance agents who consider environmental risks as a business threat; and
- stock market analysts who study the environmental performance of a company as a management quality.

The environmental damage resulting from increasing greenhouse gas emissions and global warming, disappearance of species, and reduction in global water supply has contributed to the deterioration of overall quality of life worldwide. This has motivated green consumers to put pressures on local and global leaders and government officials to construct and enforce new laws and regulations for business practices that are of utmost importance for the sake of our planet's future.

**See Also:** Consumer Behavior; Consumer Culture; Consumer Ethics; Ethically Produced Products; Green Communities; Locally Made; Morality (Consumer Ethics); Quality of Life; Simple Living.

## Further Readings

Autio, M., et al. "Narratives of 'Green' Consumers—The Antihero, the Environmental Hero and the Anarchist." *Journal of Consumer Behaviour*, 8/1:40 (2009).

Chan, R., et al. "Applying Ethical Concepts to the Study of 'Green' Consumer Behavior: An Analysis of Chinese Consumers' Intentions to Bring Their Own Shopping Bags." *Journal of Business Ethics*, 79/4:469–81 (2008).

Diamantopoulos, Admanantios, et al. "Can Socio-Demographics Still Play a Role in Profiling Green Consumers? A Review of the Evidence and an Empirical Investigation." *Journal of Business Research*, 56/6:465 (2003).

D'Souza, Clare, et al. "Green Decisions: Demographics and Consumer Understanding of Environmental Labels." *International Journal of Consumer Studies*, 31/4:371 (2007).

Grønhøj, Alice. "Communication about Consumption: A Family Process Perspective on 'Green' Consumer Practices." *Journal of Consumer Behavior*, 5/6:491–503 (2006).

Haanpää, Leena. "Consumers' Green Commitment: Indication of a Postmodern Lifestyle?" *International Journal of Consumer Studies*, 31/5:478–86 (2007).

Hopkins, M. and C. Roche. "What the 'Green' Consumer Wants." *MIT Sloan Management Review*, 50/4:87–9 (2009).

Jain, Sanjay K. and Gurmeet Kaur. "Role of Socio-Demographics in Segmenting and Profiling Green Consumers: An Exploratory Study of Consumers in India." *Journal of International Consumer Marketing,* 18/3:107–46 (2006).

Klintman, M. "Participation in Green Consumer Policies: Deliberative Democracy under Wrong Conditions?" *Journal of Consumer Policy*, 32/1:43–57 (2009).

Mahenc, P. "Signaling the Environmental Performance of Polluting Products to Green Consumers." *International Journal of Industrial Organization*, 26/1:59 (2008).

McDonald, Seonaidh and Caroline J. Oates. "Sustainability: Consumer Perceptions and Marketing Strategies." *Business Strategy and the Environment*, 15/3:157–85 (2006).

Moisander, Johanna. "Motivational Complexity of Green Consumerism." *International Journal of Consumer Studies*, 31/4:404 (2007).

Moisander, Johanna and Sinikka Pesonen. "Narratives of Sustainable Ways of Living: Constructing the Self and the Other as a Green Consumer." *Management Decision*, 40/4:329–42 (2002).

Nyborg, Karine, et al. "Green Consumers and Public Policy: On Socially Contingent Moral Motivation." *Resource and Energy Economics*, 28/4:351 (2006).

Pedersen, Esben Rahbek and Peter Neergaard. "Caveat Emptor—Let the Buyer Beware! Environmental Labelling and the Limitations of 'Green' Consumerism." *Business Strategy and the Environment*, 15/1:15–29 (2006).

Rivera, Jorge. "Assessing a Voluntary Environmental Initiative in the Developing World: The Costa Rican Certification for Sustainable Tourism." *Policy Sciences*, 35/4 (2002).

Schlegelmilch, Bodo B., et al. "The Link Between Green Purchasing Decisions and Measures of Environmental Consciousness." *European Journal of Marketing*, 30/5:35–55 (1996).

Shrum, L. J., et al. "Buyer Characteristics of the Green Consumer and their Implications for Advertising Strategy." *Journal of Advertising*, 24/2:71 (1995).

Straughan, Robert D. and James A. Roberts. "Environmental Segmentation Alternatives: A Look at Green Consumer Behavior in the New Millennium." *Journal of Consumer Marketing*, 16/6:558–75 (1999).
Thogersen, John. "Media Attention and the Market for 'Green' Consumer Products." *Business Strategy and the Environment*, 15/3:145–56 (2006).
Wagner-Tsukamoto, Sigmund and Mark Tadajewski. "Cognitive Anthropology and the Problem-Solving Behavior of Green Consumers." *Journal of Consumer Behaviour*, 5/3:235–44 (2006).

*Mousumi Roy*
*Penn State University, Worthington Scranton*

# Green Consumerism Organizations

President John Kennedy identified six basic consumer rights: choice, safety, courteous and respectful service, being heard, information about product and service characteristics, and education about consumer concerns. The modern consumer movement has focused on the rights to safety, information, and being heard. Green consumerism has somewhat different emphases. Green consumers particularly desire the availability of a greater number of environmentally friendly, sustainable, or green products; information to help determine that products and services are more sustainable than traditional alternatives; and tips on how to live and consume more sustainably. Unlike the traditional consumer movement, which has tended to see consumers' interests as opposed to businesses' interests, the green business and green consumer movements appear to be much more compatible.

Several major purposes of green consumerism organizations and business organizations serving the interests of green consumers appear to exist: advocating more green products and services; promoting green businesses, locally produced products, and Fair Trade; establishing standards and promoting and/or certifying green products; ensuring the validity of claims made by companies; identifying instances of greenwashing; educating consumers regarding sustainable living; and advocating less consumption. Few organizations are solely aimed at serving green consumers. This article therefore briefly discusses those activities that advance the interests of green consumers—several of which are examined in more detail in other articles—and identifies some of the organizations engaging in these activities.

## Advocates of Green Products and Services

Although supportive of environmental issues, many organizations spawned by the U.S. consumer movement of the 1970s—for example, Public Citizen—have tended to focus on public advocacy, paying little attention to green consumers' special concerns. Traditional environmental organizations such as the Sierra Club and Environmental Defense Fund have been among the major advocates of increasing product sustainability. The Sierra Club gave awards to Honda for its original Insight and Toyota for its Prius. The Environmental Defense Fund has, since 1990, worked with many companies to green their practices, including McDonald's (reducing fast food packaging waste and antibiotic use in chicken), FedEx (with hybrid delivery vans), Wegmans supermarkets (with an ecofriendly farmed shrimp purchasing policy), and S.C. Johnson (using environmental assessment software for product and packaging design).

In addition to its corporate partnerships, the Environmental Defense Fund has advocated for greener products and services by championing market incentive–based government policies—cap-and-trade systems for sulfur dioxide and carbon dioxide, and "catch shares" giving fishermen incentives to conserve fish stocks—establishing costs for harms ("negative externalities") caused by pollution or resource depletion. Under such policies, companies have an economic incentive to green products and services by finding innovative ways to reduce harmful environmental impacts and avoid raising consumer prices. Such schemes mean that consumers need not be informed of products' environmental attributes. Because acting more sustainably actually reduces businesses' costs, more sustainable performance should be reflected in lower prices than those offered by competitors acting less sustainably. Although traditional regulation establishes an upper limit on negative behavior, carefully crafted market-based incentive policies can encourage continued green innovation as companies seek to further reduce costs. However, such market incentive–based policies do increase prices, at least initially.

Absent such policies, consumers must rely on voluntary adoption of greener production practices. Green America (formerly Co-op America), formed in 1982, has been a major advocate of business greening. The organization's stated mission is "to harness economic power—the strength of consumers, investors, businesses, and the marketplace—to create a socially just and environmentally sustainable society." Through its Website and publications, Green America encourages consumers to urge companies to adopt more sustainable practices, particularly in the areas of green energy and labor rights. Although many other organizations do this as well, Green America offers additional consumer resources—notably "Responsible Shopper," a Website section with three components: Learn, Act, and Live. The "Learn" section opens access to information about companies' social and environmental practices, searchable by company and industry, and the "Act" section provides information about boycotts and other campaigns designed to influence corporate behavior, thus enabling consumers to become advocates for greener products and services. The "Live" section gives tips for green consuming and living. Green America also indirectly encourages companies' sustainable practices by encouraging and providing substantial information on social investing, including a "Socially Responsible Mutual Fund Performance Chart."

## Promoters of Green Businesses

Green America also encourages consumers to patronize specialized green businesses, providing online and print information through its "Green Pages" about thousands of green products and services. GreenPeople.org and the Organic Consumers Association provide another similar online directory. Many other Websites provide green business directories, but most appear to be oriented more toward advertising than promoting the growth and development of green business per se.

An emerging trend in the promotion of green businesses is the development of metropolitan or regional associations, such as the Chicago Sustainable Business Alliance. The alliance attempts to help member companies become more environmentally and economically sustainable through activities such as networking, member education, actively promoting sustainable practices and policies, and encouraging information exchange and learning. As such organizations foster the successful development of green business, green consumers will have more choices.

## Promoters of Local Products

Other efforts to encourage consumers to buy locally rather than from national and multinational companies connect green businesses and consumers, including the [San Francisco] Bay Area Green Business Program, Green Directory Montana, Boise [Idaho] Green Living, and the Bainbridge Island [Washington] Green Business Directory. Although the buy local movement is not strictly a green consumer movement, many local businesses spearheading this movement attempt to act in a more sustainable manner. Furthermore, shorter supply chains can produce lower carbon footprints for local companies.

Local business networks in over 60 communities have joined under the umbrella of the Business Alliance for Local Living Economies (BALLE). Created in 2001, with its roots in the Sustainable Business Network of Greater Philadelphia, BALLE attempts to foster Local Living Economies throughout North America to facilitate vibrant communities, environmental health, and prosperity. BALLE's activities include fostering networks of independent locally owned businesses, educating member companies, engaging in community economic development, and promoting public policies conducive to successful local businesses. Although not strictly a green business organization, BALLE has spotlighted local entrepreneurs' green practices, including sustainable agriculture, green building, renewable energy, and zero-waste manufacturing. Local BALLE networks largely aim at promoting successful business practice, but they also engage in outreach to consumers through their Buy Local campaigns.

## Promoters of Fair Trade Products

The Fair Trade movement, which has expanded dramatically in recent years, seems in some ways diametrically opposed to "buying locally." Despite the distance Fair Trade goods travel, they can still be considered a form of green consumption because the movement promotes more sustainable practices among small producers in developing nations. Organizations facilitating, promoting, and/or selling Fair Trade products are thus aligned with green consumerism. The Fair Trade Federation is a member association of retailers and importers who deal with Fair Trade products. In addition, Green America educates about and encourages Fair Trade.

## Product Certification and Ecolabeling Organizations

Green consumers often question whether a product's environmental claims are accurate. The lack of such assurance helped to short-circuit a wave of green consumerism in the early 1990s. Certification of a product's environmental friendliness gives green consumers confidence and helps expose regular consumers to green products. As labels and certifications gain credence, they encourage existing companies to green their products, increasing the number of green alternatives. Developers of labels and certification schemes are therefore central to the green consumer movement. In some nations, ecolabels are issued by the government—the U.S. Environmental Protection Agency issues Energy Star and WaterSense labels. Other labels are issued by private organizations, notably the Forest Stewardship Council (FSC) and the Carpet and Rug Institute. Some certification schemes' governing organizations also act as the certifiers of products under the scheme—for example, Green Seal and Greenguard Environmental Institute. In other cases, a third party certifies: SmartWood certifies wood products for the FSC, and NSF International certifies furniture products for the "level" program of the Business and Institutional Furniture Manufacturers Association. Some certifying organizations, such as Scientific Certification Systems, conduct certification for certification schemes such as FSC while also developing their own certification systems and labels. Certification is also an essential component of Fair Trade. About 20 organizations such as TransFair USA are engaged in independent certification of fair trade products and are members of Fair Trade Labeling Organizations International. Although they are not consumer products, green buildings are certified as meeting different levels established by the U.S. Green Building Council's Leadership in Energy and Environmental Design (LEED) standards.

As ecolabels and certification schemes have proliferated, particularly with the development of "store brand" ecolabels, there is a risk that information will become too confusing or, worse, that meaningless ecolabels will mislead consumers. *Consumer Reports* now provides an online "ecolabels center," explaining the meaning of various ecolabels and certifications and assessing their reputability.

## Other Independent Validators of Green Claims

Other means exist to validate companies' green claims. For example, companies wishing to be listed in Green America's Green Pages must submit a detailed questionnaire for staff review. A second type is independent certification that a facility meets an environmental management standard, such as International Standards Organization's (ISO) 14001. Certification by accredited third parties such as Det Norske Veritas and Certification International that a facility meets ISO 14001 standards assures companies that their suppliers follow high environmental management standards. Unlike product certification schemes, ISO standards assume that following a set of specific environmental management processes will generate good environmental performance. A third form of independent assurance is "verification" or "audit statements" of the claims made in corporate sustainability reports. Many large public accounting firms, environmental engineering firms, and environmental consulting firms issue conduct such verification. Sustainability reports are available to consumers and are reviewed by environmental, socially responsible investment, and green consumer organizations such as Green America and used as a basis for the information they give to constituents and the public. Other types of assurance statements identified by the environmental consultancy SustainAbility include "professional evaluations," "expert statements," and "stakeholder comments." These are less binding but do offer opinions that may help green consumers make informed decisions.

## Identifiers of Greenwashing

Unfortunately, most environmental claims are not independently certified or verified. "Greenwashing" occurs when false or misleading claims are made in an attempt to develop an environmentally strong reputation, and green consumers need to be aware of these attempts. Environmental marketing agency TerraChoice identifies different types in "the seven sins of greenwashing" on its Website. As noted above, *Consumer Reports'* Ecolabels Center aids in identifying ecolabeling greenwashing. Many other organizations identify and publicize instances of greenwashing, including the Center for Media and Democracy's PRWatch.org, Greenpeace, and EnviroMedia Social Marketing's Greenwashing Index. The latter aims to help consumers evaluate advertisements' environmental claims, hold companies accountable for such claims, and stimulate demand for sustainable company practices. Green consumers can submit questionable ads to the site, and readers assign them a greenwashing score.

## Promoters of and Educators Regarding Green Lifestyles

Education about various aspects of green lifestyles is provided by many media and environmental organizations, most frequently as stories about and lists of general or specific topics. Several organizations have developed ongoing columns that focus on very specific issues. The Sierra Club e-mails its Green Life daily tip. Grist's "Ask Umbra" and Slate.com's "The Green Lantern" advice columns both answer readers' specific green lifestyle questions. *National Geographic* publishes *The Green Guide* magazine and maintains a Website of the same name, providing buying guides to various product categories, answers to reader-submitted questions, quizzes on greening, and links to local green guides. These independently produced guides focus on identifying green consumer alternatives in local communities. Green America also has a site on "living green."

Consumers Union, the publisher of *Consumer Reports* magazine, first covered environmental issues in the early 1970s, including in an award-winning 1974 three-part series on water pollution. Although a variety of its articles on green products are archived online, *Consumer Reports* has created a separate, extensive "Greener Choices" Website, with sections on products and green ratings; tools such as different types of energy calculators and information about toxins in products; "hot topics" including information on electronics use and recycling, ecolabels, and global warming; and community pages including blogs, campaigns, reader forums, and links to resources from other organizations.

## Advocates of Less Consumption

A unique sector of the green consumer movement is one of anticonsumption. Many organizations exist whose missions are to encourage individuals to consume less. Generally referred to as promoters of voluntary simplicity (VS), the specific focus of these organizations varies. Most VS organizations emphasize not being influenced by advertising and the consumer culture but, rather, engaging in "mindful" or "conscious" consumption. To some extent, a tension exists between green consumerism organizations and VS organizations, and even among the different VS organizations themselves, as to whether the goal is to make the products consumed more sustainable or to limit consumption of products, regardless of how green they may be. However, Green America serves as the fiscal agent for one of the major VS organizations, the Simplicity Forum.

VS advocates emphasize the importance of personal choice in developing simple and sustainable lifestyles, and a number engage in lifestyle education, generally taking a highly intensive and personal approach. Many of these organizations use "simplicity circles"—an approach illustrated in *The Circle of Simplicity: Return to the Good Life* by Cecile Andrews. Simplicity circles are discussion groups in which individuals meet together to discuss simpler lifestyles and support one another in simplifying their lives. The Simple Living Network Website provides information on simplicity circles and other study groups, including a "simplicity study group database search" page, which allows individuals to locate or attempt to start a study group. A 2009 search of this list indicates the existence of approximately 100 active study groups in the United States. Most of these appear to follow the approach outlined by Cecile Andrews, use the book *Your Money or Your Life* as a study guide, or use a course developed by the Northwest Earth Institute, an organization that develops courses and training programs that help individuals explore their values and behaviors regarding lifestyles and the environment. Some of these courses have been held in corporate workplaces.

Other VS organizations include Alternatives for Simple Living, Green Heart Institute, Simple Living America, and the Simple Living Network. Although most of these organizations focus on providing encouragement and information to individuals, the Center for a New American Dream also engages in some policy research and advocacy aimed at making products more sustainable. This national organization's mission is to help "Americans consume responsibly to protect the environment, enhance quality of life, and promote social justice." It offers information on greening organizational purchasing, offices, and schools; being a conscious consumer; personal action campaigns in areas such as bottled water, energy, and paper use; and petition campaigns.

Although reducing consumption has clear environmental benefits, VS advocates have additional motivations for their actions. These include saving money (Your Money or Your Life), increasing family time (Take Back Your Time Day), and promoting local communities (Community Solution). Other advocates of lower consumption strongly oppose the influence of corporations on societal culture. Buy Nothing Day encourages people to avoid any purchases on Black Friday, the day after Thanksgiving, which is the busiest U.S. shopping day of the year. Adbusters describes itself as "a global network of culture jammers and creatives working to change the way information flows, the way corporations wield power, and the way meaning is produced in our society." Adbusters produces a magazine, creates advertisements that make fun of commercial ads, critiques the corporations behind such ads, and hosts various anticonsumption and anticorporate videos. A major VS community has developed online, and one site, The Compact, encourages individuals to pledge not to make any new purchases for a year aside from food, medicine, and utility services.

Product sharing and product reuse organizations also help reduce consumption. Organizations such as Zipcar and tool rental companies promote the sharing of products rather than individual purchases. Secondhand outlets and thrift stores such as Deseret Industries, Goodwill, and Salvation Army facilitate buying used consumer goods. Web-based organizations such as Craigslist and eBay facilitate direct sales, and networks such as YahooGroups' Freecycle facilitate free distribution of such products. All of these organizations and groups help reduce consumption of new products.

## Conclusion

There are few organizations that can truly be considered to specialize in green consumerism, but Green America appears to be the major such organization, engaging in most of the activities discussed above. Existing consumer organizations, with the exception of Consumers Union through its Greener Choices Website, appear to have paid relatively little attention to green consumer issues. It is clear, however, that a multiplicity of organizations, some of them no more than Websites, are engaging in the specific activities needed to provide green consumers the kinds of information that they need. Although the resources exist, it is not yet apparent that the general public is aware of and uses them to the degree needed to encourage a massive voluntary greening of products. This suggests that policy changes are still needed. Those green consumerism organizations engaging in policy advocacy, such as Green America and the Center for a New American Dream, will likely need to continue to work with environmental organizations such as the Environmental

Defense Fund and the activist consumer movement organizations formed in the 1970s, such as Public Citizen, before green products and services can become the norm.

**See Also:** Affluenza; Certification Process; Certified Products (Fair Trade or Organic); Consumer Activism; Consumer Boycotts; Consumer Culture; Consumerism; Consumer Society; Downshifting; Ecolabeling; Ethically Produced Products; Fair Trade; Frugality; Government Policy and Practice (Local and National); Green Consumer; Green Marketing; Greenwashing; Lifestyle, Sustainable; Locally Made; Materialism; Overconsumption; Product Sharing; Regulation; Secondhand Consumption; Simple Living; Sustainable Consumption; Websites and Blogs.

### Further Readings

Adbusters. http://www.adbusters.org (Accessed June 2009).

Center for a New American Dream. http://www.newdream.org/ (Accessed June 2009).

Environmental Defense Fund. "Corporate Partnerships." http://www.edf.org/page.cfm?tagID=56 (Accessed June 2009).

Green America. http://www.coopamerica.org (Accessed June 2009).

Greener Choices.org. http://www.greenerchoices.org (Accessed June 2009).

GreenMuze. "Anti-Consumer Organizations." http://www.greenmuze.com/waste/consumption/354-anticonsumer-organizations.html (Accessed June 2009).

*Gordon P. Rands*
*Pamela J. Rands*
*Western Illinois University*

## Green Design

Environmental effects were ignored during the design stage for new products and processes in the past. Waste was common in material production, manufacturing, and distribution. Hazardous wastes were dumped in the most convenient fashion possible, ignoring possible environmental damage, and inefficient energy use resulted in high operating costs. It was realized in the 1970s that to achieve a real effect on reducing product- and process-related environmental impacts, environmental considerations should be built into product development at the earliest opportunity, and customers should be steered toward the greener options. Rapid industrialization, regulatory pressure, higher requirements from customers, open markets, increased demands to defend or expand market share and to create the ability to attract foreign investments, and an increase in competitiveness between companies locally and globally have paved a way for this paradigm shift in the design process.

Green design is intended to improve environmental quality through better design and management in the production and consumption activities of economic enterprises. This green approach involves the smart design of products, processes, systems, and organizations and evolves a particular framework for the implementation of relevant analysis, synthesis methods, and smart management strategies that effectively harness technology and ideas to avoid environmental problems before they arise.

Various aspects of green design involve improving energy efficiency, safeguarding water, managing waste, planning a sustainable site, and creating healthy indoor environments using environmentally preferable materials.

Three goals for green design in pursuit of a sustainable future are to reduce or minimize the use of nonrenewable resources, manage renewable resources to ensure sustainability, and reduce, with the ultimate goal of eliminating, toxic and harmful emissions to the environment.

The objective of green design is to pursue these goals in the most cost-effective fashion. To achieve this aim, we need to adopt the following green design principles: minimize or design away the extraneous, integrate design aspects for multiplicity of function, design for all aspects of climate at all levels, think about the unintended consequences of maintenance and renewal, select materials that use their base resource most efficiently, and design to use maximum local and regional resources.

## Rules of Thumb

Rules of thumb proposed for Design for Sustainability can also be applicable to green design:

- selection of low-impact material such as cleaner, renewable, recycled, recyclable materials and materials available in the local region;
- reduction of material usage in terms of weight and volume;
- optimization of production techniques through alternative productive techniques, few production steps, lower/clear energy production, less waste production, fewer/cleaner consumables use, safety, and cleanliness of the workplace;
- optimization of distribution system through less/cleaner/reusable packaging, energy-efficient transport mode, energy-efficient logistics, and involvement of local suppliers;
- reduction of impact during use by low energy consumption, use of clean energy source, and reduction of waste of energy and other resources;
- optimization of initial lifetime involving reliability and durability, maintenance and repair, modular product structure, classic design, strong product–user relations, and involvement of local maintenance and service systems; and
- optimization of an end-of-life system through efficient reuse of product, remanufacturing/refurbishing, recycling of materials, and safer incineration, and involving local pollution prevention systems.

## Product Development Process

The ecoproduct or green product development process generally involves the following steps: idea generation, concept development, evaluation and testing, manufacturing, launching, product management, and end-of-life management. There are two contemporary applications of green development—green infrastructure (GI) and green building (GB).

## Green Infrastructure

GI consists of an integrated/interconnected network of multifunctional, high-quality, open green spaces provided across all spatial planning levels (regional, subregional, local, and neighborhood) that support native species, maintain natural ecological processes, sustain air and water resources, and contribute to the health and quality of life, in addition to ensuring sustainable livelihoods. This requires that GI be more proactive and less reactive. GI is a new term, but it is not a new idea. It has roots in planning and conservation efforts that started 150 years ago.

GI is central to promoting smart opportunities for facilitating nature conservation/ecorestoration and protecting habitat/ecosystem/biodiversity/life support systems (soil, air, and water) through improvements in environmental quality; better air, soil, and water quality; local climate control and noise reduction; and sustainable drainage and flood mitigation. Setting aside vegetation buffers along rivers and streams can protect them against flooding, which will sustain land values and provide enhanced environmental surroundings to encourage ecofriendly business investments. It also repositions open space protection from a community amenity to a community necessity. Thus, it is an answer to habitat degradation/fragmentation as a consequence of conventional "development" strategies.

## Green Building

GB applies principles of resource and energy efficiency, healthy buildings and materials, and ecologically and socially sensitive land use to achieve "an aesthetic sensitivity that inspires, affirms, and ennobles." GB requires an integrated, multidisciplinary design process and a "whole-building" systems approach that considers the building's entire life cycle from planning, design, and construction to operation and maintenance, renovation, and demolition or building reuse. Together, these provide the means to create solutions that optimize building cost and performance. As a result of these innovative design principles/strategies, GBs consume at least 40–50 percent less energy and 20–30 percent less water vis-à-vis a conventional building. This comes at an incremental cost of about 2–8 percent, with a payback period of three to seven years.

## Some Green Design Methods and Tools

Following are definitions of some terms common to green methods and tools.

*Mass balance analysis* involves tracing the materials or energy in and out of a process or an analysis area, such as a manufacturing station or a plant. Ideally, mass balances are based on measurements of inflows, inventories, and outflows including products, wastes, and emissions.

*Life cycle assessment* is used to determine the total environmental effect of a product throughout its life cycle (i.e., from cradle to grave, or the more recent cradle to cradle).

*Green indices* or ranking systems attempt to summarize various environmental impacts into a simple scale. The designer or decision maker can then compare the green score of alternatives such as materials, processes, and so on, and choose the one with the minimal environmental impact.

*Design for disassembly and recycling aids* means making products that can be taken apart easily for subsequent recycling and parts reuse. Design for disassembly and recycling aids software tools generally calculate potential disassembly pathways, point out the fastest pathway, and reveal obstacles to disassembly that can be "designed out."

*Ecoperformance profile* is an identification of energy and material-related environmental impacts generated by the company and along the products' life cycle.

*Risk analysis* is a means for tracing the chances of different effects occurring in a particular area or process or product.

*Material selection guides* attempt to guide designers toward the environmentally preferred material and green alternatives. Manuals are intended to provide information for users about how to use, maintain, and dispose of products. Companies need management information systems that reveal the cost to the company of decisions about materials, products, and manufacturing processes.

*Environmental management subsidies* are intended to provide subsidies for the development and marketing of cleaner products including the creation of know-how, methods, product development processes, greener purchasing, and waste/recycling systems.

*Green tax* refers to a policy that introduces taxes intended to promote ecologically sustainable activities via economic incentives.

*Ecodesign* has been used in companies as a tool to incorporate environmental considerations into product design and development process of enterprises.

*Ecolabeling* refers to a label that identifies overall environmental preference of a product or service within a specific product/service category based on life cycle considerations. These labels are awarded by an impartial third party in relation to certain products or services that are independently determined to meet environmental leadership criteria.

*Shared responsibility* is the sharing of different tasks by different stakeholders along the product's life cycle. The field of green design is more in favor of this approach, rather than passing all the responsibility to the producer.

*Supply chain management* refers to technology partnerships in which reuse or recycling relationships and supplier evaluation can also be used to manage supplier chains for environmental quality.

With recognition of *information technology* in this field, various supportive databases and software can also be used for product design.

## Benefits and Future Perspectives

Developing and marketing green products is a concrete step toward sensible and sustainable economic development. Green products imply improved worker productivity, health, and satisfaction; more efficient resource use; reduced emissions; reduced waste; and environmental protection. Greener designs promise greater profits to companies by reducing material requirements, the social cost of pollution control such as disposal and environmental cleanup fees, and raising revenues through greater sales and exports.

Designing and manufacturing green products requires appropriate knowledge, tools, and production methods. To enable the development of an appropriate organizational infrastructure will demand a paradigm shift from the existing engineering mindset of ecodesign to a broader stakeholder network that captures, processes, stores, and disseminates information from and to formal and informal systems. There is a need for development of new training/capacity building approaches and the knowledge delivery systems to enable faster performance.

Sustainability is today's buzzword in design. The green market is expanding rapidly, and ecofriendly design is helping companies to stand out from the competition. Green designers—a new breed of environmentally conscious engineers and architects—are rethinking entire product life cycles, from the industrial manufacturing processes to what happens at the end of the life of the product. They aim to build nonpolluting factories that make products that are safe for the environment and 100 percent recyclable by designing new industrial methods and scrutinizing every raw material that goes into fabrication. Some products created according to these principles now carry a new certification mark: Cradle to Cradle. The hard-headed Fortune 500 companies have already started working in this area. However, for wider adoption of green design principles, appropriate policy/legal/ administrative/financial support is essential in the form of a system of incentives and disincentives such as subsidies, soft loans, taxes, and so on to encourage companies/ industries to incorporate environmental and social factors into product development throughout the life cycle of the product, throughout the supply chain (preproduction, production, and distribution, including packaging, sale, use, and disposal), and with respect to their socioeconomic surroundings. In addition to research and development funds for promoting green design, GI, and GB, a specific green investment fund—a competitive grant program that offers funds to industrial, commercial, residential, and mixed-use projects—may be promoted.

Changing design procedures is particularly difficult because designers face many conflicting objectives and uncertainties and a work environment demanding speed and cost effectiveness. Environmental concerns must be introduced in a practical and meaningful fashion into these complicated design processes.

Although still a fledgling movement, this holistic, ethical, pragmatic, and wildly inventive mode has the potential to redirect design toward progressive ends. This shift drives toward sustainable ways of living and working. A key to successful, integrated GD is the participation of people from different specialties of design: general architecture, lighting and electrical, interior design, and landscape design. By working together at key points in the design process, these participants can often identify highly attractive solutions to design needs. If companies are to take advantage of the business opportunities and minimize the effect of changes that may emerge from a greener and/or more sustainable world, then new expertise and skills need to be developed. This will mean the need to rethink with a new vision and to redesign products and industrial processes to make them safer, environmentally friendly, and more efficient. The real question is not whether we can afford to do this, it is whether we can afford not to.

**See Also:** Biodegradability; Carbon Credits; Carbon Offsets; Certification Process (Fair Trade or Organic); Durability; Ecolabeling; Ecotourism; Energy Efficiency of Products and Appliances; Environmentalism; Environmentally Friendly; Ethically Produced Products; Green Communities; Green Consumer; Green Consumerism Organizations; Green Discourse; Green Gross Domestic Product; Green Homes; Green Marketing; Green Politics.

## Further Readings

Armstrong, Tom. "Design for Sustainability." http://www.indeco.com/Files.nsf/Lookup/dfs/$file/Dfs.pdf (Accessed April 2009).

Fairs, Marcus. *Green Design: Creative Sustainable Designs for the Twenty-First Century.* Berkeley, CA: North Atlantic Books, 2009.

Hendrickson, Chris, et al. "Introduction to Green Design." http://www.gdi.ce.cmu.edu/gd/education/gdedintro.pdf (Accessed April 2009).

World Green Building Council 2007. "WorldGBC News." http://www.worldgbc.org (Accessed June 2007).

*Gopalsamy Poyyamoli*
*Rasmi Patnaik*
*Pondicherry University*

# Green Discourse

In defining green discourse, it is first necessary to introduce the role and meaning of the term *discourse*. We live in a complex world, one in which we do not understand all aspects of our natural and social environment, yet are constantly bombarded with information aimed at shaping the way we understand issues and act in response. It is within this context that we are constantly required to make decisions, from the mundane, (e.g., should I drive or walk to the store?) to the exceedingly complex (e.g., is my current lifestyle sustainable?).

Acting in such an environment can feel overwhelming, so we have a tendency to develop a framework (albeit subconsciously) of ideas, perceptions, assumptions, judgments, and attitudes that we apply to help us understand the world and make decisions. It is this framework that is referred to by the term *discourse* (you might like to think of it as the lens through which we view the world). As John Dryzek stated, discourses "construct meanings and relationships, helping to define common sense and legitimate knowledge."

Each discourse is shaped by key influences that affect us, including ideology, ethics, history (including our own experiences), political institutions, public perspectives, and societal trends and demographics.

There are almost an infinite number of discourses possible, as no two individuals are subject to the same experiences or interpret events and ideas in the same way. Nevertheless, because people have similar attitudes in a few core areas, we are able to group these discourses together to form a shared way of apprehending the world.

Just as individuals develop discourses to view and understand the world, so do institutions such as corporations, government, and other stakeholder groups. Indeed, it is necessary for them to do so, as without an overarching framework to guide the work of the individuals that make up the institution, it could not progress or develop along a predetermined path. Through this discussion it may be becoming apparent that there are multiple discourses that exist at any one time, and that these discourses need not be compatible. To understand the interaction between different discourses and between their proponents we can draw on the work of M. Foucault, who examined how discourses change over time within society, institutions, and disciplines. In examining why and how discourses evolve, Foucault focused attention on the role of power (who holds it and how it was applied to shape how issues are viewed).

## How Do Discourses Apply to Environmental Decision Making?

The concept of discourses can be applied to the case of environmental decision making. As R. Alexander identifies, it is increasingly accepted that the ways in which we view the "natural" world are socially constructed. In this sense, ecological issues become social and political problems as individuals and institutions view and frame the environment and our relationship to it in alignment with the discourse to which they subscribe. Proponents of different discourses vie for dominance in an attempt to colonize and dominate environmental decision making so as to narrowly frame decision making such that it aligns with their discourse. This is done in an attempt to ensure that environmental decision making occurs in a way (and produces outcomes) that is in alignment with their environmental discourse.

Recently, there has been an emergence (or resurgence) of what can be termed the *green discourse* in contemporary environmental decision making. Proponents of the green discourse are challenging the currently dominant neoliberal environmental discourse and seek to shape the debate and decision making concerning the environment in accordance with the (namely, nature-centric) key ideas that make up the green discourse. The focus of this article is to examine and explain the term *green discourse*, but it is also important to understand that this discourse is but one competing for dominance in policy debates regarding environmental decision making (see Dryzek 1997 for a discussion of discourse typologies).

The literature examining discourses in environmental decision making has predominately examined key decisions in an attempt to highlight how the dominant discourse has evolved over time and how this has been a result of changes in institutional and societies' contextualization of the environment and our relationship to it. These works provide valuable insight into the relationship among key stakeholders, their discourses, and their relative power in shaping decision making. However, they do not characterize or define discourses into typologies, making it difficult to determine the meaning and role of the green discourse in environmental decision making.

Some authors have worked to define the core ideas that comprise different discourses in environmental decision making. In examining the meaning of green discourse, it is this work, and particularly the typologies posed by John Dryzek in 1997, that prove most valuable.

## What Is a "Green Discourse"?

Within the literature there is no consensus on the core ideas that make up the green discourse. Indeed, there is not even consensus on the terms used to encompass these ideas, with variations including *alternative discourse*, *green radicalism*, *conservation discourse*, *ecological discourse*, *green movement* (namely, in politics), and more broadly, *environmental consciousness*. As such, it is important to develop a clearer understanding of the concept.

Some feel that green discourse has its roots in the development of an environmental consciousness that was a product of the 1960s—a time when, as part of the counterculture critique of the "technocratic society" and the widespread questioning of the dominant values of consumer culture—environmentalism emerged as a new political cause. These ideas form the foundation for what now can be classified as a green discourse. Over the following three decades, the green discourse has been somewhat marginalized from environmental decision making, as concepts such as sustainable development and ecological modernization came to dominate environmental policy making—reflections of shifts in society as a whole away from the protection of the environment to the greening of our own human societies. However, this discourse has never disappeared and may be gaining power in environmental decision making.

Broadly speaking, the green discourse can be defined as an approach in which environmental protection finally takes its place alongside the economy and social issues in decision making that affect the environment. In his 1997 analysis of environmental discourses, Dryzek describes the green discourse as being both radical and imaginative. It is radical in the sense that its proponents reject the basic structure of industrial society, economic growth as societies' main objective, and the way the environment is conceptualized within decision making. It is imaginative in its proposed solutions. The green discourse challenges typical views of, and responses to, environmental problems.

Those who subscribe to the green discourse believe that the world is facing a multifaceted social and ecological crisis. Proponents identify that the world's resources are finite and that current levels of consumption (driven by materialism and population growth) are consuming resources at a rate far in excess of what the Earth can sustain. As such, proponents of the green discourse reject the (currently dominant in global politics) notion that economic growth should form the dominant goal of government and society, citing that there are limits to industrialization and that continued growth will lead to exhausting the Earth's resources, resulting in a collapse of the natural system and society. Industrial society is seen to induce a warped conception of persons and their place in the world; namely, one that de-emphasizes how human existence is inextricably linked to the health of the Earth. Proponents believe that the environment should become the primary consideration in decision making (as opposed to economic growth).

Required to remedy this situation are new kinds of human sensibilities—ones that are less manipulative and destructive of nature. There is the need for a reconsideration of the human/nature relationship, such that humans develop a more humble attitude toward the natural word and toward each other. In essence, it reflects the need to link human systems with natural systems. At the center of this thinking is the concept of biocentrism, or ecocentrism, which presumes that nature has inherent or intrinsic value and should be protected for these reasons, not simply because of its value to human society.

Proponents of the green discourse seek to remake society in line with these beliefs through collective action (reflecting a democratic rather hierarchical approach) to change both the way people think, and so behave, on the one hand, and social institutions and collective decisions on the other (including governments and their policies, international organizations, and corporations). Sometimes these two foci advance in tandem, and sometimes they lead to distinct approaches to decision making, which has led Dryzek to further subdivide the green discourse along these lines. Within the green discourse, the agency to enact change rests with each individual, corporation, and government. Remaking society along the lines proposed within the green discourse will require changes within all facets and levels of decision making.

It is this element that critics cite as a barrier to the widespread adoption of the green discourse. Despite having an identifiable set of core beliefs, the green discourse proposes no real blueprint for how society should be restructured. Instead, it focuses more on trying to reshape the way we make decisions and allowing these new decision-making processes to shape society rather than prescribing a particular end point. To achieve modification, each individual, corporation, and government must reexamine their behavior, reject materialism and overconsumption, and reorder priorities such that the environment is the main priority in decision making. Current trends indicate that society may be unwilling to sacrifice immediate gratification for long-term sustainability. To counter this critique, proponents of the green discourse advance education and tools such as ecological economics and life cycle analysis to better understand the effect and unsustainable nature of society.

Although this article outlines the core ideas that form the foundation of the green discourse, it is important to keep in mind that, as with competing discourses, there are factions within each discourse that challenge and vie for power to shape the discourse. Although this is the case for all discourses, it is especially so for the green discourse, as proponents, who are united in their rejection of industrialization, hold diverse and often divergent views on other issues, such as the role of gender politics and the preferred course of action to change decision-making structures. This diversity has led some authors to further subdivide proponents of the green discourse into subcategories such as deep ecology, ecofeminism, and eco-Marxism.

**See Also:** Consumer Activism; Consumer Society; Environmentalism; Government Policy and Practice (Local and National); Green Consumer; Green Politics; Sustainable Consumption.

### Further Readings

Alexander, R. *Framing Discourse on the Environment: A Critical Discourse Approach.* New York: Routledge, 2009.

Benton, L. and J. Short, eds. *Environmental Discourse and Practice: A Reader.* Hoboken, NJ: Wiley-Blackwell, 2000.

Bøgelund, P. "Opportunities and Barriers for the Green Discourse to Penetrate Policy-Making in Sweden and Denmark." *Ecological Economics*, 63/1:78–92 (2007).

Dryzek, John S. *The Politics of the Earth.* Oxford: Oxford University Press, 1997.

Foucault, M. *The Archeology of Knowledge.* London: Routledge, 1994.

Foucault, M. *The Birth of the Clinic.* New York: Pantheon Books, 1973.

Foucault, M. *Madness and Civilization.* New York: Pantheon Books, 1971.

Foucault, M. *Power/Knowledge: Selected Interviews and Other Writings 1972–1977*, Colin Gordon, ed. Brighton: Harvester, 1980.

Hajer, M. *The Politics of Environmental Discourse: Ecological Modernization and the Policy Process.* Oxford: Clarendon, 1995.

Jamison, A. *The Making of Green Knowledge: Environmental Politics and Cultural Transformation.* Cambridge: Cambridge University Press, 2001.

Litfin, K. *Ozone Discourses.* New York: Columbia University Press, 1994.

Toke, D. *Green Politics and Neo-Liberalism.* New York: St. Martin's Press, 2000.

Torgerson, D. *The Promise of Green Politics: Environmentalism and the Public Sphere.* Durham, NC: Duke University Press, 1999.

*Daniel Murray*
*Independent Scholar*

## GREEN FOOD

The recent growth in the interest in green consumer goods has made extensive inroads into food production and consumption patterns. This growth has accelerated over the past 20 years, with global sales of green food now in the billions of U.S. dollars every year. This article briefly considers what green food is, the reasons people are motivated to purchase and consume green food, and some of the multiple ways green food is grown. These categories often overlap, reflecting the complexity that inheres to the ecosystems in which green food is grown and in how such food is marketed, sold, and transported.

Green food can be considered to be any food product that is grown with environmental concerns at the forefront, outlined later. A farmer may be motivated to grow food in this way for any of the following reasons, whether for the individual merit of the reason or a mix of any of the possible motivating reasons: because of personal environmental or health commitments, for greater profit, to receive farming subsidies, because of changing legal requirements, because of religious/spiritual beliefs, and/or because of consumer demand. Consumers may be motivated to purchase green food for similar environmental, health, political, religious, and/or ethical reasons.

To say green food is food grown in an environmentally friendly manner can mean that a mix of any of the following green farming practices are used by a farmer in the growth and production of such food. The agricultural practices that appear on this list occur within an overall goal of striving toward soil and farm health and environmental sustainability:

- rotational grazing of livestock
- cover cropping
- applying compost to fields
- free-range grazing of livestock
- humane slaughtering of livestock
- provision of habitat for wildlife, birds, and soil organisms at various on-farm sites
- leaving some farm fields fallow for extended periods of time
- using natural, nonsynthetic, nonpolluting herbicides and pesticides that are not petroleum based (with the use of petroleum-based chemicals as a last resort, and then only strategically and in minimal doses)
- attempts to conserve on-farm water and soil
- paying farm workers a livable and humane wage
- use of seeds adapted to local ecosystems, where such seeds are available
- preservation of endangered seed and livestock species
- attempts to grow seasonal produce
- various other organic farming practices consistent with U.S. Department of Agriculture (USDA) or other international or third-party organic certification guidelines

Furthermore, green food and the farmers who grow green food, as well as the consumers who purchase green food products, in large part oppose the following industrial farming practices. These practices tend to be practiced by the overwhelming majority of farmers in the United States and other Western countries, and increasingly are being adopted in developing countries such as China and India:

- genetic engineering of crops, seeds, and livestock
- commodity farming of single, large-scale monocrops
- patenting of traditional seed varieties and indigenous farming methods by transnational corporations, called biopiracy by critics
- dependence on government subsidies that encourage farmers to grow monocrops, especially corn and soybean
- use of groundwater and depletion of healthy topsoil at unsustainable levels
- use of agrochemicals that critics claim lessen soil fertility, contribute to increased rates of cancer, and leech into waterways, eventually contributing to "dead zones" in various international bodies of water
- large-scale breeding and confinement of livestock that depend on increasingly larger doses of antibiotics and that generate multiple tons of manure waste daily, affecting the local water and air quality
- inhumane slaughtering of livestock
- loss of wildlife habitat and diverse ecosystems, as these are transformed into farms that grow monoculture crops or livestock for export markets

All of these practices are considered by proponents of green food farming practices to be environmentally damaging, are claimed to lead to the loss of farming communities as a result of the costs and needed scale of production required by these methods, and are claimed to result in food that is devoid of nutritional and health-giving qualities.

The growth of interest in and demand for green food in the United States, and the development of the various methods for growing green food listed above, can be traced to a confluence of cultural tributaries that began to gain ascendancy beginning with the back-to-the-land movement of the 1960s and 1970s. These tributaries include, but are not limited to, the following:

- the publication of Sir Albert Howard's classic treatise on "organic" agriculture
- the rise of "New Agrarianism" in the United States, especially associated with the writings of Wendell Berry
- the publication of Rachel Carson's *Silent Spring* in 1962, which helped educate the public about the dangers of unregulated use of agrochemicals, especially pesticides, in agriculture
- the legacy of romanticism
- the introduction of Asian cuisine and diets to the United States with the onset of increased immigration from and easier travel to Asia, as seen, for example, in macrobiotics
- the celebration of Earth Day, beginning in 1970
- the growth in marketing acumen and concomitant proliferation of green food products in the retail health food industry, as seen, for example, in the history of the Fortune 500 company Whole Foods
- the rise of the animal rights and animal welfare movement and associated rise in vegetarian and vegan diets (with recognition that such diets are not necessarily practiced for ethical reasons but are also practiced for health reasons)
- the lobbying and marketing efforts of early organic-certifying bodies, such as the Northeast Organic Farming Association and California Certified Organic Farmers
- the rise of both the science of ecology and the development of metaphysical holism, as traced by the environmental historian Donald Worster

The ideals of the back-to-the-land movement and the growth of Green Party politics in Western Europe and Australia, coupled with cultural overlap with the United States on some of the above tributaries, has also contributed to the respective growth of green food consumerism in these areas of the world. Japan and other parts of Asia are also undergoing current growth in the demand for green food for both similar and cultural-specific reasons.

Given this diversity of goals that constitute the growing of green food and the diversity of cultural tributaries that have led to the demand for green food, it should come as no surprise that there are a variety of agricultural methods that are practiced to produce green food. Some of these methods and a quick summary of what motivates their adoption by green food producers are described here.

Veganic farming is a type of agricultural production that relies solely on plant-based materials to provide nutrients to agricultural soils. Veganic farmers are motivated by vegan ideology and lifestyle practices so they do not use animal products, such as manure, to help build soil health, nor do they raise animals as livestock. Rather, veganic farmers depend on "green" manure in the form of cover crops and composted leaves and grasses that they apply back to fields to build soil fertility. Most veganic farmers farm organically and use companion planting and rotational planting of crops.

Permaculture farming is a type of agricultural production that fits into a larger "permaculture" worldview. This worldview and the farming practices associated with it were in large part developed and popularized by Bill Mollison. Permaculture is the attempt to create sustainable, place-based permanent cultures that are able to live in perpetuity within a bioregion. Permaculture includes agricultural practices that recognize various "zones" in which each zone, measured by its distance from the home of a farmer, has zone-specific crops planted within it. Permaculture agriculture also attempts to make use of "edge" spaces in which nature's creative forces and productivity are said to be at their peak. Such farming practices include polycultures of various crops and plant species, as well as animal husbandry.

Biodynamic farming is a method of agriculture directly associated with the work of the anthroposophist Rudolph Steiner. His lectures on agriculture became the basis of biodynamics, and this method of agriculture views a farm as a complete organism in and of itself. All nutrients are recycled within this organism, and plant and animal species that evidence good health are selectively bred so they can continue to flourish within the farm organism over succeeding generations. Biodynamics takes into account what Steiner calls "etheric" forces. Such cosmic forces are believed to affect the growth of plants and animals, so that biodynamic farmers plant according to a biodynamic calendar put together by Maria Thun, which factors in the perceived ethereal effect various stars and planets have on the growth of plants. Biodynamics also creates homeopathies for the soil, collectively known as BD preps (biodynamic preparations) 501 to 508; biodynamic farmers also create BD compost using some of these preps. These preps and the compost are then applied to agricultural fields. Biodynamic farmers, similar to farmers using organic, veganic, and permaculture methods, also do not depend on petrochemical fertilizers and pesticides, and they also tend not to plant monocultures.

Organic farming is currently the most popular form of green food production and is also the most popular form of green food purchased and consumed by the public. The USDA has codified into law what counts as acceptable organic farming practice. An Internet link to the USDA "organic" Website is included in the recommended reading listed for this article. This governmental process of codifying and certifying organic farming practices has been met with much criticism from organic farming and consumer "purists," as these parties feel mainstream agribusiness interests have tried to weaken organic standards so that these interests can capitalize on what has become a multibillion-dollar annual global market.

In addition to this critique of what is perceived to be governmental intervention into organics and the lessening of organic standards, numerous other critiques of green food production and consumption exist. Some critics charge that green food produced by large growers for national and international markets is still a form of large-scale monoculture farming. Other critics point out that green food is still shipped long distances and that these "food miles" are dependent on petroleum and help contribute to climate change. Various critics claim that consumers of green food are elitist and that the prices of green food make such food cost-prohibitive, especially for those in lower income brackets. Supporters of green food claim that this charge is disingenuous and point out that the U.S. government heavily subsidizes conventionally farmed foods so that the costs of such foods are kept artificially low. Another charge levied against green food is that the health claims—in terms of soil, human, and animal health—made by its proponents are unfounded and that tests do not show a significant difference in health-inducing qualities of green food. Supporters of green food counter this charge by claiming that large-scale agribusiness interests have corrupted the U.S. Food and Drug Administration, USDA, and the land-grant universities where such tests are carried out, so that these tests are biased and faulty.

Recent developments in the consumption of green foods points to increasing sophistication in marketing and distribution of green food and to an increasing sophistication in consumer knowledge about food issues. Some of these developments, in both marketing/distribution and consumer knowledge, are the rise in popularity of the slow food and local food movements, the growth of farmers markets that sell regionally and seasonally produced green foods, the rise in popularity of community-supported agriculture, and the burgeoning growth of Fair Trade–certified green food products. Many retail outlets and restaurants now offer Fair Trade, organically produced, and animal cruelty–free products. Some green food producers and retailers are now offsetting the carbon dioxide emissions of their operations by purchasing wind credits or using photovoltaics to power their operations. Some are also using biodiesel and other nonpetroleum-based fuels to power their shipping fleets. Such trends in green food consumption and distribution appear to be growing in popularity, so that even large corporations like Wal-Mart carry green food items in their grocery, dairy, and meat sections. Current trends suggest that green food production and consumption are processes that will have longevity and will most likely continue to grow in national and international popularity and demand.

**See Also:** Composting; Ethically Produced Products; Fair Trade; Food Miles; Genetically Modified Products; Greenwashing; Organic; Slow Food.

### Further Readings

Acres U.S.A.: A Voice for Eco-Agriculture. http://www.acresusa.com/magazines/magazine.htm (Accessed April 2009).

Bird, Christopher and Peter Tompkins. *Secrets of the Soil: New Solutions for Restoring Our Planet*. Anchorage, AK: Earthpulse, 1998, 2002.

Faulkner, Edward. *Plowman's Folly*. Norman: University of Oklahoma Press, 1944.

Fick, Gary. *Food, Farming, and Faith*. Albany: State University of New York Press, 2008.

Fukuoka, Masanobu. *The One Straw Revolution: An Introduction to Natural Farming*. Goa: Other India, 2004, 1978.

Holthaus, Gary. *From the Farm to the Table: What All Americans Need to Know About Agriculture*. Lexington: University Press of Kentucky, 2009.

Howard, Albert. *An Agricultural Testament*. Emmaus, PA: Rodale, 1972, 1943.

Kimbrell, Andrew, ed. *The Fatal Harvest Reader: The Tragedy of Industrial Agriculture*. Washington, D.C.: Island, 2002.

King, Franklin Hiram. *Farmers of Forty Centuries: Organic Farming in China, Korea, and Japan*. Mineola, NY: Dover Publications, 2004.

Mollison, Bill. *Permaculture: A Practical Guide for a Sustainable Future*. Washington, D.C.: Island Press, 1990.

Organic Consumers Association. http://www.organicconsumers.org (Accessed April 2009).

Patel, Rajeev. *Stuffed and Starved: The Hidden Battle for the World Food System*. Brooklyn, NY: Melville House, 2008.

Pollan, Michael. *The Omnivore's Dilemma: A Natural History of Four Meals*. New York: Penguin, 2006.

Seed Savers Exchange. http://www.seedsavers.org (Accessed April 2009).

Shiva, Vandana. *Stolen Harvest: The Hijacking of the Global Food Supply*. Boston, MA: South End, 2000.

Sorensen, Neil, et al., eds. *The World of Organic Agriculture: Statistics and Emerging Trends 2008*. London: Earthscan, 2008.

Steiner, Rudolph. *Agriculture: An Introductory Reader*. Forest Row, UK: Sophia Books, 2003.

Sustainable Agriculture Coalition. http://sustainableagriculture.net (Accessed April 2009).

U.S. Department of Agriculture Organic Farming Portal. http://www.usda.gov/wps/portal/!ut/p/_s.7_0_A/7_0_1OB?navid=ORGANIC_CERTIFICATIO&navtype=RT&parentnav=AGRICULTURE (Accessed April 2009).
Ward, Ronald and Philip Wheeler. *The Non-Toxic Farming Handbook*. Metairie, LA: Acres U.S.A., 1998.
Wirzba, Norman, ed. *The Essential Agrarian Reader: The Future of Culture, Community, and the Land*. Lexington: University Press of Kentucky, 2003.
Worster, Donald. *Nature's Economy: A History of Ecological Ideas,* 2nd Ed. New York: Cambridge University Press, 1994.

*Todd J. LeVasseur*
*University of Florida*

# Green Gross Domestic Product

GDP is an abbreviation for gross domestic product. It is a term economists use to state the economic value of a country's economic activity. It is the sum of all of the goods and services produced in a country in a year. Green GDP is the gross domestic product with environmental costs and benefits added into the estimate of the GDP.

Wealth, similar to many economic terms, has been the subject of different theories and disputes through the centuries. When Adam Smith published his monumental economic study, *The Wealth of Nations: An Inquiry Into the Nature and Causes of the Wealth of Nations,* he was discussing a controversy of his time. What causes a country to be wealthy? Smith's answer was that it was the productivity of the people of a country. It was how many goods and services they produced in a year. His claim was that the GDP was what counted when calculating the wealth of a nation. This answer was in stark contrast to the view of the economic theory of mercantilism. This theory of economics said that a country was wealthiest when it had the most gold and silver.

All of economics is political. The mercantilist economic theory urged governments to engage in policies that help it to garner the most gold and silver. Smith's political economic theory is often called market capitalism. He refuted the policies of governments that hinder or promote economic production in ways that distort the economy.

In the 19th century, Great Britain, the United States, and a few other countries adopted at the least portions of Smith's free market idea. In opposition to the ideas of various economic statists, who want the government of a country to tightly control its economy, Smith proposed the idea of an economy of many individual producers who were free to engage in economic activity under a policy of *laissez faire, laissez passer*. This was a policy of economic liberty, according to which, producers should not have to brook government interference in their legitimate economic activities.

The development of unregulated economic activity in the 19th and 20th centuries was economically very productive; however, the great outpouring of goods from factories and farms was often accompanied by environmental attitudes that disregarded the damage done to the environment. For example, sheep grazing in many rangelands in the western United States abused the range. Stripped of grass, noxious weeds such as sagebrush supplanted the grasses, making the range less productive. It was like farming in the East, which wore out the land and rendered it insufficiently productive.

The consumption of resources and the closing of the frontier by 1900 were matched by concerns for preserving wild spaces and conserving resources. The development of national parks such as Yellowstone and Yosemite under a preservationist philosophy such as that espoused by John Muir and his allies at the University of California–Berkeley and elsewhere was an ecological step that was part of the conservation movement of that era.

## Multiple Uses and the Conservation Ethic

Preservationist strategies for ecological management were paralleled by a different conservation philosophy that sought to manage some natural resource such as timberlands. A major exponent of the philosophy of multiple uses was Gifford Pinchot, who used the term *conservation ethic*. He eventually was able to have a major role in the development of the forestry schools at Yale University and at the Biltmore Estates in North Carolina. He also was to lead the U.S. Forest Service as its first chief during its foundational years.

Muir, Pinchot, and others were very influential in promoting changes in American attitudes toward natural resources and their exploitation. The changes in attitudes convinced many businessmen, farmers, miners, industrialists, and others that conservation was a form of environmental stewardship. The changes in attitudes also prompted legislation promoting conservation programs. All manner of legislative changes were opposed by many on ideological grounds developed from free-market economic studies. The usual basis for opposition was that the conservation laws favored rich corporations and other important players such as government bureaucracies. The view of opponents was that the freedom of individuals and small business people was threatened by regulations.

By the 1970s, conservation programs had become settle policies; however, new developments in chemistry and other sciences had created powerful new tools for exploitation of natural resources. Many of these, such as dichlorodiphenyltrichloroethane (DDT), emerged to give people a way to protect themselves from mosquito-borne diseases. Other chemical developments paved the way for the use of chemical fertilizers. These opened new lands that had been marginal for use in farming with chemical fertilizers, allowing crop production to soar. However, the effects of the use of these powerful chemicals slowly became evident, and a new environmental protection movement emerged from the writings of Rachel Carson, especially after publication of her book, *Silent Spring* (1962). The new environmental movement became national, with the first Earth Day event in 1970. It was to usher in a new global environmental movement that among other things taught the public a new awareness and understanding of the problems of the environment caused by human activities.

One of the outgrowths of the environmental movement was the development of new tools for measuring the effect of human activities on the environment. The activities included agriculture, industry, military, and even tourism. Oceanography emerged as a new science that was able to describe the very depths of the oceans. It also began to provide data on the effects of pollution and overfishing. Biological studies sparked studies of the value of natural resources.

By putting a price on the value of clean air and clean water, or of national parks and forests, a common standard developed for comparing the price of these values with the value of goods and services in the economy. In effect, the economy was expanded to include the natural resources that were used to make goods and service. The cost of pollution was also measurable by its effect on the environment and on people. For a firm to dump its wastes, especially its toxic wastes, was a cost savings to the firm; however, this externalization of costs was simply a process for passing the costs on to the environment or to other people.

The development of environmental thinking concerning the effect of human activity on the environment was aided by advances in science that provided increasingly precise answers to environmental questions. The growth in education in recent decades has provided the armies of researchers and technicians who work to describe the environment and the effect that human activities are having.

## Changes Since World War II

The period dating from the end of World War II has been one of enormous change. The world population grew from about 3 billion people to about 7 billion people. This growth has required ever-greater industrial and agricultural development to feed, house, clothe, and otherwise provide for these billions. Huge parts of the rainforests have been logged, vast new areas have been farmed, and cities have grown to the level that the world's population has become more urban than rural for the first time in human history. The huge consumption of fossil fuels to provide energy for electrical and transportation grids has also had huge environmental effects. As a consequence, the growth in human population has also raised questions about the carrying capacity of the Earth.

The discovery of increases in the atmosphere of greenhouse gases has evoked controversy. These gases, if unregulated, will contribute to climate change by creating global warming with seriously negative effects on the environment of the entire Earth, such as it has been experiencing for the last several hundred years. Some scientists have disputed the claim that there is greenhouse gas–generated global warming. Regardless of the answer to this dispute, the fact is that human exploitation of the planet has reached a scale much greater than at any time in human history.

With all of these changes, and with a growing global ecological understanding, the move to develop an economics of the environment, matched by lifestyles and consumption, has become the cause of many people who want a green economy. Green economics means that the lifestyle, the tourism, the consumption, and the other effects of humans on the environment have a smaller footprint than has been the case in the latter decades of the 20th century.

The world since 2000 has become a globally connected world. Whether through live television broadcasts seen around the world or via the Internet or telephones, people around the globe are increasingly aware of what others are doing. The effect has been to enlarge the common understanding of the global economy to one that is no longer measured on a national scale but, instead, on a global scale. GDP was originally called gross national product; however, the development of a global economy has shifted economic perspectives to seeking to understand the vastly expanded global economy and its environmental impact. As a result, green GDP is now seen as a statistic that will represent economic growth and its environmental consequences.

Although there have been a number of attempts made at trying to measure the relationship between economic activity and the degradation of the environment, which also includes the depletion of natural resources, sustainability is an important part of green GDP.

In 2004, the premier of China, Wen Jiabao, told the world that China was developing a green GDP index. The goal sought was a measurement that would describe how well the economy was doing, and at what environmental price. The green GDP index, it was hoped, would give the government of China and its Communist Party a performance measurement.

In 2006, the Chinese government published the world's first green GDP, which was a report of the year 2004. The report showed the cost to the economy of ecological damage to be about 3 percent of the economy, or a price tag of $66.3 billion. The attempt to use a green GDP measure for the year 2005 failed in 2007 because including the environmental costs leveled the growth rate of the Chinese economy to nearly zero. The fact is that green measures can show that environmental degradation exceeds the economic returns. To respond to this kind of outcome would lead to the conclusion that the economic activity was too costly and that it should be shut down. In many cases, including China, the political costs would be very high as people moved to expel political leaders supporting environmental concerns over the jobs of workers and owners.

Efforts to create a green GDP are still in their infancy. The development of true measurements that calculate environmental costs of losses in biodiversity or the damage done by global warming is still a work in progress. To develop concepts that would measure the amount of waste per year or the carbon footprint of production is similar to development in other fields of human activity—a matter of slow progress until a breakthrough occurs.

**See Also:** Consumerism; Ecotourism; Lifestyle, Sustainable; Materialism; Overconsumption; Resource Consumption and Usage; Sustainable Consumption.

### Further Readings

Brekke, Kjell A. *Economic Growth and the Environment: On the Measurement of Income and Welfare*. Northampton, MA: Edward Elgar, 1997.

Cato, Molly Scott. *Green Economics: An Introduction to Theory, Policy and Practice*. London: Earthscan, 2009.

Goleman, Daniel. *Ecological Intelligence: How Knowing the Hidden Impacts of What We Buy Can Change Everything*. New York: Broadway Books, 2009.

Hecht, Joy E. *National Environmental Accounting: Bridging the Gap Between Ecology and Economy*. Washington, D.C.: Resources for the Future, 2005.

Jones, Van. *Green Collar Economy: How One Solution Can Fix America's Two Biggest Problems*. New York: HarperCollins, 2008.

Makower, Joel and Cara Pike. *Strategies for the Green Economy: Opportunities and Challenges in the New World of Business*. New York: McGraw-Hill, 2008.

Murty, M. N. *Environment, Sustainable Development, and Well-Being: Valuation, Taxes, and Incentives*. Oxford: Oxford University Press, 2009.

Uno, Kimio and Peter Bartelmus, eds. *Environmental Accounting in Theory and Practice*. New York: Springer, 2002.

*Andrew Jackson Waskey*
*Dalton State College*

## Green Homes

A green home is one that aims to minimize environmental and health impacts. More specifically, building a green home means using a holistic design and construction process that takes into consideration the size of the home, appropriate site selection and orientation, energy- and water-efficient design, and the effect of materials used, as well as minimizing waste and optimizing the environmental health of the home, such as indoor air quality. Some proponents of green homes argue that a better term for a green home is a "healthy home"—healthy for the planet and healthy for the household.

Over the past 50 years, the average size of housing in North America increased from 1,100 square feet in 1950 to 2,300 square feet in 2003. Larger houses mean more resources are used in terms of energy, water, and building materials.

*Source:* iStockphoto

Homes also give us a sense of place and are a reflection of who we are—as an individual, a neighborhood, and a society. The architectural literature on green buildings discusses the relationship between the design of a home and the environment. On one end of the continuum, a home can be designed to coexist with nature, and on the other end, man-made objects and nature act as opposing forces. The design of a home is a holistic reflection of who we are and how we see the world. Although green home design often focuses on materials and technological innovations that reduce the environmental impact of the home, the larger relationship between a home and its surrounding ecosystem should not be overlooked.

## Environmental and Health Impacts of Homes

The environmental and health impacts of homes are both numerous and far-reaching. In Canada it is estimated that 17 percent of all energy used goes toward running homes, whereas the U.S. Department of Energy reports that 20 percent of all fossil fuel use in the United States is from the residential sector. The implications of this are increased greenhouse gas emissions and other air quality issues. According to Natural Resources Canada, homes that are more than 25 years old have the potential to save an average of 35 percent of their energy use.

The effect of buildings in the residential sector is also increasing as a result of the increased average size of houses in North America. Larger houses mean more resources are used in terms of energy, water, and building materials. Over the past 50 years, the average size of housing has more than doubled, from 1,100 square feet in 1950 to 2,300 square feet in 2003. Furthermore, as the number of people per household decreases, average square footage per person has tripled from 290 square feet in the 1950s to 893 square feet in 2003.

From a health perspective, both new homes and older ones can have an effect on the individuals living in the home in terms of indoor air quality (e.g., from toxins in building materials, poor air flow, and lack of natural light). The impacts on the environment include waste from construction (estimated at 25–30 percent of the municipal waste stream), deforestation for wood products, and energy used in construction. A typical wood-frame house produces somewhere between three and seven tons of solid waste from the construction process alone.

## Emerging Trends in Green Homes

Although green home designs have been around for some time, it has only been in the last decade that they have gained popularity among multiple stakeholders such as planners, engineers, developers, builders, architects, and homeowners. In the 1960s and 1970s, green homes were seen as outliers, often built away from other homes, and their construction and design were not always aesthetically pleasing from an architectural point of view. The 1990s produced homes with green features that were making a design statement. These features were designed purposely to stand out from the house, and public awareness of green homes began to increase.

The gap between aesthetics and greenness is now being bridged. Green features are more integral to the design and less obvious visually. Increasingly, green homes are emerging as focal points of neighborhoods—there are even entire developments that feature all green homes. The popularity of green homes tours and green living shows are further indications that green homes are moving toward the mainstream.

Although North America has made great leaps in green home design, Europe is still much more advanced in this area. Stricter building codes, more innovative design from more years of experience in green homes, and a higher expectation among the general population have contributed to these advancements. Meanwhile, countries like Japan have developed efficient, innovative home designs using a very small footprint. Although some of the motivations of this movement toward small, well-designed houses are the result of economics and the high cost of available land, the outcomes have resulted in innovative designs and have created an opportunity to build well-designed homes in a very small space.

In the United States, green homes are expected to make up 10 percent of new home construction by 2010, up from 2 percent in 2005. The green home market in the United States currently exceeds $7 billion per annum and is projected to increase by 40 percent over the next few years.

## Features of a Green Home

There are varying shades of green homes. Many would argue that the greenest home is the one already built—that is, it is better to retrofit or renovate an existing home to be greener than to tear it down to start over building a completely green home. The notion of reduce, reuse, recycle applies to homes as it does to other materials, and it is important to adopt a flexible and holistic approach to designing a green home (or greening an existing home) at every stage in the process.

Although innovations in available technology products for designing green homes are continually evolving, the basic principles of green design remain the same.

### Size

The size of a home is a fundamental, yet often overlooked, aspect of green design. The greenest 4,000-square-foot home is no match for an 1,100-square-foot home in terms of construction materials needed, waste generated, and energy consumed. Well-designed homes can provide effective living space without needing a lot of room. An extreme example of this is the Tumbleweed Tiny Home, which has gained notoriety across North America and beyond. S. Susanka's series of books on "The Not So Big House" has also demonstrated smart design on a smaller footprint.

## Building Design, Site Selection, and Location

Creating green homes starts with proper site selection. Consideration should be given to renovating an existing building before starting a new one. A green home takes advantage of existing sun/shade on-site and minimizes damage to what is already there. The design of the home also greatly affects the potential for water and energy efficiency, as well as the comfort, or livability, of the home. Important elements to consider in the design are orientation (with respect to sunlight and wind), amount of natural light, shading devices, and light-colored or green roofs to reflect heat/sun.

Closely related to site selection is the location of the lot in relation to services, community, and surrounding ecosystems. An ideal location for a green home in an urban area is one to that is close to work, school, shopping, and public transport (e.g., how many errands can you do without your car?). Higher-density areas and the use of infill lots (such as former parking lots, rail yards, or subdivision of larger urban lots) are ideal, and new green homes and neighborhoods should not be built on environmentally sensitive sites like wetlands and prime farmland.

## Building Materials

Materials used in the building process are chosen with careful consideration of their life cycle. This includes choosing local products and using products made from renewable resources, as well as choosing those that minimize waste, toxins, and other environmental impacts. Emphasis is placed on building materials that either have significant recycled content or are reclaimed materials. The ability to recycle the materials is also important. Some materials are considered "green" because of their durability or long life span.

## Energy Efficiency

The energy efficiency of a house is a key factor in reducing the home's ecological footprint over the long term. The design of the house in this regard has an effect on both greenhouse gas emissions and energy bills for years to come. Energy-efficient homes include those with high R-value insulation, high-efficiency heating and air conditioning systems and water-heating systems, energy-rated windows and exterior doors, and efficient lighting and appliances. Homes aim for self-sufficiency by using renewable energy technology (such as through solar or geothermal heating systems that provide potential for houses to be "off-grid"). These are known as net-zero energy homes and are capable of producing an annual output of renewable energy that is at least equal to the amount used.

## Water Efficiency

A green home uses water efficiently and also implements proper design for site runoff. When feasible, water is reused or recycled on-site. Green water-saving features include water-conserving irrigation systems such as rainwater collection and storage systems and water-efficient kitchen and bathroom fixtures.

### Indoor Environmental Quality

Green homes are also healthy homes for the people living in them. Important considerations are the amount of natural light, natural ventilation and moisture control, and avoidance of materials with high–volatile organic compound emissions.

### Landscaping

Landscaping is appropriate for the climate/region, with native and drought-tolerant plants, appropriate drainage, rainwater barrels, and minimal need for lawn mowing and watering. Homes provide appropriate shading in summer and wind protection in winter. Landscaping has a significant effect on the comfort of the home, as well as on its energy and water efficiency. For example, large-canopy trees and other landscaping shades exterior walls, the driveway, and patios to minimize heat islands. In most regions, yards are landscaped with drought-tolerant rather than water-intensive plants.

## Green Home Certification Programs and Incentives

In recent years, several programs have developed that offer environmental certification and energy ratings of new and existing homes. Many of these programs offer varying levels of achievement such as "bronze," "silver," and "gold" certification, depending on the number of green features that have been included in the design. Some of these programs are independently run, whereas others are government sponsored; however, most programs promote third-party independent assessment of green home achievements. Examples include

- Leadership in Energy and Environmental Design (LEED) certification;
- Energy Star (U.S. Environmental Protection Agency);
- R-2000, EcoEnergy, and EnerGuide for New Homes (Natural Resources Canada);
- National Green Building Standard (recently approved by American National Standards Institute as their green building rating system); and
- Built Green (for Canadian homebuilders).

This list is just an example of some of the many programs that currently exist. In the United States alone there are nearly 50 regional and national green home labeling programs. Most of these programs have developed standards for most of the green features that were described earlier, although some of them place more emphasis on the energy efficiency of the building's envelope. Government incentive programs for green homes are also on the rise, and some banks are offering green mortgages and green renovation loans.

## The Benefits of a Green Home

Building or owning a green home can have profound effects not only on the environment but also on the health and comfort of its occupants, as well as on the long-term financial costs of owning a home:

- Significantly reduced greenhouse gas emissions over the lifetime of the house (the U.S. Environmental Protection Agency estimates that these savings equate to 4,500 pounds of greenhouse gases per house per year)
- Long-term savings in energy and water costs (average reduced energy usage for green-certified homes ranges between 30 and 60 percent)

- Health improvements for home occupants
- Lower insurance premiums and lower mortgage rates (in some instances, depending on availability)
- Possibility of rebates, tax breaks, and other incentives, depending on where you live

## Barriers to Green Homes

At this time, the biggest barrier to building green is cost or, in some cases, perception of cost. When looking at costs, most people consider the capital costs of building the house without taking into consideration the potential for long-term savings through reduced energy and water bills and maintenance. Furthermore, there are the indirect environmental and health benefits that are hard to define financially. A study commissioned by the state of California on LEED-certified buildings suggests that it costs about $4 more per square foot to build green, with potential savings over 20 years of between $48 and $67 per square foot. For a 2,000-square-foot house, that would be an extra initial cost of $8,000 with the potential for savings of between $96,000 and $134,000 over the lifetime of the house.

Some people find it hard to build green because of lack of knowledge, lack of extra funds, or lack of contractors in their area who know how to build green homes. Some homeowners have had difficulty in using reclaimed materials in their design because of building code restrictions. Planning and zoning laws in some communities need to catch up to trends and to make it easier for innovative green home ideas to be implemented. Some municipalities currently require homes to be of a minimum size, which is not encouraging people to build smaller houses. As green homes become more mainstream, these barriers are beginning to decrease.

As we look forward to the next era of green homes, perhaps we also need to look backward. Perhaps the greenest homes are the ones of yesteryear—simple designs, local materials, and of reasonable size and use of amenities. It all comes back to thinking of a green home as a reflection of who we are.

**See Also:** Environmentally Friendly; Floor and Wall Coverings; Heating and Cooling; Insulation; Lawns and Landscaping; Windows.

### Further Readings

Duran, S. C. *Green Homes: New Ideas for Sustainable Living*. New York: Collins Design, 2007.

McGraw-Hill Construction. *Residential Green Building SmartMarket Report*. New York: McGraw-Hill Construction, 2006.

Stang, A. and C. Hawthorne. *The Green House: New Directions in Sustainable Architecture*. New York: Princeton Architectural, 2005.

Susanka, S. *The Not So Big House: A Blueprint for the Way We Live*. Newtown, CT: Taunton, 2001.

U.S. Green Building Council. http://www.usgbc.org (Accessed August 2009).

Whole Building Design Guide. http://www.wbdg.org/design/sustainable.php (Accessed August 2009).

Wilson, A. and J. Boehland. "Small is Beautiful: U.S. House Size, Resource Use and the Environment." *Journal of Industrial Ecology*, 9/1–2:277–87 (2005).

*Jennifer K. Lynes*
*University of Waterloo*

# GREEN MARKETING

Green marketing, sometimes referred to as "environmental marketing" or "ecological marketing," is a recent and emerging system of labeling and advertising for products and services that are promoted by claims of either a reduction or elimination of negative ecological impact. A constituency of businesses and individual consumers subscribing to this trend is rapidly developing for motives both financial and of genuine environmental concern. Although the number of businesses that advertise their products and services as "green" has continued to grow, this marketplace transformation has been delayed by a lack of clarity in what constitutes being green. Although corporate interests often remain hesitant to expose themselves to the financial liabilities of civil or regulatory reprimand as a result of perceived embellishment, the individual consumer often regards advertised claims with mistrust. The phenomenon of green marketing oftentimes necessitates radical alterations in the way products function and are manufactured, packaged, and advertised. Those interested in green marketing must consider the conceptions and history of green marketing, specific types of green marketing, and regulations and consumer protections related to green marketing

One major concept closely associated with green marketing involves the promotion of a service's or product's purported lack of ecological debasement despite evidence to the contrary. When an institution promoting such claims demonstrates blatant disregard for environmental concerns in another area of its commerce, this can inflict serious damage to a company's market share. These contradictions and fabrications are known as "greenwashing." Use of this terminology usually pertains to accusations of corporations defrauding the community, but the term has been applied to nongovernmental agencies (NGOs) and government agencies, as well as to politicians and individuals. The allure, and analogous menace, of greenwash is its potential to sway favorable opinions of regulators, investors, employees, competition, and the general public based on one's desire to engage in ecologically responsible behavior. Politicians might attempt to appeal to the green market within their constituencies by promoting themselves, their views, and their actions as environmentally sound. Similarly, government agencies tasked with enforcing or regulating environmental issues have an innate motive to rely on greenwash, rather than confess inaction or inefficacy to taxpayers. A problematic, recurring, and oftentimes contentious issue within the greater discussion of green marketing on the whole is the demarcation or distinction between green marketing and greenwashing. Indeed, most legitimate action aimed at the reduction of negative environmental impact is motivated largely in part by profit incentives. In a system in which profit incentive is the main factor in determining which products, services, and businesses are successful and will continue to exist, it seems hypocritical for the public to demand that those products, services, and businesses regard moral or ethical considerations more highly than the financial considerations that their very existence necessitates.

## The Birth of Green Marketing

As early as 1975, the American Marketing Association held a workshop on ecological marketing. This workshop was spurred by increased interest in green lifestyle and products that became popular in the 1970s. Although interest in a more sustainable lifestyle, and the products that supported this, began to attain a foothold in that decade, green marketing first became significant during the 1980s. Certain companies, such as Nike and Ben and Jerry's Homemade, began to issue corporate social responsibility reports. Corporate social responsibility reports augmented traditional financial reports to explore a company's environmental impact. This decade also saw a split in green marketing, in which some advertisements focused on the environmental benefits of certain products while others concentrated on the green sensibilities of the manufacturer. Advertisements focusing on products, on the one hand, tended to emphasize the positive attributes of a specific manufactured good, especially in comparison with preexisting products. Marketing that dwelled on the green sensibilities of the corporation, on the other hand, stressed corporate good deeds and initiatives that were beneficial to society. In the early 1970s, chlorofluorocarbons (CFCs) were identified as chemical substances that, despite their usefulness, can deplete the Earth's protective ozone layer in the upper atmosphere. This led to a near-total banning of the use of CFCs as propellants in consumer aerosol products in 1978. Though some manufacturing industries still make use of these chemicals, this use is gradually being phased out entirely.

During the 1980s, most green marketing focused on the necessity of the current population to meet their needs in a manner that did not compromise the ability of future generations to meet their own needs. This approach stressed widespread thinking about sustainability and focused on getting the public to consider ways in which they could alter everyday activities to foster sustainability. Green marketing began to be seen as something that could be integrated into all aspects of the sales process: new product development, analysis, packaging, consumer trials, communication, and advertising. Other advocates of sustainability also were enlisted to promote green products, including suppliers, retailers, consumers, educators, members of the community, regulators, and NGOs. This approach greatly broadened interest in and support for green marketing and the products it promoted. Environmental issues came to be balanced with primary consumer needs as the realization was reached that environmental goods needed to support primary consumer needs.

Since May 2000, the Federal Trade Commission (FTC) has taken legal action against only three companies for violating the guidelines it established to guide companies marketing green claims. A problem arising in the enforcement of these regulations is that the FTC is an independent and underfunded agency within the federal government that does not have the financial capability to provide massive national oversight or enforcement of its regulations regarding green marketing.

After 2000, green marketing evolved in its focus. Scientific environmental points of view were disseminated to a wider public, supported by a broad range of authors and reports, including former U.S. Vice President Al Gore, the 2005 United Nations Millenium Project report and the U.K.'s *Stern Report*, all of which proposed moving to a low-carbon global economy. At this point, green marketing also began suggesting a variety of different approaches that society could take to achieve a more sustainable culture—not surprisingly, individual marketing campaigns tended to support the approach taken by the specific manufacturers that underwrote the advertising campaign.

## How Green Marketing Works

One example of green marketing that has garnered much criticism is Clorox's new line of GreenWorks cleaners, which feature prominent green-colored labels adorned with large sunflowers. Clorox GreenWorks products are marketed as being biodegradable, nontoxic, plant-based, non–animal-tested, nonallergenic, having full ingredient transparency, and made with recyclable packaging. However, many consumers are reluctant to trust a company's "green" claims when the company continues to sell other products viewed as hazardous to individuals' health and the environment in general. This paradox is reminiscent of the criticism directed toward tobacco manufacturer Phillip Morris's promotion of health awareness and "light" cigarettes, despite their continued sale and revenue from the products that, by virtue of their own promotional claims, are in fact hazardous to the health of consumers. When some statements by a manufacturer are contradicted by their other actions, it becomes difficult for consumers to place faith in green marketing declarations made by that manufacturer.

Philips Lighting, a division of Royal Philips Electronics, initially had difficulty marketing a compact fluorescent light (CFL), which it had designed as a replacement for incandescent bulbs. Philips Lighting met with initial consumer resistance to the CFL bulbs, which retailed for over $15 each—much more than the incandescent bulbs they replaced, which sold for approximately 75 cents each. Although the CFLs had some appeal to green consumers, Philips Lighting had difficulty making the product attractive to ordinary consumers. Only after the bulbs were reintroduced under the Marathon label was the product successful. Philips Lighting's Marathon advertising campaign emphasized the CFL bulbs' extended life and stressed that consumers would save over $26 in energy costs over each bulb's five-year life span. Philips Lighting also sought and obtained the U.S. Environmental Protection Agency's (EPA) Energy Star label for the CFL bulbs—a move that added to the product's credibility.

The marketing efforts of multinational corporations such as Clorox and Philips, which have mature and nongreen product lines in addition to their sustainable offerings, are very different from niche companies that were formed exclusively to manufacture and market green products. Burt's Bees, for example, was started as an environmentally friendly natural personal care company in 1984. Its products are all intended for personal hygiene, such as hair care, lip care, skin care, and outdoor remedies. Burt's Bees uses natural products for all its products and uses minimal processing to ensure the purity of its products. In keeping with the company's motto to work for "The Greater Good," Burt's Bees developed a business model that sought to ensure that all company practices were socially responsible. To that end, Burt's Bees uses all natural ingredients, engages in environmentally friendly packaging and product containers, and seeks to engage in humanitarian practices. Ironically, Burt's Bees was purchased by Clorox in 2007.

In contrast, there are many smaller ecofriendly companies that have not opted for acquisition by a major manufacturing company. One such smaller company is Seventh Generation, Inc. With its very name stemming from the Iroquois law that states that one must consider the ramifications of every action as it impacts the next seven generations, Seventh Generation, founded in 1988, markets natural household cleaners, personal care, and paper products consisting of nontoxic, biodegradable, hypoallergenic, and recycled materials containing no phosphates or chlorine. Although some public criticism has been directed at the company for a perceived inefficacy of their dish-cleaning products, defenders claim that much of this is a result of mineral-heavy hard water that is resistant to natural cleaners. Nonetheless, Seventh Generation has experienced a reduced market share since Clorox's GreenWorks line of cleaners was introduced. Despite this fact, the company continues its attempts to independently revamp its products aimed at a loyal constituency of consumers who prefer to support these smaller, nonmultinational alternatives over larger competitors. Small companies earnest in their attempts to bring responsibly manufactured products to a large market face much difficulty when competing with large corporations.

An example of a highly successful green marketing campaign is the Toyota Motor Corporation's Prius. The hybrid vehicle made its debut in the Japanese market in 1997 and has rallied significant, positive critical acclaim in Asian and North American markets, as well as among agencies such as the U.K.'s Department for Transport, the EPA, and the California Air Resources Board. The Prius, which is sold in both four- and five-door models, combines an internal combustion gasoline-powered engine with a battery-power backup system to increase mileage per gallon of gasoline, which is now in the range of 50 miles per gallon. The Prius has been heavily marketed as a green automobile, and Toyota continues to design vehicles with increasing appeal to the green market.

## Regulation of Green Marketing

In the United States, the FTC works in cooperation with the EPA to regulate green marketing claims made by manufacturers, merchants, and vendors. These government agencies have developed guidelines to ensure that advertised environmental marketing claims do not mislead consumers. On receiving complaints of companies violating these guidelines, the FTC enters consumer complaints into the Consumer Sentinel Network, a secure online database and investigative tool used by hundreds of civil and criminal law enforcement agencies in the United States and abroad. The Consumer Sentinel Network terms itself "the unique investigative cyber tool" that provides member law enforcement agencies with access to millions of consumer complaints regarding matters including identity theft, do-not-call registry violations, computers, the Internet, online auctions, telemarketing scams, advance-fee loans, credit scams, sweepstakes, lotteries, prizes, business opportunities, work-at-home schemes, health and weight loss products, debt collection, credit reports, and financial matters.

Part of the confusion stems from the variety of terms used to indicate environmentally friendly practices. Some ecoclaim labels currently used on product packaging include environmentally safe, environmentally friendly, environmentally preferable, ecosafe, Earth smart, recyclable, (bio)degradable, ozone friendly, preconsumer, postconsumer, (essentially/practically) nontoxic, compostable, CFC-free, the three-chasing-arrows recycling symbol, and the Society of the Plastics Industry's symbol/code that indicates the type of plastic from which a particular product is made.

The FTC and EPA guidelines for green marketing claims assert that labels must evidence their claims and state the specifics unless they are obvious to consumers. For example, if a product is labeled "nontoxic," the manufacturer must have reason to believe that the products will not pose any significant risk to people or to the environment. A product that is labeled with a broad claim such as "ecosafe" can be misleading because all products, packaging, and services have at least some environmental impact. Even something labeled "biodegradable" could be considered misleading, as laws that mandate that landfills be impervious to excess sunlight, air, and moisture to reduce pollution also ensure slow decomposition. In a landfill, even biodegradable materials like paper and food may take decades to decompose. If a product is labeled "CFC-free," this can be misleading, as the product could potentially contain volatile organic compounds, such as alcohols, butane, propane, or isobutane, which are also damaging. In addition, the product could cause ozone formation, which, although beneficial in the upper atmosphere, could create smog and exacerbate breathing problems for some people if this formation were to occur at the ground level. Therefore, if a product is labeled "ozone safe," the company making that assertion should have substantial reason to believe that the product is not harmful to the atmosphere at any layer of altitude.

All of these issues underscore just a few of the many complications associated with ensuring that green marketing claims are based on something more substantial than financial consideration. In a free-market economy, profit drives many decisions regarding marketing products, whether the products are green or not. In an atmosphere in which environmentally friendly products have great appeal, marketing products as "green" can be highly profitable. Businesses that seek to tailor their products to the environmentally conscious consumer, along with the government agencies and NGOs that regulate and monitor these claims, must take adequate precautions to ensure that the information they provide to consumers is clearly articulated, evidenced on fact, and in no way misleading to the end consumer. Otherwise, consumers will have no way to distinguish green marketing from greenwashing, causing great confusion. This confusion could hinder the advancement of responsibly manufactured products and services, causing the market for these to falter. As the evolution of green marketing continues, clearer definition of what constitutes green products is necessary. Transparency in decisions and processes determining these matters on the part of all involved, along with swift, decisive enforcement of corporate greenwashing policies, are essential components that will ensure security and stability in the green market.

**See Also:** Advertising; Biodegradability; Composting; Consumerism; Ecolabeling; Environmentalism; Environmentally Friendly; Ethically Produced Products; Greenwashing; Regulation.

### Further Readings

Esty, Daniel and Andrew Winston. *Green to Gold: How Smart Companies Use Environmental Strategy to Innovate, Create Value, and Build Competitive Advantage.* Hoboken, NJ: Wiley, 2006.

Grant, John. *The Green Marketing Manifesto*. Hoboken, NJ: Wiley, 2008.

Levinson, Jay Conrad and Shel Horowitz. *Guerrilla Marketing Goes Green: Winning Strategies to Improve Your Profits and Your Planet*. Hoboken, NJ: Wiley, 2010.

Makower, Joel. *Strategies for the Green Economy: Opportunities and Challenges in the New World of Business*. New York: McGraw-Hill, 2008.

Ottman, J. A. *Green Marketing: Opportunity for Innovation*. New York: BookSurge, 2004.

Peattie, K. *Environmental Marketing Management: Meeting the Green Challenge.* Philadelphia, PA: Trans-Atlantic Publications, 1995.

*Stephen T. Schroth*
*Jason A. Helfer*
*Daniel O. Gonshorek*
*Knox College*

## Green Politics

*Green politics* refers to the political sphere of the Green movement (this article will capitalize Green for clarity), an international movement that often overlaps with other political movements but tends to describe itself as more than merely political and concerned with more than just specifics of governance. Though especially prioritizing environmental concerns, the Greens tend to intersect with social liberal concerns, such as feminism, peace movements and nonviolent protest, social justice, the protection of civil liberties, and a strong focus on participatory democracy and grassroots organization. This is an important consideration because it is not an automatic consequence of environmentalism: Although an individual or group could theoretically call for a dictatorial regime that curtails business and individual activity to protect the environment—a sort of martial law environmentalism—this would not be considered a Green position. Although there are a range of positions that can be taken in relation to green consumption (from consuming less to consuming green products as alternatives), green politics can both inform and facilitate the actions of consumers, organizations, and governments in relation to key themes of ecological wisdom, social justice, participatory democracy, nonviolence, sustainability, and respect for diversity.

Green politics developed out of 19th-century conservation movements that were a reaction to the rapid expansion, industrialization, mining, and urbanization seen in the later years of the Industrial Revolution. Complaints about the effects of urban living—contamination of bodies of water, air pollution, noise—had been made since the ancient world, and efforts to create nature reserves and to introduce responsible, sustainable methods of crop management are just as old. Human civilizations had weathered global climate changes in the 6th and 13th centuries without having the vocabulary to talk about them or the scientific awareness to conceive of climate as a complex system. Events like the spread of plague in 14th-century Europe led to a sort of embryonic understanding of the connections between public health (and public good) and the interaction between man and the natural world; it was the squalid conditions and dense population of urban living that permitted the epidemiology of the Black Death. The long-term effects of industrialization on the countryside and the environment catalyzed the conservation movement in the United States, which borrowed from the French and German "scientific forestry" that had begun a century or so earlier. Scientific forestry was designed to preserve the state of the forest, to safeguard it against what we now recognize as climate change, as well as from wildfires and artificial deforestation. Though no one was using the word sustainability yet, the conservation movement wanted exactly that—sustainable forests, active efforts at planting trees to make up for those being lost to industry, and lands set aside that industry would not be allowed to touch.

## Green Politics and the National Parks

Though Green parties in Europe have had considerably more success in politics, the first major win for Green politics was in the United States, with the creation of the national parks. In Europe, as in most of the "Old World," parks and other vast areas of land set aside and preserved from development are generally donations from one noble or another, or the property of the monarch. In the United States, in contrast, at a point in its history when it was still mostly unsettled but the age of the wild frontier was clearly over, certain lands were specifically set aside as public property and protected against development, often for the sake of its own natural wonder. The United States is so vast, so geologically diverse, that the radicalness of this is often taken for granted. Yellowstone National Park was the first true national park, established in 1872; the lands that became Hot Springs, Arkansas, and Yosemite National Park had been put aside by earlier acts of government but were not specifically designated as national parks. Yellowstone was directly managed by the federal government and represented what American writer Wallace Stegner called "America's best idea," putting lands of natural beauty aside for the enjoyment of all, "reflect[ing] us at our best, not our worst."

Conservationism spread through Europe and British-controlled India as well as the United States—anywhere that the resource needs of the post–Industrial Revolution world were leading to or threatening deforestation. In time, concern also arose for the species of animals put at risk by local and global climate change and ecosystem disruption; in the early days, the forests were the main area of concern. The oldest environmental group in the United States—and still the largest, with a 2009 membership of 730,000—is the Sierra Club, founded in San Francisco in 1892 with the mission "to explore, enjoy, and protect the wild places of the earth; to practice and promote the responsible use of the earth's ecosystems and resources; to educate and enlist humanity to protect and restore the quality of the natural and human environment; and to use all lawful means to carry out these objectives." Founded by professors, the club in its early days pursued the transfer of Yosemite Valley—now Yosemite National Park—from the possession of the California government to the federal government, the establishment of the Glacier and Mount Rainier national parks, and the protection of California's coastal redwood forests, which remain a club concern. In addition to these early victories, the Sierra Club supported the 1916 creation of the National Park Service—a federal agency devoted to the parks' administration. In addition to forest protection, the club pursued waterway protection by opposing many dams, an effort with which it had less success. At some points in the past, the club's policies on population control have intersected with the eugenics movement and anti-immigration nationalism, although in the present day the club has largely distanced itself from such intersects.

## Foundations of Green Political Thought

In the early 1970s, various Green parties formed in countries around the world, responding to the social changes of the 1960s and the new awareness of anthropogenic environmental damage. Green political thought became more formalized over the course of the decade, expressed by the German, Australian, and Swedish Greens as "the four pillars": ecological wisdom, social justice, participatory democracy, and nonviolence. In 2001, the Global Greens Charter was drafted by 800 delegates from Green parties in 72 countries and added to the four pillars two more principles: sustainability and respect for diversity. Individual national parties may interpret these values differently, but all agree on the broad strokes.

Ecological wisdom recognizes the need to limit, and when possible undo, the anthropogenic harm done to the environment. In many cases this calls for learning from the natural world to understand what is healthy for the ecosystem and to determine humankind's place in it. In particular, the philosophical school of "deep ecology" is based on the core principle that the living environment has the same inherent right to life and health that the human race does, and that therefore a course of action should be pursued (and when necessary, discovered) that does not put human health at odds with the ecosystem's health. Deep ecology considers irrelevant the religious question of whether humans have a soul—that is, whether there is a supernatural case for human exceptionalism. Humans are, rather, part of a larger living world, gifted with agency, conscience, and the ability to consider consequences. Deep ecology is at its core more metaphysical and spiritual than inductive and particularly draws on Spinoza as an antecedent. Both deep ecology and the idea of ecological wisdom overlap with the animal rights movement.

Social justice refers generally to policies that pursue economic egalitarianism and is one example of overlap between the Greens and the Catholic Church, which has long condemned significant disparities of income and means and pursued justice for the poor. The interpretation of this principle varies from party to party, with some describing it in somewhat socialist terms and the American Greens considering it to represent the right of all people to benefit equally from social and natural resources—not exactly a market-fundamentalist interpretation, but certainly capitalist-minded. Social justice also includes the need for equal and affordable access to healthcare for all people.

Participatory democracy is a democracy in which citizens get directly involved. Although this can include direct democracies, it is not at odds with the more common representative democracy. Though in representative democracy most of a citizen's votes are for candidates who will legislate and govern in their stead, such citizens can participate more by becoming informed about the issues, "getting out the vote," publicly and actively advocating the issues that are important to them, and helping others become informed. In addition, most districts have ballot issues as well as elections for local, county, and state concerns. Also key to Green issues is anticipatory democracy, which seeks to inform political decisions (by both voters and officials) by taking into account reasonable expectations for the future, such as the long-term consequences to the environment or the benefit of passing laws today about the issues that will arise tomorrow. Anticipatory democracy would include, for instance, seeking to reduce carbon emissions or other sources of environmental damage before they reach critical levels rather than dealing with the damage once it occurs, or calling for a thorough safety inspection of nanoparticles in use in consumer goods to avoid dealing with the human and environmental consequences after the fact, as happened with dichlorodiphenyltrichloroethane (DDT) and asbestos.

Nonviolence rejects the use of violence in the pursuit of social change. Nonviolence is not inaction, nor pacifism—pacifism rejects violence on the personal level, nonviolence rejects it as the agent of change. One reason the Greens embrace nonviolence is to distance themselves from many of the radicals on the left in the 1970s who otherwise pursued the same goals—the extremists who, it could be argued, were hurting the cause by alienating the general public. The principle of nonviolence, or at least some of its supposed adherents, is sometimes criticized because it is not always applied to property damage, which when intentional and widespread certainly seems as aggressive and inappropriate as personal assault.

Sustainability is the long-term capacity to continue. A logging industry that cuts trees down and does not plant them is not sustainable—it will eventually run out of trees, even if it does so long after its current captains are dead or retired. The oil industry, similarly, is unsustainable because new fossil fuels cannot be created (except in the truly long term, measured in geological time). Sustainability is a particularly important concern in capitalist societies, in which market forces are not likely to address the issue without social pressures: If the negative consequences of an action will not be felt until so long into the future that the actors will not be subject to it, there is less motive to avoid it. Indeed, were that not the case, the 19th and 20th centuries would not have dealt the pollutive blow with which the 21st century is now dealing. American and Canadian Greens sometimes cite the Great Law of the Iroquois, which states that "In every decision, we must consider the impact on the seventh generation."

Respect for diversity includes not only biodiversity (the vastness of life on Earth, and even elsewhere) but also human diversity in ethnic, gender, religious, and sexual identities.

In the United States, the Green Committees of Correspondence in 1984 expanded the Four Pillars into their Ten Key Values, adding decentralization, community-based economics, feminism, respect for diversity (later added to the Global Greens Charter), global responsibility, and future focus. Though sustainability is not specifically listed, this has more to do with the difference in language currency between 1984 and 2001 than a difference in values.

Green parties have had noteworthy success outside of the United States. In 1995, Finland's Green League received nine seats out of 200 in the election and joined the coalition-cabinet of which the Social Democrats were the majority party, naming a Green (Pekka Haavisto) as Minister of Environment and Development Aid—Europe's first Green to achieve a cabinet position. The Finnish Greens rose in popularity in subsequent years, particularly in the country's urban centers. In Germany, the "Red-Green Alliance" between the Greens and the Social Democrats ruled from 1998 to 2005, though many Greens outside Germany opposed the German Greens' compromise with their political rivals to support the war in Afghanistan.

**See Also:** Consumer Activism; Consumer Ethics; Disparities in Consumption; Ecotourism; Environmentalism; Finance and Economics; Government Policy and Practice (Local and National); Green Communities; Greenwashing; Regulation.

### Further Readings

Carter, Alan. *A Radical Green Political Theory*. London: Routledge, 1999.

Dobson, Andrew. *Green Political Thought*. London: T & F Books, 2009.

Radcliffe, James. *Green Politics: Dictatorship or Democracy?* New York: Palgrave Macmillan, 2002.

Torgerson, Douglas. *The Promise of Green Politics: Environmentalism and the Public Sphere*. Durham, NC: Duke University Press, 1999.

*Bill Kte'pi*
*Independent Scholar*

## Greenwashing

The belief that consumer demand for "environmentally friendly" products and services will encourage industry to perform in a more sustainable manner has been much touted by both government and industry alike. The strategy is simple: place ecolabels and/or ecosymbols on products deemed environmentally superior to their counterparts to allow consumers to make more environmentally conscious purchasing decisions. Consumers face a growing selection of adverting and labeling encouraging them to buy products or services that claim to reduce their personal ecological footprint. Ecolabels and ecosymbols can be found on a plethora of products, with certain retailers claiming exclusive use of their own copyrights or trademarks. However, on closer scrutiny, it is clear that many ecolabeling schemes have more to do with greenwashing than bona fide environmental protection. *Greenwashing* is a term derived from "whitewashing" by environmentalists, who claim that some corporations want to present an environmentally responsible public image by misleading consumers regarding their environmental practices or the benefits of their products or services.

There are many factors that account for the general failure of ecolabels to promote the sale of environmentally superior products, but this failure can ultimately be attributed to the misuse and abuse of ecolabels on the part of industry, the lack of government regulations covering these labels, and too few government-certified environmental labeling schemes based on the findings of objective life cycle assessment studies. Furthermore, the relative ignorance of consumers regarding ecolabeling is a major problem. As such, there exists much potential for the improvement of ecolabeling schemes.

Environmental labeling was adopted in principle by governments at the 1992 United Nations Conference on Environment and Development to "encourage expansion of environmental labeling and other environmentally related product information programs designed to assist consumers to make informed choices." The Canadian Standards Association, along with the International Organization for Standardization established a technical committee to develop international environmental labeling standards. An environmental claim can be any statement, graphic, or symbol that refers to or creates the general impression that it reflects the environmental aspects of any product or service through packaging labels, product literature, technical bulletins, advertising, publicity, telemarketing, and digital or electronic media including the Internet. The three types of ecolabels identified by CAN/CSA-ISO 14021 include the following:

*Type 1:* Labels from independent third parties who award them to the best environmental performers in various product categories. For example, in Canada the Environmental Choice Ecologo (ecologo.org) is awarded to companies by the federal government through TerraChoice Marketing, Inc. Companies must pay licensing fees after the submission of life cycle assessments—studies that pass a peer review. Similarly, in the United States, Green Seal (greenseal.org) is a nonprofit, third-party certifier that has been advocating a life cycle assessment approach since 1989. Other schemes include the European Union Eco-label (flower), the German Blue Angel, and the Nordic Swan of Sweden, Denmark, Finland, Iceland, and Norway.

*Type 2:* Self-declared labels used by manufacturers to make environmental claims about their products. These represent the vast majority of ecolabels, likely because there are few government regulations and no licensing fees to pay.

*Type 3:* A much less common label licensed by independent organizations. This label serves as a report card, providing information on the possible environmental impact of a product and leaving it to the consumer to decide which product is best.

The true value of ecolabels rests on faith by consumers that the environmental information provided is credible, objective, and easily understood. Unfortunately, research indicates this is generally not the case, and a recent study found most ecolabels contain some form of greenwashing. A 2007 study of type 2 self-declared labels by TerraChoice Marketing, Inc., examined labels on products claiming to be "environmentally friendly" in six leading big box retail stores in the United States and Canada. Green claims were examined against accepted practices as defined by the International Organization for Standardization and Consumer Affairs Canada. The researchers found 1,018 consumer products bearing 1,753 environmental claims, and they found that 1,017 products made claims that are demonstrably false or that risk misleading intended audiences. Six categories of greenwashing were identified, including the following and their frequency:

1. *Hidden Trade-Offs* (57 percent): Suggesting a product is "green" based on one or a narrow set of desirable attributes without acknowledging other important or more important environmental issues. These claims may not be false; instead, they create a "greener" image for the product than a life cycle assessment would support through empirical evidence. For example, paper that promotes its recycled content or sustainable harvesting practices while ignoring the relatively poor performance of emissions, effluents, and greenhouse gases associated with that manufacturer.
2. *No Proof* (26 percent): Claims that cannot be verified by easily accessible supporting documentation (point of purchase or company Website) or by a reliable third-party certification.
3. *Vagueness* (11 percent): Claims that are poorly defined or too broad, including: green, environmentally friendly, nontoxic, all-natural, and the ubiquitous use of the mobius loop symbol. The U.S. Federal Trade Commission and the International Organization for Standardization consider most of these terms to be too vague to be meaningful to consumers unless qualified with empirical evidence, which often they are not.
4. *Irrelevance* (4 percent): Claims that may be accurate but that are unimportant or irrelevant, such as claims that relate to chlorofluorocarbons that have been illegal for over three decades (i.e., no products are manufactured with chlorofluorocarbons).
5. *Lesser of Two Evils* (1 percent): These are claims that may be true but that risk distracting the consumer from more serious environmental effects of the product category, such as organic cigarettes and green insecticides/herbicides.
6. *Fibbing* (1 percent): Claims that are untrue and/or the illegal use of certified ecologos.

Despite the prevalence of greenwashing, Futerra Sustainability Communications points out that this problem exists mainly because of ignorance and/or sloppiness, as opposed to corporations demonstrating malicious intent. Despite the lack of government regulations concerning ecolabeling, the U.K. Department for Environment, Food and Rural Affairs points out that businesses and advertisers can prevent greenwashing in self-declared labels by taking into account the quality of the actual information being communicated (i.e., content), the way in which the information is presented (i.e., presentation), and the steps and methods taken to verify its accuracy (i.e., assurance of accuracy). The U.K. Department for Environment, Food and Rural Affairs suggests that the content of the claim should be accurate and truthful, relevant, specific, and unambiguous. The presentation of the claim should ensure that the claim uses plain language, that all relevant information is presented together, and that the meaning of any symbols or pictures is clear and relevant. To ensure accuracy, claims should be substantiated and verifiable, updated as necessary, based on the best agreed standards available, and supported by information needed to verify its accuracy.

Unfortunately, greenwashing exists because most countries do not regulate ecolabels with the same vigor that they regulate food labels. It is caveat emptor for the time being.

**See Also:** Advertising; Ecolabeling; Green Consumer; Green Marketing.

## Further Readings

Futerra Sustainability Communications. "The Greenwashing Guide." http://www.futerra.co.uk/downloads/Greenwash_Guide.pdf (Accessed January 2009).

Ramus, Catherine. "When Are Corporate Environmental Policies a Form of Greenwashing?" *Business and Society*, 44/4 (2005).

TerraChoice Environmental Marketing, Inc. "The Six Sins of Greenwashing: A Study of Environmental Claims in North American Consumer Markets" (2007). http://www.terrachoice.com/files/6_sins.pdf (Accessed January 2009).

U.K. Department for Environment, Food and Rural Affairs. "Green Claims—Practical Guidance: How to Make a Good Environmental Claim" (2003). http://www.defra.gov.uk/environment/consumerprod/pdf/genericguide.pdf (Accessed January 2009).

*Chuck Hostovsky*
*University of Toronto*

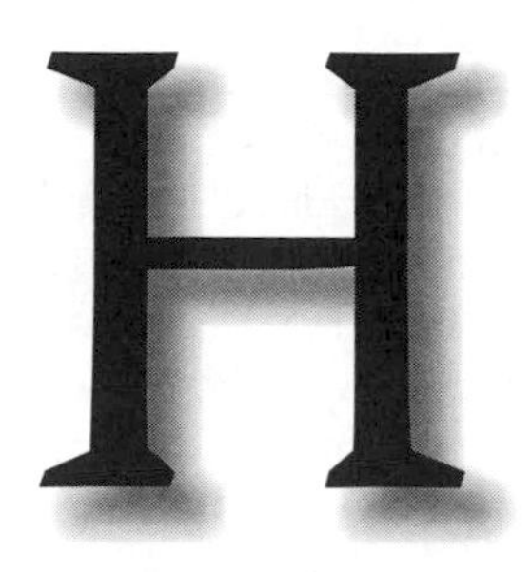

## Healthcare

The term *healthcare* refers to maintaining or restoring health. It comprises a variety of care such as care of skin, eyes, teeth, heart, kidneys, lungs, liver, bones, blood, and mental state, among others. Aside from practitioners of forms of alternative and natural forms of medicine, numerous green healthcare practices are emerging in the United States and elsewhere. These practices focus on a range of activities, from the environmental impacts of building and administering medical and healthcare services to promoting community and environmental health. In addition, there has been a proliferation of healthcare products, particularly vitamins, minerals, and herbal supplements—a consideration that forms the focus of this entry. These products are also known as dietary supplements and are widely consumed. In the United States, the Dietary Supplement Health and Education Act defines a dietary supplement as a product that is intended to supplement the diet; contains one or more dietary ingredients (including vitamins, minerals, herbs or other botanicals, amino acids, and other substances) or their constituents; is intended to be taken by mouth as a pill, capsule, tablet, powder, or liquid; and is labeled as a dietary supplement. This article gives an overview of the dietary supplement aspect of healthcare; enumerates critical reviews and details regulations of dietary supplements; and identifies trends in dietary supplement consumption.

Herbal supplements are plants or plant parts that are valued for their medicinal properties, flavor, and scent. An example is St. John's wort, seen here in its natural form.

*Source:* Ciar/Wikipedia

## An Overview of Dietary Supplements

The most commonly dietary supplements used to maintain or restore health are vitamins, minerals, herbal supplements, and other substances. They are described briefly here.

Vitamins are organic compounds required in small amounts to maintain the normal physiological functions and to restore health. On the basis of their solubility, vitamins are classified as fat-soluble vitamins—A, D, E, K—and water-soluble vitamins—thiamin, riboflavin, niacin, pyridoxine, pantothenic acid, biotin, folic acid, cobalamine, and vitamin C. Vitamins are not synthesized in our body in an amount adequate to meet the normal physiological need; however, they are naturally present in foods, and therefore, a well-balanced diet that includes a variety of foods should meet the daily requirements for various vitamins. Fat-soluble vitamins are stored in the body if consumed in excess and can cause toxicity, whereas water-soluble vitamins are not stored in the body, and the excess is generally excreted in the urine.

Similar to vitamins, minerals are not made by the body and must be obtained through the diet or supplements. Minerals are inorganic molecules that are essential for health and the maintenance of body functions. Minerals required by adults in excess of 100 mg/day are known as macrominerals—calcium, phosphorus, magnesium, potassium, sodium, chloride, and sulfur, whereas minerals for which less than 100 mg/day is required are known as microminerals or trace elements—iron, zinc, iodine, selenium, manganese, fluoride, molybdenum, copper, chromium, cobalt, and boron. Minerals constitute about 4–5 percent of total body weight.

In addition to vitamins and minerals, herbal supplements, also known as botanicals, are widely used as health supplements. Herbal supplements are plants or plant parts (e.g., leaves, flowers, seeds, bark, etc.) valued for their medicinal properties, flavor, and scent. These supplements may contain a single herb or mixtures of herbs. Commonly used herbal supplements include echinacea, ginseng, gingko biloba, garlic, ginger, flaxseeds, and St. John's wort.

Other products, such as protein powder, fish oils, amino acids mixes, fiber supplements, and enzyme preparations, among others, are also used as supplements to maintain or boost health.

## Critical Review of Dietary Supplements

Critical review of a dietary supplement is important to assess its safety and efficacy. Before deciding to use any dietary supplement, one should be aware of certain information about the product, such as how the product works, and what are the adverse effects, if any; any available clinical trials on animal versus human subjects, conducted by reputable, unbiased scientists/authorized institutions using a valid study design and methodology; the confirmation and/or duplication of the result of clinical trials by other reputable scientists; the short-term and long-term safety and efficacy of the product; and the population most likely to benefit from the product.

## Regulations of Dietary Supplements

In the United States, dietary supplements do not require approval from the U.S. Food and Drug Administration (FDA) for their safety and effectiveness. Manufacturers and/or distributors of the dietary supplements are responsible for ensuring the safety and label claims of their products. Once a supplement is on the market, the FDA monitors its safety, label claims, and package inserts. If a product is found to be unsafe, FDA can take action against the manufacturer and/or distributor by issuing a warning or removing the product from the marketplace. The Federal Trade Commission regulates the product advertising. It requires that all information be truthful and not misleading.

## Trends in Consumption of Dietary Supplements

Consumption of the aforementioned health or dietary supplements has increased among the geriatric age group, white race, cancer patients, females, and athletes. A U.S. national survey conducted in 2007 found that 17.7 percent of American adults had used natural products (dietary supplements other than vitamins and minerals) in the past 12 months. The most popular products used by adults for health reasons in the past 30 days were fish oil/omega-3/DHA (37.4 percent), echinacea (19.8 percent), flaxseed oil or pills (15.9 percent), and ginseng (14.1 percent). Another U.S. national survey (2007) covering all types of dietary supplements reported that approximately 52 percent of adult respondents used some type of supplements in the last 30 days. The most commonly reported supplements were multivitamin/multimineral supplements (35 percent), vitamins E and C (12–13 percent), calcium (10 percent), and B-complex vitamins (5 percent).

Data regarding the use of dietary supplements among infants and children are scanty. J. M. Eichenberger and colleagues (2005) discovered that 11.5 percent of newborns and 42.8 percent of infants (20–24 months) were given supplements, though their use was intermittent in 22 percent. Consumption of dietary supplements is expected to rise as supplement manufacturers target children by selling the supplements in more desirable form such as vitamin gummy candies and gumballs.

Research indicates that some health supplements are effective in preventing or treating diseases. For example, folic acid (a vitamin) can prevent certain birth defects, a regimen of vitamins and zinc can slow the progression of age-related eye disease such as macular degeneration, and calcium and vitamin D supplementation can prevent and treat bone loss and osteoporosis (i.e., thinning of bone tissue). Further, some dietary supplements may improve certain health conditions. For example, omega-3 fatty acids may reduce coronary artery disease, and probiotics may improve gastrointestinal health.

## Conclusion

Different people consume dietary supplements for different reasons—to compensate for inadequate diets, treat or support medical conditions, enhance work performance, improve mental status, and maintain overall health, among others. Recent years have seen increased marketing of "green" dietary supplements, with their promotion as alternatives to supplements produced by major pharmaceutical companies. It is common for producers of these supplements to indicate that these "green" products have "natural" or "organic" ingredients and to indicate that they are nonaddictive and/or chemical free. Though there are protections in many countries to ensure that products are safe and that their health claims are not misleading, regulation is not uniform. The use of dietary supplements to support and improve health may be desirable; however, long-term use of some supplements can be harmful when they are consumed in high amounts. Further, some supplements may potentially interact with foods and other drugs regardless of their sources—natural or chemical.

Hence, one should consult with healthcare providers when deciding to use any dietary supplement or healthcare product.

**See Also:** Beverages; Bottled Beverages (Water); Dairy Products; Fish; Food Additives; Organic; Overconsumption; Pharmaceuticals; Vegetables and Fruits; Water.

### Further Readings

Eichenberger, J. M., et al. "Longitudinal Pattern of Vitamin and Mineral Supplement Use in Young White Children." *Journal of American Dietetic Association,* 105/763 (2005).

Food and Drug Administration (FDA). "Consumer Health Information." http://www.fda.gov/consumer (Accessed April 2009).

Food and Drug Administration (FDA)/Center for Food Safety and Applied Nutrition/Office of the Nutritional Products, Labeling and Dietary Supplements (December 2004). http://www.cfsan.fda.gov/~dms/ds-take.html (Accessed April 2009).

U.S. Department of Health and Human Services, National Institute of Health, National Center for Complementary and Alternative Medicine. "Using Dietary Supplements Wisely." http://nccam.nih.gov/health/supplements/wiseuse.htm (Accessed April 2009).

*Meera Kaur*
*University of Manitoba*

## Heating and Cooling

In the United States, buildings use 72 percent of the nation's annual electricity consumption. The residential sector constitutes 21.4 percent of the total annual U.S. electricity consumption, and a substantial portion of residential energy consumption is used to heat and cool the built environment. The average U.S. household uses approximately 16 percent of its total energy consumption for air conditioning, 10.1 percent for space heating, and 9.1 percent for water heating. These figures vary throughout the different climate zones of the country based on seasonal thermal characteristics. There are three general types of climates within the United States: humid, arid, and temperate. Within various geographical regions, there are zones of extreme cold and extreme heat.

Residences within humid climates generally use air conditioners and dehumidifiers to remove humidity from the air. Typical current systems for these practices are energy intensive. Residents situated in arid climates use evaporative coolers and other water additive solutions to condition the environment, thus eliminating or reducing the need for air conditioners. This process is typically less energy intensive, and there are a myriad of passive methods available. Temperate climates generally have relatively mild temperatures in winter and summer and require less energy use to heat and cool internal spaces. This climate is more amenable to passive strategies. Areas that experience a combination of the various thermal typologies, such as Washington, D.C., which is both humid and temperate throughout the year and has fairly cold winters, require mixed modes of air conditioning strategies and typically are more energy intensive.

To maintain an optimal thermal environment in a given space, various solutions can be employed that range in quantity of embodied energy, energy consumption, noise pollution, temperature fluctuations, and occupancy comfort probability. Passive thermal design strategies generally have the lowest embodied energy, minimum electricity consumption, and allow for greater occupancy comfort and comfortable temperature ranges. These are the most sustainable heating and cooling strategy options and can contribute a significant, if not complete, portion of heating and cooling needs, depending on the local climate. Passive thermal design strategies, depending on the method employed, can cost more or less than traditional thermal conditioning systems in regard to up-front and life cycle costs. When passive design strategies require augmentation to provide optimal thermal environments, active systems are employed. Active systems generally have greater embodied energy, greater electricity consumption, and reduced temperature ranges of comfort. They can also contribute to sick building syndrome and trap volatile organic compounds within the building envelope. These systems may have a lower initial cost, but they typically cost more over the lifetime of the building when electricity consumption rates are factored into the analysis.

## Strategies and Housing Types

Strategies for developing a cost-efficient, minimal–energy-consuming thermal system depend on the type of housing. Existing homes can be modified inexpensively to attain significant cost savings and energy efficiencies with rapid payback timetables. Improving building insulation and fixing thermal leaks and bridges are highly effective—and usually the cheapest—strategies. A professional energy audit can quickly help owners determine the most cost-effective strategies to achieve maximum cost savings and efficiencies by pinpointing the major sources of thermal bridges and leaks. Some power companies or state energy offices will perform an energy audit for free, whereas other companies offer these services at relatively low prices. A more capital- and labor-intensive strategy is replacing windows with double- or triple-paned windows with optional low-E glazing for increased thermal performance. Air filters for air conditioning systems, if not replaced regularly, contribute to poor indoor air quality and increased energy consumption. Plants can be used to improve indoor air quality, as various plants break down toxic materials found in common household materials. Plants can also be used for passive cooling, both as shading devices for the external walls of the building and for adding humidity to interior arid environments through transpiration. Shading the exterior of the building both reduces internal temperatures by reducing thermal gain and has been proven to increase building materials' lifetime by up to 50 percent as a result of the reduction in direct ultraviolet exposure of building materials. Trellises, vegetated walls, outdoor patios, and stand-alone trees and shrubs vary in effectiveness based on the way they are used. When replacing an active system for heating, electric heating is the least effective and the most cost intensive, whereas natural gas is the most thermally and cost-efficient and is becoming increasingly common throughout the country. Electric heating is currently the most common system used in rental apartments, as they are preferable for owners because of the relatively inexpensive initial cost.

New homes allow for the comprehensive design of thermal systems based on the local climate conditions. The use of air to warm or cool a space is very inefficient when compared with the use of water, as air is roughly four times the weight of water and 3,500 times its volume. Thus, water is more efficient to condition spaces, and air should be used as a fresh ventilation source. This strategy reduces the level of volatile organic compounds within the built environment and allows for greater opportunities to employ passive thermal design strategies, which can be designed to be more cost-efficient and are inherently less energy intensive. Primary problems with natural ventilation strategies are noise and air pollution infiltration. The first step to maximizing passive thermal conditioning strategies is the design of the building envelope and façade on the site, focusing on thermal gain and natural ventilation potential. The building orientation, window-opening sizes and locations, form, and height are the greatest contributing factors. Existing site-shading devices, such as contextual buildings, trees, and so on, should inform the building's attempt to maximize the devices' shading potential in the summer and to minimize their effect in the winter.

Care should be taken to reduce the prospect of the building obstructing surrounding buildings' natural ventilation potential, while also considering the effect of the building's height and form in terms of winter and summer shading on surrounding buildings. Plants also can be used as windbreaks—they can channel wind from various directions, increase air speed to increase natural ventilation potential, filter the passing air, and serve as shading devices in the summer. Once the building form has been determined, space planning for the various programming should be situated on the basis of thermal impacts and interior wind flows. For example, kitchens act as a heat source during specific daily times and should be located where this excess heat can be used in the winter and removed easily in the summer. High ceilings allow hot air to rise above the occupant level and allow greater daylight access and promote a sense of being in a more open space. Cooling systems should first employ cost-effective passive strategies and integrate active, cost-intensive strategies to supplement the passive systems.

As previously stated, water conditioning systems, such as radiant slab cooling and heating, are preferred to air conditioning systems, although water condensing on the slab can be an issue if the building is not properly ventilated. These systems can use geothermal temperatures as the water thermal source, but the up-front costs of these systems are considerable and vary based on the geological typology of the site. Thus, the payback periods vary with different sites. Concrete air ducts can be used as air intake sources and can be cost-efficient when integrated with the building foundation. However, this system can promote mold growth in humid climates if not properly designed and maintained. Within arid environments, plants can be used to humidify the air, as well as passing intake air over pools of water. When the built environment requires supplementary heating, passive design strategies can maximize cost- and energy efficiency.

## Passive Strategies

There are two general passive heating strategies: thermal mass and solar greenhouses or sun porches. Within these broad categories are a myriad of successful strategies that range in cost- and thermal efficiency. The most effective solution is designing multivalent building elements, such as building slabs and walls that can be used as thermal masses, light diffusers, and spatial boundaries. Rock gardens and pools of water can be used as thermal masses and can be combined with a solar greenhouse. In general, it is more effective to avoid placing the thermal mass in direct contact with the external environment to minimize heat loss to the exterior. Thus, using solar greenhouses or sun porches with a thermal mass is effective and provides an amenable interior space for plant growth and social space. Typical active systems for new homes are the same as existing home systems. However, new homes can implement innovative active systems more fluidly into the design, such as radiant heating and cooling systems.

Innovative thermal systems are constantly being developed and employed. One promising technology is solar absorption chiller technology, sometimes referred to as solar air conditioning. This system uses fewer moving parts, has less embodied energy, and greatly reduces electricity consumption compared with a typical air conditioning system. The limitation for this system, as with all solar-driven systems, is that it requires solar access. During days when solar access is inadequate, this system must be augmented with a secondary air conditioning system, such as traditional air conditioning or geothermal systems. Another interesting development is the use of composting piles for heating a residence's water supply. It has been shown that a compost pile that is exposed to the sun provides enough heat through solar gain and the inherent decomposition process to heat water passed through the pile through a pipe. This system can potentially use domestic organic waste to heat one's water supply and increase the cost-efficiency of constructing and maintaining a compost pile, while promoting residential composting.

Passive thermal conditioning, innovative active system design, and existing home thermal efficiency improvements, although not currently widely employed, have the potential to greatly reduce annual rates of energy consumption, the need for new power plants, and costs for electricity in the United States, while simultaneously developing a healthier, nontoxic built environment.

**See Also:** Composting; Electricity Usage; Green Homes; Insulation; Resource Consumption and Usage.

### Further Readings

Chiras, Daniel D. *The Solar House: Passive Heating and Cooling*. New York: Chelsea Green, 2002.

Environmental Information Administration. "EIA Annual Energy Outlook." http://www.eia.doe.gov/oiaf/aeo/electricity.html (Accessed May 2009).

Kibert, Charles. *Sustainable Construction: Green Building Design and Delivery*, 2nd Ed. Hoboken, NJ: Wiley, 2007.

Lechner, Norbert. *Heating, Cooling, Lighting: Sustainable Design Methods for Architects*. Hoboken, NJ: Wiley, 2008.

U.S. Department of Energy. "End-Use Consumption of Electricity 2001." http://www.eia.doe.gov/ emeu/consumption/index.html (Accessed May 2009).

*Giancarlo Mangone*
*University of Virginia*

## Home Appliances

Public policies in many countries seek to steer manufacturers and consumers toward using more energy-efficient goods. This article examines one of the largest uses of electrical goods—electrical appliances for home—that are produced and purchased in large and increasing quantities.

A home or domestic appliance is a machine used to accomplish routine housekeeping tasks, such as food preservation (e.g., refrigerators or freezers), cooking (e.g., stoves, ovens, microwave ovens), cleaning (e.g., dishwashers, washing machines, dryers). Household appliances consume a lot of energy—on average, a quarter of all the energy used in the house. Furthermore, they can affect the environment through their "associate" consumptions, for example, the use of water and soaps, as well as their disposal.

Home appliances are usually referred to as major household appliances (such as stoves and refrigerators) and small appliances (such as stereos, computer equipment, and television sets). Major appliances are large, difficult to move, and generally fixed in place to some extent. Small appliances are portable or semiportable. They look small and appear not to be heavy users of energy compared with major appliances. Nevertheless, as the International Energy Agency warns, by 2010 there will be over 3.5 billion mobile phone subscribers, 2 billion televisions in use around the world, and 1 billion personal computers.

All appliances are intended to perform, enable, or assist in performing a job or changing a status; for example, the temperature of a room. Most are not necessities but, instead, make our lives easier, save us time, or provide entertainment. In any case, they are a growing part of our lives, from major items like televisions to a host of small gadgets. As these small devices gain in popularity, they became an important and growing portion of household energy consumption.

In developed countries, appliances that previously used the majority of electricity, that is, major appliances, are close to saturation levels. The stabilization of ownership rates and the improvement in efficiency of these appliances has made their share of residential electricity consumption fall. In contrast, the ownership of small electronic equipment (as well as the use of lighting equipment) has increased in several world regions. In developing countries, although the ownership level for major appliances is already high in urban areas, increased access to electricity in rural areas and growing urbanization is driving up overall ownership levels, and hence electricity consumption.

The result is that household appliance energy efficiency has increased by one-third in the last 20 years, but household energy consumption has decreased by only about 2 percent. This is because of the increase in the rate of equipment purchase. Furthermore, even when efficiency improvements have been made to small appliances, any savings have been cancelled out by the demand for equipment that provides more functionality or is larger or more powerful, and therefore uses more electricity.

This increase in the equipment purchase rate also brings to the forefront the problem of discarding home electronics. Electronic waste, which encompasses loosely discarded, surplus, obsolete, and broken electrical or electronic devices, has become more and more of a problem and causes serious health and pollution issues. Electronic equipment can contain serious contaminants, such as lead, cadmium, beryllium, and brominated flame retardants. Recycling and disposal of this waste involves significant risk to workers and communities, and great care must be taken to avoid both unsafe exposure in recycling operations and materials such as heavy metals leaching from landfills and incinerator ashes.

Is concentrating on energy efficiency the solution to rising home appliance electricity consumption? Energy consumption is an aggregated indicator that does not do justice to the diversity of practices related to its use. These practices are articulated around objects and sectors of activity and are not necessarily coherent with attitudes toward environment or energy consumption. Attitudes appear to have less influence on the consumption than the facts and reality of possessing and using appliances. Energy-consuming practices are fueled by appliances.

An effective energy policy needs to take into account the factors shaping user perceptions about the use of energy. In the case of home appliances, the increase in energy consumption in this sector shows that the problem cannot be faced only through energy efficiency measures. In fact, rebound effects lurk around the corner: more households (at constant population) have an increasing number of appliances, which—used more often—can drive energy consumption up, even if appliances are built to be more energy efficient.

Savings can be achieved through better equipment and components, but the largest improvement opportunity must come from making hardware and software work together more effectively to ensure that energy is only used when and to the extent it is needed. To deliver these savings, strong public policies are needed. In particular, given that new devices increasingly offer a variety of functions, each of which may have differing energy needs, policies are needed that set maximum energy budgets for each function.

**See Also:** Audio Equipment; Ecolabeling; Energy Efficiency of Products and Appliances; Green Consumer.

### Further Readings

Bittman, Michael, et al. "Appliances and Their Impact: The Ownership of Domestic Technology and Time Spent on Household Work." *British Journal of Sociology*, 55/3:401–23 (2004).

International Energy Agency (IEA). *Gadgets and Gigawatts: Policies for Energy Efficient Electronics*. Paris: IEA, 2009.

Jackson, T. "Motivating Sustainable Consumption: A Review of Evidence on Consumer Behaviour and Behavioural Change." A Report to the Sustainable Development Research Network (2005). http://www.comminit.com/en/node/219688 (Accessed July 2009).

*Marco Orsini*
*Institut de Conseil et d'Etudes en Développement Durable*

## Home Shopping and Catalogs

Green consumerism is viewed by some in the green movement as an oxymoron. For others, it is the easy way to save the planet. Green consumers now have a huge variety of goods available for purchase that vary in the depth of the "greenness" of the product. Their consumer choices can make a difference in environmental effects. One effect is the "greening" of a growing array of products, although how green these products are varies significantly from one catalog product line to another.

Retailers spend billions of dollars each year distributing 17 billion catalogs through the mail because they believe that this form of advertising works. The catalogs come at a rate of 55 per person per year, which amounts to millions of tons of garbage, particularly during the holiday season. They are environmentally expensive, and most are not recycled. They require specially treated papers and special inks that may be environmentally unfriendly. Attempts to switch to catalogs that are much more environmentally friendly have not been very successful because consumers prefer slick-papered and colorfully printed catalogs to browner ones. Some environmentalists have sought to limit catalog mailings with "Do Not Mail" lists, but with little success. The other alternative is home shopping via the Internet, which is cheap and may be the best "green catalog."

The business of producing home catalogs as a marketing tool began in 1872. Aaron Montgomery Ward was a traveling salesman. He and many other salesmen would carry samples to many small towns and cities, where they would call on businesses to sell their products in bulk at wholesale prices, or go door-to-door to take retail sales orders. Other salesmen would visit farms to sell; however, farms were often difficult to reach, and the terrain or the weather could present many challenges to this type of personal marketing to homes.

Ward realized that printed catalogs could be mailed to homes as a marketing tool and that people would be able to shop from home. Customers would order from the catalog by mailing in the order form. They could prepay, pay cash on delivery, or pay after delivery. A large variety of goods was for available for sale in those early catalogs, which gave millions of people a way to access goods from home that were previously unknown or unavailable to them.

Sears, Roebuck and Company developed a catalog that eventually grew to be several inches thick. Its array of goods included toys, tools, farm equipment, kitchen equipment, bedding, linens, personal items, and virtually anything consumers wanted. The Sears catalog was often referred to as the "wish book." Old copies also served as toilet tissue.

The invention of the automobile and paved highways allowed people to travel to towns, but shopping was still difficult in certain seasons, such as Christmas, winter, or during vacations. It became a convenience to order products from catalogs. For retailers, catalog sales were a welcome addition to store sales and greatly expanded their businesses. It was also a boon to the postal service, which handled more packages, increasing its business.

As the mail-order business grew, competition also grew. Although many general catalogs were mailed to the public by large retail stores, shoppers were soon presented with numerous specialty catalogs. Specialty mail-order catalogs usually arrive in increasing numbers as the holiday buying season approaches.

Mail-order catalogs are currently supplemented with information about companies posted on their Websites. Many companies put their entire catalog of goods online and invite home purchases. Online booksellers Barnes & Noble, Amazon, and others have expanded their catalogs to include many other goods, thereby moving in the direction of becoming online department stores.

Television shopping channels also allow for shopping from home. The goods are displayed, described, and often demonstrated to cable television home-viewer consumers. With a toll-free phone call to an 800 number, these consumers can become buyers of the products, paying with a credit card. Television shopping channels appeal to people who are unable to leave home or who are reluctant to go to shopping centers where there are at times crowds of people. Online shopping, shopping channels, and mail-order catalogs are mechanisms for marketing goods to a wide variety of people who may not have the interest or the time to shop at retail outlets. Shopping from home allows consumers to be green because they do not burn gasoline driving to stores.

**See Also:** Green Consumerism; Internet Purchasing; Lifestyle, Rural; Shopping.

### Further Readings

Cherry, Robin. *Catalog: An Illustrated History of Mail-Order Shopping*. Princeton, NJ: Princeton Architectural, 2008.

Mars-Proiette, Laura, ed. *Directory of Mail Order Catalogs 2010*. Armenia, NY: Grey House, 2009.

Montgomery Ward. *Montgomery Ward & Co. Catalogue and Buyers' Guide 1895*. New York: Skyhorse, 2008.

Oharenko, John. *Historic Sears, Roebuck and Co. Catalog Plant, Illinois. Images of America Series*. Charleston, SC: Arcadia, 2006.

*Andrew Jackson Waskey*
*Dalton State College*

# Homewares

Selecting green homewares is principally a matter of buying those that have been made of the right materials. This includes using recycled or recyclable materials and using renewable materials. Green homewares are often the product of "design for environment," an approach to product and service design that reduces the environmental impact of a product across its life cycle. For instance, such products will be made easy to disassemble, when applicable, to make recycling easier. They will also be designed through source reduction—the reduction of the toxicity of any product's final stage of life; that is, the point at which it becomes solid waste. Source reduction looks not only at whether a product is toxic to consumers while in use—which, of course, manufacturers have done everything to avoid all along—but whether it will be toxic to soil and groundwater when it becomes part of a landfill, and whether when it degrades it will release any toxic chemicals into the air, even decades from now. When possible, source reduction therefore favors biodegradable materials.

Very similar, almost congruent, to design for environment is sustainable design, which goes a step farther by eliminating the use of nonrenewable resources entirely. Sustainable design has been a movement in architecture since the 1970s, and before long was considering those objects that fill the home as well as the design of the home itself. Sustainable design is also known as ecodesign, environmental design, green design, or "the triple bottom line" (adding "people" and "the planet" to the bottom line of profit—a succinct way of addressing the matter of externalities).

Many green homewares are listed in the bimonthly *The Green Guide* published by the National Geographic Society, available either in print or online. (The print version is printed on "sustainable paper," made from wood from Forest Stewardship Council–certified forests and recycled material.) A sort of Green *Consumer Reports, The Green Guide* helps consumers find green, environmentally conscious versions of the household products they commonly purchase, from beer to air conditioners to fertilizer, with articles on the relative merits of products and the environmental hazards of common household goods.

Green dishes are generally made from recycled material, clay, or glass. Glass is both easily recyclable and food-safe, and several lines of recycled-glass dining sets exist. As a natural product, clay is also environmentally conscious, as are china, earthenware, and stoneware. In all of the above cases, consumers are encouraged to buy new goods, not buy from yard sales, antique stores, and the like, because of the possibility of their containing lead—a standard practice until recent years. The disadvantage these materials present is their heavy weight and, in the case of glass, its fragility—which in today's transport-addicted economy means that many more units of carbon emissions involved in bringing the product from the factory to the consumer. Whether this is balanced by the gains of not using plastic depends on the specific circumstance and on the consumer's proximity to the point of origin.

When it comes to cookware, consumers are advised to avoid nonstick pans from Teflon and other brands. The fluoropolymers used in nonstick finishes degrade at high temperatures into various toxic substances and have been implicated in the presence of perfluorinated acids in the blood of nearly the entire population—a Johns Hopkins University study in 2004 found perfluorooctanic acid (PFOA) in the umbilical cord blood of 99 percent of 300 babies, for instance. The widespread diffusion of such chemicals in our blood is partly the result of industrial air pollution and partly of now-discontinued substances used on consumer products. PFOA is considered carcinogenic, and although manufacturers have agreed to eliminate 95 percent of PFOA emissions by 2010, its use will continue in non-stick cookware finishes.

Earlier studies showed that overheated nonstick pans release chlorofluorocarbons, the emission responsible for much of the ozone layer depletion, and the manufacturer warning identifies fumes that can be fatal to birds. Green alternatives to nonstick pans include stainless steel—which requires the addition of fat in many applications to avoid sticking—or a well-seasoned and well-used cast-iron pan. Part of the popularity of nonstick pans has been their light weight compared with heavy cast iron, and their ease of cleaning without scrubbing; cast iron also performs best when used frequently (in which case it will last literally generations with minimal basic care), and the rise of nonstick cookware coincided with cooking becoming more of a casual activity in America than an everyday chore, in part because of TV dinners, microwaved food, take-out, and drive-through restaurants.

The fast food chain Chipotle has recently introduced a sustainable cutlery plan to their restaurants, on the heels of their wind-powered green restaurant in Illinois. Recycled bioplastic cutlery is manufactured by Cereplast. Similar recycled/recyclable plastic utensils are available to the public. A number of brands of biodegradable cutlery are available as well, including "Spudware," made from potato starch instead of petroleum products.

**See Also:** Advertising; Baby Products; Cleaning Products; Consumerism; Disposable Plates and Plastic Implements; Home Appliances.

### Further Readings

Brower, Michael and Warren Leon. *The Consumer's Guide to Effective Environmental Choices: Practical Advice From the Union of Concerned Scientists.* New York: Three Rivers, 1999.

Garlough, Donna, et al., eds. *The Green Guide: The Complete Reference for Consuming Wisely.* New York: National Geographic, 2008.

Rogers, Elizabeth and Thomas Kostigen. *The Green Book.* New York: Three Rivers, 2007.

*Bill Kte'pi*
*Independent Scholar*

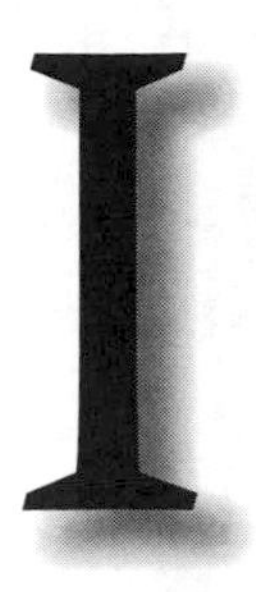

## INSULATION

Insulation is considered here in relation to the application or incorporation of specific materials into domestic building and maintenance practices. It can be integral to the whole building process itself and can be highly effective. For example, new domestic dwellings have been built wherein the insulation, as part of other environmentally sustainable design considerations, is so effective that it obviates the need for heating technologies or appliances. More commonly, insulation refers to a range of material add-ons to already built dwellings. The ability to insulate is the prime quality of the materials in question and the provision of insulation is offered as the principle justification for the consumption of these specific materials.

Both approaches rest on a common apprehension of insulation as diminishing heat loss. However, within the context of green consumerism, the emphasis is usually more on saving energy, and this itself is often advocated to the consumer via the slightly different prospect of saving money. While insulation could, then, be understood in relation to energy consumption, building design, and dwelling habits, such considerations are often downplayed in favor of the economic language of costs and savings. However, within the context of green consumerism, insulation is not so easy to actually sell. The materials advocated can be expensive, their fitting time-consuming, and the benefits, especially when promoted in terms of saving money rather than saving energy, may be quite long-term. Furthermore, it can also be unattractive to consumers, due to its increasingly regulated nature. Many of the materials are highly technical and/or have quite complex technological demands. The specialized energy rating information they carry can also be difficult to decipher and compare. However, there are some clear and increasingly meaningful benefits to be had for some: for example, an increase in the thermal efficiency of homes is seen to add economic value, something that is now officially recognized through obligatory energy ratings reports that are a feature of house sale and purchases in many areas.

The efficiency or thermal resistance of common bulk insulation materials associated with domestic consumption is designated by their R-value. The higher the value rating, the better the insulating properties of the material. In general, increasing the thickness of insulating material increases the R-value proportionally. In addition, enclosing materials, which may be a safety requirement—for example, in relation to fire prevention legislation—will also increase R-values. However, these types of materials do not eliminate all heat loss, as

Commercially available thermal insulation materials are generally efficient at disrupting the flow of acoustic energy if they are good at inhibiting conductive heat flow.

*Source:* iStockphoto

they are focused on preventing convective heat loss through air movement. They do not prevent convective heat loss through other building structures, such as studs or windows, nor do they prevent heat loss through conduction or radiation. The insulating effectiveness of various materials also has to be balanced against a range of further considerations, including ease of installation, durability, toxicity, fire retardation, energy costs of manufacture, and sustainability. Moreover, many insulating materials are prone to various problems, including dramatic drops in efficiency due to poor fitting, deterioration (both physical and chemical), weathering, escape of insulating gases, and ingress of air and moisture. Consequently, the materials offered for public, domestic consumption may have relatively low R-values but be advantageous in other ways. For example, fiberglass-type insulation tends to have middle-range R-values but is easy to fit, cheap, and increasingly incorporates recycled content.

Safety, in installment and consequent use, is of increasing importance in relation to domestic insulation. Some materials, including asbestos (an excellent insulating material), are now outlawed due to their carcinogenic potential. Urea-formaldehyde foams are also widely banned due to their leakage of formaldehyde gas post-installation. Common insulating materials with high R-values (for example, polyurethane foams), are toxic when burned, need fire retarding cladding, contain greenhouse gases, require protection from sunlight and solvents, and are made from scarce petrochemicals via energy-intensive processes. While more ecofriendly versions are in ongoing development, the advocacy and use of alternative materials and alternative conceptions of the problem of heat loss have gained pace. For example, cotton insulation has a higher R-value than most fiberglass alternatives, is made from recycled cotton scraps, contains no formaldehyde, and consists of a much less energy-intensive manufacturing process. Similar claims can be made for loose-filled cellulose cavity wall insulation, which contains up to 80 percent recycled newspaper. Indeed, alternatives with very high environmental, technical, and safety ratings (such as Warmcel in the United Kingdom, which is made from 100 percent recycled newspaper), are increasingly available. Another U.K.-branded alternative to fiberglass-type wools (Thermafleece, which is made from sheep's wool) is also notable here. Although expensive, such branded alternatives may fare better in the context of green consumerism as compared to other ecoalternatives, such as straw bale insulation, which is generally thought to be impractical, to have undesirable cultural associations, and to be technologically unrefined. Indeed, straw bales have a relatively low R-value and rely on very thick walls for their insulating effect.

The context of insulation as a concept can also be usefully expanded. For example, "eco-minimalism" is concerned with saving and retaining energy in buildings, and runs counter both to approaches that emphasize novel energy generating technologies and those advocating insulation as a panacea. Insulation should also be used in concert with other material modifications to be most effective. Triple glazing has a 30 percent energy saving over double glazing, for example. When such factors are seen as prerequisites for the potential maximum thermal efficiency that insulation promises, it can be argued that expansion of the concept of insulation to include design considerations, energy generation, and usage and dwelling patterns and behaviors has advanced considerably. The best energy practices, broadly conceived, would render insulation as a superfluous consumer add-on.

**See Also:** Consumer Culture; Finance and Economics; Floor and Wall Coverings; Green Homes.

### Further Readings

Hopkins, Carl. *Sound Insulation: Theory into Practice*. Oxford: Butterworth Heinemann, 2008.

Pfundstein, Margit, et al. *Insulating Materials: Principles, Materials, Applications (Detail Practice)*. Basel, Switzerland: Birkhäuser, 2008.

U.S. Department of Energy. "End-Use Consumption of Electricity 2001." http://www.eia.doe.gov/emeu/consumption/index.html (Accessed May 2009).

*Neil Maycroft*
*University of Lincoln*

## International Regulatory Frameworks

International regulatory frameworks provide "hard law" obligations and rules of treaties and "soft law" standards and guidelines about safe and sustainable state and corporate practices. They are typically derived from normative consensus that arises through interaction among states, market actors, and civil society proponents of a green agenda. These legal regulations—an outcome of interstate negotiations—cover a wide variety of conservation issues. This is because many environmental problems extend across national boundaries and affect the life chances of various species as well as the preservation of land, air, and water on which human livelihood depends. All such efforts at regulatory structures, then, have, at their core, the attempt to not just mitigate the violation against human rights but also enable the enjoyment of human rights.

The key international regulations derive from the normative ideals enshrined in the international law—the 1948 Universal Declaration of Human Rights. These rights cover civil and political rights as well as social, economic, and cultural rights. Key to both sets of rights is the right to life and the right to a standard of living adequate for health and well-being. Obviously these issues strike deeply into the heart of the key concern of the green agenda—sustainable development.

International regulations could be enshrined in the charters of international institutions like the United Nations (UN), the World Bank, and the World Health Organization. They could also become part of bilateral treaties. Typically, the adoption of such regulations tends to be voluntary, and hence not all states necessarily are part of such treaty frameworks. Moreover, the states could sign on to an international treaty regulation, but the treaty still has to be ratified by the domestic governing institutions. This means that many states do not move beyond structural facades to genuine efforts at processes to implement the regulations and ensure just outcomes.

Hence, regulations could be codified into laws within the domestic arena, or they could be left to be auto-accomplished through market and civil society pressures. For example, civil society activism focuses on naming and shaming delinquent actors and their activities. Hence, actors such as Amnesty International or Oxfam copiously document human rights violations by states and businesses and publicly propagate them through various media. These entities have successfully unearthed the problem of "blood diamonds" and associated civil wars related to the extraction of such resources. As citizen watchdogs, they have also recorded toxic environmental pollution activities by international oil and gas companies in resource-rich, but regulation-poor, developing countries.

International institutions like the World Bank, UN Development Programme (UNDP), and International Labour Organization also document state performances on various indicators of justice. For instance, the International Labour Organization's labor standards cover issues of antidiscrimination, antislavery, safe and hygienic working conditions, and minimum wage. The World Bank and UNDP reports cover issues such as gender equity and health. The public records of such data are available in international reports such as the Human Development Report. Hence, indicators like the Human Development Index or the Environmental Performance Index (which assesses a country's commitment to environmental and resource management) become critical in raising awareness of sustainable issues and progress made on that front. The UN's Millennium Development Goals (to be achieved by 2015) also seek to mobilize states in an international partnership to implement concrete plans on the reduction of human travails of burgeoning population, degrading environment, increasing poverty, and so on, and bring about global equity.

Springing from a similar concern, the UN Food and Agriculture Organization has established a "red alert" list of more than 50 pesticides that have been banned in many countries. It requires the manufacturer of such pesticides to inform importing countries about why they have been banned elsewhere. This is because food produced with these pesticides in the importing countries and exported to other countries that have banned them continues the vicious cycle of jeopardizing human health as a result of lax practices. Moreover, crop production with pesticides results in enhanced greenhouse gas emission and reduces "healthy" precipitation in comparison with the depletion of water tables that provide quality water.

## Some Key Regulations

One major treaty that has been the source of huge controversy was the Kyoto Protocol (2005). This treaty seeks to reduce greenhouse gas emissions, particularly carbon dioxide emissions, through the provision of operational rules and legally binding timetables. It is hoped that a temporary rise in economic costs will be offset in the long run, given the reduction in costs incurred from damage to coastal infrastructure, ecosystems, and human health.

However, the United States, the largest consumer of fossil fuel, has refused to join the Kyoto Protocol. It considers the treaty an arbitrary imposition on its sovereign right to growth. This is because the treaty requires the imposition of a tax or permit system to reduce emissions. Such measures can result in increased economic cost through higher energy costs, which burdens growth. Emerging countries like China and India also refuse to join the treaty convention for the same reason. They claim that the Northern countries, especially the United States, have been responsible for large-scale pollution of the environment during their development phase, and hence, it is unfair to demand such regulations from newly industrializing countries of the South. It is the North's burden to undertake environmental stewardship.

In anticipation of the post-2012 climate regime, further critique of extant practices related to the treaty has started to emerge. For example, because of the system of permits or assigned amount units, the carbon market has grown, rather than abated, thereby reducing net atmospheric benefit. This is particularly because rich regions like Japan and the European Union have been buying up carbon allowances from cash-strapped Eastern European states (which have pollution permits to spare because of their industrial decline, rather than real emission cuts) to meet their treaty requirements before the Copenhagen Summit. The permits, which are cheaper and more certain than the treaty's Clean Development Mechanism credits, act as a perverse incentive. Rich countries buy fewer international offset credits generated from genuine emission cuts from developing country projects undertaken through the protocol's Clean Development Mechanism. However, there is hope that such constant scrutiny of state activities will nudge the states toward more altruistic behavior in the forthcoming treaty negotiations.

International scrutiny has also resulted in beneficial treaties like the Basel Convention (1989), developed by the UN Environment Programme. This treaty restricts the international transport of hazardous wastes. The 1995 amendment to the treaty bans the export of any hazardous waste from industrialized to developing countries. Similarly, the Stockholm Convention on Persistent Organic Pollutants (2004) seeks to protect human health from chemicals that disrupt the endocrine system, cause cancer, and generally harm developmental processes of humans. It also seeks to protect the environment from the 12 most toxic chemicals, including dichlorodiphenyltrichloroethane (DDT). However, an exception is made for poor countries that do not have access to alternatives and require DDT to be administered in small quantities to control mosquitoes and associated malaria disease.

In 2004, the International Treaty on Plant Genetic Resources for Food and Agriculture went into effect. This treaty seeks to limit the genetic materials that agribusinesses are allowed to patent. This enables farmers, particularly those from developing countries, to save, use, exchange, and sell farm-saved seeds and ensures their food autonomy. Similarly, to ensure that human technological engineering efforts at creating sustainable seed varieties do not result in the decimation of biodiversity, the Cartagena Protocol on Bio-safety went into effect in 2003. This treaty tries to provide appropriate procedures for the handling and use of genetically modified organisms to ensure that the threat of gene transfer from them to their wild relatives is lessened. International treaties have also come into effect to ensure continued marine food stocks. In 1986, the International Whaling Commission imposed a global ban on commercial whaling in international waters. Issues of food safety have also been covered by product labeling standards such as the Agreement on Sanitary and Phytosanitary Measures administered under the aegis of the World Trade Organization.

## Preservation of Biodiversity

Several conventions have become critical for the preservation of biodiversity. The Convention on Biological Diversity produced by the 1992 Earth Summit requires signatory nations to build a detailed plan for managing and preserving biodiversity. The Convention on International Trade in Endangered Species of Wild Fauna and Flora (1975) seeks to ban the hunting, capturing, and selling of endangered species and wants to regulate the trade of organisms considered to be potentially threatened. Agenda 21, adopted in 1992 by the UN General Assembly, includes a nonbinding Statement of Principles for the Sustainable Management of Forests. The World Heritage Convention helps to protect natural sites like national parks and cultural sites deemed as important for international heritage by a nation. The UN Convention to Combat Desertification, adopted in 1994, spells out practical measures to be undertaken to combat desertification and mitigate the effects of drought in specific ecosystems.

The UN Convention on the Law of the Sea (1994) helps protects its mineral resources. Multilateral agreements have also been reached over control of ocean pollution and protection of wetlands in 1975. Similarly, the Environmental Protection Protocol to the Antarctic Treaty (1990) puts a moratorium on mineral extraction and development. This helps to ensure that a vital ecosystem that plays a key role in regulating many aspects of the global climate is preserved for purposes of peace and future security.

The World Trade Organization implemented the legal Agreement on Trade-Related Intellectual Property Rights in the 1990s. There are negative and positive aspects of such an agreement. There are undesirable consequences in terms of negative incentives for invention and production of medicines required for the majority of the world population that lives in developing countries. This is because these people cannot afford to pay for healthcare, and hence medicine-related patents are not profitable for large pharmaceutical organizations. However, popular activism has resulted in the amendment of the treaty to ensure access to vital medicines for public health crises, such as AIDS.

The protection of geographical indications in the Agreement on Trade-Related Intellectual Property Rights has also become a source of powerful arguments for proponents of Fair Trade. The Fair Traders consider it a long-term national public right that helps to develop niche export markets for high-value food and nonfood agricultural products and to protect indigenous knowledge and biodiversity as a public good. It also protects against biopiracy, counterfeiting, and free riding by third parties; reduction of transaction costs in terms of information search by consumers; and prevention of market failures resulting from asymmetric information and lack of transparency. Furthermore, they help to empower a network of production and consumption actors. Consumers do not simply choose products for their benefits but also become citizens in collective political choices related to economic rules and their environmental consequences.

It is in light of the above regulations that it becomes critical to understand that corporate social responsibility (CSR) has become international rules in the global commodity production and value networks. As stakeholders, the relationships among consumers, retailers, and producers are increasingly being regulated by standards, codes of conduct, and certification systems to enhance ethical practices through building reputation and credibility. The World Economic Forum in its Global Competitiveness Report of 2006 also pointed out the competitive relevance of CSR being part of a firm's core business strategy. The International Standards Organization has created a vital standard for CSR in ISO 14001. This standard demands managerial commitment to specific environmental policy measures like reduction in emissions.

International regulatory frameworks are complex. Hence, increasingly there are regional attempts to recodify them for more effective implementation. Such regional attempts can be seen at the level of regional institutions like the EU and the Association of Southeast Asian Nations. Regional treaties such as the 2003 African Convention on the Conservation of Nature and Natural Resources have also emerged to add to the increasing pool of global attempts at conservation.

**See Also:** Environmentalism; Green Communities; Green Politics; Kyoto Protocol.

### Further Readings

Boiral, O., et al. "The Action Logics of Environmental Leadership: A Developmental Perspective." *Journal of Business Ethics,* 85 (2009).

Dubuisson-Quellier, S. and C. Lamine. "Consumer Involvement in Fair Trade and Local Food Systems: Delegation and Empowerment Regimes." *GeoJournal,* 73 (2008).

Friel, S., et al. "Global Health Equity and Climate Stabilization: A Common Agenda." *The Lancet,* 372:9650 (November 8–14, 2008).

Grote, U. "Environmental Labeling, Protected Geographical Indications and the Interests of Developing Countries." *The Estey Centre Journal of International Law and Trade Policy,* 10/1 (2009).

Policy Briefing International. "UNFCC: 'Hot Air' Could Threaten Climate Policy Credibility." Ends Report. (March 2009).

Rockwood, L. L., et al., eds. *Foundations of Environmental Sustainability: The Coevolution of Science and Policy*. New York: Oxford University Press, 2008.

Shestack, J. J. "60th Anniversary of the Universal Declaration of Human Rights." *International Law News,* 38/1 (2009).

*Pallab Paul*
*University of Denver*

## Internet Purchasing

Internet purchasing is the act of buying products or services via the Internet, as differentiated from purchasing by some other means, such as retail. From a consumer standpoint, purchasing items online is an increasingly attractive option. Internet shopping can be done in the comfort of one's own home without the need to face traffic, crowds, limited business hours, or checkout lines. Online, consumers can buy nearly anything. Popular stores sell items such as electronics and home furnishings, and some offer nonperishable—or even perishable—foodstuffs if the consumer happens to live in an area served by such a provider. In addition, in the United States, the extra cost added for shipping fees is often completely offset by immunity from out-of-state sales taxes, although this might not be the case in the near future as legislatures move to tax online purchases. Nonetheless, it comes as no surprise, given those benefits, that online shopping has been rapidly gaining market share from other forms of purchasing.

Internet purchasing is a logical extension of home shopping that arose when entrepreneurs realized that the business model of mail-order catalogs and television shopping channels could be applied to the World Wide Web. Online stores have evolved since then to mimic traditional shops by giving their customers the ability to browse catalogs and product information. However, not all online shops adhere to the traditional business model. Online auction stores permit customers to bid on products, wholesale and overstock shops assist in selling off bulk or surplus products, and new digital distribution methods bypass the shipping of music or video discs entirely. The only things they all share in common are the medium through which they communicate with their customers and the methods by which they transfer physical items to customers.

## Costs

Charting the total environmental cost of Internet purchasing can be difficult. First, there are the heavy-metal mining and manufacturing costs of the involved computers and support equipment. Then, power plants must provide electricity to operate those devices. Data centers must be constructed and air conditioned to keep the Website equipment running smoothly, and switching stations must be added to the Internet to support the increased amount of commercial traffic. One could even go as far as to chart the damage to the environment caused by the laying of hundreds of thousands of miles of cable to carry the data around the world. However, all these damages pale in comparison to the environmental harm that is caused once the purchasing has been completed. The products themselves must be stored and shipped, requiring warehouses to be lit and packages to be routed across the country before they can be delivered to the doorstep. However, all of those costs are shared with other forms of purchasing as well.

## Benefits

The environmental benefits are made evident when the damages of purchasing online are compared with those of purchasing by some other means, such as retail. Such activity involves the same amount of computer and Internet use, as retail stores purchase from their suppliers, as well as a similar degree of housing and shipping of the products. On top of that, however, the shops must also be lit and air conditioned. Then one must take into account the additional use of fuel for the consumer to drive from their home to the store. Because the extra environmental costs in retail shopping are primarily the fault of the consumer driving to and from the store, Internet purchasing becomes a very appealing alternative to people who care about their impact on the environment but do not have the time or the means to walk or make use of public transportation to do their shopping.

Businesses are also starting to realize the benefits that could be brought both to themselves and to their customers. A study by the RAND Corporation found that the online bookstore Amazon.com will spend roughly half as much on energy as a single retail store per square foot of building space, and 16 times less per sale on upkeep costs. The overall benefit of applying Internet technologies to the traditional shopping mindset has been a net decrease in energy use of 1 to 2 percent annually since 2005 in the use of energy in some markets, despite the continued growth of sales. A separate research study compared the environmental costs of two identical DVD rental services, one operating a nationwide network of retail stores for rentals and returns and the other operating solely online and processing its storage, rentals, and returns through three regional warehouses and the U.S. postal system. The online store consumed roughly one-third less energy and produced two-fifths less carbon dioxide emissions per disk rented than the average disk rented. Though the measurable environmental benefits of Internet purchasing are great, it is very difficult to measure the effects of browsing and shopping online without making a purchase compared with doing the same via a retail store.

## Application

The environmental benefits of Internet purchasing cannot be applied as efficiently to all purchases. Large purchases that involve a significant amount of packaging, products that must be refrigerated, and small purchases that must be packaged into boxes larger than the products themselves will all increase the amount of environmental waste caused by shipping. In addition, the cost of shipping those packages increases with the distance from the local post office or shipping center, and the remoteness of those locations can also be a factor. In totality, it is most efficient to walk to a retail store and carry the purchased items home without the assistance of plastic bags that are immediately disposed, but online purchasing is more environmentally friendly than driving.

**See Also:** Packaging and Product Containers; Shopping; Websites and Blogs.

### Further Readings

Mayo, Donna T., Marilyn M. Helms and Scott A. Inks. "Consumer Internet Purchasing Patterns: A Congruence of Product Attributes and Technology." *International Journal of Internet Marketing and Advertising*, 3/3:271–98 (2006).

Romm, Joseph. "The Internet and the New Energy Economy." *Proceedings of E-Vision 2000*. Washington, DC: RAND Corporation, April 25, 2009. https://www.rand.org/pubs/conf_proceedings/CF170.1-1/CF170.1.romm.pdf (Accessed September 2009).

Sivaraman, Deepak, et al. "Comparative Energy, Environmental, and Economic Analysis of Traditional and E-Commerce DVD Rental Networks." *Journal of Industrial Ecology*, 11:77–91 (2007).

*Anthony R. S. Chiaviello*
*Craig Blaylock*
*University of Houston-Downtown*

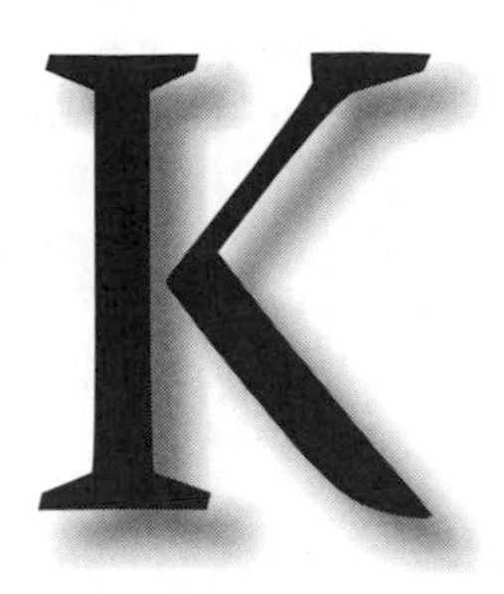

## Kyoto Protocol

The Kyoto Protocol is a protocol to the United Nations Framework Convention on Climate Change that establishes legally binding commitments for the reduction of four greenhouse gases (GHGs) and two families of gases produced by industrialized nations. As of May 2009, 183 parties have ratified the protocol, which was initially adopted for use on December 11, 1997, in Kyoto, Japan, and which entered into force on February 16, 2005. Under Kyoto, industrialized countries agreed to reduce their collective GHG emissions by on average 5.2 percent compared with 1990. National limitations range from 8 percent reductions for the European Union and some others to 7 percent for the United States, 6 percent for Japan, and 0 percent for Russia. The protocol limits GHG emission increases to 8 percent for Australia and 10 percent for Iceland.

On May 9, 1992, the United Nations Framework Convention on Climate Change was adopted. The ultimate objective of the Convention (Article 2) is to stabilize concentrations of GHGs in the atmosphere at a level that would prevent dangerous anthropogenic interference with the climate system. The convention divided countries into two groups: those listed in Annex I (Annex I Parties) and those not listed (Non-Annex I Parties). Annex I Parties are industrialized countries that have historically emitted the most GHGs. Their per capita emissions are higher than those of most developing countries, and they have more financial and institutional resources to address the problem. The principles of equity and of "common but differentiated responsibilities" set out in the convention require these parties to take the lead in changing emissions trends. To this end, the Annex I Parties agreed to adopt policies and measures with the (legally nonbinding) objective of stabilizing their emissions at 1990 levels in 2000. Non–Annex I countries are mainly developing countries. However, there are also some that would now be categorized as newly industrialized countries, such as South Korea, China, Mexico, and South Africa. The convention recognizes that financial assistance and technology transfer are essential to enable developing countries to cope with global warming and to adapt to its effects.

The parties meet annually at the Conference of Parties, the supreme body of the convention. At these meetings, the parties make the necessary decisions to promote the effective implementation of the convention and pursue dialogue on the best measures to fight global warming.

At the first Conference of the Parties, which took place in Berlin in 1995, the parties agreed that the specific commitments of the convention for the Annex I parties were not adequate because they were too vague. The parties then launched a new round of discussions to achieve tougher and more specific targets for Annex I parties. After two and a half years of intense negotiations, the Kyoto Protocol was adopted at the Third Conference of the Parties on December 11, 1997, in Japan.

The Kyoto Protocol to the convention entered into force on February 16, 2005, or 90 days after the date of deposition of the instrument of ratification by Russia. Russian participation was essential, following the refusal to ratify by the United States, because a prerequisite for the protocol's entry into force is that ratifying parties account for at least 55 percent of the total carbon dioxide ($CO_2$) emissions of all Annex I Parties of the convention, which Russia enabled it to do. On December 3, 2007, the new prime minister of Australia Kevin Rudd signed the instruments of ratification by Australia, making the United States the only industrialized country that has not deposited its instrument of ratification.

## Innovative Steps

Under international law, this protocol is innovative for several reasons. First, inspired by the success of the Montreal Protocol, the negotiators decided to define measurable and binding targets, moving away from the declarations of intent that often characterize international environmental law. Furthermore, this protocol introduced three market-based mechanisms (the Clean Development Mechanism [CDM] and Joint Implementation), thereby creating what is now known as the "carbon market."

The Kyoto Protocol commits the Annex B parties to a binding target for reducing or limiting their emissions of GHGs. The Protocol recognizes six gases or groups of GHGs: $CO_2$, methane, nitrous oxide, hydrofluorocarbons, perfluorocarbons, and sulfur hexafluoride. The commitment period extends from 2008 to 2012 (the parties who have ratified the protocol have to meet their commitments relating to this five-year period).

Reduction targets vary from an 8 percent reduction in GHG emissions for the period 2008–12 compared with 1990 levels for some countries, to a limit on the increase for others (e.g., Iceland is allowed an increase of its emissions by 10 percent over the period).

The best approximation for the baseline is the level of emissions in 1990 for the six gases or families of gases addressed in the protocol. However, there are quite a few exceptions, and the determination of the baseline is not as straightforward as one might expect. The definition of the cap and the actual extent of reductions are more complex because some activities related to change of land use (e.g., deforestation or reforestation), which emit or capture $CO_2$ in the atmosphere, are also covered. The estimate of these changes in land use and the effect of $CO_2$ equivalent can have significant consequences for national targets and has therefore been a controversial issue within Kyoto negotiations. In addition, GHG emissions from fuels used in international aviation and marine transportation are not accounted for within the targets of the Kyoto Protocol. This is partly because of accounting difficulties, which are widely recognized as a weakness of the Kyoto Protocol.

The Kyoto Protocol and subsequent decisions under the conferences/meetings of the parties recognize four types of emissions allowances or credits. First, assigned amount units (AAUs) are the allowances allocated to parties (based on historical emissions and emissions targets, as explained earlier). An AAU is equal to 1 metric ton of $CO_2$ equivalent.

The Kyoto Protocol recognizes three other types of units that can be used instead of AAUs to comply with its emission target:

- certified emission reduction, issued under the CDM;
- emission reduction unit (ERU), issued under Joint Implementation; and
- removal unit, issued pursuant to the relevant provisions of the modalities concerning increasing the capacities of sinks.

The original allocation of AAUs was made on the basis of historical emissions. This approach, called grandfathering, favors industrialized countries because it generally leads to the allocation of more allowances to big emitters. This allocation would be very unfavorable for developing countries.

## Determining Effectiveness

The Kyoto Protocol's effectiveness depends on two critical factors: whether parties comply with their commitments, and whether the emissions data used to assess compliance are reliable. Recognizing this, expert review teams check annual inventories to make sure they are complete and accurate and conform to the guidelines. The annual inventory review is generally conducted as a desk or centralized review. However, each Annex I Party is subject to at least one in-country visit during the commitment period. Annex I Parties must also provide national communications on activities they undertake to implement the protocol.

Registries record the holdings of Kyoto units and any transactions involving them through a structure of accounts. They record and monitor all transactions in AAUs, ERUs, certified emission reductions, and removal units. This is similar to the way that banks record balances and movements in money using accounts allocated to individuals or to other entities. Whenever a national registry undertakes a transaction that affects the number of units held by a Kyoto party, the register communicates with the International Transaction Log, which then verifies the compliance of the transaction in accordance with the general rules of accounting for Kyoto units.

Next to the trading of allowances, the CDM is a flexibility mechanism that helps parties achieving their emission targets. Through a CDM project, an Annex I Party invests in a non–Annex I Party for the purpose of reducing GHG emissions, along with the promotion of sustainability principles in developing countries. For every ton of $CO_2$ reduced or absorbed through the project, the investor will receive a certified emission reduction. The calculation of the emission reduction is based on a comparison with a baseline scenario without the project.

The Kyoto Protocol itself gives very little guidance on the practical implementation of this mechanism. This was clarified in the Marrakesh Accords (Conference of Parties 7 held in November 2001), which defined the practical features of the CDM. These agreements included the establishment of a CDM executive board and detailed the different steps leading to the issuing of certified emission reductions.

Additionality is the fundamental criterion for the recognition of a project. Under this criterion, the project developers must, from a business-as-usual (BAU) scenario, show that their project will result in GHG emissions reductions that would not occur otherwise. The difference between the level of emissions in the BAU scenario and in the scenario with the CDM project determines rights to certified emission reductions. The additionality test is essentially composed of three elements: environmental additionality (does the project reduce emissions below the BAU scenario?), investment additionality (does access to certified emission reductions make the project viable?), and technological additionality (does the project lead to a transfer of technologies in the host country?). The Marrakesh Accords implicitly recognize that investment additionality can be a means of proving the environmental additionality of the project, but it is not necessarily the only way (other barriers to investment such as technological barriers or availability of capital may also prove the environmental additionality of a project). Finally, economic additionality is also considered. This implies that the capital provided by the developed countries is not used as a substitute for traditional aid to developing countries.

If the CDM has been successful in promoting thousands of carbon reduction projects in developing countries, the mechanism also faced some criticisms resulting from uneven geographical or technological spreads and concerns about the additionality of some registered projects. In practice, projects are concentrated in three countries (China, India, and Brazil) and in the most profitable activities. Economically and environmentally this is desirable, and the foundation for the establishment of an emissions market. However, the CDM was also intended to deliver on the third and least tangible pillar of sustainable development; that is, improving the social conditions in local communities. Although the majority of projects developed in the energy (small hydropower, wind farms, etc.) or waste (waste recycling, recovery of methane, etc.) sectors improve social conditions around the project site, this is not true for large industrial projects involving the destruction of F-gases or large dams. Unfortunately, these large projects represent the lion's share of the issued credits so far.

Joint Implementation is a flexibility mechanism similar to the CDM but in which projects are carried out in an Annex I Party. To avoid double-counting, the issuance of ERUs must correspond to a cancellation of a corresponding amount of AAUs. By requiring Joint Implementation credits to come from a host country's pool of AAUs, the Kyoto Protocol ensures that the total amount of emissions credits among Annex I parties does not change for the duration of the Kyoto Protocol's first commitment period. To illustrate, suppose the United Kingdom finances a project for the reduction of 10,000 tons of $CO_2e$ in Russia. The United Kingdom then receives 10,000 ERUs. Thanks to this project, Russia needs fewer AAUs (its real emissions are reduced by 10,000 tons, so that is 10,000 AAUs not needed by Russia). To avoid double-counting the reduction (i.e., the fact that two countries benefit from the same reduction project), ERUs issued by Russia must come from its reserve of AAUs. Given the link between ERUs and AAUs, ERUs can only be issued during the commitment period; that is, between 2008 and 2012 inclusive.

Although the trading provisions are likely to stay in the successor to the Kyoto Protocol, they were initially a controversial policy. The scope of the initial cap reflects a political negotiation, rather than an optimization of environmental outputs. Often criticized for its lack of ambition, Kyoto should, however, be considered a first step, as requirements to meet the convention will be modified until the objective is met. Issues such as "hot air," concerns about additionality for some CDM or Joint Implementation projects, and the lack of transparency and liquidity of the carbon market itself will need to be considered carefully. This is a prerequisite for the next protocol to be more environmentally effective.

**See Also:** Carbon Credits; Carbon Emissions; Carbon Offsets; International Regulatory Frameworks; Regulation.

### Further Readings

Brohé, Arnaud, et al. *Carbon Markets. An International Business Guide*. London: Earthscan, 2009.

Kyoto Protocol. "Toward Climate Stability." http://www.kyotoprotocol.com (Accessed December 2009).

McGovern, Joe. *The Kyoto Protocol*. Pittsburgh, PA: Dorrance Publishing, 2006.

Stewart, Richard B. *Reconstructing Climate Policy: Beyond Kyoto*. Washington, D.C.: AEI, 2003.

Victor, David G. *The Collapse of the Kyoto Protocol and the Struggle to Slow Global Warming*. Princeton, NJ: Princeton University Press, 2004.

*Arnaud Brohé*
*Amandine Bourmorck*
*Université Libre de Bruxelles*

# Lawns and Landscaping

A well-groomed lawn is the hallmark of the suburban context. Although familiar, comforting, and healthy at first glance, most individually owned and maintained lawns are actually ecological deserts that have little resemblance to the native local landscape.

Suburban green space, despite providing very few of the ecological services associated with the natural habitat that it replaces, is much more resource intensive to maintain than naturally occurring flora. Domestic irrigation exacerbates water scarcity in many regions, and the overuse of chemical fertilizers and pesticides is a primary cause of declining water quality in most impaired watersheds. Heightened awareness of the environmental cost of maintaining the typical suburban lawn has given rise to the growing proliferation of organic lawn and garden practitioners and products. Xeriscaping—the use of locally adapted, native plants to reduce the need for irrigation—is becoming a household term in arid regions, and this approach to landscaping will become even more common as a changing climate intensifies water scarcity issues. In addition, the current trend away from suburban living in many regions has home buyers demanding, and developers supplying, higher-density residential development in which individually owned and maintained lawns are replaced by larger, shared open spaces that often include elements of the natural landscape.

## Environmental Impact of Conventional Lawn and Landscape Practices

In many regions of the United States, over 50 percent of residential water consumption is used for the irrigation of lawns and gardens. In the United States, rainwater harvesting for domestic use is only just creeping back into the public consciousness in all but the most drought-stricken regions. In addition, building codes in many communities actually make low-tech rainwater harvesting—even for irrigation—illegal or overly complicated by requiring the use of engineered systems. As such, almost all water used for irrigation in suburban America is potable water, meaning that it has been treated and is fit for human consumption. In addition to creating or exacerbating water scarcity issues, the allocation of such a high proportion of potable water to irrigation represents a great inefficiency because the treatment and distribution of potable water require expensive infrastructure that is resource- and energy-intensive to run and maintain.

Except for in the most drought-stricken regions of the United States, rainwater harvesting for domestic use is only just reentering the public consciousness.

*Source:* iStockphoto

Lawn care and landscaping are multibillion dollar industries built largely around chemical products and resource-intensive practices aimed at supporting ornamental species that are grown out of their native context. Because of the accessibility of chemical fertilizers and pest control agents, their overuse in domestic applications is rampant. Nitrogen, phosphorus, and potassium are the three main components of most fertilizers. Nitrogen, in the form of water-soluble ammonium nitrate and ammonium sulfate, is typically the largest constituent in inorganic fertilizers. Nitrates become available to plants as soon as water is applied but may leach through the root zone to contaminate groundwater. In contrast, nitrogen from manure and other natural organic sources tends to remain in place but is not biologically available until it is broken down by natural microbial activity in the soil. Nitrogen is the limiting nutrient in most estuarine and marine systems, and unabsorbed ammonium nitrates and sulfates, which are carried to these water bodies in storm water runoff, can cause algal blooms that eventually lead to low dissolved oxygen, with grave consequences for fish and other aquatic organisms. Phosphorus is the limiting nutrient in rivers, lakes, and other freshwater systems, so phosphorus from excessive fertilizing can be similarly detrimental to these habitats. Phosphates—phosphorous combined with oxygen—are immobilized in soil and are thus of little threat to groundwater supplies, although phosphorus from organic sources may be less tightly bound to soil and therefore be more prone to leaching or runoff.

Many pesticides sold for domestic application are broad-spectrum chemicals that may have unintended consequences for human and environmental health. Broad-spectrum chemical insecticides may wipe out beneficial parts of balanced and healthy terrestrial ecosystems, including bees, birds, and other insects that may be natural predators of other undesirable pests. As a result, the use of these chemicals often engenders a vicious cycle that suppresses natural ecological processes and necessitates the continued use of such products.

## Greener Alternatives for Lawn Care and Landscaping

Plant selection determines the need for supplemental irrigation. Xeriscaping is a broad strategy by which the need for supplemental irrigation can be reduced. Native plants provide habitat for local fauna and may have symbiotic relationships with, or at least be more tolerant of, local insects, thus reducing the need for supplemental pest control. Native plants are also more likely to tolerate indigenous soil conditions, thereby reducing the need for supplemental fertilizers.

If required, the source of water for supplemental irrigation and the means by which it is delivered are principal determinants of overall water usage on a site. The collection and use of rainwater for irrigation is an efficient and cost-effective way to offset the demand for water from off-site sources. Rooftop runoff can be directed, via gutters and downspouts, into simple rain barrels, cisterns, or other storage devices. Even without the use of these storage devices, rooftop runoff can be directed onto lawns or into gardens or flowerbeds to reduce storm water runoff and reduce the need for supplemental irrigation. Drip irrigation, or microirrigation, delivers water from a pressurized water source directly to plant roots through a combination of tubes, valves, pipes, and in some instances, microsprayers. The timing of water delivery is also an important consideration in practicing water conservation. Water delivered in the early morning is less apt to evaporate than water delivered during midday heat. Last, the use of mulch or an alternative helps stabilize soil temperature and slow evaporation, further reducing the need to irrigate landscaped areas.

Organic lawn care and gardening, which employs a holistic, ecology-based approach and excludes the use of all chemical fertilizers and pesticides, signifies the green standard on the continuum of environmental sustainability. A wide variety of natural sources of organic nutrients can be used in lieu of chemical fertilizers, including composted manure and organic compost, which can be generated on-site from the collection of food waste. Freshly mowed grass clippings, if left in place rather than bagged and removed, serve as natural compost and are an excellent supplemental source of organic nutrients for the yard. In situ composting of clippings also saves landfill space that would otherwise be occupied by bagged yard waste. In addition to these and other natural sources of organic nutrients, there are a myriad of organic fertilizers available in the marketplace. Most organic fertilizers contain easily identifiable ingredients, including vegetable matter such as kelp or alfalfa, minerals such as zinc and gypsum or lime, and animal byproducts such as manure or blood meal. Other common ingredients include molasses, whey, potash, and corn gluten meal. Most organic fertilizers are categorized as slow-release—a characteristic that is preferable in terms of minimizing unintended environmental impact, even when selecting chemical fertilizers.

Holistic consideration of landscaped habitats has given rise to integrated pest management, a systematic approach by which information about the life cycle of particular pests is used to identify targeted management strategies that are cost-effective and minimize hazard to people, pets, property, or the environment. As with fertilization, accurate diagnosis of the problem is an integral first step in identifying targeted pest management practices. Very low impact natural solutions, such as sprinkling black pepper to deter rabbits, releasing ladybugs to control aphids, or employing nematodes to control grubs, are typically the first actions taken along the continuum of integrated pest management. Only if these strategies are unsuccessful might more intensive practices be used. This incremental approach ensures that appropriate, targeted techniques are used to minimize unintended environmental impacts.

Although the use of locally adapted native species and natural ecological processes is the foundation of greener lawn care and landscaping in the conventional suburban context, arguably the ultimate in sustainability is the modification of suburbia itself. The construction of the typical suburban neighborhood, with single-family homes isolated as islands in a sea of green lawn, necessitates nearly complete modification of the natural landscape, which has a big effect on both the local ecology and the project budget. In a lower-impact model, individual dwelling units are clustered together, enabling larger sections of the native landscape to be preserved as shared, multifunctional amenities that provide recreational opportunities, aid with assimilating storm water runoff, buffer adjacent natural waterways, and provide natural habitat for birds and wildlife. In addition, preserved natural open space requires little or none of the care and maintenance associated with conventional suburban green space, making it far and away the most sustainable alternative.

**See Also:** Composting; Gardening/Growing; Lifestyle, Suburban.

### Further Readings

Robbins, Paul. *Lawn People: How Grasses, Weeds, and Chemicals Make Us Who We Are.* Philadelphia, PA: Temple University Press, 2007.

U.S. Environmental Protection Agency. "GreenScapes Program." http://www.epa.gov/waste/conserve/rrr/greenscapes/index.htm (Accessed August 2009).

U.S. Environmental Protection Agency. "Xeriscape Landscaping: Preventing Pollution and Using Resources Efficiently" (April 1993). http://yosemite.epa.gov/water/owrccatalog.nsf (Accessed August 2009).

*Jeffrey Pollack*
*Independent Scholar*

## Leisure and Recreation

The leisure industry, with its steady proliferation of goods and services, constitutes a major force in society that influences economies, cultures, social identities, and nature itself in cities and regions throughout the world. As the ecological footprint of the constantly expanding leisure industry has grown, so too have concerns about environmental sustainability, global climate change, and green, ecofriendly business strategies.

Beginning in the early 20th century, three interrelated developments transformed the way people engaged in leisure and recreation in the United States: organized pasttimes and activities replaced spontaneous and informal forms of leisure; recreation activities became increasingly commercialized and commodified; and leisure and recreation shifted from the public to the private sphere. Entering the second decade of the 21st century, these tendencies have intensified and become global in nature.

### Sustainability and the Greening of Leisure, Recreation, and Tourism

Sustainability has become a ubiquitous feature of leisure, recreation and tourism business and development strategies. Following the publication of the 1987 Bruntland report, ideas of sustainability—development that seeks a balance between economic growth, environmental protection and social equity—became widely used and began to inform discussions about the environmental consequences of economic growth. At the core of sustainable development is the notion that economic growth and environmental protection are compatible. Such a notion replaced the prevailing view that held that the economic logic of capitalism was inherently environmentally destructive, and that there might be limits to continuous economic growth and development with a more pro-growth and pro-technology stance.

Tourism is one of the largest industries in the world, and both contributes to global climate change and is impacted by it. Life cycle analysis tools are shedding light on the true environmental cost of tourist travel. One response to the hidden environmental, social, and economic costs of travel has been the proliferation of various types of ecotourism. According to the International Ecotourism Society, ecotourism is "responsible travel to natural areas that conserves the environment and improves the well-being of local people."

The dominant approach to dealing with GCC impacts generated by industries such as tourism encourages voluntary efforts on the part of businesses and consumers to become ecologically efficient and sustainable, and downplays calls for government regulation and mandates. Increasingly, various responsible travel websites are using life cycle analysis tools to rate the sustainability of tourist packages. One popular voluntary measure is purchasing carbon offsets, which in theory is meant to cancel out the emissions generated by activities like air travel by directing money to programs that reduce emissions elsewhere, such as planting trees in Africa. This approach has been criticized because it may enable people to feel less guilt about their environmental impact and does little to change the behavior causing the problem in the first place. A mantra of the corporate responsibility discourse is that going green is a smart, win-win proposition that helps both the environment and business profits. Time will tell if corporate greening is a genuine business strategy or a form of greenwashing. There are indications that corporate responsibility and greening will need to be helped along by responsible government action.

## Tourism, Parks, Outdoor Recreation, and the Experience of Nature

Wildlife-related recreation is a major leisure activity and a powerful economic engine. In 2001, more than 87.5 million Americans enjoyed outdoor recreations like fishing, hunting, bird watching, or wildlife photography, spending approximately $122.3 billion in their pursuits. In 2007, almost 300 million recreational visits were made to U.S. national parks. The United States has more than 778 million acres of publicly owned recreation lands—more than any other nation.

Historically, parks and wildlife recreational preserves have been a perennial source of conflict over commercialization and privatization. In the 1940s, Aldo Leopold, a prophet of the modern ecology movement, warned that the steady barrage of motorized tourists and the rise of the recreational gadget industry were wreaking havoc with the natural environment. Since the 1980s, the multi-billion dollar recreational industry has responded to environmental issues and concerns by developing a host of green and sustainable recreational products and services. With the steady decline in public revenues, park agencies have increasingly come to rely on public/private partnerships schemes to help pay for park maintenance and upkeep. Many fear that this is exacerbating the trend toward commercialization and tipping the balance away from low-impact, sustainable forms of nature-based recreation toward unsustainable, corporate-backed, technologically intensive ones.

### Cities as Spaces of Consumption, Leisure, and Sustainability

Increasingly, consumption associated with leisure, recreation, and tourism has become a primary economic base for many postindustrial cities throughout the world. For the last three decades, planning for leisure spaces and tourism have become key components of cultural regeneration and branding strategies undertaken by governments eager to attract tourists and high income residents and businesses. In many cities, commercial strategies have transformed downtowns into urban entertainment districts, complete with museums, shopping centers, stadiums, and other leisure destinations in an attempt to stimulate consumption. Increasingly, leisure and recreational amenities are key selling points for mixed-use residential developments.

In an attempt to reduce energy consumption, carbon emissions, and other forms of environmental pollution, urban regeneration strategies have taken a green turn, and have begun to adopt smart growth and sustainable urban design principles. In many cities, transforming environmentally damaged areas or brownfield sites into leisure spaces and parks has become a major smart growth strategy. New Urbanism's emphasis on building medium-density towns and neighborhoods where houses, offices, shopping, and leisure activities would all be within walking distance is a sustainable development strategy that lessens the input of fossil fuels and reduces carbon missions. Another approach to dealing with some of the negative social and environmental impacts of leisure and recreation can be found in voluntary simplicity movements that question excessive consumerism and call for a new leisure ethic that emphasizes unharried and noncommodified leisure time, such as gardening, artistic endeavors, hiking, visiting libraries, parks, or museums.

The recent global economic downturn has led many people to rethink how they spend their free time and question the trend toward the increasing commercialization and commodification of leisure. Economic necessity is forcing many people to engage in "time intensive" over "goods intensive" forms of leisure. Only time will tell if this form of green leisure, with its light ecological footprint, will become a common practice.

**See Also:** Consumer Culture; Ecotourism; Greenwashing; Lifestyle, Urban; Simple Living; Sustainable Consumption.

#### Further Readings

Gossling, Stefan and Michael Hall. *Tourism, and Global Environmental Change: Ecological, Social, Economic and Political Interrelationships*. London: Routledge. 2006.

McLean, Daniel, Amy Hurd, and Nancy Brattain. *Kraus' Recreation and Leisure in Modern Society*, 8th Ed. London: Jones and Bartlett, 2007.

Smith, Melanie. *Tourism, Culture and Regeneration*. Cambridge: CABI, 2007.

*Steven Lang*
*LaGuardia Community College*

## Lifestyle, Rural

Rural lifestyles may be characterized by the term *simplicity*: just as the level of interaction and impact on the environment is limited, it manifests a high degree of variability.

Notwithstanding the varying definitions of rural life, a relatively low population, occupations dependent on working the land, lower-than-average per capita income, and a general inadequacy or outright lack of basic facilities and services are some of the distinguishing features of rural lifestyles. These features are also important determinants of the level and pattern of consumption of environmental resources, as well as fundamental environmental conservation and preservation factors.

One rural consumption pattern consists of subsistence agriculture, particularly farming, fishing, and livestock rearing. Here, the owner of a lodge in Ecuador picks maize from her garden.

*Source:* Karen Williams/Agricultural Resource Service/USDA

## What Is a Rural Dweller?

Globally, there is great variation in what it means to be a rural dweller, both within and between nations, and experiences of rural dwelling vary according to age, gender, sexuality, ethnicity, and culture. In many "developed nations," rural dwellers may be dependent on rural areas for their livelihoods (e.g., they are employed in primary industries such as agriculture, horticulture, or resource extraction, or craft and/or service industries in rural communities). They may choose to live in rural communities for lifestyle reasons, working from home for an employer based elsewhere, or commuting to urban areas for employment. In developing countries rural dwellers may be involved in agriculture or petty trading and may walk or cycle to daily work and recreational activities. In rural communities with strong social and family networks, rural dwellers can exist in communities in which every other member is known, and the emphasis for many may be on material survival, rather than on spending disposable income. Many rural dwellers in developing countries now earn a salaried income that is often a consequence of employment as unskilled labor in an urban area.

There are many forms of rural lifestyles, ranging from the most simple or rather primitive to the most modern (equaling urban lifestyles with respect to facilities and services). The various forms reflect the relative location and accessibility of the rural community to the nearest city, as well as the way in which the rural locality, economy, and culture are linked to urban life. The remainder of this article reflects primarily on rural communities in developing countries.

## Rural and Urban Lifestyles Compared

If we conceive of lifestyles as a continuum, rural and urban lifestyles can be two extremes. Rural lifestyle is essentially a land-based or land-related culture, in which land (including natural resources) and the family (its size and composition) form the major assets. This differs from the urban lifestyle, including ownership of means of production, movable and disposable assets, and a living and livelihood revolving around home, work/business, and recreation. Although the rural lifestyle can be characterized by a high level of interpersonal relationships, and well-defined individual and group roles and responsibilities, the urban lifestyle is often characterized by alienation, freedom from traditional bonds, and anonymity. The social network in the rural lifestyle is simple and defined largely by kinship relationship; in urban lifestyles, it could be a dense pattern defined along occupational, class, religious, and even ethnic lines. The majority of rural lifestyle culture depends on consumption of firewood and coal as sources of residential energy, whereas its urban counterpart relies on electricity and household gas and oil. Although urban lifestyle is more economically successful, it is more environmentally destructive, with a large ecological footprint and cultural artifacts.

## Features of Rural Lifestyles and Consumption Patterns

Rural dwellers in developing countries live in homes built with indigenous materials that may have a lesser impact on the environment and, in many cases, lack basic facilities such as indoor plumbing and electricity. Land is an important asset in rural lifestyle, and it is a common heritage and basis of generational socioeconomic prosperity that demands great commitment and deep family bonds to jointly use, manage, and secure it for the future. The importance of family size and composition often becomes an asset in rural living. In traditional African culture, for instance, large family size is greatly cherished, as many wives and children constitute the labor force in farming activities. Preference is usually shown for a male child, not only to sustain the family name and lineage but also to guarantee rights to family inheritance. Recreation is an important aspect of rural lifestyles, as many communities engage in such popular spectator sports as wrestling (popular among Igbos of southeastern Nigeria), horseracing, and many indoor and outdoor games. Among the Kazakhs of Kazakhstan, traditional horseback games are as old as the villages that practice them.

The rural consumption pattern is closely related to the prevailing economic structure and prominently features subsistence agriculture—particularly farming, fishing, and livestock keeping. The dietary structure is based on the consumption of locally produced agricultural products, largely carbohydrates, and is generally aimed at ensuring survival. Protein-rich products such as fresh milk and meat are sold in urban markets in exchange for urban goods like clothes. The generally low income realized from the sale of farm produce and the low ability to obtain loans to increase and diversify economic productivity generally leads to widespread poverty.

## Rural Lifestyles and Brown Environment

There are many aspects of the rural lifestyle that contribute significantly to the initiation and wide spread of brown environment. This perspective is based on the experience of rural dwellers in developing countries; in particular, Nigeria.

### Rural Physical Environment

Most rural communities have sprung up without being planned and have grown without guidance and control; they have thus acquired a morphology that is simple, but largely with no sound physical planning. Houses adjoin one another without setbacks. Access to most homes is possible only by footpaths. Facilities such as toilets and bathrooms are not within houses, where available, so residents use nearby bushes and open spaces as their bathrooms and cook in huts outside their houses. The quality of the indoor and neighborhood environment is a significant risk to many rural inhabitants, as livestock and birds are often kept within the family home.

### Economic Activities

The land-based rural lifestyle focuses on primary industries such as farming, fishing, small-scale animal husbandry, hunting, and trading. However, the use of bush clearing or burning, either to acquire more land for farming or to stimulate the growth of fresh herbage for grazing, removes vegetative cover, leaving the soil exposed to the sun's intense radiation and with a propensity to erode, increasing the frequency and severity of floods in many places. The use and or misuse of chemical fertilizers have deleterious effects on aquatic animals when they contaminate rivers and streams. The habit of raiding bush animals has not only destroyed many animal species but also may lead to the extinction of several endangered species.

The absence of sound planning coupled with a lack of basic facilities and services, particularly waste disposal facilities, have contributed to rural environmental degradation, though not as severely as suburban or urban environments. In many places, the level of degradation is largely within the carrying capacity of the natural environment, which makes a rural environment a growing lifestyle option for many urbanites.

**See Also:** Environmentally Friendly; Lifestyle, Suburban; Lifestyle, Sustainable; Lifestyle, Urban.

### Further Readings

Bascom, W. *The Yoruba of South Western Nigeria.* Austin, TX: Holt, Rinehart and Winston, 1969.

Duncan, Susan J. "Urban and Rural: Lifestyles Clash Over Differing Views of Open Space" (2007). http://www.newwest.net/topic/article/urban_rural/c57/l35 (Accessed August 2009).

Fadipe, N. A. *The Sociology of the Yoruba.* Ibadan: Ibadan University Press, 1970.

*Nigerian Compass.* http://www.compassnewspaper.com (Accessed August 2009).

*Adeboyejo Aina Thompson*
*Ladoke Akintola University of Technology, Ogbomoso*

## Lifestyle, Suburban

While the term *lifestyle* has become synonymous in popular culture with terms such as *way of life*, in academic circles the term finds its strongest elaboration in debates on consumption and consumerism. Returning the focus to consumption enables us to assess more clearly both the ways in which lifestyles contribute to environmental harm as well as how

Suburbs often form from the transformation of farmland into tract housing, like these new homes replacing farmland outside Des Moines, Iowa.

*Source:* Lynn Betts/Natural Resources Conservation Service/USDA

they might be redirected toward lessening environmental impact. Moreover, orienting lifestyle consumption within a larger vision of green ways of life may strengthen the ways in which more sustainable living practices can be developed. The suburban environment can be seen as both the location for many environmentally harmful consumption practices as well as a fertile one for the development of more environmentally beneficial alternatives. Indeed, there are specific elements of traditional suburban patterns of life and culture that lend themselves to such developments.

## Lifestyle and Consumption Practices

As a term most fully developed in relation to consumption practices, lifestyle is not simply reducible to personal consumption by individuals. Various promotional industries may present consumption in that manner, constantly orienting it around images of personal consumption, but this is somewhat misleading. First, individuals are usually part of households, and household consumption decisions are significant in terms of their environmental impact and the prospects for lessening that impact. Consumption results from both the addition of personal consumption habits, practices, and plans, and also from household consumer practices which may not be specifically oriented to any one individual in a household.

Moreover, individual consumption is not all oriented toward lifestyle. Much individual, and most household, consumption can be best defined as ordinary consumption, and is concerned with matters including convenience, habit, and individual responses to changing social contexts. Much ordinary consumption is either necessary or obligatory; food, fuel, travel, insurance, and, increasingly, services that were often formally wholly provided by the state and funded through taxation, for example, paid elements of health and social care. Lifestyle consumption, on the other hand, is generally defined as consumption in excess of, or in addition to, that ordinary consumption oriented toward satisfying basic needs.

Lifestyle consumption is also usually represented as being oriented to certain kinds of goods, including clothing, cosmetics, and the products of the culture industries. As such, lifestyles are regarded very much as self-oriented sets of consumer practices forming part of a reflexive, biographical project of identity-formation and self-presentation and based particularly upon the consumption of the symbolic dimensions of commodities. Lifestyle is seen as a sensibility—something that helps make sense of consumer choice, reduces the anxiety that flows from having to make such choices, and provides a consistent framework within which consumption decisions are made so that there is a consonance between the objects, services or experiences chosen and consumed. This kind of consumption is seen by some environmentalists as particularly wasteful and harmful, not simply because of the resources it uses, but because it is seen as frivolous and superfluous.

## A More Sustainable Lifestyle

There is, then, a consequent need to change patterns of both ordinary and lifestyle consumption, both in relation to individuals and households, if environmentally sustainable ways of living are to be developed. In a sense, the aim is to transform lifestyles into greener ways of life. That said, the prospect of individual and meaningful green consumption, or ways of consuming that lessen harmful environmental impacts, is attractive to many. For example, one of the advantages of lifestyle within the context of green consumerism is that the market generates a myriad of choices of goods and services from which to construct a personalized green lifestyle.

These rest on the idea and assumption that replacing environmentally harmful goods and services with green ones will lead to the greening of consumption, generally through a kind of trickle-down or sidewards-creep effect. Consequently, a range of green lifestyles are available, based around cleverly designed and attractively promoted products. Many of these are aimed at affluent consumers. As suburban developments, certainly in the developed world, are home to many of those affluent consumers, the congruence between green lifestyles and suburban life are clearly discernible. While superficially attractive, such options are problematic; for example, there is already an identifiable and environmentally harmful duplication of very similar products.

One consequence of this may be environmentally damaging competition between manufacturers all vying for a relatively small market. Another unanticipated result may even be increased levels of consumption of goods, energy, packaging, and distribution services overall, as well as a concomitant increase in defunct, unrepairable, nonrecyclable products and waste. And, of course, such high-value, high-cost green lifestyles present enormous entry barriers for those not as well off.

## Suburban Perceptions and Developments

Initially, one might be forgiven for thinking that the suburbs represent a barrier to greener individual and household consumer practices. They are often seen as an undesirable and unnecessary proliferation of expensive and environmentally damaging urban infrastructure and housing without the advantages of the city centers. That is, they consume roads, sewers, telecommunications, and water and fuel infrastructure, but exist without the shops, restaurants, businesses, and other facilities, redolent of urban life. Urban population density confers many environmentally advantageous factors, including the possibilities for face-to-face transactions and exchange of goods and services, reduced traveling distances manageable by foot, bicycle, or public transportation, as well as a lower overall environmental impact of populations relative to rural settlements.

The suburbs, in contrast, are marked by lower population densities, relatively larger traveling distances that often require the use of private transportation in the absence of public alternatives, and a lack of both local amenities and the social density that allows the development of sustainable, informal networks. Suburbs seem to have neither the advantages of urban centers or village cultures. Such, at least, are the common perceptions of the suburbs. Indeed, this is part of the attraction of the suburbs for many people: they are an escape from the pandemonium of urban life or, conversely, an escape from isolated small town life without having to make a full commitment to urban living. Suburbs are also where much of the bigger, more expensive housing stock can be found and thus they represent a material and social aspiration for many people.

Suburbs, however, are also dynamic, and a fixed definition or characterization of their content is not possible. For example, in the United Kingdom, government policies geared toward urban and suburban infill of housing, rather than expansion into protected greenbelts, has led to dramatically rising land prices. This, combined with rapidly increasing housing prices, has led to a situation in which suburban population density has actually increased as land has been redeveloped, often replacing one large house and garden on a single plot with three or four smaller houses and gardens. One result of this has been a more diverse social mixing of various income groups, replacing the overwhelmingly affluent character of many suburbs. The increased social density, the rising number of suburban disposable incomes, as well as an expanding range of personal consumer preferences has led to a marked regeneration of many suburban areas. This raises the prospect of developing the environmentally friendly consumption practices associated with urban density; local shops and amenities, the meeting of basic needs, organized around short journeys by walking or cycling, increased demand for suburban public transport, and so on.

However, the same developments can also be viewed negatively. The example of food consumption is instructive in this regard. The practices previously described, "garden grabbing," and the infill of vacant lots have resulted in a greater number of smaller suburban plots that not only may have a deleterious effect on suburban wildlife habitat and species diversity but also reduces the amount of potentially productive space available for homeowners. If urban centers offer provision of consumer goods by others, the suburbs have often featured a higher degree of self-provision of food via productive domestic gardens and allotments. The loss of productive suburban space thus not only reduces the potential for self-provision, but also increases the demand for alternative provision. The two solutions that have developed in order to service the suburbs are, first, the use of private transportation to drive to out-of-town supermarkets and, second, using telecommunications technologies in order to procure food via home delivery services. These two "solutions" are mirror images in terms of environmental impact. The environmental impact of one is simply displaced to the other, from private provision to that of the service sector. While either option may provide a preferable choice of particular products to consumers, for example, offering seasonal and locally produced food, they are a poor match environmentally speaking compared to self-provision. Indeed, supermarkets are expert at obliterating the times and spaces of global food production via their cornucopia displays of de-contextualized foodstuffs available year round.

A further environmentally damaging consequence of suburban lifestyles also results from the changes previously discussed. Again, particularly in relation to the United Kingdom, houses are now seen by many consumers less as homes and more as economic assets to be traded regularly in order to make money. Families move much more frequently than ever before. The result has been an enormous increase of consumption related to setting up or changing homes. These include material costs, for example, furniture and decorating goods, service costs, such as real estate agents fees, and obligatory costs, including the consumption of legal services. Many environmentalists argue that the climate of viewing homes as readily disposable financial assets needs to change in favor of a stable, more place-bound, homemaking sensibility.

## Sustainable Solutions

In order to lessen the impact of suburban lifestyles on the environment, particular practices immediately present themselves; car sharing, domestic sorting of household waste, use of neighborhood recycling facilities, domestic food growing, the encouragement of a culture of reuse, repair, and social recycling of still useful goods. Patronizing local shops may, despite their sometimes higher prices, be cheaper overall than paying the cost of transportation, if one is shopping in out-of-town facilities, or paying for delivery via home shopping services.

A further benefit of such practices results from bringing people together, a result that the promotion of lifestyle consumption does not do, with its emphasis on personal meaning and identity as functions of individualized consumption. One beneficial consequence of people coming together is the potential growth of the collective provision that is currently irrationally consumed on a private basis: for example, sharing infrequently used tools, garden appliances, or the informal redistribution of small food surpluses. In previous centuries, it was precisely this type of arrangement, supplemented by neighborhood waste traders, including "rag and bone men," that worked in concert to informally regulate and sustain household consumption at relatively low rates.

How we consume has been neglected in relation to analyses of what is consumed, but it is becoming clear that sustainable lifestyles depend on consuming different goods for different purposes and in novel ways. Sustainable "communities" of consumption may prove to be as significant as attempts to reengineer, either by ideological means, behavior-steering initiatives, legal enforcement, or individual consumption practices.

These practices all resonate with the spatial scale and social composition characteristic of suburban life. This does not imply that they may be the most appropriate solutions for densely packed urban centers. Different solutions and different fulcrums of change, in terms of lifestyle and consumption, are likely to be most appropriate for each. A sustainable suburban lifestyle is likely to have different temporal and spatial dimensions to its urban equivalent, though it will be oriented via differing means to common ends.

**See Also:** Conspicuous Consumption; Consumerism; Durability; Lifestyle, Urban; Overconsumption.

### Further Readings

Crocker, David A. *Ethics of Consumption: The Good Life, Justice, and Global Stewardship (Philosophy and the Global Context Series)*. Lanham, MD: Rowman & Littlefield, 1997.

Hamilton, Clive. *Growth Fetish*. London: Pluto Press, 2003.

Herring, Horace and Steve Sorrell. *Energy Efficiency and Sustainable Consumption: The Rebound Effect (Energy, Climate and the Environment)*. New York: Palgrave Macmillan, 2009

Jackson, T. "Live Better by Consuming Less?" *Journal of Industrial Ecology*, 9/1–2:19–36, 2005.

Robbins, Paul. *Lawn People: How Grasses, Weeds, and Chemicals Make Us Who We Are.* Philadelphia, PA: Temple University Press, 2007.
Ward, Colin and David Crouch. *The Allotment: Its Landscape and Culture.* Nottingham, UK: Five Leaves Publications, 1997.

*Neil Maycroft*
*University of Lincoln*

# LIFESTYLE, SUSTAINABLE

The focus of conventional development policies and production, consumption, and trade patterns on the improvement of the social and economic conditions of individuals, societies, and nations has causal relationships with environmental maladies such as depletion or degradation of natural resources, destruction of the ecosystem, creation of heat islands, pollution, and increasing large-scale climate change. Within the context of available technology, capital, and culture (including dietary habit), man's current lifestyles emphasizes optimal utilization and, in many places, an exploitation of Earth's resources for the creation of wealth and an enhancement of social and economic conditions. The resultant effect is now feasible not only in precarious environmental conditions but also in the declining quality of life and economic fortunes that humankind sought to improve in the first place.

## Sustainable Lifestyle: Meaning, Scope, and Content

A sustainable lifestyle is a way of life that minimizes the use of energy, water, and food, as well as curbs the production of both air and water pollution. It is a lifestyle that reduces the consumption of energy resources, water, and food to a minimum, as well as reducing and recycling waste output. A sustainable lifestyle revolves around, or relies more on, renewable energy resources. Sustainable lifestyles put a heavy emphasis on progressive improvement and advancement in social, economic, and environmental conditions in the present and, more important, in the future. The concept emphasizes improvement in the quality of life of individuals and the defense of the integrity of the ecosystem. The goal of a sustainable lifestyle is to create a minimal ecological footprint and to reuse or convert waste to resources.

Sustainable lifestyles view the well-being of humankind and of the planet as two inseparable spheres and thus posit that lifestyle improvement, as well as developments in social and economic conditions, must or should be pursued not only in an environmentally and ecologically sound manner but also in a way that ensures continual renewal and availability of natural resources for future generations. There are two closely related major dimensions of sustainable lifestyles: environmentally sound social development and environmentally sound economic development.

Environmentally sound socioeconomic development implies, first, the ability of individuals to have access to an income-generating activity to support self and family, and second, the right of societies and nations to pursue industrial development and to participate in the global economy so as to increase national income and achieve economic growth and development. The social and economic well-being of the individual and the society are closely related to the well-being of their immediate environment. Sustainable lifestyles place priority on protecting the Earth, safeguarding natural resources, and conserving biodiversity for future generations. The main objective is to progressively improve the living conditions of humankind while pursuing social and economic developments in an environmentally friendly and ecologically sound manner.

There are three main lifestyles in which sustainability is required: urban, suburban, and rural lifestyles.

## The Unsustainability of Urban Lifestyles

The loss of biodiversity; the pollution of air, water, and soil; and the creation of local heat islands are some of the direct effects of the creation and urbanization of cities. As a result of a growing urban population, an increasing demand for housing, food, water, and energy imposes a heavy toll on Earth's resources. The rates of waste generation and associated poor management practices in many places have come to overwhelm the Earth's carrying capacity. In many cities in developing countries, for instance, mountains of refuse are piling up and continue to grow uncontrollably, posing a serious threat to people's health and putting a strain on urban management. The urban dwellers are indeed the greatest consumers, as well as destroyers, of Earth's resources. The pattern of urban production, distribution, and consumption of goods and services leads to a heavy, growing, and insatiable demand for energy—fossil fuels, electricity, and gas. In the urban production process, industries continue to release millions of tons of carbon by burning coal, gas, and oil. In the distribution of goods and services, the transport-dependent urban lifestyles emit more carbon dioxide compared with other sources of carbon emission. The attendant increase in local and global temperature has serious implications on man, living organisms, and indeed, the entire ecosystem.

## Unsustainable Practices of Suburban Lifestyles

In many countries in developed and developing nations, the increasing rate of urbanization, coupled with the congestion of inner cities and improvements in transport systems, have resulted in the creation of transport-dependent suburban areas of cities. This development has brought about dispersal of workplaces (particularly in the developed countries) and, in some cases, residential areas to outer parts of the urban areas. The result is increasing commuting distances and emissions of carbon from vehicles. Suburban areas are some of the most perplexing areas of human agglomeration, particularly with respect to the pattern and pace of development. Urban workers flock to the suburbs for cheaper housing. The immediate result is a preponderance of hastily developed buildings that in many cases pay no regard to any set standards. Houses and developments often lack basic facilities and services, such as roads, drainage, potable water, sanitation, and electricity. Facilities for waste disposal are often lacking. Some of the most visible environmental consequences of the suburban lifestyle include, first, the conversion of prime agricultural land to residential and industrial uses, changes in the ecological balance, pollution of the peri-urban areas in which urban wastes are deposited, and population morbidity, a state of ill health arising from the high incidence and prevalence of environment-related communicable diseases.

## Unsustainable Practices in Rural Lifestyles

The land-based rural lifestyle is, relative to urban and suburban lifestyles, a simple lifestyle, particularly in regard to morphology—the land use pattern, the size of the population, the level and pattern of consumption of Earth's resources, and the implications of man's interaction with the natural environment. Nevertheless, the ubiquitous presence of rural settlements, the total population of all rural populations, and the combined impacts on the environment call for a careful consideration of aspects of the rural lifestyle that may threaten the sustainability of the environment. Although activities are largely limited to agricultural practices including livestock and fisheries, the technology adopted is relatively simple and in most cases rudimentary, involving the use of simple implements. However, farming practices such as bush clearing and burning, cultivation of erosion-prone areas, hunting endangered animal species, and the use of wood as an energy source for cooking, among other habits, pose a serious threat to the stability and viability of the ecosystem.

## How to Create a Sustainable Lifestyle

Creating sustainable lifestyles implies reducing ecological footprints. This is a major challenge for both individuals and nations. For the former it implies, first, education on the implications of current consumption and behavioral patterns, and information and access to sustainable goods and services. For nations and corporate organizations, it implies provision of adequate infrastructure for the production of sustainable goods and services, as well as creation of appropriate conditions, particularly through policy instruments for the consumption of these goods and services.

There are several strategies for creating sustainable lifestyles. The starting point in all cases is a change in mental attitude and consumption behavioral patterns. Claudia Barner's 2009 research detailed some guidelines, which revolved around individual choices along four directions: animal, plant, mineral, and human.

- Sustainable guidelines for the animal world centers on the preservation of endangered species, the treatment of domestic animals with respect, and the consumption of animals native to the area.
- In the plant world, emphasis is on tree planting, urban agriculture, protecting ecosystems, and using recycled plant materials.
- With regard to minerals, emphasis is on the consumption of organic materials, a rational utilization of nonrenewable resources, and recycling mineral products.
- For humans, the concern is overpopulation and a need to reduce family size to two children. Consuming fewer goods and services is stressed, and a call is put out to enlist in programs and movements to protect endangered animal and plant species.

For all intents and purposes, the strategies center on consumption of only what is necessary; a reduction of waste and pollution arising from production and consumption of goods and services, as well as from human socioeconomic activities; consumption of organic foods rather than processed food materials; reuse of products before recycling (none of the byproducts of human activity should be considered a waste product to be disposed of but, rather, a resource to be used in another process); and use of sustainable products and the propagation of the message of sustainable lifestyles. However, research must be conducted into the various ways of reusing and recycling waste.

The choices people make in their production of goods and services, as well as in the consumption and distribution of food, leave an ecological footprint and the environmental effects of pollution and diminishing and degraded natural resources.

**See Also:** Environmentally Friendly; Lifestyle, Rural; Lifestyle, Suburban; Lifestyle, Urban; Resource Consumption and Usage.

### Further Readings

Adeboyejo, A. T. and Olajoke Abolade. "Household Response to Urban Encroachment on Rural Hinterland in Ogbomoso Urban Fringe, Nigeria." *Proceedings of Conference on Urban Population, Development and Environment Dynamics in Developing Countries.* CICRED of Columbia University, Nairobi, Kenya. http://www.cicred.org/Eng/Seminars/Details/Seminars/PDE2007/PDEpapers.htm (Accessed September 2009).

Barner, Claudia. "Guidelines for Creating Sustainable Lifestyles." DEER Tribe Metis Medicine Society. http://www.dtmms.org/readingroom/aligning-earth/sustainable-lifestyles.htm (Accessed March 2003).

Wann, David. *Simple Prosperity: Finding Real Wealth in a Sustainable Lifestyle.* New York: St. Martin's Griffin, 2007.

Wright, Liz. *Natural Living: The 21st Century Guide to a Sustainable Lifestyle.* Berkeley, CA: GAIA, 2010.

*Adeboyejo Aina Thompson*
*Ladoke Akintola University of Technology, Ogbomoso*

## Lifestyle, Urban

The urban environment can provide a fruitful location for the development of sustainable lifestyle practices and a means of lessening of environmental impact. Lifestyle as a practice is intimately connected to consumption, but is not simply based on personal consumption by individuals. Individuals are generally part of households, and household consumption decisions not only make a significant environmental impact, they also have the ability to lessen that impact. More significantly, individual consumption is not all concerned with lifestyles. As Tim Jackson has pointed out, much consumption can be best defined as ordinary consumption and is concerned with matters including convenience, habit, and individual responses to changing social contexts.

Much ordinary consumption is obligatory; food, fuel, travel, insurance, and, increasingly, services that are often provided by the government and funded through taxation. Lifestyle consumption, on the other hand, is generally defined as consumption in excess of, or in addition to, that ordinary consumption oriented toward satisfying basic needs. Lifestyle consumption is often represented as being oriented to certain kinds of goods, including clothing, cosmetics, and the products of the culture industries. Lifestyle is presented as a reflexive, biographical project of identity-formation and self-presentation, based particularly upon the consumption of the symbolic dimensions of commodities. Some kind of project or strategy is implied or explicitly

Urban planning measures include the creation of green belts within the city limits, like this restored wetland in Chicago. Green space helps reduce the so-called heat island effect of urban areas.

*Source:* Lynn Betts/Natural Resources Conservation Service/USDA

claimed, and a lifestyle is seen as something that makes sense of consumer choice, reduces the anxiety that flows from having to make such choices, and provides a consistent framework within which consumption decisions are made, such that there is a consonance between the objects, services, or experiences chosen and consumed. This kind of consumption is seen by many environmentalists as particularly wasteful and harmful, not simply because of the resources it uses, but, because it is regarded as frivolous and superfluous.

## Urban Needs and Infrastructure

There is, then, a consequent need to change patterns of both ordinary and lifestyle consumption if environmentally sustainable ways of living are to be developed. At first glance, urban centers do not seem to provide the most propitious locations for such desirable change. The sheer scale and density of many cities often works against a sanguine attitude toward meaningful environmental change. The physical expansion of urban centers demands enormous consumption of resources, including land, building materials, infrastructural elements, labor, energy and capital. The provision of energy, communications, healthcare, transport, education, and other social goods and services is, out of necessity, intensified as cities either physically expand or grow through increased population density.

Added to these are the myriad consumption practices, both ordinary and lifestyle, of the individuals, households, and communities occupying urban areas. Cities draw in people, raw materials, and finished goods in enormous amounts and expel pollution, waste, unused heat, and light and noise on a comparable scale. Such a malign view seems even more apposite with the ever-expanding nature of the urban environment that is now home to more than half of the world's population. Growth in the slums and shanty towns of the southern hemisphere is particularly rapid and marked.

## Opportunities for Sustainability

However, there are reasons to be sanguine. While some ordinary consumption is obligatory, much of it coincides with necessary consumption aimed at satisfying basic needs. Individuals and households can make informed decisions in order to lessen the environmental impact of their consumer practices. One potent change would be for consumers to reject products that have built-in obsolescence. Andre Gorz has referred to the accelerated obsolescence that characterizes an increasing number of consumer goods. One particular pernicious consequence is that many products are deemed to wear out stylistically before they do materially, due to the superfluous elements of fashion they embody.

Other goods lack the durability that modern manufacturing processes could guarantee. For those who can afford them, higher purchase prices of many goods bring with them improved materials and stylistic durability as well as better function. It is, however, pertinent to point out that for unemployed and low-income individuals and families, the relative costs of meeting basic needs are higher than for the even moderately well off. This extra burden is manifested in many ways, including the pressure to lease pre-owned or refurbished goods that may function inefficiently, as well as having a higher environmental impact when disposed of. Access to the world of well-designed, functionally efficient goods is often on the basis of expensive credit agreements many times in excess of the purchase price of the goods in questions. Fuel may also be more expensive due to the necessity of having to prepay or the use of meters calibrated disadvantageously toward the less well off.

A further way in which the environmental impact of consumption could be reduced would be to take advantage of the density and proximity that urban centers offer. Increased urban population density often lowers the aggregate environmental impact of human settlements compared to rural living. Cities, especially those with high rates of residency, tend to involve face-to-face consumption services, including shops and restaurants. Walking, cycling, trams, and other forms of public transport make environmental sense in terms of accessing these densely packed sites of consumption. Choosing seasonal and locally produced food would be another advantageous practice.

Focusing on changing ordinary urban consumption practices, along with reducing wasteful institutional consumption in cities (unnecessary shop, bank, and restaurant refurbishments, replacing company car fleets frequently, and not leaving shops and offices illuminated all night), lessens the significance of lifestyle consumption. Indeed, Conrad Lodziak points out that many life-satisfaction surveys reveal that once basic needs have been met, most people give priority to noncommodifiable values, such as love, friendship, harmonious family relationships, mutuality, and so on. Lifestyle consumption has little to offer in this regard, whereas the social density of cities does. Nevertheless, one of the advantages of a lifestyle lived within the context of green consumerism is that the market generates a myriad of choices of goods and services from which to construct a personalized green lifestyle.

When lifestyle consumption is seen as one element of urban consumption overall, along with ordinary and institutional consumption, the prospects of a greener urban way of life become clearer, including the necessity of identifying the most appropriate methods of change for each area of consumption.

**See Also:** Conspicuous Consumption; Consumerism; Durability; Lifestyle, Suburban; Overconsumption.

### Further Readings

Gorz, Andre. *Reclaiming Work: Beyond the Wage-Based Society*. Queensland, AU: Polity Press, Cambridge, 1999.

Jackson, T. "Live Better by Consuming Less?" *Journal of Industrial Ecology*, 9/1–2:19–36, 2005.

Lodziak, C. "On Explaining Consumption." *Capital & Class*, 72:111–35 (2000).
Newman, Peter and Isabella Jennings. *Cities as Sustainable Ecosystems: Principles and Practices*. Washington, D.C.: Island Press, 2008.
Owen, David. *Green Metropolis: Why Living Smaller, Living Closer, and Driving Less are the Keys to Sustainability*. New York: Riverhead, 2009.
Pinderhughes, Raquel. *Alternative Urban Futures: Planning for Sustainable Development in Cities throughout the World*. Lanham, MD: Rowman & Littlefield Publishers, 2004.
Portney, Kent E. *Taking Sustainable Cities Seriously: Economic Development, the Environment, and Quality of Life in American Cities (American and Comparative Environmental Policy)*. Cambridge, MA: 2003.

*Neil Maycroft*
*University of Lincoln*

# Lighting

In many countries (including the countries of the European Union, the United States, Australia, Ireland, Brazil, the Philippines, Argentina, Venezuela, the United Kingdom, and Russia), there exists a scheduled phasing out of incandescent lamps in favor of compact fluorescent lamps (CFLs). Fluorescent lighting uses between one-fifth and one-quarter of the electricity that the former bulbs use and can therefore contribute to the reduction of greenhouse gas emissions. Despite this strong argument, there is resistance toward the adoption of these energy-efficient products, notably for environmental reasons. At this time, completely green lighting does not exist.

Lighting is inherently part of our history, and along with heating it is one of the aspects of fire. The advent of electrical lighting in the second half of the 19th century has been celebrated under the umbrella of the "Electricity Fairy." The electrical network is indeed linked to the provision of light—first public and then private. Lighting then completely changed the space and time frame of human activities; for example, by transforming nightlife in cities. Electrical lighting is now integrated into daily life and is everywhere. However, this modern comfort is now challenged by the threat of climate change.

Emitting 1,900 Mt of carbon dioxide in 2005, lighting is one of the biggest sources of climate change (7 percent of all emissions). North America alone represented 29 percent of estimated residential building lighting energy consumption in 2005. China, with 20 percent of the world's population, accounted for only 9 percent of estimated residential building lighting energy consumption in 2005. The International Energy Agency foresees a global growth of 60 percent of artificial lighting demand between 2005 and 2030. This growth is mainly expected from the increase of lighting demand in non–Organisation for Economic Co-operation and Development countries.

The efficiency of the residential lighting sector is around two and a half times lower than for commercial lighting and is the lowest of any of the major lighting end-use sectors. This is the reason for the current implementation of policies discouraging the use of incandescent bulbs, known to be the most inefficient lamps. In conventional incandescent (or filament) lamps, around 95 percent of the electricity is transformed into heat, and only

5 percent into light. Their efficiency is typically 10–15 lm/W. (Energy efficiency of lighting is usually measured in lumen/Watt, with lumen being the physical unity of light.) Halogen lamps, which are also filament lamps filled with gas, have an efficiency between 15 and 25 lm/W. Gas discharge lamps such as CFLs have a typical efficiency of 45–60 lm/W; namely, four to five times higher than incandescent lamps. Despite higher lamp prices, the overall cost of lighting from CFLs is cheaper than that from incandescent lamps because their operating costs are much lower when properly used.

Many resist a change from incandescent lighting for several reasons. The first is aesthetics—the way people feel about their old lighting and the beauty of light itself. Many customers are still reluctant to buy CFLs not only because of their higher prices but also because of their supposed lower quality of light (especially when first lit), even though huge improvements in the quality of these bulbs have been made, sometimes making their life span shorter. The second reason concerns the mercury content of CFLs, which means that they have to be carefully discarded. This mercury content is regarded as acceptable when compared with the reduction of mercury emission in coal power plants that the use of CFLs entails. Indeed, the decrease of mercury emissions resulting from energy savings outweighs the need for mercury in the lamps. However, this implies that CFLs must not be thrown away in the waste bin. The third, lesser-known reason is the emission of electromagnetic radiations by CFLs (resulting from the integrated ballast); as a result of this radiation, some suggest not using CFLs in bedside lights.

One option is light-emitting diodes (LEDs), becoming available on the market with increasing efficiency and increasing life span, with their theoretical limit being around 150 lm/W. However, LEDs make use of rare raw materials (gallium and indium) that are also used in many other high-tech applications (photovoltaic panels, monitors, liquid crystal displays with coatings of indium tin oxide). This will probably impede the development of inorganic LEDs. Research is now focused on organic LEDs, which could be well-suited for indoor area illumination and could appear as "glowing wall paper" without the need for "luminaries." However, this lighting is still "not yet available technology."

Today there is no satisfactory solution for artificial lighting that could be both sustainable and highly energy efficient. If such a technology were developed, rebound effects could be substantial: if new devices were more energy efficient, users could simply use them more, or leave them on for longer. In the meantime, other possibilities exist to tackle the important question of green lighting. There is a trend toward modulation of lighting according to the user's presence and ambient light. Lighting is a system, a mix of natural and artificial lights, and architecture plays a large part by designing buildings that are naturally illuminated most of the time.

**See Also:** Consumer Behavior; Ecolabeling; Electricity Usage; Energy Efficiency of Products and Appliances; Windows.

### Further Readings

Bertoldi, Paolo, ed. "Proceedings of the 4th International Conference on Energy Efficiency in Domestic Appliances and Lighting." EEDAL06, London, 2007.

International Energy Agency. *Light's Labour's Lost: Policies for Energy-Efficient Lighting.* Paris: IEA/OECD, 2006.

Lefèvre, Nicolas, et al. *Barriers to Technology. Diffusion: The Case of Compact Fluorescent Lamps*. Paris: International Energy Agency/Organisation for Economic Co-operation and Development, 2006.

*Grégoire Wallenborn*
*Université Libre de Bruxelles*

# Linen and Bedding

Humans have used a variety of materials for bedding since the days when cavemen put branches and leaves down to sleep on. Eventually covered with animal skins, these bedding materials blocked the cold of the cave floor and created a softer bed than the hard ground. The principles of materials that provide protection from the cold ground and comfort from its hardness have been universal in the development of human bedding. With the development of higher levels of culture, bedding materials were also made with elegance that gave warmth, romance, and esthetic pleasure.

Beds have traditionally been made from a wide range of organic and inorganic materials such as wood, metal, rock, rope that have differing ecological impacts. Modern beds can contribute to forest destruction when made from nonsustainable wood, or they may be upholstered with textiles that can be toxic when discarded. It is possible to create a more ecofriendly bedding environment. Beds can be made with reclaimed or recycled materials, with certified sustainable wood, or rapidly renewable materials like bamboo, rattan, or straw and wheat-based particleboard. Consumers can seek out beds made with natural latex rubber, low-VOC paints and stains, and nontoxic fabrics. Buying used or antique furniture keeps usable items out of landfills, while often saving the consumer money as well.

Bed frames hold mattresses and use hard steel coil springs (recyclable) as the cushioning material for comfortable sleep. Mattress covers can be synthetic, but are often a blend of natural and synthetic fabric that will resist absorbing odors or harboring pathogenic organisms. The use of synthetics can reduce the presence of germs or insects that are attracted to natural fibers, but there is a health and environmental tradeoff in using these synthetic materials, as many are processed with harsh chemicals.

Bed sheets, pillow covers, and other bedding or bath textiles are called "linens," even when they are made of cotton, hemp, synthetic, or other nonflax fibers. Flax is usually woven to a very high thread count, which creates a fabric with a fine feel ("hand"). Bed linens and blankets continue to be made from wool, cotton or synthetic blends. Comforters may be filled with artificial fibers that create insulating dead air spaces, or may use down or feathers as the insulating material. Eco-fi, a polyester fiber made from 100 percent post-consumer recycled plastic bottles, is also used as stuffing in pillows and comforters.

Because it is a natural fiber that is also a renewable resource, green consumers can also purchase flax products to reduce their carbon impact. Cotton has traditionally been produced on farms that use large amounts of water, as well as chemical fertilizers and pesticides. Although linen is usually a much greener choice, cotton is also grown organically. According to the Organic Trade Association, organic cotton has a low environmental impact. They state that "organic production systems replenish and maintain soil fertility, reduce the use of toxic and persistent pesticides and fertilizers, and build biologically diverse agriculture." Organic cotton is grown without using pesticides and synthetic fertilizers, and no genetically engineered seeds can be used. The total acres of organic cotton planted increased 152 percent during the 2007–08 crop season.

Linen is a fiber that comes from the flax plant and has been used since ancient times. It is the strongest of vegetable fibers and has been used for centuries in many fine fabrics, including lace. It continues to be used extensively for bed sheets because it feels cool and fresh even in hot, humid weather. Bamboo can also be made into a soft, durable fabric and consumers can now find bamboo sheets, towels, blankets, and duvet covers.

**See Also:** Consumer Culture; Green Consumer; Organic; Simple Living.

### Further Readings

Esty, Daniel and Andrew Winston. *Green to Gold: How Smart Companies Use Environmental Strategy to Innovate, Create Value, and Build Competitive Advantage.* Hoboken, NJ: Wiley, 2009.

Hann, M. A. *Innovation in Linen Manufacture. Textile Progress Series,* 3/37. Cambridge: Woodhead, 2005.

Johnston, David R. and Kim Master. *Green Remodeling: Changing the World One Room at a Time.* Gabriola Island, British Columbia, Canada: New Society Publishers, 2004.

Seo, Danny. *Conscious Style Home: Eco-Friendly Living for the 21st Century.* New York: St. Martin's Press, 2001.

Tanqueray, Rebecca. *Simply Green: Sustainable Style for Every Room in the House.* London: Carlton Books, 2002.

Wilson, Alex T. *Your Green Home: A Guide to Planning a Healthy, Environmentally Friendly New Home (Mother Earth News Wiser Living Series).* Gabriola Island, British Columbia, Canada: New Society Publishers, 2006.

*Andrew Jackson Waskey*
*Dalton State College*

## Local Exchange Trading Schemes

Local Exchange Trading Schemes or Systems (LETS) are local, not-for-profit, community-based networks for exchanging goods and services using a local trading system. The LETS system of exchange sits outside the mainstream use of a national currency. Instead, LETS schemes typically employ a locally negotiated system of credit or exchange for the value of goods and services found within the boundaries of the community. LETS serve as a new form of economic organization that is intended to support the sustainable development of a community.

Around the world, LETS initiatives are growing in popularity and are seen as a localized response to the rise of multinational trade, rapid globalization, and the role of institutions such as the World Trade Organization. LETS programs operate like a supplementary currency and trade system, whereby alternative value systems are introduced into the community and operate like a community banking project and, as a result, insulate the local system from fluctuations in the labor conditions and mainstream economic markets. As grassroots programs, LETS initiatives are all inclusive, allowing for people of all ages, skills, employment status, and aptitudes to participate, as well as nonprofits, associations, small business, local government, and others.

In its most basic format, LETS programs operate via a systemized directory of "wants" or "needs" and "offers" or "bids." A community "bank" acts as an escrow-like entity and monitors and tallies the flow of exchanges made between wants and offers. In addition, LETS systems typically account for time at a standard rate to equalize the value of services between parties. Community currency is made when credits move from one person to another between the exchange of wants and offers made. Other LETS currency systems also use tokens or other forms of credit vouchers to account for the distribution of local capital.

In more sophisticated LETS programs, local communities, usually under the guidance of their national treasury, print their own local money, often known as scrip. With this, individuals are typically able to exchange their national currency for the local currency at community banks at a 1:1 ratio. Once exchanged, this local currency can be used the same way people use national currency, but with the exception that it is only accepted within the local municipality.

As a mutual credit system, the benefits and rationale for LETS initiatives are twofold: first, individuals, organizations, business, and local government are able to decouple themselves from traditional economic constraints, and instead allow LETS programs to offer an alternative approach to money, work, trade, and economic development. Second, LETS programs bring people together, particularly in poorer areas, by creating a community system of value and exchange that relies on mutual engagement, trust, and reliance on each other. Moreover, LETS programs offer services that may be desired but sit outside the financial reach for people engaged in a traditional system of an exchange of national currency for goods and services.

Membership in a LETS initiative allows local individuals or organizations to buy and sell goods and services without using traditional cash or credit cards. LETS initiatives operate outside the traditional money and trade economies and aim to keep local economies thriving by encouraging people to invest in their own backyards, as well as to take a fresh look at the concept of money and currency.

In addition, many LETS programs allow local community members to open local accounts to earn and accumulate credits by providing a direct service in exchange for the local currency scrip or credit voucher. In doing so, people can later spend LETS credits for other goods and services found within the community's marketplace of offerings; goods and services may include items such as food, childcare, alternative healthcare, car repairs, or cleaning services. LETS programs offer interest-free credit generated at the point of purchase or exchange to develop a free trading local economy.

Although LETS systems supplement traditional cash flow, this does not always mean they are exempt from regional or national taxation. Taxation of LETS initiatives varies, depending on the country and its tax laws. The general rule of thumb, however, has been for community members to pay taxes on a professional exchange of services, but not for people who are engaged in a hobby or personal/social exchange. As an example of a taxable service, a licensed mechanic who offers a professional service to another as an exchange for an equal professional service or LETS credit would create an exchange that is subject to taxation. However, neighbors who are not professional mechanics and who offer to fix the other's car for an equal or lesser exchange may not be taxable.

Overall, LETS initiatives are becoming one of the many tools for change that people and communities are choosing to revitalize the notion of consumerism and to energize their local regions in efforts to build a more socially just, sustainable world.

**See Also:** Affluenza; Consumer Activism; Consumer Boycotts; Consumer Culture; Consumer Ethics; Disparities in Consumption; Downshifting; Fair Trade; Green Communities; Green Consumerism Organizations; Locally Made.

### Further Readings

Aldridge, Theresa, et al. "Recasting Work: The Example of Local Exchange Trading Schemes." *Work, Employment & Society*, 15 (2001).

Douthwaite, Richard. *The Ecology of Money. Dartington.* Schumacher Briefing 4. Devon: Green Books, 1999.

Zelizer, Vivian. *The Social Meaning of Money: Pin Money, Paychecks, Poor Relief and Other Currencies*. Princeton, NJ: Princeton Paperbacks, 1997.

*Jeff Leinaweaver*
*Fielding Graduate University*

## Locally Made

The term *locally made* is typically applied to locally grown food or locally manufactured goods. Instead of items being centrally grown, manufactured, or outsourced, items are created and grown within a local proximity. The phrase *locally made* means different things to different retailers. Currently in the United States, there is no official definition of the term from either the Federal Trade Commission or U.S. Department of Agriculture.

Unlike organic standards (entailing specific legal definitions, labels, and inspections), locally made foods (sometimes referred to as locally grown or local) have no overarching regulation. Food defined as locally made depends on where consumers live, the time of year, and the item in question. In different cases, it could stand for items created within the same state, items that have been trucked into stores within one day, or items created within 100 miles. It is not uncommon to see produce, for example, carrying an organic label (indicating adherence to federal standards) as well as a locally grown label (noting the number of miles traveled to deliver the product). In recent years, however, studies have shown consumer preferences for local over organic food.

The new trend toward locally supported lifestyles encompasses not only locally made food, but locally crafted and manufactured goods and products. Some countries that mass-produce consumer products have less stringent safety standards, resulting in toxic or hazardous materials in end products. Recent toy recalls have illustrated this weakness to consumers, resulting in legislation regarding manufacturing, as well as an increase in purchases of locally made or handmade items. Consumers that are tired of the chain store culture and global manufacturing have embraced the market of locally made goods and products to rebel against conformity. Buying locally made items allows consumers to purchase creative, one-of-a-kind goods and products, and to support independent creators.

The benefits to shopping locally are many. Local businesses tend to be more accessible than large corporations, mainly because they are smaller in size and located near the consumers. Independent businesses, choosing products based on their local customers needs and desires, rather than a national sales plan, guarantee a more diverse range of product and service choices. Asking questions about where and how locally made items were produced can help build community. According to sociologists, consumers have 10 times more conversations with vendors in farmers markets than in grocery stores. When buying directly from the producer, consumers establish a relationship that goes beyond exchange of money, creating a sense of trust and mutual respect.

The environmental impact of locally made food and goods is also appealing to today's green consumer. According to a study by the Leopold Center for Sustainable Agriculture, "food miles" (the distance food travels from where it is grown to where it is ultimately purchased or consumed by the end user) for nonlocal items are, on average, 27 times higher than good purchased from local sources. Most of these nonlocal food and goods (often requiring costly refrigeration during shipping) contribute to global warming and air pollution through the burning of irreplaceable fossil fuels.

In addition to the environmental impact, health and nutrition greatly concern today's green consumer. Locally made food can be harvested at the peak of ripeness, instead of being picked and shipped before it is fully matured. Harvesting and ingesting local food at the proper state of ripeness stage generally increases nutritional content and reduces exposure to unnecessary and potentially harmful chemical substances used to artificially ripen produce. Many green consumers state that the taste of locally grown food is significantly better than commercial, large-scale farm products. This may be due to the fact that while in transit, certain aspects of nonlocal foods change (for example, sugars turn to starches, and plant cells shrink). Food grown in local communities are typically sold within two or three days.

Studies show that buying locally made foods directly from the farmer sends 90 percent of those food dollars back to the farm. Instead of spending money on processors, packagers, distributors, wholesalers, truckers, and other middlemen, locally made purchases result in localized economic support, with communities benefiting from support of local jobs and businesses and tax monies staying local. The benefits of purchasing locally made food and goods have introduced the concept of "buy local" campaigns in cities and states around the country. These campaigns typically involve farmers markets, local farms and co-ops, independent producers, and crafters. Portrayed as stimulating for the local economy, "buy local" campaigns have the ability to strengthen communities and further the locally supported lifestyle.

A survey of restaurant chefs by the National Restaurant Association found "locally grown" food to be the hottest industry trend for 2009. With more and more consumers interested in knowing the location and conditions of their food and goods, the designation of "locally made" risks becoming a new marketing strategy with little regulatory power to back up such a claim. Recent years have seen proposed legislation in support of local agriculture in a number of states. In 2007, Illinois created a task force to develop strategies to increase local, organic buying programs for public institutions, and the U.S. Department of Agriculture Marketing Service administers grant programs to help a variety of local-organic initiatives around the country.

**See Also:** Consumer Ethics; Consumer Society; Ecological Footprint; Ethically Produced Products; Food Miles; Green Food; Lifestyle, Sustainable; Markets (Organic/Farmers); Quality of Life; Seasonal Products; Services; Shopping; Vegetables and Fruits.

## Further Readings

Global Exchange. "Buy Local Day: Local Business Directory." http://www.buylocalday.org/article.php?list=type&type=9 (Accessed April 2009).

Institute for Local Self-Reliance. "Key Studies on Wal-Mart and Big-Box Retail." http://www.newrules.org/retail/key-studies-walmart-and-bigbox-retail (Accessed March 2009).

LocalHarvest, Inc. http://www.localharvest.org/ (Accessed March 2009).

Natural Resources Defense Council. "Eat Local." http://www.nrdc.org/health/foodmiles/default.asp (Accessed April 2009).

Sustainable Table. "What Is Local?" http://www.sustainabletable.org/issues/eatlocal (Accessed December 2009).

*Erin E. Dorney*
*Millersville University*

# M

## Magazines

Published periodically, magazines have been issued as serial publications since 1731, when *The Gentleman's Magazine* was first published in London. They are often financed by newsstand purchase, subscriptions, and advertising. The distribution costs for magazines are similar to those of newspapers and are normally paid for by the retail price of the magazine. The bulk of the operating costs and profits from magazines are met by advertising revenues.

Many shoppers' magazines are distributed free of charge. Shoppers' magazines can include periodicals devoted to real estate sales and automotive sales. Some are targeted to people seeking tourist information. The goal is to reach consumers who seek the goods or services offered in the advertising information. This can be thought of as a green activity because it provides consumers with information that would be difficult to obtain otherwise, and it saves them the cost of searching; however, the costs, material inputs, and potential energy invested in producing and distributing such magazines need to be considered.

The cost of the periodicals is not insubstantial, but the sales generated are usually much greater. Environmental critics have pointed out that the average magazine on the newsstands—for example, a weekly news magazine—creates carbon dioxide emissions that have been calculated to amount to over two pounds. This figure is based on the emissions from logging, paper production, shipping, and consumer use. However, recycled paper, especially old newspapers, is used to make the glossy paper used in magazines. The several processes for making glossy paper use recycled newsprint and some virgin material, so recycled newspapers, which are scrap paper, are returned to the consumer as a useable product.

Magazines today can be print magazines or online "e-zines." A great many magazines are profit-making businesses. These provide customers with all manner of special-interest stories and information. Some magazines are published by nonprofit organizations that may or may not have a public education agenda.

One of the oldest magazines about the environment is *National Geographic*. It prints stories about physical and human geography. Many of its stories in recent decades have been about environmental problems. Other nature magazines that deal with the environment include publications such as *Wildlife,* published by the National Wildlife Federation, which also publishes the *Ranger Rick*, *Just for Fun*, and *Animal Baby* nature magazines for children.

Recent years have seen the emergence of numerous magazines aimed at the green market. Green magazines seek to increase societal awareness of the need to live sustainably, act in an environmentally sensitive way, and consider the social and environmental impacts of existing social practices. A number of these magazines have been targeted at consumers with an interest in hunting and fishing. Many hunting organizations are active as conservationists of habitats needed by game animals or fish. Their concerns also may include the introduction of invasive species that threaten native species. In addition to magazines encouraging individuals and households to live and consume sustainably, there are magazines aimed at builders and architects who are seeking to green the built environment.

Green magazines may also be published by government agencies such as the National Park Service. *National Parks* is a magazine that educates the American public about the history, nature, and conservation needs of the country's national parks. Articles may focus on potential environmental threats from mining, logging, building, or industrial activities.

Nonprofit green magazines include the environmental magazine *E - The Environmental Magazine,* which has been published for 20 years. It is a bimonthly magazine that seeks to be a clearing house on the environment. It provides its readers with environmental information that ranges from global issues to backyard issues.

*The Green Guide* magazine is produced by the National Geographic Society, which purchased the publication in 2007. The society is using the magazine to fulfill its mission to inspire people to care about the planet. *The Green Guide* is used by many consumers to aid in purchasing decisions that will be ecologically responsible. It is a much-used resource because it seeks to be very practical and informative. Its staff is dedicated to seeking out the facts about products and services to aid consumers who are consciously seeking to be "green." In its reporting, it strives as a matter of policy to be objective, fair, and factual.

Topics found in green magazines include personal, business, and global effective waste reduction. Others deal with topics such as forest conservation, greenhouse gases, protecting the ozone layer, endangered animals, organic foods, and choices about food, sustainable agriculture, or the impact of urbanization on the environment. Green sources of energy from wind, waves, sun, or biomass are topics for stories featured in other green magazines.

The increasing availability and popularity of green magazines may help bring issues connected with forest conservation, greenhouse gases, protecting the ozone layer, endangered animals, organic foods, sustainable agriculture, land-use impacts, and energy use to the forefront, with the consequent hope that consumers will not only be better informed but also act on the information and ideas contained within these publications.

**See Also:** Advertising; Green Consumer; Leisure and Recreation; Sustainable Consumption.

### Further Readings

Blanchette, Rick and Jayne Vandervelde, eds. *Green Profit on Retailing*. Chicago, IL: Ball, 2000.

*The Green Guide: The Complete Reference for Consuming Wisely*. Washington, D.C.: National Geographic Society, 2005.

Lockwood, Charles. *The Green Quotient: Insights from Leading Experts on Sustainability*. Washington, D.C.: Urban Land Institute, 2009.

*Andrew Jackson Waskey*
*Dalton State College*

# MALLS

The growing interest in sustainable consumption and green products over the last decade has led to a concern with the environmental impact of places where those products are sold—shopping malls. Increasingly, mall developers and operators are beginning to factor ecological and sustainability issues into their planning, marketing, and development schemes. The International Council of Shopping Centers, the global trade association of the shopping center industry, has recently established a sustainability portal for its Website and sponsored a "green retailing" conference that addressed "Sustainability, Energy and Environmental Design." Given that mall operators are in the business of selling products and making profits, it is not surprising that a common refrain in this new discourse of sustainability places great emphasis on the fact that green shopping centers are good both for the environment and for business.

The self-enclosed mall design, an icon of retail development that has had a profound cultural and economic influence throughout the world for almost 50 years, began to decline in the 1990s.

*Source:* iStockphoto

Malls are both highly efficient and rationalized forms of retailing that are structured in ways that almost compel people to shop, as well as "cathedrals of consumption" that are structured to have an enchanted and sacred character so that a trip to mall is akin to a religious pilgrimage. Along with products, malls sell experiences. For example, the Mall of America, the largest mall in the United States, is not only an assemblage of stores but is a major entertainment and tourist destination as well. Although malls, a ubiquitous feature of the modern U.S. retail landscape, have been around since the 1920s, they really began to proliferate after the Vienna-born architect Victor Gruen, referred to as the "mall maker," entered the scene in the 1950s. Gruen's Southdale Center in Edina, Minnesota, built in 1956—a fully enclosed, weather-controlled mall anchored by department stores with lots of public and pedestrian spaces surrounded by parking lots all under one management—became the standard model that has been replicated all over the world. For Gruen, malls were more than places to shop, they were community spaces much like the ancient Green Agora and the medieval marketplace. Despite the widespread adoption and economic success of his mall concept, Gruen became a harsh critic of shopping malls and regarded their worldwide triumph as a dismal failure. He often claimed that real estate interests, in their blind pursuit of profits, had hijacked his concept of malls as community centers and reduced them to "machines for selling." In the later part of his life, the pioneer of the shopping mall began to address the environmental problems and issues of the sustainable city. Reflecting on his legacy, Gruen criticized shopping malls for encouraging sprawl, destroying nature, and degrading a sense of community.

## Malls and Suburban Sprawl

In the post–World War II years, U.S. malls have become synonymous with low-density, automobile-dependent suburban sprawl—a land use pattern that is detrimental to the environment and to public health. Along with contributing to the loss of wetlands, farmlands, and wildlife, sprawl is linked to global warming because of high levels of carbon emissions from increased automobile usage. Runoff that spills into streams and waterways and diminished infiltration for groundwater are also major problems because of the paved surfaces that dominate the landscapes of sprawl.

Abandoned or underperforming malls, often referred to as "greyfield" sites or "ghost boxes" because they leave behind a sea of empty asphalt, have become a common feature of the modern suburban landscape. Several new urbanists and smart growth–oriented groups and organizations have become involved in searching for creative and environmentally sound ways to retrofit failing shopping centers and malls, claiming that the greening of abandoned malls could help to reverse the problem of suburban sprawl and its associated negative environmental impacts. Because malls occupy large, flat, well-drained, developable space linked to existing infrastructure and transportation routes, they offer opportunities for more sustainable, mixed-use, less auto-dependent places that could relieve traffic, help reduce pollution, and provide residents with a downtown. The demise of malls built on wetlands before environmental legislation slated them for protection has led many to advocate the restoration of the landscape and the creation of green and open spaces.

The enclosed mall, an icon of retail nirvana that has had a profound cultural and economic influence throughout the world for almost 50 years, began to decline in the 1990s. Despite their decline, there has been a profusion of different types of specialty malls—factory outlet malls, lifestyle malls, megamalls, power centers (e.g., groups of big box retail stores)—and malls like the Easton Town Center in Columbus, Ohio, which is essentially a town unto itself.

## Can Malls Be Made Green?

In recent years, amid growing concern about issues of environmental sustainability and energy conservation, several mall developers and operators have began to explore ways to make their sites more ecofriendly and energy efficient. Some of their green initiatives include installing green roofs and solar panels, relying on renewable energy sources, and developing their own green power sources. Some are encouraging their retailers to adopt green strategies such as ecofriendly packaging. Megamall growth is particularly strong in Asia, which is home to 7 of 10 of the world's largest malls. Many new Asian malls are claiming to be green.

Greening the infrastructure of malls has limitations. Combating the environmental problems stemming from suburban sprawl will require more than better design. Creating or retrofitting a green development site does not really address the structural conditions that create and promote sprawl in the first place. Better architectural design cannot, in and of itself, change the larger pattern of social, economic, and environmental devastation developed by tax-subsidized growth machines that profit from sprawl. Real sustainability gains will require addressing the carbon footprint of the products that are being sold. Wal-Mart, the world's largest retailer, in addition to greening its buildings, has begun to address the wider issue of sustainable consumption by focusing on the carbon footprints and the ecological and social impacts of the products being sold in its stores. Their focus on the life cycle of consumer products and the environmental cost associated with their manufacture, distribution, consumer use, and post use is potentially far-reaching. Wal-Mart, because of its immense power and influence over manufacturing and retailing, might help to green both processes of production and consumption.

These developments raise several questions. Are mall operators and big box stores like Wal-Mart jumping on the sustainability bandwagon because they view greening efforts as a way to reduce cost through increased efficiency? Is it a response to changes in consumer demand? Or is it a sophisticated form of greenwashing, a public relations scheme to project a favorable image? Although time will be needed to answer these questions, one thing is clear: Wal-Mart and other mall developers regard adopting sustainable retail strategies as a sound business decision. On another level, greening the cathedrals of consumption and stocking them with all sorts of ecofriendly products reinforces and legitimizes a culture of excessive conspicuous consumption. If malls are places where people go to practice their consumer religion, than perhaps what is needed is a new environmental and consumer ethic that promotes purchasing less as well as green consumption.

**See Also:** Consumer Culture; Consumer Ethics; Green Consumer; Green Design; Greenwashing.

### Further Readings

Dunham Jones, Ellen. *Retrofitting Suburbia: Urban Design Solutions for Redesigning Suburbs*. Hoboken, NJ: John Wiley & Sons, 2008.

Hardwick, Jeffrey. *Mall Maker: Victor Gruen, Architect of an American Dream*. Philadelphia: University of Pennsylvania Press, 2004.

Hayden, Dolores. *Building Suburbia: Green Fields and Urban Growth, 1820–2000.* New York: Pantheon Books, 2003.

International Council of Shopping Centers. "Sustainability Portal." http://www.icscseed.org (Accessed August 2009).

Ritzer, George. *Revolutioning the Means of Consumption: Enchanting a Disenchanted World*. Philadelphia: University of Pennsylvania Press, 2006.

*Steven Lang*
*LaGuardia Community College*

## Markets (Organic/Farmers)

Markets are places where the exchange of goods and services occurs. Specifically, in relation to green consumerism they are places where organic foods are sold, but they may also be associated with notions of buying local, supporting local firms, lowering food miles, and slow food. Originally, in the 1920s, organic food tended to be marketed through small-scale, informal links directly between farmers and consumers. Direct marketing continued to be the main way in which organic food was sold up until the 1950s and 1960s. By the 1970s, in both Europe and the United States, retail health-food outlets specializing

Farmers markets have burgeoned in both the United States and European Union countries. The sale of organic produce encourages a closer connection between producers and consumers.

*Source:* iStockphoto

in organic products started to appear. These were often run by people who were part of alternative movements and were concerned with a variety of issues such as the industrialization of farming and social justice within developing countries.

However, by the mid-1980s and early 1990s the production of organic produce increased rapidly. Small-scale shops were no longer able to handle the volumes involved, and a number of other market outlets developed. These included conventional supermarkets, which have increasingly stocked organic produce, encouraged by the premium prices available. The 1990s and early 2000s have also seen a growth in large-scale box schemes, such as Abel & Cole and Riverford Organics in the United Kingdom and the Good Food Box in Canada. At the same time, farmers markets have burgeoned in both the United States and Europe as a medium that includes the sale of organic produce in a way that encourages the connection of producers and consumers. Small-scale vegetable box schemes have also grown, as have specialized organic food outlets—especially in countries such as Germany and the Netherlands.

The net result of these processes is that there is now a wide range of markets for organic food. It is also evident that there is some kind of division between those that involve a degree of direct marketing—wherein there is contact between the producers and consumers of the products involved—and outlets where the produce is more anonymous and distanced, whereby institutional organic regulations become more important.

## Starting Small

Specialized independent retail outlets have been an important market for organic produce since the 1960s, when in many cases they may have been run as not-for-profit food cooperatives. Initially, the emphasis was on retaining direct links with the place of production and on encouraging the consumption of locally grown food, but throughout the 1970s and 1980s, they often became more specialized and urban based. Indeed, in countries such as the Netherlands and Germany, specialized shops selling only organic products were the most important market outlet for organic produce. However, by the mid-1980s and early 1990s, in part following deliberate attempts by certain players within the organic movement (such as the Soil Association in the United Kingdom), the scale of organic food production increased markedly and the role of independent retail outlets declined, relative to other markets. Nevertheless, they still have an important role to play in certain countries (e.g., Germany), as well as in the guise of farm shops, which continue to be a significant market outlet in some countries.

An important example of a direct marketing outlet is small-scale box schemes. Originally, these involved local producers selling a limited range of their own produce (usually vegetables, but sometimes meat as well) directly to consumers living within the vicinity. These box schemes continue to be an important outlet for organic produce, although they are limited in terms of the volumes they can handle. They are also limited to the produce that the producer has available at any one time (although in some respects this is heralded as part of their appeal, wherein there is a direct connection to the seasonality and localness of production). Partly as a result of these factors, a number of large-scale box schemes have emerged in recent years, selling, in some cases, in excess of 30,000 boxes a week. Usually these are coordinated via the Internet and, although attempting to source produce from as locally as practically possible, also source produce from other countries. This allows them to increase the range they can offer, as well as to overcome seasonal shortfalls. Inevitably, there is then a greater distance between the producer and the consumer, although the lack of direct contact is countered by information supplied in the weekly boxes, as well as via the operators' Websites.

## Growing Bigger

The Internet has also grown in importance in recent years for individual producers and is effectively a more interactive version of "mail order." Clearly, this form of marketing is not limited to organic producers, but it does enable producers to explain in detail why they feel their product is distinctive and worthy of a price premium. It also provides a relatively low-cost market opportunity for producers who may be geographically isolated from their potential consumers. Modern courier systems have facilitated this process.

Sales through the Internet can also work well in tandem with farmers markets, whereby consumers can directly assess the quality of both the product and the producer at the market, subsequently buying it online. Farmers markets have grown in terms of both numbers and popularity across Europe and the United States over the last 10 years. Although few are exclusively reserved as an outlet for organic produce, many have at least some organic producers. Although there are differences in emphasis between countries, farmers markets are concerned about reconnecting the producers of food with the end consumers. They do this by insisting that someone who has been involved in the production process is actually selling the produce (thereby allowing the consumer to ask detailed questions about the produce being sold), and that the producer comes from within a specified distance of the market. Farmers markets are an important outlet in terms of the low costs involved in participation, as well as their accessibility, especially for smaller-scale organic producers. They also allow for product experimentation and direct consumer feedback, which can be useful to organic producers of all sizes. As such, they may be the principal market for some smaller organic producers, but simply one of a number of outlets in the overall marketing mix of larger producers.

## Supermarkets Join in Sales

Growing consumer interest in organic produce in the late 1980s and 1990s, as well as an increase in supply, led the major multiples, or supermarkets, to become interested in stocking it as part of their overall offer. Initially, this involved pioneering supermarkets (such as Tegut in Germany or Waitrose in the United Kingdom), but subsequently more mainstream chains (such as Tesco in the United Kingdom, or the Whole Foods Market and Wal-Mart in the United States) also started to stock organic produce. Conventional supermarket interest was in part driven by the premium prices and higher margins available, but also by the institutional and international regulation of organic standards. In the European Union this was through EC Regulation 2092/91, which established uniform rules for organic produce throughout Europe (in 1991 for plants and in 1999 for animal products). Similar standards were subsequently introduced in Japan in 2001 and the United States in 2002. The introduction of these regulations facilitated the international trade in organic produce, in that all those involved were then working to comparable definitions of what is meant by "organic produce." Since the late 1990s, supermarkets have played an increasingly significant role in the marketing of organic produce, mainly because they are the place where most consumers do most of their shopping. They are now responsible for a high percentage of organic sales. For example, in the United Kingdom they are responsible for 75 percent of organic sales, whereas in Switzerland 80 percent of organic products are sold through just two retail chains. Supermarkets have been instrumental in increasing the scale and scope of the organic market, but at the same time they have also taken much of the power from those who were originally involved, such as specialized, independent organic stores.

Before the 1980s and early 1990s, organic food was largely associated with alternative food outlets such as local, farm-based, vegetable box schemes and independent retailers. However, as the volume of organic produce increased rapidly throughout the 1990s and 2000s (until the worldwide recession starting in 2008), it started to be sold through a wider range of market outlets. These outlets range from large-scale, corporate supermarket retailers through national-level box schemes and the Internet to farmers markets, farm shops, and local box schemes. In reality, in most instances, both producers and consumers of organic food are likely to use a range of market outlets for their sales and purchases, respectively.

**See Also:** Food Miles; Organic; Slow Food; Vege-Box Schemes; Vegetables and Fruits.

### Further Readings

Holt, Georgina and Matthew Reed. *Sociological Perspectives of Organic Agriculture: From Pioneer to Policy*. Wallingford, UK: CABI, 2007.

International Federation of Organic Agriculture Movements (IFOAM). "The World of Organic Agriculture 2009." http://www.ifoam.org (Accessed April 2009).

Kremen, Amy, et al. "Organic Produce, Price Premiums, and Eco-Labeling in U.S. Farmers' Markets." Electronic report from the Economic Research Service, USDA. http://www.ers.usda.gov/publications/VGS/Apr04/vgs30101/vgs30101.pdf (Accessed April 2009).

Lockeretz, William, ed. *Organic Farming: An International History.* Wallingford, UK: CABI, 2007.

Soil Association. "Organic Market Report 2009." http://www.soilassociation.org/marketreport (Accessed April 2009).

*James Kirwan*
*Countryside and Community Research Institute*

# MATERIALISM

Materialism is defined as the importance a consumer attaches to worldly possessions. At the highest levels of materialism, possessions assume a central place in a person's life and are believed to provide the greatest sources of satisfaction and dissatisfaction. Russell Belk has framed materialism as a higher-order construct with three second-order dimensions; that is, possessiveness, nongenerosity, and envy. Marsha Richins approaches materialism as the belief in the desirability of acquiring and possessing things, "a value that guides people's choices and conduct in a variety of situations, including, but not limited to, consumption arenas." Her measure of material value has three subscales, measuring possession-defined success, acquisition centrality, and acquisition serving as the pursuit of happiness. In her study, respondents are asked whether they admired people who owned expensive homes, cars, and clothes (success); whether the things that they owned were important to them (centrality); and whether they would be happier if they could afford to buy more things (happiness). Overall, whether for pleasure seeking, self or relationship definition/expression, or status claiming, materialism is an excessive reliance on consumer goods to achieve these ends—a consumption-based orientation to happiness-seeking and placing a high importance on material possessions.

## Materialism and Perceived Well-Being

Studies of material values have consistently shown that a high degree of materialism ultimately affects consumers' well-being and quality of life. Nevertheless, the criteria for well-being are likely to vary cross-culturally. Consumers low in materialism are more satisfied with their socioeconomic status than are those who place a higher value on material things. In turn, one's perceived socioeconomic status is an important life domain that affects one's evaluation of overall life satisfaction. Related to perceived socioeconomic status effects, the state of the country's economy within which consumers live also appears to influence their level of materialism. A value survey of 50,000 people in 40 countries reveals that individuals in poorer nations appear to be more materialistic than those from wealthier countries.

There is evidence substantiating the negative correlation between materialism and happiness in life. In a Singapore study, those with a higher level of materialistic inclination were significantly less satisfied with life compared with the group with a lower level of materialistic inclination. Other studies have demonstrated negative correlation between materialism and psychological well-being, happiness, or life satisfaction in various regions around the world.

A debate continues in the popular press over whether reducing such materialistic tendencies are indeed the secret to a contented life. Some point out that the negative consequences of such materialistic tendencies are creating higher debt, less savings, more bankruptcies, more stress, and less time for family and friends, and thus a lower quality-of-life satisfaction. Others argue that consumerism is not against our better judgment—it is our better judgment. Our life satisfaction is predicated on the fulfillment of our desires to indulge in luxury spending habits.

## Quantity of Television Viewing, Perceived Realism, and Materialism

As access to advanced media, technology, and advertising accelerates exponentially throughout the developing world, it seems particularly relevant to investigate the consequences of such exposure on the evolving cultures of these countries, as well as on individual perceptions regarding their socioeconomic status and relative life satisfaction.

The power of media exposure has been partially explained by the "cultivation theory," the effect of the quantity and consumerist content of television programming and advertising. The viewer's perceived reality of television content, along with repeated exposure, strengthens acceptance of the messages in that television content. Past research also has shown that as posited by cultivation theory, frequent viewing of the distortions of reality will increasingly lead the consumer to believe that these distortions reflect reality.

For those who perceive television commercials to be a realistic portrayal of consumers, research further indicates that the correlation between hours spent watching television and the extent of materialism is significant. Others identifying the link between being materialistic and watching large amounts of television include findings that also suggest that the cultivation effect is a function of the content in television programs. Thus, increased exposure to mass media is likely to fuel consumption and materialism.

Media images, along with peers and aspiration groups, tell consumers what they should want or strive to obtain, with respect to one's possessions, lifestyle, and status, and serve as a standard of comparison. The perceived realism of programming and advertising content influences the values held by consumers, and ultimately, that of society as a whole. Media images, as a source of information on which elevated standards of comparison are based, affects not only the level of materialism in a culture but also consumers' perspectives on their general socioeconomic status, and creates satisfaction or dissatisfaction with their lives.

As posited by L. Festinger's social comparison theory, individuals learn about and assess themselves in comparison to other people. Although such comparisons can potentially be "downward," that is, with some perceived as worse-off than others, or "upward," with others perceived as better-off than others, James Twitchell suggests that humans are consumers by nature, and that we proactively seek out media messages that reinforce our desire to live and create ourselves through things. We often depend on such material objects for meaning. Consumers are not misled by advertising, packaging, and branding; instead, they actively seek out and surround themselves with what they enjoy, especially when they are young. Cross further noted how attractive media images created a consumerism that moved even farther away from reality and "toward an enveloping personal fantasy."

## Materialism vis-à-vis Religiosity

The *Random House College Dictionary* definition of materialism provides support for the relationship between religiosity and materialism: "attention to or emphasis on material objects, needs, and considerations, with a disinterest in or rejection of spiritual values." This definition posits that materialism is in direct opposition to spiritual values.

All major religions have long criticized excessive materialism as being incompatible with religious fulfillment, or at least distractive to spiritual development and—more typically—contrary to the attainment of happiness and true joy and peace of mind. Criticisms that the excessive pursuit of material goods at the expense of "higher" pursuits stem from many organized religions (e.g., Buddhism, Hinduism, Islam, Judaism, and Christianity), all of which condemn concentrating on building excessive material wealth.

Each of these faith perspectives concurs that one's primary goal should be spiritual, while only the minimum level of material wealth is necessary to maintain a content life.

Although religion can be a barrier to the adoption of consumer culture (e.g., the Amish), it is not always so. Recent research finds that some modern-day religious sects view material gain as compatible with spirituality—even seeing material success as a blessing from God. A study based in Singapore finds that the level of materialistic inclination differs significantly between respondents with different religious affiliations. In general, research shows that although income and some aspects of materialism are positively related to the subjective well-being of low-religiosity consumers, these variables are negatively related to the subjective well-being of high-religiosity consumers. The relationship between the cultural values of religiosity and materialism can, indeed, be complex. On the basis of the findings of the majority of the research discussed earlier, we would expect religiosity to be a negative influencer of materialism and a positive influencer of life satisfaction.

As the studies on material values suggest, increasing materialism has not yielded higher levels of happiness and life satisfaction in the United States. In addition to a rising level of life dissatisfaction, others have already noted the negative ramifications of consumer resource misallocation motivated by the growing materialism in the non-Western world. Such messages may help to deemphasize the role played by material possessions in realizing happiness and life satisfaction. Further research into how to address this challenge seems paramount, given the important role played by materialism in influencing the well-being of societies worldwide.

**See Also:** Advertising; Affluenza; Consumerism; Disparities in Consumption; Environmentalism; Needs and Wants; Overconsumption; Quality of Life; Resource Consumption and Usage; Shopping.

### Further Readings


Belk, R. W. "Materialism: Trait Aspects of Living in the Material World." *Journal of Consumer Research*, 12/3:265–80 (1985).

Burroughs, J. E. and A. Rindfleisch. "Materialism and Well Being: A Conflicting Values Perspective." *Journal of Consumer Research*, 29/3:348–70 (2002).

Cross, G. *All Consuming Century: Why Commercialism Won in Modern America.* New York: Columbia University Press, 2000.

De Graff, J., et al. *Affluenza: The All-Consuming Epidemic.* San Francisco, CA: Berrett-Koehler, 2001.

Festinger, L. "A Theory of Social Comparison Processes." *Human Relations*, 7/2:117–40 (1954).

Ger, G. and R. W. Belk. "Accounting for Materialism in Four Cultures." *Journal of Material Culture*, 4/2:183–204 (1999).

Ger, G. and R. W. Belk. "Cross-Cultural Differences in Materialism." *Journal of Economic Psychology*, 17/1:55–77 (1996).

Griffin, M., et al. "A Cross-Cultural Investigation of the Materialism Construct: Assessing the Richins and Dawson's Materialism Scale in Denmark, France and Russia." *Journal of Business Research*, 57/8:893–900 (2004).

Kasser, T. *The High Price of Materialism.* Cambridge, MA: MIT Press, 2002.

Keng, K. A., et al. "The Influence of Materialistic Inclination on Values, Life Satisfaction and Aspirations: An Empirical Analysis." *Social Indicators Research*, 49/3:317–33 (2000).
Richins, M. L. and S. Dawson. "A Consumer Values Orientation for Materialism and Its Measurement: Scale Development and Validation." *Journal of Consumer Research*, 19/3:303–16 (1992).
Rindfleisch, A., et al. "Family Structure, Materialism, and Compulsive Consumption." *Journal of Consumer Research*, 23/4:312–25 (1997).
Ryan, L. and S. Dziurawiec. "Materialism and Its Relationship to Life Satisfaction." *Social Indicators Research*, 55/2:185–97 (2001).
Schwartz, B. *The Paradox of Choice: Why More Is Less*. New York: HarperCollins, 2004.
Shrum, L. J., et al. "Television's Cultivation of Material Values." *Journal of Consumer Research*, 32/3:473–79 (2005).
Sirgy, M. J. "Materialism and Quality of Life." *Social Indicators Research*, 43/3:227–60 (1998).
Speck, S. K. S. and A. Roy. "The Interrelationships Between Television Viewing, Values and Perceived Well Being: A Global Perspective." *Journal of International Business Studies* (November 2008).
Twitchell, J. B. *Living It Up: Our Love Affair With Luxury*. New York: Columbia University Press, 2002.
Watson, J. J. "The Relationship of Materialism to Spending Tendencies, Saving, and Debt." *Journal of Economic Psychology*, 24/6:723–40 (2003).

*Abhijit Roy*
*University of Scranton*

# Meat

We term *meat* the organic matter of which animals are made, with particular reference to muscle tissue, soft tissue, and some edible internal organs (e.g., brains and liver). In industrial jargon, the term is generally used to signify foods deriving from the butchering of reared animals (bovines, pigs, sheep, goats, rabbits), with the exception of poultry.

The earliest evidence for farm animal domestication comes from Asia and is dated between 6000 and 5000 B.C.E. Up to the beginning of the 19th century, domestic animals had been chiefly reared with traditional extensive methods; therefore, numbers and production were conditioned by the availability of grass and winter feedstocks. Beginning from the second half of the 19th century, however, animal breeding systems have become increasingly more intensive—animals have been kept inside, in extremely artificial environments, and breeders have begun to feed them with high-calorie feedstocks.

The progressive expansion of intensive breeding has led to the growth of a meat-processing industry. Between the second half of the 19th century and the first half of the 20th century, great abattoirs were built where, thanks to refrigerated railcars, animals were brought to distant places (frequently they arrived covered with excrement), so much so that as a result of urban residents' protests and of the complaints of people sensitive to the treatment of animals, governments had to resort to suitable legislation aimed at improving the entire process of meat processing.

In the 1960s, the Iowa Beef Packers (IBP) introduced a fundamental innovation in the processing of meat: boxed beef. Because by then refrigerators with freezers existed in most houses, IBP thought it more advantageous to make available to consumers a number of selected cuts of meat. Pushed by supermarkets and by a great demand from consumers, beginning from the early 1980s, the boxed beef market has expanded rapidly in such a way that the system adopted by the IBP has been extended to all other types of meat.

At present, the global meat processing industry butchers around 56 billion animals: of these 39 percent are pigs, 24 percent bovines, and 7 percent sheep and others; the remaining 30 percent is represented by poultry.

World production of meat has grown from 44 million tons in 1950 to 280 million tons in 2008. Mean production per person has increased from 17.2 kilos in 1950 to around 42 kilos in 2008; in developing countries, an average consumer eats around 30 kilos of meat per year, whereas in industrialized countries, the average consumer eats about 80 kilos of meat.

Worldwide production of meat grew from 44 million tons in 1950 to 280 million tons in 2008. The global meat processing industry butchers around 56 billion animals per year, including pigs, cattle, poultry, sheep, goats, and others.

*Source:* Brian Prechtel/Agricultural Research Service/USDA

The continuous growth of meat consumption and, consequently, the number of animals reared for slaughter, has caused several environmental problems. The two main problems are represented by the considerable quantities of water needed during the various phases of rearing (the production of one kilo of beef reared with cereals requires around 16,000 liters of water) and by the complexity of the disposal of butchered animal waste.

Precisely the intrinsic difficulties in the disposal of the considerable quantities of waste produced by the butchering industry have induced the food industry to use such waste for producing feedstock destined for the rearing of herbivores. Feeding such animals with feedstock containing animal proteins has been conducive to the spread of bovine spongiform encephalopathy (BSE, commonly known as mad cow disease), which, discovered for the first time in Great Britain in 1986, has spread to other countries, leading to the prohibition of feeding animals with animal proteins. Having established that such disease is directly connected with the Creutzfeldt-Jakob syndrome, which affects the human brain, BSE has quickly become a serious epidemic global problem, causing the destruction of many animals and, during the peak of the epidemic, the collapse of the meat market in some European countries.

A third problem concerns soil and water pollution caused by significant amounts of nitrates, antibiotics, and heavy metals present in animal feces, and in particular those of pigs and bovines. The question of antibiotics is especially important: livestock uses an average quantity of antibiotics eight times above that of human beings, which encourages the development of antibiotics-resistant microbes, thus making the fight against disease harder for both humans and animals.

A fourth problem concerns the emission of considerable quantities of greenhouse gases: animal breeding produces, in fact, 37 percent of world emissions of methane (a gas whose global warming potential is about 20 times above that of carbon dioxide) and 18 percent of global greenhouse emissions.

Last, the large, intensive animal-rearing farms are conducive to the spread of *Salmonella* and other agents that produce food infections. Furthermore, these farms may encourage the spread of diseases such as avian flu and pig fever, as these viruses quickly spread among animals confined to narrow, overcrowded spaces.

Given concerns around animal husbandry and ethical treatment from rearing to slaughter, and about the presence of chemicals in the food chain for meat, there has been a growing market for organic and chemical additive–free meat. A number of major meat producers and packers now market "green" meat products to consumers, and new commercial operators promoting green/healthy/sustainable and/or humane products and production processes have emerged. There are also a number of practices that consumers can engage in to reduce their carbon footprint, including cooking at home, buying less processed meats, and buying direct from meat producers.

**See Also:** Consumer Society; Poultry and Eggs; Resource Consumption and Usage.

### Further Readings

Ferriéres, Madeleine. *Sacred Cow, Mad Cow: A History of Food Fears*. New York: Columbia University Press, 2006.

Lee, Paula Young, ed. *Meat, Modernity, and the Rise of the Slaughterhouse*. Durham: University of New Hampshire Press, 2008.

Nierenberg, Danielle. *Happier Meals: Rethinking the Global Meat Industry*. Washington, D.C.: Worldwatch Institute, 2005.

Niman, Nicolette Hahn. *Righteous Porkchop: Finding a Life Beyond Factory Farms*. New York: Collins, 2009.

Perren, Richard. *Taste, Trade and Technology: The Development of the International Meat Industry Since 1840*. Aldershot, UK: Ashgate, 2006.

Worldwatch Institute. *Vital Signs 2009: The Trends That Are Shaping Our Future*. Washington, D.C.: Worldwatch Institute, 2009.

*Federico Paolini*
*University of Siena*

## Mobile Phones

In the United States, typical cell phone contracts can provide the customer with a new phone every 18–24 months. Many consumers may choose to switch phones more often than that in response to the rapid improvement of products and releases of premium offerings like the iPhone and the Blackberry. The potential electronic waste of discarding these millions of old cellular phones a year is considerable, as is the amount of raw materials used to manufacture the new ones, and even the valuable gold and copper included in small amounts in each phone. State and federal agencies have encouraged initiatives to recycle cell phones and cell phone accessories, as they do with all electronic waste; in some states, such as California, such recycling is mandatory. Although often ignored, such laws have succeeded in increasing the recycling rate somewhat; the Environmental Protection Agency reported that the recycling rate for cell phones was 10 percent from 2006 to 2007 compared with 18 percent for electronic waste in general. Of the 140 million cell phones that were disposed of that year, 14 million were recycled, and the remaining 126 million were simply thrown away. A number of charities accept donated cell phones, which have their data wiped and are refurbished for reuse.

One reason electronic waste is such a growing concern is that there is more to a discarded cell phone that just its plastic casing. The small amounts of metals used, although harmless when inside a phone carried in your pocket, are dangerous when they build up in the environment, as they do when hundreds, thousands, or millions of phones are thrown away. Lead, which is used to solder the circuit boards inside the phone, is extremely toxic to humans if it enters the water supply and can interfere with the development of young children. Mercury, present in some batteries, is deadly to the nervous system and the development of unborn babies. Cadmium and beryllium can contribute to cancer, as well as damage the organs. And although they are present in our phones in small amounts, the highly toxic poisons antimony and arsenic used in cell phone batteries and battery contacts can poison our soil and water supply. Simply from an antiwaste perspective, cell phones also include precious metals gold, silver, platinum, palladium, rhodium, copper, tin, brass, and zinc, which can be extracted and recycled. Four tons of gold alone could be retrieved from the cell phones that are discarded every year, which according to the Environmental Protection Agency would take so much mining to extract from the Earth that over six million tons of soil, sand, and rock would be displaced—and that is just gold: a typical cell phone contains 500 times more copper than gold.

As cell phones have become more ubiquitous, they have also become greener—even those that do not specifically market themselves as such. Newer phones are smaller, using not only less plastic but less packaging and fewer trucks to transport; considering the sales volumes of cell phones, such small differences really do add up. More phones are designed with "end-of-life management" in mind, making it easier to extract metals from them and to separate different recyclable components. More recycled materials, both metals and plastics, are used in the manufacture of many cell phones, and toxic materials (particularly outside of the circuitry, such as toxic flame-retardant coatings on the phone body itself) have been reduced or eliminated.

There are a small number of phones specifically designed and marketed as green, though they are being introduced and used more slowly than expected because of the industry's perception that the global financial meltdown that began in 2008 will dissuade consumers from spending extra money, even for added value. Samsung's Reclaim phone, a 3G phone with a slider QWERTY keyboard and one-click access to popular social Websites, is made of 80 percent recyclable materials—half of which is corn-based bioplastic. The charger meets Energy Star standards for energy efficiency. Motorola's Renew is billed as the market's first carbon-free cell phone; made from recycled water bottles. Purchase and use of the phone confers carbon credits to offset the emissions resulting from the phone's manufacture and use over two years. The packaging is made of recycled paper and comes with a shipping envelope so the phone can be recycled when the consumer is done with it. In Jamaica, the first mass-produced solar-powered cell phone is marketed by ZTE, the Coral-200-Solar. Although the "greening" of mobile phones is frequently premised on companies' promotion of the recyclability of their products, demand for phones with greener components is likely to increase with a growing consumer awareness of the ecological impacts of using and discarding mobile phones.

**See Also:** Audio Equipment; Carbon Offsets; Energy Efficiency of Products and Appliances; Ethically Produced Products; E-Waste; Recycling.

### Further Readings

Brower, Michael and Warren Leon. *The Consumer's Guide to Effective Environmental Choices: Practical Advice From the Union of Concerned Scientists*. New York: Three Rivers, 1999.

Garlough, Donna, et al., eds. *The Green Guide: The Complete Reference for Consuming Wisely*. New York: National Geographic, 2008.

Rogers, Elizabeth and Thomas Kostigen. *The Green Book*. New York: Three Rivers, 2007.

*Bill Kte'pi*
*Independent Scholar*

# Morality (Consumer Ethics)

Morality, with reference to consumer ethics, refers to the codes of conduct that are followed by members of a consumer society to avoid causing harm to other human beings. Moral decisions made by each consumer, either directly or indirectly, are responsible for protecting the well-being of all members of society. In a business transaction, both the buyer and the seller are expected to act according to their own economic interests; however, a mutual trust in the morality and ethical behavior of both parties is essential for establishing a long-term and successful business relationship that is beneficial to both parties of the consumer society.

Corporate scandals involving companies such as Enron, Global Crossing, WorldCom, and many others have caused concern and suspicion among consumers about business ethics and corporate social responsibilities. However, business behavior is not always independent of consumers' own immoral, unethical behavior and acceptance of dishonesty. A holistic approach of shared responsibility of business and consumers on issues such as Fair Trade and social and environmental sustainability is growing in popularity in the 21st century. Moral and ethical consumers are motivating companies to recreate their policies to be sustainable in three areas—the planet, people, and profit. Positive buying, such as purchasing energy-saving lightbulbs and household appliances, is encouraged by the consumer media, and negative buying, such as driving gas-guzzling vehicles, is positioned as less desirable to consumers.

According to past research, consumers in Western and Eastern cultures have different interpretations and tolerance for morality and ethics; that is, what is right and what is wrong. A consumer learns his moral values from the culture in which he has been brought up and behaves accordingly. In Western cultures, such as in Germany, under certain circumstances a questionable behavior may be considered "good" and an individual may avoid the consequences of such a behavior. For example, morality was not a consideration for a German company when it sent 95,000 tons of household plastic waste materials to Pyongyang in North Korea without considering safety and environmental sensitivity. In France and Germany, companies are allowed to treat bribes of foreign officials as a business expense. If a situation permits, ignoring moral values is not uncommon to Austrian consumers. In Eastern cultures, integrity, fairness, and high standards are generally the social customs; however, they are preached but not always practiced by high-ranking officials and leaders of the communities. For example, officers in Indonesia did not object to taking bribes from Wal-Mart, the giant U.S.-based retailer, and the senior minister of Singapore, Lee Kuan Yew, accepted a discount offered by a real estate developer.

There are criticisms of multinational corporations (MNCs) for exploiting natural and human resources in the countries in which they operate. Some of the MNCs, such as Wal-Mart, earn significantly more revenues than the gross domestic product of entire countries, such as Pakistan, Israel, and Romania. Often these MNCs are criticized for using their power to manipulate the governments of these countries to gain special privileges. To avoid controversy, many MNCs are adopting codes of corporate social responsibility to foster labor rights, protect the environment, and work against corruption, bribery, and other similar issues. However, whether such codes are genuine or part of corporate greenwashing is important to ascertain.

A challenge for business professionals operating internationally across markets is negotiating differences in the application of moral values in regulations and legislation of different countries and/or trading blocs.

Conflicts of interest may arise in these areas:

- *Utilitarian ethics:* It must be determined how and to what extent the constituent groups will be benefited or harmed by the action.
- *Rights of the parties:* Actions must respect the fundamental rights of the individuals involved.
- *Justice or fairness:* Actions must be fair and just to all parties involved.

The United States has led the campaign against international bribery, whereas European firms and institutions are putting effort and money into the promotion of "corporate social responsibility." A study in Europe, in cooperation with the European Institute of Business Administration, found that there is a strong link between firms' social responsibility and European investors' choices for equity investments.

Researchers are finding the link between consumer products, corporate and government policies, and the sustainability of our planet. Deterioration of our environment resulting from global warming, the failure of the financial sector in the United States and the Western world, and the subsequent slowdown of the world economy, as well as an increase in life-threatening diseases such as diabetes, cancer, and so on, are a few examples of recent phenomena that are threatening the survival of humanity in the long run. Concerned consumers are eager to find out information about the companies they do business with, their products, and their policies. Widespread Internet access, along with the general media, is greatly helping consumers to make the right choices for their purchases. For example, a list of the top-100 environmentally-friendly (green) large corporations has been created by *Newsweek* magazine. There are also many Web communities where ethical consumers share their knowledge and provide information about companies and products causing harm to the public, wildlife, and the environment.

Green consumers are increasingly able to express their moral views through their shopping purchases. They may do so by supporting companies that engage in socially and environmentally responsible production, marketing, distribution, and recycling practices, or they may avoid companies seen to engage in practices or to produce products that are detrimental to society and the environment.

**See Also:** Consumer Culture; Consumer Ethics; Consumerism; Consumer Society; Disparities in Consumption; Ecological Footprint; Environmentally Friendly; Fair Trade; Green Design; Green Marketing; Green Politics; Lifestyle, Sustainable; Materialism; Quality of Life; Simple Living; Sustainable Consumption.

### Further Readings

Arjoon, S. "Reconciling Situational Social Psychology With Virtue Ethics." *International Journal of Management Reviews*, 10/3:221–43 (2008).

Aupperle, K. "Moral Decision Making: Searching for the Highest Expected Moral Value." *International Journal of Organization Theory and Behavior*, 11/1:1–11 (2008).

Baker, J. "Virtue and Behavior." *Review of Social Economy*, 67/1:3 (2009).

Behnam, M. and A. Rasche. "Are Strategists From Mars and Ethicists From Venus?—Strategizing as Ethical Reflection." *Journal of Business Ethics*, 84/1:79–88 (2009).

Bell, C. and J. Hughes-Jones. "Power, Self-Regulation and the Moralization of Behavior." *Journal of Business Ethics*, 83/3:503–14 (2008).

Brinkmann, Johannes. "Understanding Insurance Customer Dishonesty: Outline of a Situational Approach." *Journal of Business Ethics*, 61/2:183–97 (2005).

Britz, J. "Making the Global Information Society Good: A Social Justice Perspective on the Ethical Dimensions of the Global Information Society." *Journal of the American Society for Information Science and Technology*, 59/7 (2008).

Caruana, R. "A Sociological Perspective of Consumption Morality." *Journal of Consumer Behaviour*, 6/5 (2007).

Cohn, Deborah Y. and Valerie L. Vaccaro. "A Study of Neutralisation Theory's Application to Global Consumer Ethics: P2P File-Trading of Musical Intellectual Property on the Internet." *International Journal of Internet Marketing and Advertising*, 3/1:68–88 (2006).

Derry, Robbin and Ronald M. Green. "Ethical Theory in Business Ethics: A Critical Assessment." *Journal of Business Ethics*, 8/7 (1989).

Dunfee, Thomas W., et al. "Social Contracts and Marketing Ethics." *Journal of Marketing* (July 1999).

Maguad, B. and R. Krone. "Ethics and Moral Leadership: Quality Linkages." *Total Quality Management & Business Excellence*, 20/2 (2009).

Rawwas, Mohammed Y. A. "Culture, Personality, and Morality: A Typology of International Consumers' Ethical Beliefs." *International Marketing Review*, 18/2:188–211 (2001).

Rawwas, Mohammed Y. A., et al. "Consumer Ethics: A Cross-Cultural Study of the Ethical Beliefs of Turkish and American Consumers." *Journal of Business Ethics*, 57/2:183–95 (2005).

Thogersen, John. "The Ethical Consumer. Moral Norms and Packaging Choice." *Journal of Consumer Policy*, 22/4:439–60 (1999).

Tsalikis, J., et al. "Relative Importance Measurement of the Moral Intensity Dimensions." *Journal of Business Ethics*, 80/3:613–26 (2008).

Vanberg, V. "On the Economics of Moral Preferences." *American Journal of Economics and Sociology*, 67/4 (2008).

Wagner-Tsukamoto, S. "Consumer Ethics in Japan: An Economic Reconstruction of Moral Agency of Japanese Firms—Qualitative Insights From Grocery/Retail Markets." *Journal of Business Ethics*, 84/1:29–44 (2009).

*Mousumi Roy*
*Penn State University, Worthington Scranton*

# Needs and Wants

The difference between a need and a want, and what (if any) difference this makes from a personal or a collective point of view, are issues that have been intensely discussed in many disciplines. The responses to this will depend on and reflect different conceptions of justice, development, and individual happiness. Debates about the differences between needs and wants are also at the heart of critiques of consumerism, which are framed in terms of overconsumption and misconsumption.

## Differences Between Needs and Wants: The Moral Dimension

For some commentators—notably mainstream economists and the political philosophers close to them—and in the consumer culture literature, there are no significant differences between needs and wants. They can be used more or less indifferently or, still better, not used at all, but replaced by the pair "preferences" and "utility." For economic liberalism, individuals behaving as sovereign consumers maximize their utility (their welfare, pleasure, happiness) by purchasing on the market the commodities corresponding to their preferences within the limits of their budget constraints. There is no moral perspective in economic liberalism: the individual alone is the sole authority over his desires and his ability to pay is the sole criterion for judging whether his desires are to be satisfied.

Therefore, from the market point of view, it does not matter whether I buy a car because I think I need it or because I just want it. The difference becomes relevant only if I cannot afford the car and claim that I have a right to some help from the state to get it, or if my buying a car threatens legitimate and more urgent needs of other people. In both cases, what will enter into the moral or political deliberation is the comparison between the harm I would endure if I lacked the car and the harm others would undergo otherwise. In this deliberation, what will be weighted is the importance of wants with respect to needs and of some needs with respect to others. In sum, from a collective point of view, the distinction between needs and wants is meaningful only in a moral and/or political context, in relation to rights, moral obligations, and claims on social entitlements or shares of some public resources. It follows that the difference between needs and wants is crucial for sustainable development. The authors of *Our Common Future,* the well-known report of the World Commission on Environment and Development, would certainly resist firmly the idea of replacing "needs" by "wants" in their definition of sustainable development as "development that meet the needs of the present without compromising the ability of future generations to meet their own needs." By talking of needs instead of wants or desires, they had in mind objective, urgent, and satiable motives (needs), not subjective, particular, and insatiable ones (wants).

Contrary to wants, needs are objective because they can be assessed by external and impartial observers. Medical doctors and psychologists can, in principle, diagnose unfulfilled physical (food, water, sleep, clothes, shelter) or psychological (autonomy, recognition, self-esteem) needs even in people unaware of their needy situation, on the basis of specific symptoms generally associated with a deficit in some need satisfaction. Thus, a need can be ascribed to individuals even in the absence of any expression or articulation of it (the anorexic's need for food, the desk-bound person's need for exercise), and there can be unwanted needs, as there are unneeded wants. Some needs are also objectively ascribed to individuals by the social, economic, and cultural norms and values of their society and by the necessity to have them satisfied to become and stay a fully participating member of this society. For example, depending on one's job or other circumstances of life, a car can be a real necessity—not a luxury or a mere convenience. The need for it could be objectively assessed by an impartial observer aware of the existing conditions of membership in our society and informed on the circumstances of living of the needing person.

Contrary to wants, some needs—those characterized as basic—are universal because they are constitutive of the biological and psychological make-up of every human being. They belong to human nature. It is useful at this stage to distinguish dispositional needs from occurrent ones. A dispositional need is one the needing beings have simply by virtue of being what they are. As such, it can only be met, not eliminated. On the contrary, an occurrent need is one the needing beings have only when in a state of lack. For example, human beings dispositionally need water and food to survive, but they are in nonneedy states when they do not experience an occurrent need for water (being thirsty) or food (being hungry) and so on. It is important to remark that the universality of human needs is totally compatible with the historic, cultural, and sociological relativity of what is considered adequate "satisfiers" for them.

Contrary to wants, needs are urgent because not satisfying them is harmful for the physical or psychological health of the person. Of course, the more vital or basic a need, the more harm thwarting it is likely to lead to. More generally, basic needs are: grave, for which the harm resulting from their nonfulfillment is very bad and may be irreversible; urgent, for which the harm will ensue rapidly; entrenched, which are determined by relatively unchangeable facts of nature; or unsubstitutable or weakly substitutable.

Finally, contrary to wants, needs are satiable. This means that if a good or service can satisfy a given need (is therefore a "satisfier" for that need), there is a threshold level of consumption beyond which that good, or its characteristics, may bring no additional satisfaction to the consumer.

All these differences explain why needs have moral preeminence over wants and why we can feel committed to help satisfy the needs, but not the unneeded wants, of people—even strangers—in a needy situation. This is the main reason why sustainable development is about needs, not wants. Introducing an additional distinction between legitimate and nonlegitimate wants can, however, be justified, in accordance with the physicist and philosopher Mario Bunge. Legitimate wants are those that can be satisfied without hindering the satisfaction of any basic need or without endangering the integrity of any social and natural system. Therefore, legitimacy is dependent on the society (or group), and more precisely, on its level of prosperity. In other words, a want can be legitimate in a prosperous society and illegitimate in a less prosperous one. The history of consumption in Western societies can therefore be read as a transformation, correlative with productivity growth, of illegitimate wants into legitimate ones and then of legitimate wants into needs. What were once unneeded (luxury) wants become needed ones either because the general living conditions make them necessary (the refrigerator and the car are examples of luxuries becoming necessities because of the way they have changed systems of provision of food and of transport) or because they are progressively integrated into the definition of the minimum standard of living.

Overconsumption can thus be defined as the satisfaction of illegitimate wants. More precisely, there is overconsumption when some people do not have access to sufficient amounts (i.e., above a specified threshold or norm) of a given resource or of resources in general to satisfy their needs (underconsumption), others enjoy levels of consumption of these resources that fall above the aforementioned threshold (overconsumption as such), and there is a causal relation between the underconsumption of the former and the (over) consumption of the latter.

In others words, there is overconsumption if, and only if, it is responsible for underconsumption elsewhere (contemporaries) or later (future generations). There can be no overconsumption of nonrivalrous, nonexcludable goods such as security, air, love, information, and so on. Only resources existing in limited supply can be overconsumed. The threshold or norm is the level of need satisfaction considered adequate in the given society—a level that, of course, changes across societies and through history.

## Differences Between Needs and Wants: The Personal Well-Being Dimension

The distinction between needs and wants or between "true" and "false" needs can also be mobilized in the assessment by individuals themselves of their quality of life and of the contribution consumption of commodities really makes to it. Critics of the consumer society maintain that, far from being really sovereign, the consumers are alienated—victims of the "hidden persuasion" from commercials and, more deeply, of the consumer culture that renders them unable to see where their true interests reside, to discriminate between genuine human needs and the false needs artificially created by firms to sustain their growth and the profits of their shareholders. This process of inefficient allocation of money and time resources by the individual consumer, which is conducive to a suboptimal level of well-being, could be called misconsumption. Thus, if overconsumption is consumption criticized from an external perspective, from a nonconsumer point of view, misconsumption is consumption assessed from inside against the criterion of genuine well-being, happiness, and flourishing.

There is another objection to the myth of the sovereign consumer: the thesis of the consumer lock-in. The problem, it is argued, is not so much marketing or advertising but the socioeconomic functioning of society, which requires higher and higher levels of income and consumption just to meet fundamental economic needs. Jerome Segal, for instance, denies that the high level of consumption in the first world means that essential needs of population are really met. On the contrary, it could just be that the costs in terms of commercial goods and services of reaching the same level of need satisfaction are steadily increasing, leading to a kind of "red queen" situation (being obliged to run faster and faster to stay at the same place), rather than to a hedonic treadmill.

At any rate, there is greater and greater statistical indication that, beyond some threshold of material needs satisfaction, additional income and consumption bring less and less additional well-being. This is confirmed by a host of statistical analysis at the aggregate level, as well as at the individual and household levels. Positive psychology, in particular, is accumulating evidences on the importance of nonmaterial needs for well-being. For instance, the influential self-determination theory in psychology posits three basic psychological needs (autonomy, competence, and relatedness), the fulfillment of which is deemed necessary for well-being. These basic needs are of the same nature as biological needs and are necessary for the growth, wellness or health, and integrity of the person. This means that a deprivation in autonomy, competence, or relatedness is associated with objective harm as "measurably degraded forms of growth and impaired integrity." These needs are to be considered "nutriments" for psychological optimal functioning in the same way that food, water, and so on are nutriments for optimal physical functioning.

In contrast, they are objective—not subjective—needs. Even if people are not aware of it, their insufficient satisfaction is conducive to impairments in growth, mental health, and well-being. It has also been observed that people react to thwarting of these psychological needs by depressive responses or by trying to adjust through substitute and compensatory activities. For instance, self-determination theory interprets strong materialism and excessive concern with social status as resulting from psychological needs deficits and insecurity resulting from insufficient nurturing in childhood or adult conditions of living. Tim Kasser, in particular, has explored in depth the "high price of materialism." He showed that people whose values center on the accumulation of wealth or material possessions face a greater risk for unhappiness, anxiety, depression, low self-esteem, and problems of intimacy. He then concluded that materialistic values actually undermine well-being and that four sets of needs are basic to the motivation, functioning, and well-being of all humans: safety, security, and sustenance; competence, efficacy, and self-esteem; connectedness; and autonomy and authenticity. These recent findings corroborate the intuitions of the most influential theorist of needs, the American psychologist Abraham Maslow. According to Maslow, needs are organized in a hierarchy that looks like a five-floor pyramid.

At the base live the basic physiological needs, such as food, drink, sleep, and sex. They must be fulfilled to allow the second level—needs for safety, security, health, and comfort satisfied by healthcare, money, and economic security—to express themselves. The third level is occupied by belongingness needs: love, affection, and acceptance, and satisfaction from being a member of a family, a community, a society. Continuing up in the hierarchy are needs for self-esteem, prestige, and status. Finally, on the top of the pyramid, Maslow put the need for self-actualization and the moral needs (truth, justice, perfection, aesthetic, meaningfulness). Humans strive to reach the fifth level, but they can succeed only after having climbed successfully all preceding stages. He called the first four levels "deficiency needs" because deprivation in these areas produces illness.

Manfred Max-Neef, a Chilean economist and environmentalist, winner of the Right Livelihood Award in 1983, objected to Maslow's assumption of hierarchy of needs. He worked out a matrix of human needs and satisfiers intended to help grassroots communities assess their well-being and define their own development path. The matrix consists of 10 human fundamental needs (subsistence, protection, affection, understanding, participation, leisure, creation, identity, and freedom) crossing four axiological categories of satisfiers (having, being, doing, interaction). The participative exercise aims at filling the 40 cells of the matrix with the satisfiers felt as lacking by the community. The method helps people identify multiple poverties, and in particular, the immaterial poverties resulting from unmet needs for affection, understanding, identity, freedom, leisure, and creation. Except for subsistence, which is a clearly a prerequisite for all others, no need has any fixed and permanent supremacy over the others. On the contrary, they are interactive and interrelated, and the process of needs satisfaction is characterized by simultaneities, complementarities, and trade-offs.

## Conclusion

As some have argued, drawing a sharp distinction between needs and wants can lead to committing the sin of paternalism. The assumption that needs are different from expressed preferences or wants may open the Pandora's box of the manipulation by experts of individual wants and of the repression, in the name of their true interests, of the expression of preferences different from officially legitimate wants. Historically, it has indeed led some political regimes to exercise a "dictatorship" over needs. However, with the notable exception of the consumer culture discourses, almost all cultures and many current scientific approaches and ethical theories maintain that the distinction is essential. It is also at the heart of the sustainable development discourse and program. The best way to avoid the risk of paternalism or of technocratic domination is therefore to make the definition and articulation of needs the outcome of a public, democratic deliberation.

**See Also:** Consumer Culture; Consumer Society; Overconsumption; Poverty; Quality of Life; Sustainable Consumption.

### Further Readings

Bunge, Mario. *Treatise on Basic Philosophy, Volume 8: The Good and the Right.* Dordrecht, Netherlands: Reidel Publishing, 1987.

Doyal, Len and Ian Gough. *A Theory of Human Needs.* Basingstoke, UK: Macmillan, 1991.

Gasper, Des. *The Ethics of Development.* Edinburgh, UK: Edinburgh University Press, 2004.

Kasser, Tim. *The High Price of Materialism.* Cambridge, MA: MIT Press, 2002.

Maslow, Abraham. *Motivation and Personality.* New York: Harper and Row, 1954.

Max-Neef, Manfred. *Human Scale Development: Conception, Application and Further Reflection.* London: Apex, 1991.

Reader, Soran. *Needs and Moral Necessity.* London: Routledge, 2007.

Segal, Jerome M. "Consumer Expenditures and the Growth of Needed-Required Income." In *Ethics of Consumption. The Good Life, Justice and Global Stewardship*, David A. Crocker and Toby Linden, eds. Lanham, MD: Rowman and Littlefield, 1998.

*Paul-Marie Boulanger*
*Institut pour un Développement Durable*

# ORGANIC

The organic movement is mainly associated with agricultural produce, including crop production and the rearing of livestock, with a primary focus on the health benefits of consuming food products and ingredients that have been produced without the aid of artificial chemical fertilizers, pesticides, antibiotics, genetic modification, and growth hormones. In this way, organic methods embody sustainable practice in that they are human-, wildlife-, and planet-friendly and aim to foster the natural production of food by allowing the supporting ecosystems to function without interference or manipulation. Born out of a reaction against intensive farming methods, the organic sector has consistently gained market share in recent years, with strong annual increases in the global sales of organic food and drink. Notable is the appearance of organic products and ingredients in other sectors such as organic pet food, but also particularly in nonfood sectors such as fashion, with organic cotton and leather, as well as health and beauty products. Despite having strong sustainability credentials, the organic sector has come under increasing pressure to maintain growth as a result of increased competition from sustainable alternatives, and some producers have been forced out of the market. The higher prices commanded by organic products have become an obstacle for expansion during the global recession, and recent scientific research has put the health and nutritional benefits of organic in doubt.

The organic movement is mainly associated with agricultural produce and focuses on the health benefits of eating foods that were produced without chemical fertilizers, pesticides, or genetic modification.

*Source:* iStockphoto

The term *organic* first appeared in the 1940s, with various observers describing the farm as a living entity that must be based on a balanced organic life. Though still a small sector in comparison with conventional farming, the shift toward organic production has grown steadily since the end of World War II and has gained ground as the detrimental effects of intensive technology-driven farming methods have become more apparent. Controversial issues; for example, the widespread use of aggressive pesticides such as dichlorodiphenyltrichloroethane (DDT) and the use of artificial petrochemical-based fertilizers that contribute to natural resource shortages, pollution, and global warming, as well as animal health scandals such as mad cow disease, have all served to promote the central ethos of the organic movement that natural is best.

As the environment and the quality of food have become more of a public concern, the sustainable credentials of organic have contributed to sector growth. National and international organizations such as the U.K. Soil Association, the Organic Trade Association, and the International Federation of Organic Agriculture Movement have established standards for organic practice to guarantee quality and sustainable practice for the public and provide certification and labeling to allow for the easy identification of organic products. Gaining certification for a farm and its produce is a rigorous and lengthy process and is subject to inspections to ensure standards are maintained. Fruits and vegetables were the earliest products to get mainstream distribution; dairy products followed, with some brands becoming quickly established. The fashion industry has quickly learned from the food sector, and organic cotton helps that sector to demonstrate ethical credentials and show more transparency in their supply chain. The cosmetics and beauty sector have also turned to organic products to compete in ethically motivated markets.

Organic standards cover a range of agricultural aspects, including avoiding the use of pesticides and additives; soil fertility is maintained through crop rotations and the planting of crops such as clover to restore nitrogen levels to soil naturally, without the use of artificial fertilizers. The maintenance of larger hedgerows and wild growing spaces around cultivated crops allows for wildlife diversity to increase and helps in the control of pests that may affect crops by providing shelter and breeding grounds for their natural predators. Animal welfare is also central to organic farming, including all facets of rearing, shelter, feeding, transportation, and slaughter. Animals have access to the outdoors, are given organic-approved feed, and are kept in smaller numbers to reduce stress. Animals are not given genetically modified products or growth hormones. Antibiotics are only given to animals when a vet decides it is necessary, and if so, a period of time must pass before their meat and other products can be sold, which is longer than for nonorganic livestock.

The higher price placed on organic goods has often been criticized for restricting the market for organic foodstuffs to those with more disposable income, but the pricing strategy has been justified by organic associations because of the need to maintain standards of care and production that require greater investment than conventional farming. As a result, much organic produce remains outside mainstream consumption, with less than 15 percent of the population accounting for two-thirds of all organic purchases in some markets.

The organic market faces some key challenges. Five consecutive years of growth in the organic food sector until 2008 has been stalled by the recent global economic recession. The premium price demanded by much of the organic food sector has led recession-hit consumers to trade down to cheaper alternatives. In addition, the organic food market has quickly become crowded with a number of alternatives, each with their own sustainability features. Labeled, branded, and certified foodstuffs now compete against the organic certification, such as weekly or monthly farmers markets, welfare food, freedom food, and

Fair Trade, as well as locally sourced produce, often sold in local shops rather than in large supermarkets. Animal welfare, the environment, food miles, and food carbon footprints are now of growing concern to consumers. Research suggests that local sourcing is now the most sought-after ethical food type, with purchase intention among consumers belonging to that segment ahead of both the Fair Trade and organic categories.

In 2009, the Food Standards Agency in the United Kingdom published research based on the results of 50 years of research into the health and nutritional benefits attributed to organic produce and found there were no significant benefits from organic sources compared with conventional farming systems. As the consistent positioning of organic food has been on health benefits, rather than on other aspects of sustainability, the health-related position of organic has now become a weakness and added to the problems of the sector. As organic sales have dropped, farmers have been unable to meet the extra costs involved in organic food production, and some have turned back to cheaper conventional methods of livestock management and cultivation and dropped their organic certification. In December 2008, the Soil Association, together with other organic certification bodies, requested an organic holiday from some of the stricter organic rules, such as the use of organic feed for animals, which costs twice the price of normal feed. This change would enable farmers to get through the difficult economic times but would mean they would lose their organic label during the holiday period; this may also confuse consumers as to the credibility and value of organic certification. In spite of the market difficulties, the organic sector is still in the growth stage, with the potential to develop in the future by repositioning and competing on the other sustainable issues at its disposal and by entering new markets.

**See Also:** Certified Products; Dairy Products; Fair Trade; Green Food; Markets (Organic/Farmers); Meat; Vegetables and Fruits.

### Further Readings

Guido, Gianluigi. *Behind Ethical Consumption: Purchasing Motives and Marketing Strategies for Organic Food Products, Non-GMOs, Bio-Fuels*. New York: Peter Lang, 2009.

Wiswall, Richard. *The Organic Farmer's Business Handbook: A Complete Guide to Managing Finances, Crops, and Staff—And Making a Profit*. White River Junction, VT: Chelsea Green, 2009.

Wright, Simon and Diane McCrea. *The Handbook of Organic and Fairtrade Food Marketing*. Oxford: Wiley-Blackwell, 2007.

*Barry Emery*
*University of Northampton*

## Organisation for Economic Co-operation and Development (OECD)

The notions of sustainable production and consumption patterns can easily be traced within international policy texts back to the Earth Summit (United Nations Conference on

Environment and Development, Rio 1992), where Principle 8 of the Declaration states that "To achieve sustainable development and a higher quality of life for all people, States should reduce and eliminate unsustainable patterns of production and consumption and promote appropriate demographic policies," Also at the Rio Conference, Agenda 21 titled its chapter 4 "Changing Consumption Patterns." The concept remained current at the influential United Nations' level 10 years later at the World Summit on Sustainable Development (Johannesburg 2002), where "Changing Unsustainable Patterns of Consumption and Production" became the title of chapter 3 in the Plan of Implementation. The same document outlines the launch of a "10-Year Framework of Programmes" (§14), known as the Marrakech process, which is still ongoing. This article considers the main features by which these initiatives are interpreted and promoted by the OECD.

The OECD is a forum in which the governments of 30 highly industrialized countries work together to address the economic, social, and environmental challenges of globalization. This organization has been working actively on sustainable consumption issues since 1994 as part of its sustainable development efforts, emphasizing the achievement of long-term economic growth that is consistent with environmental and social needs. Sustainable consumption policies take into account the ecological impacts of the products through a holistic view of their life cycle (from cradle to grave). They also consider, to a lesser extent, certain social and ethical dimensions of product production and use.

The OECD member countries are Australia, Austria, Belgium, Canada, the Czech Republic, Denmark, Finland, France, Germany, Greece, Hungary, Iceland, Ireland, Italy, Japan, Korea, Luxembourg, Mexico, the Netherlands, New Zealand, Norway, Poland, Portugal, the Slovak Republic, Spain, Sweden, Switzerland, Turkey, the United Kingdom, and the United States. The Commission of the European Communities takes part in the work of the OECD.

The OECD facilitates comparisons between the policies and practices of its members and promotes guidelines that address country-specific challenges. The organization categorizes its sustainable consumption policy instruments as follows:

- *Standards and Mandatory Labels:* These tools are the most direct policy instruments for eliminating unsustainable products from the market. The most common sustainability-related performance standards are those aimed at reducing energy use; for example, promoting energy efficiency in household appliances, effectiveness of home insulation, and fuel economy in motor vehicles. These standards have become stricter and more widespread as climate change concerns intensify.
- *Taxes and Charges:* By raising prices on less sustainable products, taxes and charges can be effective in shifting consumer behavior toward sustainability. These tools help internalize negative externalities, which in theory helps the market play the critical role of changing purchasing patterns. Taxes on motor fuels are applied in all OECD countries and often form the bulk of environmental tax revenues. In Europe, taxes on motor fuels are 40–60 percent of the sales price compared with 20–25 percent in the United States. The European car fleet is more energy efficient, with up to two to three times lower unit emissions of carbon dioxide from transport than the United States, suggesting that energy taxes have incentivized more efficient vehicle use. However, according to the OECD, too often taxes and charges designed to promote sustainable consumption are not set at a sufficiently high level, and they may be more effective as part of a wider tax reform strategy. For example, countries such as Austria, Finland, Germany, and Sweden have tax-shifting programs to introduce more environmental taxes in place of those on capital and labor—a policy also known as "double dividend" (reducing both pollution and labor costs).

- *Subsidies and Incentives:* These include monetary grants or fiscal incentives in the form of tax reductions for purchasing less-polluting products (house insulation, energy-efficient cars, etc.). However, altogether the amount of these sustainable consumption incentives is small when compared with environmental or social subsidies directed to production sectors.
- *Communications Campaigns:* In this domain, public authorities face tough competition with the private sector for consumer attention. More recent communications campaigns are sophisticated in their focus on single issues, advice on practical actions, and use of multimedia.
- *Education:* UNESCO has designated 2005–14 as the "Decade of Education for Sustainable Development," and many OECD countries are now teaching or developing curricula on sustainable development, which generally include material on promoting sustainable consumption.
- *Voluntary Labeling:* The most viable labels are those for which environmental or social claims are verified by a third party, including governments and nongovernmental organizations. However, the weaknesses of labels include low levels of consumer awareness, criteria differences across products, or a confusing competition between various labeling schemes.
- *Corporate Reporting:* In most OECD countries, sustainability reporting is voluntary, and approaches vary by country, sector, and company. In general, these reports emphasize direct production impacts, and not many examine how the company's product range supports sustainable consumption in terms of life cycle and social impacts.
- *Advertising:* Whereas at one time the sole function of advertising was to make people buy more, advertising is now responding to new demands from consumers looking for greater significance, transparency, and ethics. With encouragement from governments, the advertising industry is starting to self-regulate in various countries with regard to sustainability claims.
- *Public Procurement:* The average share of public procurement in gross domestic product in OECD countries is about 11 percent, reaching 16 percent in the countries of the European Union. Although most OECD countries have adopted green procurement practices that emphasize the environmental characteristics of products and services, these vary in extent and coverage.

Within these categories, there is considerable variation and application in different countries and contexts. However, one can indeed find common initiatives orientation within OECD countries. For example, the European Union's "Renewed Strategy for Sustainable Development" (2006) identifies its key challenge number 3: "Sustainable consumption and production." This includes decoupling economic growth from environmental degradation, improving the environmental and social performance for products and processes, aiming to achieve by 2010 a European Union average level of green public procurement equal to that currently achieved by the best-performing member states, and increasing the European global market share in the field of environmental technologies and eco-innovations.

Governments have to consider which instruments or combinations of instruments are likely to achieve the desired consumption changes within groups of consumers for particular products and services. However, the OECD advocates specifically for sustainable consumption programs that promote coherence and realize synergies across a range of policies. It claims that without coherent approaches to sustainable consumption in terms of sectors (food, energy), actors (households, women, youth), and instruments (regulations, taxes, communications), initiatives may have inconsistencies or significant gaps, producing generally ineffective results. Moreover, in designing effective sustainable consumption policies, general consumer behavior (awareness, rationality), as well as attitudinal variables, should be taken into account. In this respect, the OECD and its member countries are conducting studies examining inter alia the influence of income, age, gender, and perception on sustainable consumption across several sectors.

Will these policies achieve their long-term goals? In the eyes of the organization itself, the nature and size of the problem in OECD countries makes achieving sustainable consumption appear daunting, even without considering the still-greater implications of a global community consuming in the style and on the scale of OECD countries. Analysis shows that environmental impacts from household activities have worsened over the last decades, and they are expected to intensify over the next decade—particularly in the areas of energy, transport, and waste—if strong and comprehensive policies are not implemented.

**See Also:** Ecolabeling; Environmentalism; Government Policy and Practice (Local and National); Regulation; Sustainable Consumption.

### Further Readings

Council of the European Union. "Review of the EU Sustainable Development Strategy: Renewed Strategy." 10117/06, June 9, 2006.

European Environment Agency. "Using the Market for Cost-Effective Environmental Policy (2006)." http://www.eea.eu (Accessed July 2009).

Organisation for Economic Co-operation and Development (OECD). "Promoting Sustainable Consumption: Good Practices in OECD Countries" (2008). http://www.oecd.org/dataoecd/1/59/40317373.pdf (Accessed July 2009).

Organisation for Economic Co-operation and Development (OECD). "Towards Sustainable Household Consumption? Trends and Policies in OECD Countries." Paris: OECD, 2002.

United Nations Environment Programme. "Planning for Change: Guidelines for National Programmes on Sustainable Consumption and Production." UNEP/UNDESA (2007), 10-Year Framework of Programmes on Sustainable Consumption and Production Patterns. http://www.un.org/esa/sustdev/sdissues/consumption/Marrakech/conprod10Yhtm (Accessed July 2009).

Zaccai, E., ed. *Sustainable Consumption, Ecology and Fair Trade*. London: Routledge, 2008.

*Edwin Zaccai*
*Université Libre de Bruxelles*

## Overconsumption

Whether one can speak of overconsumption and what constitutes notions of excess depends on whether consumption is assessed from the consumer's point of view. The answers depend on whether consumption is assessed from the consumer's point of view (the "self-regarding" perspective) or from another point of view (the "others-regarding" position). The self-regarding perspective can be stated as follows: "Is my consumption behavior really good for me?" whereas the "others-regarding" perspective can be stated as "Does my consumption behavior harm someone?" Following Thomas Princen, a negative answer to the first question would better be called "misconsumption," reserving the term *overconsumption* for patterns of consumption directly or indirectly incompatible with other people's welfare. However, both those defects matter for sustainable development: misconsumption because it raises questions about the effectiveness of a development model, and overconsumption because it calls into question its sustainability. The distinction is relevant because if objections to consumption from a consumer point of view are certainly relevant and have to be taken in consideration in an overall assessment of a development model, they are insufficient to legitimate pressing demands for more responsible behavior from the consumer ("Live simply so that others may simply live") or public interventions for fostering such behavior, which should be based not on—necessarily relative—conceptions of the good life but on ethical reasons, that is, on arguments about justice and fairness.

Paul Ehrlich and John Holdren's equation $I = PAT$ helps understand what can be wrong with overconsumption and why. It expresses the environmental impact ($I$) of any human collectivity as the product of three factors: $P$ (for population, the size of that collectivity), $T$ (for technology), and $A$ (for affluence, or the level of consumption; i.e., consumption per capita). To take an example, the greenhouse gas emissions ($I$) of a given transportation mode (car, train) in a given country during a given time interval can be analyzed as the product of the population ($P$) of that country times the average miles traveled per capita during the time interval ($A$) times the greenhouse gas emission per traveled mile ($T$). According to the $I = PAT$ framework, there is overconsumption of some good or service if its environmental impact exceeds some limit or threshold, $I^*$, beyond which unwanted deleterious consequences begin to show up that are likely to affect negatively other people's (including future generations') life chances. $I^*$ can denote the carrying capacity of an entire ecosystem or the maximum sustainable yield for a single renewable resource. Relevant indicators of overconsumption defined in this way are, for example, the ecological footprint, sustainability gaps, or material intensity of consumption, expressed in per capita units.

As such, the diagnostic that $I > I^*$ says nothing about the cause of the environmental problem: It can be any one of the three identified factors or some combination of them. Considering that the $T$ factor (technology) depends more on production patterns than on consumption ones, the scope of the overconsumption notion could be restricted to the two other factors: $P$ (population) and $A$ (affluence). Controlling for technology, the environmental overload of consumption depends on these two parameters only: the size of the population of consumers and/or the average consumption per capita. The two factors are in inverse relation: the more consumers, the less consumption per capita allowed, and the fewer consumers, the more each of them can consume. There is, therefore, a fundamental tension between two objectives of social and economic development: extending the size of the population of prospective consumers (be it simply by population growth or by increasing the size of the group of people entitled to some consumption) and increasing per capita consumption in a given, stable population of consumers.

The verdict of overconsumption implies that some limit has been exceeded—that a threshold has been crossed in the use of some valued thing or system, which threatens its reproduction or adequate functioning, necessary for human well-being. In addition to the ecosystems composing the natural environment, other systems and institutions contribute to human well-being. It is therefore logically possible and legitimate to extend the notion of overconsumption to those systems as well, using the same criterion as for the natural one. Indeed, there exists also an economic concept of overconsumption linked to the distinction between consumption and investment. As is well known, not all of what is given by nature or produced by man during one time interval (say one year) is consumed. A part of it might (and ought to) be saved and invested to guarantee future consumption. Therefore, when an economist asks "Are we consuming too much?" what he/she has in mind is "Are we saving enough to maintain our productive capacity and guarantee our own future welfare and that of our successors?" In other words, does consumption today jeopardize consumption tomorrow? Saving and investing consist of maintaining or improving the productive assets indispensable for generating future income and consumption. The "weak sustainability" standpoint assumes that four kinds of productive assets, between which substitutions are always possible, are used in well-being production. They are classified as natural (Earth's biophysical resources), produced (techno-economic capital), human (workforce skills, talents, know-how), and social (social norms and values that foster cooperation, trust, etc.). The World Bank has worked out a measure of savings that more or less takes into account all these kinds of assets; that is, not only produced (economic) capital, as usual, but also human and natural capital. It is called "genuine saving." On the contrary, the "strong sustainability" perspective denies that significant substitutions between man-made and natural resources are always possible and maintains that the ultimate limit is set by natural resources. It follows that overconsumption should be principally evaluated against its environmental impacts.

Hitherto the focus was on aggregate consumption. However, responsible and ethical consumption is in need of individual norms of consumption and of a criterion to assess overconsumption at an individual scale, not just at the aggregate scale. A sensible norm of overconsumption at the individual level could be the following: there is overconsumption of a given good or service if the consumed quantity of that good or service exceeds the quantity objectively sufficient for adequately satisfying the individual's needs, or the generalization of that consumption pattern to all prospective consumers would overshoot the sustainability ($I^*$) threshold.

The microperspective brings forth an additional threshold into the discussion: the adequate, sufficient level of needs satisfaction. From a sustainable development standpoint, a consumption level corresponding to the objective needs of an individual cannot be illegitimate and considered overconsumption. Nor can it be called misconsumption. Only consumption in excess of objective needs can be characterized as such.

In sum, from a sustainability perspective, there is overconsumption when

1. some people do not currently have—or will probably not have in the future—access to sufficient amounts (i.e., above a specified threshold or norm) of a given resource or of resources in general (underconsumption),
2. others currently enjoy levels of consumption of these resources above the aforementioned threshold (overconsumption as such),
3. there is a causal relationship between the deprivation of the former and the (over) consumption of the latter.

Misconsumption corresponds to a situation in which condition 2 holds, but condition 3 does not. It follows there can be no overconsumption of nonrivalrous, nonexcludable goods such as security, air, love, information, and so on. Only resources existing in limited supply can be overconsumed. Nor is there overconsumption if the consumption behavior has no effect on someone else, as was stated earlier. Likewise, only collective or public resources can be overconsumed. In the absence of adverse effects, legitimately privately owned resources can only be misconsumed, not overconsumed.

**See Also:** Consumer Behavior; Durability; Ecological Footprint; Final Consumption; Needs and Wants; Sustainable Consumption.

## Further Readings

Crocker, David A. and Toby Linden, eds. *Ethics of Consumption. The Good Life, Justice and Global Stewardship*. Lanham, MD: Rowman and Littlefield, 1998.

Ehrlich, Paul R. and John P. Holdren. "Impact of Population Growth." *Science*, 171/3977:1212–17 (March 26, 1971).

Princen, Thomas. "Consumption and Its Externalities." In *Confronting Consumption*, Thomas Princen, et al., eds. Cambridge, MA: MIT Press, 2002.

Zaccaï, E., ed. *Sustainable Consumption, Ecology and Fair Trade*. London: Routledge, 2007.

*Paul-Marie Boulanger*
*Institut pour un Développement Durable*

# Packaging and Product Containers

Packaging and product containers create a great deal of the landfill generated in the United States. Packaging and product containers also contribute significantly to the cost of finished goods sold commercially. As a result, many consumers, producers, and activists search for green alternatives that will make packaging and product containers more environmentally friendly while also reducing the amount spent on these items. The market for green packaging and product containers is estimated to be as much as $45 billion per year by 2013. Sustainable packaging and product containers focus on three related criteria: functionality, cost-effectiveness, and support for long-term human and ecological health. Advances for sustainable packaging and product containers are driven by an increased awareness of environmental issues, coupled with the elevated costs of raw materials and fuel that have occurred in the past several years. This confluence of interest in green packaging and product containers thus stems from both consumers and producers, making it a highly popular, prevalent, and powerful market force.

Making packaging and product containers more green entails the development and use of casings that result in improved sustainability. Sustainable packaging and product containers are those that exist when manufacturers and consumers consider economic, social, and environmental concerns and then take actions to ensure that present needs are met in a way that does not compromise those of future generations. Although each of these terms, of course, is used in a variety of contexts, within the framework of sustainability, special considerations help guide the assessment of each element when considering packaging and product containers.

Classical economic theory, for example, considers the costs and availability of land, labor, and financial capital when making decisions about packaging and product containers. In terms of sustainability, natural capital must be considered as well, which includes those natural resources and ecosystem services that maintain and nurture society as a whole. Natural capital includes such resources as hydrocarbons, minerals, water, and trees, as well as more intangible assets such as cultures and human intelligence. Those interested in natural capital consider the economic costs of packaging and product containers not just in terms of the resources consumed but also in terms of the ecological footprint left behind. Proponents of sustainable packaging and product containers are therefore concerned with the conservation of resources through more effective manufacturing, the reuse of materials found in natural systems, a change in focus from quantity to quality, and the restoration and sustainability of natural resources.

## Green Packaging and Social Considerations

Social considerations also guide those who work to develop green packaging and product containers. Centered on defining responsible global citizenship, social considerations involve matters of individual and local lifestyle choices, national and international law, matters of transport and urban planning, and other aspects of ethical consumerism. Decisions related to packaging and product containers also focus on the social relationships such choices affect—associations that are often inherently dysfunctional and unjust. Green packaging and product container decisions as a result concentrate on the well-being of all life on Earth and on preserving the diversity and celebrating the richness of these life-forms. As a result, decisions regarding packaging and product containers consider workers' human needs as well as the quality of life such choices make possible or imperil. Developed nations must adopt more sustainable behaviors that curtail their patterns of production and consumption while also respecting the aspirations of those who live in developing regions.

Environmental concerns also shape determinations made regarding packaging and product containers. Sustainable packaging and product containers are those that

- reduce the use of underground metals, fossil fuels, and minerals;
- decrease dependence on synthetic chemicals and other unnatural substances;
- lessen encroachment on nature, during both manufacturing and disposal; and
- respect human needs efficiently and fairly in terms of pay, working conditions, and with regard to a safe and healthy environment

Because the Earth has limited ability to absorb human waste and supply natural resources—a capacity that has already been strained—packaging and product containers must be changed to become more sustainable. Improvement of environmental concerns will occur only when economic, ideological, and technical structures are changed in a way that will support an improved quality of life, rather than increasing the standard of living.

## Green Packaging Materials and Markets

Relatively simple changes in the materials used in packaging and product containers can have a great effect on sustainability. The plastics that are commonly used in packaging and product containers, for example, take centuries to decompose. Using biodegradable polyesters in place of these plastics reduces the decomposition cycle to mere weeks. European and Asian markets have used polyesters, which are synthetic resins, for years because their rapid breakdown after disposal makes them ideal for packaging and product containers. North America has recently begun to acknowledge biodegradable polyester's potential in packaging, as shown by the 30 percent annual growth in its use. Biodegradable polyester's biggest obstacle to widespread adoption is its high cost, ranging in 2009 from $1.50 to $2 per pound.

At this level, biodegradable polyester is more expensive than alternatives such as paper or plastic. Suppliers had expected low cost of disposal, its marketability as an environmentally friendly substance, and the increased popularity of composting to increase the resin's commercial viability. The infrastructure required for sorting and composting organic waste, however, greatly impeded biodegradable polyester's marketability. In addition, its higher cost relative to other alternatives has proven a difficult obstacle to overcome in the market. Biodegradable polyesters have seen market share increase for rigid and flexible packaging. The most popular of these are highly amorphous polyesters—flexible and clear, these are often used as display casings in packaging. Semicrystalline polyesters, which are more rigid, are also popular and are used for product containers. Produced from petrochemical feedstocks in converted poly(ethylene terephthalate) (PET) polymerization plants, these resins have several unique benefits. First, these polyesters break down in approximately 12 weeks when water and microbes are present, unlike petrochemical-based polymers that take centuries to break down. Second, the polyesters decompose into carbon dioxide, thus complying with composing standards in Europe, Japan, and the United States.

The market for green polyesters focuses on paper coating and garbage bags. Biodegradable polyesters have also seen a growing market for use as thermoformed packaging, for which their superior functional features, such as moisture resistance, make them an attractive alternative to cheaper biodegradable materials. Blending biodegradable polyesters with starch, organic wastes, and natural-fiber reinforcements has also been a means of mitigating the resin's high costs. Ecoflex copolyesters have also seen their market share grow.

Ecoflex, a film that is clear, strong, and clings well, is a great substitute for vinyl in food wraps for vegetables, meats, and fruits. Other polyesters, which have a strong resistance to moisture and grease, are suitable for lawn and garden bags, paper coatings, and netting. Other resins have been used for food trays, disposable cutlery, traffic cones, and hairbrush handles. Synthetic biodegradable polyesters have also proven highly useful for food packaging. In addition to being used for disposable plates, cups, and bowls, synthetic biodegradable polyesters are used for thermoformed trays for fruits, vegetables, and meat sold in supermarkets. Although three times more expensive than nongreen alternatives, their sustainable appeal is augmented by other qualities that make them attractive to merchants and consumers. Using synthetic biodegradable polyester thermoformed trays, for example, improves the shelf life for meat to six to nine days—an increase of 50 percent. As a result, synthetic biodegradable polyester thermoformed trays are replacing expanded polystyrene trays in many supermarkets.

## Recycling

Recycled content packaging also provides a fertile ground for growth in sustainable packaging and product containers. Recycled plastic packaging, for example, is an area that has shown healthy growth in recent years. Plastic recycled content packaging needs several components to become a fully viable alternative. First, efforts to boost collection volume of used plastics are necessary, as many localities do not provide convenient or practicable means for end users to drop off used plastic items for recycling. Second, an increased focus on the development of food-contact-approved resin grades is required, as certain regulations impede the use of recycled plastic packaging. Finally, further sustainability initiatives by plastic processors and manufacturers of food and other products are necessary to build awareness of more recycled plastic packaging's benefits and potential success. Recycled glass packaging, because it benefits from an established system of recycling used materials and does not require food-contact regulation, is poised to continue increasing at an above-average rate. Industry and governmental initiatives to boost recycling rates and to include an increasingly higher percentage of recycled glass in packaging and product containers have proven highly successful to date.

Reusable packaging and product containers continue to grow as a market force as well. Historically, many beer and soda bottles were reusable, with their return encouraged by a deposit collected at the point of sale. Until the 1930s, most beverages were sold in glass bottles, which were returned when the product was consumed so that the bottle could be reused. Beginning in the 1930s, however, disposable steel beverage cans began to displace bottles, and by the 1960s, half of all beer sold was in cans, whereas 95 percent of soda pop remained in glass bottles. In an effort to reduce pollution caused by this shift, in 1971 the state of Oregon passed what became known as the Oregon Bottle Bill, which required all cans, bottles, and other beverage containers to be sold with a minimum refund value. The Oregon Bottle Bill is estimated to have encouraged a return rate of over 90 percent, thereby reducing the amount of roadside litter and increasing container recycling. Other states, including Vermont (1972), Maine (1976), Michigan (1976), Connecticut (1978), Iowa (1978), Massachusetts (1982), New York (1982), Delaware (1982), California (1986), and Hawaii (2002), have passed similar bills, although none have been as successful as Oregon in achieving such a high return rate (the others average 28 percent).

Advocates of green packaging and product containers have recently received support from some unanticipated allies. In 2006, Wal-Mart Stores, Inc., launched an initiative whereby it charged its 60,000 worldwide suppliers with reducing the amount of packaging used in products Wal-Mart carries. Specifically, Wal-Mart wanted the amount of packaging reduced to the lowest levels possible, which would save shelf space and shipping costs, as well as reduce the materials that ultimately require disposal. Early estimates of the annual savings engendered by the Wal-Mart plan are estimated to include 323,000 tons of coal and 67 million gallons of diesel fuel, as well as a reduction in carbon dioxide emissions by 677,000 tons. Wal-Mart continues to work with suppliers to design smaller, or renewable, packaging and product containers for its stores. Other retailers, such as Home Depot, Inc., have worked to make recycling drop-off easier for customers, providing a convenient place for this necessary step to occur. Retailers have also encouraged detergent manufacturers to supply more concentrated versions of their products, thereby reducing the packaging necessary to sell the merchandise and reducing transportation expenses greatly. Smaller product containers have benefits beyond the reduction in materials and transport costs—retailers can accommodate more products on the shelves, the shelves need to be restocked less often, and employee injuries resulting from the stocking process are reduced.

## The Future of Green Packaging

Appealing to consumers' green sensibilities or offering a product with improved performance over its conventional counterpart, increases sales of sustainable packaging and product containers. For greater acceptance of green packaging and product containers to occur, however, greater use of governmental regulation and economic incentives seems necessary. These regulations and incentives would cover several stages that affect the process of packaging and product container manufacture, including raw material extraction and processing; manufacturing; purchase, use, and disposal; and waste management. With regard to raw material extraction and processing, policies supporting sustainable packaging and product containers could include regulating mining, petroleum, and nonhazardous solid wastes pursuant to the U.S. Resource Conservation and Recovery Act, whereas financial incentives favoring the extraction of virgin materials could be eliminated and the production of virgin materials taxed.

Similar regulations could support the manufacture of these products, including the tightening of regulations under the Clean Air Act, the Clean Water Act, and the U.S. Resource Conservation and Recovery Act, mandating the use of recycled content in packaging and product containers, direct manufacturer and retailer return and recycling of products, and regulation of product composition, while establishing tradable recycling credits and taxing the use of virgin materials. Purchase, use, and disposal of green packaging and product containers would be facilitated by mandating consumer separation of materials for recycling and establishing weight/volume-based disposal fees, taxing hazardous products, and establishing deposit-refund systems for these items. Waste management policy steps might include tightening regulations of waste management facilities, banning the disposal of certain products in landfills, mandating recycling diversion rates for various materials, and banning construction of new landfills and incinerators while taxing emissions of effluents from waste management facilities and establishing surcharges on wastes delivered to landfills or incinerators.

**See Also:** Green Design; Lifestyle, Sustainable; Recyclable Products; Recycling; Sustainable Consumption.

### Further Readings

Fullerton, D. and W. Wu. "Policies for Green Design." *Journal of Environmental Economics and Management,* 36:131–48 (1998).

Green, K., et al. "Green Purchasing and Supply Policies: Do They Improve Companies' Environmental Performance?" *Supply Change Management: An International Journal,* 3/2:89–95 (1998).

James, S. and T. Lahti. *The Natural Step for Communities: How Cities and Towns Can Change to Sustainable Practices*. Gabriola Island, British Columbia, Canada: New Society, 2004.

United Nations General Assembly. "Millennium Declaration of the United Nations." Resolution 2, Session 55. New York: United Nations, 2000.

*Stephen T. Schroth*
*Jason A. Helfer*
*Michael Tam*
*Knox College*

## Paper Products

Paper was invented by the Chinese, perhaps as early as the Han dynasty, and its uses have been growing ever since. It is a very versatile and renewable product that is usually made from wood. The modern paper industry consumes large volumes of trees that are cut for the purpose of manufacturing paper. The largest timber companies practice sustainable forestry operations that allow them to plant, grow, and then harvest trees on tree farms for the sole purpose of producing wood pulp for paper mills. They also partner with

The making of paper is an energy-intensive operation that requires large quantities of water. The industry consumes large volumes of trees, piles of which are seen here, in front of a paper mill.

*Source:* iStockphoto

private landowners who plant trees, especially fast-growing pine trees, which when grown are cut for pulp. In the United States, large portions of the national forests are also logged for timber, a large portion of which becomes pulp.

The making of paper is an energy-intensive operation. It also requires large quantities of water. Although most paper is made from wood pulp, it can also be made from rags or grasses. Mechanized paper production began in the 19th century. It is now a major industry because of the huge demand for paper as writing or printing material and for the capacity it has to be used in many other products.

Paper as a writing material is universally available today. Offices around the world use it. Places where it's used also range from a child's book bag or a college student's desk to the largest commercial or government offices in the world. Paper is used for handwritten or printed materials, which are produced by computer printers, copiers, or the printing industry, including the publishing industry. Green paper produced from sustainable forests and paper products made from recycled paper are becoming more widely available. In many countries, there are schemes and programs to promote the reuse and recycling of paper arising from business and household consumption.

Writing paper is manufactured in large quantities as envelopes, ruled notebook paper, and copier paper and is used in publishing by printers. Huge volumes of paper are consumed for writing messages, letters, or reports among other productions. The development of computers was supposed to create "paperless" offices. In reality, this has not happened because electronic storage has been found to be imperfect. Systems "crash" and other problems occur, so a paper backup is a safe strategy for preserving valuable data. In addition, many people prefer to use a printed copy of a long work such as a book, rather than read it on a computer screen.

Grades of paper can be very inexpensive or high-quality bond paper with cotton content. In addition, money is often printed on very-high-quality paper with silk as part of its content materials. Large quantities of paper are used to make currency, postage stamps, bonds, and other securities. Some books are printed on high-quality and expensive paper that preserves photographs and text for centuries. In contrast, huge numbers of cheap entertainment books—so-called pulp fiction—are printed on cheaper paper. These books often will disintegrate in 50 or fewer years as the paper is exposed to oxidation from the atmosphere.

Paper is used in a variety of products for domestic use including toilet paper, tissues, napkins, and kitchen towels. Many of these paper products have been bleached; however, unbleached, dye-free, and chlorine-free ecobrands are available for domestic consumers. In the case of rubbish/trash bags, storage containers, and disposable plates, paper products can provide green alternatives to plastic receptacles. Soft paper material is generally biodegradable, but it requires processing in the sewage systems of the world. In some countries, the energy and water consumption needed to process toilet paper in the sewage mix exceeds the system's capacity, so it is deposited into a trash container for disposal rather than flushed down the toilet. This is a green issue for some, but not for most, people today.

Paper products in kitchens and dining rooms include the use of paper napkins, paper towels, paper lunch bags, and other products. Meals can also be served on disposable paper plates that are green products because they can be recycled in gardens or compost piles. Paper towels are commonly used in kitchens, and some companies have introduced "half" sheets that can be used in smaller quantities for numerous jobs.

Supplies for homes and kitchens that come from grocery stores display a huge number of uses for paper: labels for products, price stickers, printed receipts, boxes for many products such as cereal or popcorn, or bags for everything from flour and sugar to cookies. Coffee in paper bags is usually shelved next to paper coffee filters in paper boxes. Bulk foods such as bird seed, dog food, and cat food come in paper bags or boxes.

Paper is also used in the manufacture of a large variety of boxes. Huge rolls of unbleached brown paper are used to make corrugated cardboard. The manufacturing process sends sheets of paper into the rollers of machines that use heat, steam, and pressure to create the precisely semifolded center of the cardboard sandwich, which is glued to top and bottom sheets to form a stiff packaging material. Some boxes are waxed or given chemical applications to make them stronger or softer, such as egg crates. Other types of paper materials are made into boxes to package any number of products from shoes to crayons to furniture. Paper is also used in decorations for homes and offices. Wallpaper, paintings, and other decorative and insulating uses are made of paper.

Given the ubiquity and importance of paper products in everyday life in contemporary societies, recycling, reuse, and sustainable production of paper become significant areas through which shifts toward greener consumption and production might be achieved.

**See Also:** Environmentally Friendly; Magazines; Packaging and Product Containers; Recyclable Products; Shopping Bags.

### Further Readings

Fix, Alexandra. *Paper: Reduce, Reuse, Recycle*. London: Heinemann, 2007.

Kostigen, Thomas M. and Elizabeth Rogers. *The Green Book: The Everyday Guide to Saving the Planet One Simple Step at a Time*. New York: Crown, 2007.

Langley, Andrew. *Paper Products*. New York: Crabtree, 2008.

Smith, Maureen. *The U.S. Paper Industry and Sustainable Production: An Argument for Restructuring*. Boston, MA: MIT Press, 1997.

*Andrew Jackson Waskey*
*Dalton State College*

## Personal Products

Recently, more people have become aware that many of the traditional personal hygiene products (including toothpaste, deodorant, soap, shampoo, conditioner, shaving cream, aftershave, facial cleanser, and feminine care products) can include a number of ingredients that are both harmful to humans and hazardous to the environment. As a consequence, there is a growing consumer market for "green" (sometimes referred to as "natural" or "organic") personal hygiene products that avoid substances known to cause harmful health effects and that minimize negative impacts on the environment. Because of this increased awareness and demand, many companies are trying to market green products. However, there is currently no universal standard for certification for these products, and consumers can easily become confused or misled about what they are purchasing.

Components of green personal hygiene products can include those that

- minimize use of harmful ingredients such as mercury; toluene; petroleum distillates; parabens; sodium lauryl sulfate; diethanolamine; 1,4-dioxane; phthalates; and triclosan;
- minimize harm to the environment when acquiring the essential ingredients or during the manufacturing process;
- are not composed of animal by-products or ingredients tested on animals, and whose final products are not tested on animals;
- use packaging that is made from recycled material and/or is biodegradable or recyclable; and
- are composed of ingredients derived from sources that practice Fair Trade and provide good working conditions.

Many green personal hygiene products are made with naturally occurring minerals from plants or other organic and natural sources instead of with synthetic chemicals. For example, in place of artificial dyes such as Blue #1 or Yellow #5, vegetable juices from red cabbage, purple carrot, or annatto can be used. Fragrances from ingredients such as green tea, rosemary, and grapefruit can be used instead of synthetic chemicals, and the petroleum-based emulsifiers and surfactants in toothpastes and mouthwash can be replaced with plant-based minerals from olives and coconuts.

In response to the increased demand for green products, many personal hygiene product companies are extolling corporate values and promoting their merchandise in ways they feel will appeal to this particular market segment. There are a number of companies that strongly claim to be making products that are natural, safe for consumers, and good for the environment. For example, Terressentials, based in Maryland, is a small company in which all products are packaged by hand. Terressentials has been certified by the U.S. Department of Agriculture (USDA) and prints the USDA Organic seal on each of its products. USDA Organic certification is defined as being reserved for products containing 95 percent to 100 percent USDA organic ingredients verified by a third-party certifier. Burt's Bees and Tom's of Maine sell many of their products in natural and organic groceries around the United States. Burt's Bees offers products using natural ingredients including beeswax, botanical and essential oils, herbs, flowers, and minerals. Each natural product carries a nature seal that is regulated by the Natural Products Association Natural Standard for Personal Care Products. These products must be at least 95 percent natural, safe, and sustainable. Tom's of Maine has created its own set of natural, sustainable, and responsible standards that it claims ensure their products are produced in a socially and environmentally conscious manner. This company is certified by the American Dental Association, and it has produced the first natural toothpaste that bears the American Dental Association's Seal of Acceptance. Their products have also been given the Kosher certification, Halal certification, and the Cruelty-Free certification through People for the Ethical Treatment of Animals. The Body Shop, a large British retail chain that can be found in shopping malls across America, offers over 700 natural personal care products ranging from body care to cosmetics. Its stated corporate values include no animal testing, protecting the planet, and defending human rights; however, its products are not certified as natural or organic. Seventh Generation makes chlorine-free feminine care products, including organic cotton tampons. It donates 10 percent of its profits to nonprofit community, environmental, health, and responsible business organizations working for positive change.

## Standards for Green Certification

Certification of organic personal hygiene products is a problematic area in the quest for green personal hygiene products. At this time, there is no global standard for certification. The USDA, through its National Organic Program (NOP) regulation, does regulate the term *organic* as it applies to agricultural products. If a personal hygiene product contains or is made from agricultural ingredients, and meets the USDA/NOP organic production, handling, processing, and labeling standards, it may be eligible to be certified USDA Organic under the NOP regulations. Certified personal hygiene products can be designated as one of three different categories of organic labeling: "100 percent organic" (this product may display the USDA Organic Seal), "organic" (containing at least 95 percent organically produced ingredients), and "made with organic ingredients" (containing at least 70 percent organic ingredients). Products made with less than 70 percent organic ingredients cannot use the term *organic* on the display panel. However, personal hygiene products not made of agricultural ingredients and those that make no claims to meeting USDA standards do not come under the authority of the USDA. Given the current situation of regulation—or lack thereof—consumers can find many products in retail outlets with "organic" labeling, but without any clear standards or regulatory underpinning for making that organic claim. Groups such as the Organic Consumers Association are lobbying the USDA to enforce organic standards in personal care products that are mislabeled as organic. These groups also encourage consumers to boycott brands that mislabel their products as "organic."

## Private and Other Labeling Standards

More recently, personal hygiene products can opt to be certified under other, private standards that are not presently regulated by the USDA, including foreign organic standards, ecolabels, and Earth-friendly labels. One example of this is the Natural Products Association Standard and Certification for Personal Care Products, which requires products to be made of 95 percent natural ingredients from natural sources. Another standard known as NSF/ANSI 305: Personal Products Containing Organic Ingredients was developed to define production and labeling requirements for American personal care products containing organic ingredients. This label indicates that the product contains at least 70 percent organic ingredients. At this time, only a small percentage of the personal care products in the United States are certified according to these newly emerging standards, and consumers tend to trust brand names rather than to look for a specific certification seal. Furthermore, discrepancies exist over which ingredients and/or what quantity of these ingredients the different standards should allow.

In Europe, the Seventh Amendment to the EU Cosmetics Directive governs safety and labeling standards for personal care products sold in the European Union. New regulations within this amendment were passed in March 2009 that currently restrict, and completely ban in 2013, animal testing on personal care products. The regulations will also affect personal hygiene products imported into EU countries that may cause the United States to adopt these standards as well.

Organizations such as Organic Consumers Association, Planet Green, and Tree Hugger educate consumers on the potential health risks of chemicals in personal hygiene products and provide consumers with examples of products that are considered safer for humans and/or better for the environment. Examples of promoted green personal care products include a "Razor Saver" that claims to sharpen used razor heads and a "Mooncup" for use during menstruation. The Mooncup is designed to cut down on the number of sanitary napkins and tampons going into landfills and to reduce the risk of health issues associated with using pads and tampons, which can contain chlorine, bleach, and carcinogenic absorbency gels, and potentially can cause toxic shock syndrome. These organizations also provide people with do-it-yourself instructions on how to make natural personal hygiene products at home.

In addition to the lack of uniform standards, there is also some concern that the green consumerism occurring in the market for personal hygiene and household products may actually be to the detriment of generating social and political action for real change. Consumers might feel that by purchasing products labeled "organic" or "natural" they are helping solve the problem of environmental degradation or protecting themselves from health hazards. However, many of these products are poorly regulated, if they are regulated at all. Furthermore, green consumerism may not address one of the fundamental causes of environmental degradation: too much consumption. Rather than trying to reduce consumption, many of these companies are encouraging responsible consumption.

Although progress has been made in providing personal hygiene products that are safe and produced in a socially conscious manner, confusion over standards for certification of these products indicates that there is a need for uniformity and industry cohesion in this process. In the meantime, consumers need to question the validity of organic labeling on their personal hygiene products.

**See Also:** Certified Products (Fair Trade or Organic); Consumer Ethics; Cosmetics; Green Consumer; Organic.

### Further Readings

Gabriel, Julie. *The Green Beauty Guide: Your Essential Resource to Organic and Natural Skin Care, Hair Care, Makeup, and Fragrances*. Deerfield Beach, FL: Health Communications, Inc., 2008.

Todd, Anne Marie. "The Aesthetic Turn in Green Marketing: Environmental Consumer Ethics of Natural Personal Care Products." *Ethics and the Environment*, 9/2 (Fall/Winter 2004).

Tourles, Stephanie. *Organic Body Care Recipes*. North Adams, MA: Storey Publishing, 2007.

U.S. Department of Agriculture, Agricultural Marketing Service. "National Organic Program: Cosmetics, Body Care Products, and Personal Care Products." http://www.ams.usda.gov/AMSv1.0/getfile?dDocName=STELPRDC5068442&acct=nopgeninfo (Accessed September 2009).

*Vanessa Slinger-Friedman*
*Kennesaw State University*

# Pesticides and Fertilizers

Pesticides are chemical agents frequently used in agriculture and industry to kill unwanted organisms such as insects, weeds, and fungi. Fertilizers are materials applied to agricultural soil to enhance its ability to produce plentiful, healthy plants, and although many are phosphate- or petroleum-based, some consist of organics, such as animal and human "biosolids" (a promotional term for toxic sludge). Both pesticides and fertilizers have been increasingly used by consumers in residential and garden applications, more commonly exposing urban and suburban dwellers to toxins that were once found mostly in rural and industrial environments. For the home gardener there is considerable information available on practices and products that enable one to garden organically, including the use of natural fertilizers, composts, and forms of pest eradication.

Heavy pesticide use over the past 50 years has led to toxic surface runoff that has been documented to have disrupted watersheds and ecosystems.

*Source:* Stephen Ausmus/Agricultural Research Service/ USDA

Significant and increasing environmental damage has occurred since farmers have relied heavily on chemical pesticides and fertilizers—mostly over the past 50 years—and this damage has expanded to threaten the health of consumers. Pesticides not only harm the environment by releasing excess chemicals but also are often highly toxic to humans, pets, and wildlife. Heavy overfertilization of the soil in an effort to ensure maximum crop yields has led to dangerously toxic surface runoff that has been documented to have disrupted watersheds and ecosystems.

## Pesticides

The two major types of pesticides used in agricultural farming are chemical pesticides and biopesticides. Chemical pesticides are manufactured synthetically from source chemicals. Organophosphate pesticides are mostly insecticides and affect the nervous system by disrupting cholinesterase, the enzyme that transfers nerve impulses across synapses. Carbamate pesticides act similarly to organophosphates by also affecting the nervous system and disrupting an enzyme that regulates acetylcholine, a neurotransmitter. Pyrethroid pesticides were developed synthetically from a naturally occurring pesticide found in chrysanthemums that has been modified to increase its stability in the environment. Organochlorine insecticides, such as dichlorodiphenyltrichloroethane, are still used around the world, but were removed from the U.S. market because of their dangerous effects on animal reproductive systems that intensify up the food chain. Recent research has also shown that organochlorine pesticides suppress nitrogen-fixing bacteria from replenishing natural nitrogen fertilizer in the soil, which results in lower crop yields and stunted growth.

Biopesticides—derived from animals, plants, bacteria, and certain minerals—have advantages over conventional pesticides in that they are less harmful, are targeted to specific pests, are effective in small quantities, and decompose quickly. Microbial biopesticides are microorganisms such as bacteria, fungi, viruses, or protozoa that serve as the active ingredient, targeted to specific pests. Plant-incorporated-protectants are genetically modified pesticidal substances produced from plants that have had a specific gene inserted. For example, a protein produced from a bacterium that acts as a natural pesticide is inserted into the plant's genes so that the plant may produce its own pesticide. Biochemical pesticides are nontoxic, naturally occurring substances used to control pests and are often made from substances such as insect pheromones that interfere with mating and also include scented plant extracts to attract pests to traps.

Chemical pesticides account for environmental damage and harm humans and animals in the process. Some pesticides do not actually reach target pests, and many insects have grown resistant. Pesticides also contribute to honeybee die-offs, developmental abnormalities in amphibians, and poor immune function in dolphins, seals, and whales. In humans, pesticides increase the risk of cancer, particularly in farmworkers, and may cause endocrine and reproductive dysfunction. When dispersed into the environment, pesticides spread via spray drift, surface and groundwater contamination, and rainfall dispersal.

Spray drift is the physical movement of a pesticide through air at the time of application, spreading to any other site other than that intended. Very small droplets sprayed by a nozzle or via aircraft are carried by air currents until they contact a surface or drop to the ground. Spray drift spreads the toxic effects of pesticides to humans and wildlife and may cause both environmental and property damage.

## Fertilizers

The three most important plant nutrients are nitrogen, phosphorus, and potassium. Because of their increased depletion of soil nutrients, agricultural fertilizers containing these substances must be applied continually. Types of fertilizers used in agriculture include the aforementioned biosolids, industrial wastes, and manure. Biosolids consist of wastewater residuals, also known as sewage sludge, that have been treated for agricultural use. Industrial waste fertilizers provide a source of zinc and other micronutrient metals and are less widely used. Manure is used as a fertilizer and may be returned to the soil as soil conditioner. The original biosolid—known as "night soil"—was human waste, intensively applied to soils by smallholder farmers in China for hundreds, if not thousands, of years.

Overfertilization has caused detrimental effects to the environment that include the accumulation of nitrogen and phosphorus, which has led to surface runoffs in watersheds and may contribute to hypoxia in bodies of water—the reduction of dissolved concentrations of oxygen, stressing and killing water-dwelling organisms. Excessive phosphorus runoff from soil erosion also contributes to eutrophication: the response by a body of water to enrichment of nutrients. Only very small concentrations of phosphorus are necessary for a body of water to eutrophy, resulting in algal blooms, heavy growth of aquatic plants, and deoxygenation.

Fertilizer production has become an environmental concern as a result of the production of phosphogypsum, a solid material that results from the chemical reaction of phosphate rock with sulfuric acid and that may contain small amounts of radium. For every ton of phosphoric acid produced, five tons of phosphogypsum are generated, along with small amounts of radium. At this time, ground-stacking of phosphogypsum is under federal regulation.

Environmental scientists say swift improvement of farming methods is necessary to maintain a sustainable and ecological farming system. To reduce phosphorus runoff, they prescribe the use of improved planting methods, better fertilizer management, and enlightened soil conservation techniques. To reduce the environmental threat, ecologists suggest the development of smaller farms, which can be run profitably using less fossil fuel, fertilizer, and pesticides; incorporating greater plant diversity; and using renewable forms of energy. Switching from conventional to organic farming can reduce reliance on agrochemicals, conserve soil and water resources, and at the same time maintain high crop yields at lower costs.

Altogether, the overuse of chemical pesticides and fertilizers is having detrimental effects on the environment, including human health and wildlife. A profusion of chemicals seep into the environment through air, water, and the food chain. Indeed, agricultural scientists agree that drastic measures are needed soon to create an agriculture that is both environmentally and economically sound. Such an outcome would benefit not only those involved in the agricultural industry but also the household consumer.

**See Also:** Gardening/Growing; Organic; Plants; Water.

### Further Readings

Adler, Tina. "Harmful Farming." *Environmental Health Perspectives,* 110/5 (2002).

Hart, Murray A., et al. "Phosphorus Runoff From Agricultural Land and Direct Fertilizer Effects: A Review." *Journal of Environmental Quality,* 33 (2004).

Jasinski, Stephen M. "Fertilizers-Sustaining Global Food Supplies." U.S. Department of the Interior, U.S. Geological Survey (1999).

Pimentel, David, et al. "Environmental, Energetic, and Economic Comparisons of Organic and Conventional Farming Systems." *Bioscience,* 55/7 (2005).

Potera, Carol. "Pesticides Disrupt Nitrogen Fixation." *Environmental Health Perspectives,* 115/12 (2007).

Pretty, Jules. *Pesticide Detox: Towards a More Sustainable Agriculture.* Toronto: Earthscan Canada, 2005.

U.S. Environmental Protection Agency. "Biopesticides." http://www.epa.gov/agriculture/ (Accessed March 2009).

U.S. Environmental Protection Agency. "Nutrient Management and Fertilizer." http://www.epa.gov/agriculture (Accessed March 2009).

U.S. Environmental Protection Agency. "Types of Pesticides." http://www.epa.gov/pesticides (Accessed March 2009).

*Anthony R. S. Chiaviello*
*Laura Joanne Robles*
*University of Houston–Downtown*

## Pets

Pets play a powerful emotional role in human's lives. They share dwelling spaces, provide a sense of attachment, and can ease loneliness. Research suggests people who share their

Spending on veterinary services is roughly $10 billion per year; another $10 billion is spent on related supplies.

*Source:* iStockphoto

lives with pets may live longer. The most common household pets are dogs, cats, birds, hamsters, rabbits, mice, gerbils, rats, and guinea pigs. About 63 percent of U.S. households and 43 percent of British households have at least one pet. Pet keeping requires financial means and has a significant effect on the economy and the environment. Americans share their lives with more than 75 million dogs and 90 million cats and spend about $500 per year on each. The $41 billion a year Americans spend on their pets is more than the gross domestic product of all but 64 countries. It is a figure double what it was 10 years ago. Many dogs now go to day-care, stay in hotels instead of kennels, get braces on their teeth, and wear jewelry. In many ways, the lives of domestic animals have improved significantly in terms of diet, medical care, and attention. In industrialized countries, increased amounts of leisure time and money are spent on animals considered to be members of our families.

## What Goes In

One of the greatest areas of environmental impact comes from feeding pets. The first dry dog food was made available in the 1860s, and its promotion mirrors what we find today—free samples, sponsorships of animal shows, and advertising. This was the forerunner of the giant pet food industry, which was soon joined by marketing of other items such as collars, leashes, crates, furniture, dishes, and toys. Purchasing food specifically for pets, versus feeding them leftovers of human diets, ebbed and flowed with economic circumstances, but by the 1950s, keeping a pet became a mainstay of the family. In 1959, the Pet Food Institute was created, which today represents 97 percent of pet food manufacturers. The pet food industry, which often mimics trends in human food production and composition, has become a specialty industry, with products marketed as "all natural"; that is, free of chemicals and fragrances. Since the 1970s, the pet-related industry has grown exponentially, reflecting increased economic prosperity, the changing role of pets in people's lives, increased longevity of animals, and the changing psychological nature of the human–animal bond. As the United States became an increasingly consumer-oriented society, driven by needs and desires and purchasing power, practices once associated with the elite such as pet keeping became democratized, and all the items deemed necessary to have a happy and healthy pet also grew. This "humanization" of dogs and cats in particular has resulted in increased spending, as well as time and attention paid to the nutritional needs of companion animals.

According to the American Pet Products Manufacturers Association, Americans spent $26.5 billion on nonfood pet supplies and services in 2008. They spent $15 billion on food—kibble, biscuits, and treats. The companies that sell pet food spent nearly $300 million advertising it. A recall during spring 2008 of 60 million packages of pet food contaminated with melamine from China brought to light the complexities of pet food—ingredients, manufacture, distribution, and quality. Consumers are increasingly aware of inhumane conditions experienced by the animals they eat; these are the same sources for what goes into pet foods. Given that cats rely almost entirely on protein and that a considerable part of the canine diet is meat-based as well, even the highest-quality pet food comes from high-volume, factory-farmed conditions. For example, nearly three tons of fish go into the making of cat food annually. Experimentation on animals to gauge the effects and qualities of pet food is a behind-the-scenes source of controversy. Companies such as Purina and Procter & Gamble hire outside researchers to conduct experiments on live animals to gauge the physiological impact and waste production of their foods. An additional area of impact is the cost of packaging, shipping, storing, and distributing those sacks and cans of pet food, much in nonrecyclable plastic pouches and metal cans.

## What Goes On

Pet care is the fastest-growing retail category. In the United States, specialty stores such as PetSmart are warehouse-sized destinations for the purchase of everything from collars to kibble. Entrepreneurial businesses have appeared such as dog-walking services, companies that remove pet waste from yards, and pet bathing and grooming boutiques. Spending on veterinary services is roughly $10 billion per year, and another $10 billion is spent on related supplies. Pharmaceutical companies have created specialty lines of blood pressure, diabetes, obesity, and antidepressant drugs for pets. Pet health insurance is a growing category of growth—$195 million in premiums were paid in 2007.

## What Goes Out

Probably the greatest environmental impact of pets is what they leave behind. In the United States alone, the 73 million pet dogs produce roughly 10 tons of solid waste. According to a U.K. study, dogs eat 765,000 tons of food annually, which becomes billions of liters of urine and 365,000 tons of excrement. The 90 million cats that bury their waste in litter produce roughly two million tons of litter. The clay (bentonite) in kitty litter, a by-product of strip mining, produces dust that adds to airborne pollution. In addition, it is questionable whether this form of cat litter degrades in landfills.

Globally, the love of pets has resulted in an overpopulation crisis. Puppy and kitty mills are cruel places in which the animals are kept in small cages, producing litter after litter. The effect of pet ownership on the quality of life for animals is significant. The odds of people keeping an animal they acquire are as poor as one out of four. Millions of animals are destroyed every year because they are undesirable or unwanted. Approximately six to eight million unwanted cats and dogs are taken to animal shelters every year, half of whom are euthanized. Many are the results of people not spaying or neutering their pets. Many animals are abandoned every year and/or released into the world, but because they are unable to care for themselves, starve or pass on diseases to other animals.

Is it possible to keep a "green pet"? Yes. A variety of methods exist for the green disposal of pet waste, the feeding of humane and healthy diets, and grooming and veterinary care methods that are environmentally conscious, and there are toys made from recycled materials. More natural forms of cat litter are becoming popular, made from plants and trees and free of artificial fragrances, unnatural chemicals, and silica dust. Adopting a pet from a humane society is another way to help lessen the carbon paw print.

**See Also:** Consumerism; Meat; Waste Disposal.

### Further Readings

Adamson, Eve. *Pets Gone Green: Live a More Eco-Concious Life With Your Pets*. Irvine, CA: BowTie Press, 2009.

Brady, Diane and Christopher Palmeri. "The Pet Economy." *BusinessWeek* (August 6, 2007).

Fudge, Erica. *Pets*. Stocksfield, UK: Acumen, 2008.

Grier, Katherine C. *Pets in America: A History*. New York: Harvest, 2007.

*Debra Merskin*
*University of Oregon*

## Pharmaceuticals

Pharmaceuticals, in their current allopathic version of synthetic chemicals, have become a major greening issue. This is a result of increasing environmental hazards and adverse effects on human health resulting from the chemical ingredients. The environmental factor, or E-Factor, devised by Roger Sheldon in 1992, is a calculation derived from dividing the total waste produced in the process of production (in kilograms) by the amount of product created, thus providing a ratio that allows one to interpret the efficiency of production processes. The higher the ratio or E-Factor, the more waste is generated. According to Sheldon's analysis, pharmaceutical companies produce more waste in manufacturing drugs than oil companies as a percentage of the material processed (though much less in real-volume terms). Fifteen years after developing his E-Factor, Sheldon argues that awareness of high E-ratios in the pharmaceutical industry has increased the sustainability of the industry, providing a boost to research, funding, and marketing of safe pharmaceuticals. This has increased the market efficiency and effectiveness of green efforts.

Increasingly, soil and water contamination from pharmaceuticals have become a critical concern among scientists, consumers, manufacturers, and policy makers. Low concentrations of pharmaceutically active compounds have been detected in surface water and groundwater around the world. These include medications for human and veterinary use (as in hormones and antibiotics). Human pharmaceuticals enter the environment through routes like human excretion sent to wastewater treatment plants, which may not be designed to remove pharmaceuticals. Because most human excreta are found in metabolized form, the toxic effects of the various internally formed metabolites need to be treated with special sewage designs. Hence, depending on the compound, existent waste treatment plants may or may not be effective. The Advanced Oxidation Process is a new wastewater management scheme that is being used in Canada. It helps to remove contaminants from water, including pharmaceuticals, personal care products, endocrine-disrupting compounds, and industrial contaminants.

The environment is also affected by various other unintended modes of human activities. These include runoffs from sewage biosolids (sludge) applied to land as a fertilizer, the loss of dermally applied medications during bathing, irrigation with reclaimed wastewater or sewage effluents, release from leaky sewers, and so on. Equally critical in environmental contamination is the improper disposal of unused and expired pharmaceuticals both from manufacturing and household use.

There is a growing concern about the detrimental effects of the interactions of various types of pharmaceutical residues on aquatic species—both vertebrates and invertebrates. Because many drugs are designed to affect specific protein targets or receptors in humans, they can affect similar functions in other species that carry that receptor. For example, synthetic estrogen in human contraceptives has led to the feminization of male fish in the streams in the United Kingdom. The ecotoxicological effect of drugs can be seen among land and avian species, too. The use of analgesic diclofenac fed to livestock has led to a significant decline in the vulture population in India and Pakistan.

Engagement by multiple levels of government and multiple stakeholders holds much promise for addressing the management of pharmaceuticals in the environment. For instance, on the law enforcement front, the United States and the United Kingdom have taken the lead in passing guidelines for environmental risk assessment, such as toxicity tests for the approval of pharmaceuticals. Local governments have also taken proactive measures. In Canada, provincial waste management regulations require pharmaceutical brand owners to fund and organize pharmaceutical return programs. The County Council in Stockholm, Sweden, has undertaken a major initiative to classify pharmaceuticals according to their environmental impacts based on factors like toxicity, bioaccumulation, and persistence. This is expected to enable awareness-building labeling of medications.

The pharmaceutical companies have also taken initiatives. For example, in Spain and other countries, pharmaceutical return programs to collect unused and expired medications have been established. Some manufacturers who allow for "reverse distribution" (which allows pharmacists to return unsold drugs back to the manufacturer) could extend their program to include unused and expired medication. This solution would also involve the companies' undertaking initiatives to educate physicians.

Consumers and environmentally conscious organizations have also resorted to active management initiatives. For example, the Teleosis Institute launched the Green Pharmacy Program in California in 2008. It offers free and environmentally safe disposal of unused and expired medications (from families and hospices). It documents the returned medications through the Unused & Expired Medicine Registry—a program developed by the Community Medical Foundation for Patient Safety in Texas. The data collected help elucidate critical factors in the estimation of the environmental impact of pharmaceuticals, such as those pharmaceuticals that are most often overprescribed, unused, and disposed of. Organizations such as Save the Bay in California also seek to regulate the pharmaceutical effect on oceanic waters.

## Pharmaceuticals and Environmental Safety

Given the growing consumer demand for issues of safety, ecological think tanks such as the Science and Environmental Health Network promote a policy known as the "precautionary principle." This means that for new drugs (and other chemicals) to get approval, they should first be proven safe for the environment. The organization Health Care Without Harm also seeks to promote ethical standards for hospitals and healthcare companies. These include developing alternatives to incineration for the two million tons of medical waste that hospitals generate each year.

Given this increasing level of stakeholder interest, there has been a recent surge in green and clean pharmaceutical production with funding from venture capitalism. Greener firms are viewed as safer investments because of the reduced likelihood of litigation, government penalties, and/or catastrophic accidents. The financial markets reward this lower level of risk by charging the firm less for its capital, thus allowing the firm to carry more debt and invest more in innovative enterprises. This has helped increase the cost-efficiency of this emerging sector. This financial benefit has been boosted by other forms of incentives for innovative research. The U.S. Environmental Protection Agency's challenge award is one example among a number of initiatives, award programs, and incentives to promote and advance innovations in green chemistry in academia and industry. Many pharmaceutical companies have also developed their own internal award programs.

Going green involves the adoption of more environmentally safe approaches to the synthesis of active pharmaceutical ingredients (APIs) to increase efficiency and reduce costs. These approaches include adopting key principles of green chemistry, such as solvent reduction and replacement; refining a chemical route; and biocatalysts that relate to sustainable processing. Sustainable processing incorporates convergency (higher process efficiency with fewer operations), reagent optimization (use of catalysis and more selective and recyclable reagents), and raw material efficiency.

For example, the elimination of unhealthy chlorinated solvents is a good activity enabled by the development of rationalized bio-catalytic processes. Such processes are generally carried out in aqueous solvent because enzymes are of biological origin. This helps to avoid the usage of environmentally harmful chemicals and solvent waste disposition. The bio-catalysts can also be retained and reused in many cycles. It has been found that the enzyme-catalyzed chemical bioconversions are more acceptable by the pharmaceutical industries as a practical alternative to chemical synthesis methods. This is because of the complex and intractable synthetic problems involved with chemical synthesis methods. In comparison with traditional chemical analysis, bio-catalysis helps to modify the entire synthetic route by generating new intermediates and reducing the number of reaction steps. It can also help to minimize energy input and waste by helping to recycle an undesirable enantiomer.

## Green Processing

Some companies such as DSM, which produce semisynthetic penicillin, start with an already-green process. This includes fermentation of a microorganism that yields an aqueous slurry. This slurry contains both the producing biomass and the required precursor for the product. To obtain the precursor in a purified form from the broth, nongreen processes such as the use of chemicals need to be applied. This is where further greening and efficiency is required.

Green processes can be done in more generic facilities than chemo-catalysis. This helps to reduce energy consumption and carbon footprint by running chemical reactions at ambient temperatures and atmospheric pressures. This helps to avoid the use of more extreme conditions and to reduce cycle time. Thus, problems of rearrangement and increased atom or mass economy are minimized, which increases yield of the final product by ensuring that it contains the maximum proportion of starting materials and reduces raw material consumption and production of waste. It also helps to increase purity, to reduce toxicity of finished products, and to increase the flow ability or compressibility factor of the tablet or capsule version of the product. There is also less dust formation, which reduces the risks of sensitization for workers handling the product. Most critically, it helps to create products with better tastes and smells.

GlaxoSmithKline has developed an "Eco-Design Toolkit" that provides chemists and engineers with green chemistry information and tools for process research, development, and manufacturing. This has significantly helped in industry performance and has helped avoid the use of hazardous materials. However, there are complaints that less-regulated API markets (those outside the United States and Europe) in which generic products are more available, are more amenable to the improved and quality green products. The biocatalysts' ability to make a product or an intermediate without any side products eliminates purification steps, saving time, energy, cost, and waste. This increases its value for generic drug producers.

The strictly regulated API markets have complex regulatory hurdles that require stricter rules of registering the bio-equivalence of final dosage. This acts as a disincentive to innovation because of the higher costs and time-consuming efforts of registration. The general recommendation of industry analysts is that regulatory authorities should take a holistic perspective and try to achieve a balance between patient safety and market easement. This could include fast-tracking innovations developed by companies with a proven track record of quality and compliance.

Auxiliary companies that serve the pharmaceutical industry have also helped to increase its green efforts. For example, the need to reduce shelf space at Wal-Mart has prodded Insight Pharmaceuticals to repackage its Anacin medicine. From cartons, it changed its packaging to cost-effective extended-text label design—an "Easy Tab" label from WS Packaging Groups' Multivision line. Eliminating the carton kept an enormous quantity of chipboard out of landfills. Similarly, Newell's Rubbermaid provides innovative solutions to waste management and safe working conditions, which helps to reduce trips to the landfill through more packaging efficiency and lower litigation costs.

## Using Natural Sources

There is also a rising demand among consumers for natural sources of medicine, as botanical extracts have fewer adverse effects. Hence, the conservation of biodiversity spots and indigenous knowledge has become critical. For example, the Madagascar rainforest has an abundance of rosy periwinkle that provides for vincristine, used for treating leukemia. The immunosuppressant drug cyclosporin is extracted from a fungus. This has dramatically transformed the outlook for people having an organ transplant. There are thousands of bacteria and fungi living in the soil that secrete antibiotic compounds, such as streptomycin and cephalosporin. This helps deal with the growing problem of the emergence of drug-resistant bacterial strains. Algae that constitutes roughly half of the Earth's photosynthetic activity is a high potential source of biofuels and specialty chemicals, as well as a source for monoclonal antibodies, blood proteins, and enzymes.

Plant-made pharmaceuticals use cost-effective cutting-edge biotechnology, as chemical synthesis cannot make complex molecules such as proteins and peptides. For instance, Medicago uses genetically modified alfalfa and produces proteins and monoclonal antibodies. These are the same as human antibodies and help fight infections such as antibiotic-resistant golden staph (*Staphylococcus aureus*), which plagues hospitals. Given such progress, even vaccine production has become an exciting possibility. However, there are regulatory hurdles for such small-scale bio-pharming. This is because environmental groups are concerned about errant pharmaceutical genes finding their way into food crops like alfalfa, which normally is grown for livestock feed.

Homeopathy is perhaps the most ecologically sustainable alternative form of medicine. The process of making homeopathic remedies involves diluting and shaking the remedies multiple times. This means that a tiny amount of the original substance can be transformed into enough medicine for the whole of the planet. Homeopathy predates vaccines and works along the same principles: What can kill an individual when delivered in substantial amounts can heal the patient if given in diluted amounts. It is an ideal way of using nearly extinct herbs from rainforests and is also an effective alternative to the mainstream petrochemical-based conventional drugs.

However, efficiency can run counter to conservation even in natural therapy sectors. When the multinational pharmaceutical companies get involved, they can devastate an ecosystem (such as those in the Amazon or in China) in the quest for profits. Holistic herbalists usually understand more about harvesting herbs sustainably so that they do not become extinct. They always leave enough to regenerate growth for the following year. Hence, green movements like the United Plant Savers promote a way for saving the lifestyle and livelihoods of local herbalists, while seeking benefit-sharing alliances with large multinationals. Overall, green pharmaceuticals and the associated movements are paving the way for a better and more secure life for the planet and its habitats.

**See Also:** Ethically Produced Products; Healthcare; Personal Products; Quality of Life.

### Further Readings

"Caribbean Big Business Profiles: Pharmaceuticals and Health Care: Newell Rubbermaid Helps Pharmaceuticals Maximize Resources, Reduce Costs." *Caribbean Business* (November 27, 2008).

Datin, J. and K. Kemmerer. "VC Opportunities Still Abound." *Financial Executive* (September 2008).

De Braal, H. "Sustainability in Green Pharmaceutical Production." *Pharmaceutical Technology Europe* (2009).

Doerr-MacEwen, N. A. and M. E. Haight. "Expert Stakeholders' Views on the Management of Human Pharmaceuticals in the Environment." *Environmental Management* (2006).

McClinton, L. "Fields of Green." *Canadian Business*, 78/7 (2005).

Sheldon, Roger A. "The E Factor: Fifteen Years On." *Green Chemistry*, 9/12:1273–83 (2007).

Van Arnum, P. "Going Green in Pharmaceuticals." *Pharmaceutical Technology* (February 2009).

*Kausiki Mukhopadhyay*
*University of Denver*

## Plants

The planet Earth is dominated by plants, and the green plant is fundamental to all other life. Charles Lewis explored human ties to the green world and concluded that people and plants are entwined by threads that reach back to our very beginnings as a species. Plants have a wide spectrum of uses, supplying our food requirements either directly or indirectly as feed for animals. Crop production—the management of useful plants—is the very basis of our civilization. Plants are used as a source of construction materials and as the raw materials in the manufacture of fabrics and paper. Plants are the basis of many of the complex substances used such as dyes, tannins, waxes, resins, flavorings, medicines, and drugs. Plants, in addition to having a direct effect on our ecological position, are used to control erosion by water and wind, provide a setting for recreation and sports, and as landscape materials that satisfy our desire for beauty. This article focuses on the use of plants in contemporary American gardens and the current awareness of using plants that provide a more sustainable approach to the environment.

An estimated 36 million U.S. households, including the White House, now participate in food gardening. Here, First Lady Michelle Obama works with children from a local elementary school to break ground for a White House kitchen garden.

*Source:* Joyce N. Boghosian/The White House

According to Bruce Butterfield of the National Gardening Association, 81.2 million households in the United States in 2008 were involved in some form of gardening. These activities all require the growing or maintaining of plants and are a form of green consumerism. His research further broke this down to more specific trends. For example, homeowners spent a record $44.7 billion in 2008 to hire lawn care and landscape maintenance services, landscape installation and construction services, tree care services, and landscape design services. Thirty percent of all households nationwide, or an estimated 34.5 million households, currently hire at least one type of lawn and landscape service. As a consequence, the most popular plant-related activities in the United States are lawn care, landscaping with ornamental plants, and vegetable and fruit production. In 2008, interest in household participation in food gardening increased. Thirty-one percent of all U.S. households, or an estimated 36 million households, participated in food gardening. This includes growing vegetables, fruit, berries, and herbs.

Those 81.2 million households in the United States pose a considerable threat to natural functioning ecosystems through the overuse of pesticides and fertilizers that can affect the water quality in nearby groundwater and surface water areas. This has been documented by a number of studies showing high levels of pesticides and nutrients, primarily nitrates, leaching into drinking-water wells and flowing into rivers that eventually reach the oceans.

Any discussion of plants ought to include a brief discussion of gardening and the garden. The garden is a product of the domestication of plants, a utilitarian place, and a place of ritual. Gardening is a practice through which the division between humans and the rest of the world can be bridged. This practice is highly personal, providing real physical places to contemplate the world through the cultivation of plants. Gardening requires active engagement with the natural world: Planting a seed, running one's fingers through the soil, or watching a plant grow are living embodiments of this personal exploration. Through cultivating a particular parcel over a series of years, a gardener develops a personal connection to that place.

Today yards and gardens are significant features of the modern landscape. Yards and gardens are special places, both physically and symbolically. Environmental philosophers Mark Francis and Rudolf Hester perceive gardens as the meeting ground of nature and culture and refer to gardening as our national art form, with the power mower and spade as the paint brushes of mass culture. They function as places for rest, rejuvenation, and beauty. Ornamental gardening did not come of age as a hobby until the 1950s, with the development of the suburbs, and growth in this area has increased yearly, resulting in gardening becoming the number one leisure-time activity in the United States today.

The fastest growing area of green plant consumerism is sustainable landscaping. The basis of sustainable landscaping is using plant material to create systems that imitate nature and turn problems into solutions. The recent rise of green roofs is an apt example, as they absorb rainwater, provide insulation, and create a habitat for wildlife. Sustainable landscaping recognizes the interconnectedness of all life in nature and is an approach to connect people back to the earth by tilling the soil, growing food, or caring about the local or global landscape, consequentially becoming stewards of the Earth. What does this have to do with the landscape and the neighborhood or watershed in which you live? Everything.

How can this be achieved through the use of landscape plants? Vast strides could be made with a few relatively simple steps. These include careful plant selection, minimizing water contamination and usage, reducing yard waste, and eliminating pesticide usage. A few examples that reinforce these concepts would be the use of low-maintenance grasses in lawns, and in arid climates, lawns can be reduced or removed.

Using native plants as much as possible eliminates the need for outside inputs of fertilizer and pesticides. Many native plants have low water and maintenance requirements and are adapted to growing in specific bioregions. Many of these plants are drought tolerant and can survive where other plants cannot. Planting native plants encourages a biodiverse garden and landscape that supports a wide variety of life from plants and animals to insects and beneficial pollinators. Planting a wide spectrum of species from local wildflowers to native plants to noninvasive plants will attract local pollinators, beneficial insects, and hummingbirds. Practice good ecology by recycling plant yard waste: Transforming yard waste into yard wealth is a practice that is nothing more than recycling all organic matter including leaves, grass clippings, and yard trimmings. Sustainable landscaping is a step forward in conserving resources and preventing nonpoint source pollution.

**See Also:** Composting; Green Consumer; Organic.

### Further Readings

Butterfield, Bruce. *Personal Communication*. South Burlington, VT: National Gardening Association, 2008.

Francis, Mark and Randolph Hester Jr. *The Meaning of Gardens: Idea Place and Action.* Cambridge, MA: MIT Press, 1990.

Lewis, Charles. *Green Nature/Human Nature: The Meaning of Plants in Our Lives*. Urbana: University of Illinois Press, 1996.

*Carl Salsedo*
*University of Connecticut*

# Positional Goods

Positional goods are goods that are desirable because they confer status on the owner. This is because the goods have a quality that makes them desirable but not easily accessible. This may be because they are expensive, in short supply, not widely available, stylish, or simply because they are new. Owning positional goods gives the feeling of social distinction, leadership, and exclusive access.

Although status seeking is characteristic of many societies, in a modern consumer society the search for status is often expressed through consumer goods, and people are judged by the goods they possess. Consumer goods become symbols of success, and some goods are better than others at indicating success.

Vance Packard, in his book *The Status Seekers,* pointed out that the use of consumer goods as status symbols was a deliberate strategy of advertisers, or "merchants of discontent," who took advantage of the "upgrading urge" that people felt. Advertisers offered buyers the possibility of gaining social status through possessing goods that were better than their neighbors. With the help of installment plans and credit, they could purchase the signifiers of success even if they were not achieving that success in their workplace. Car manufacturers, particularly, exploited people's desire for status, and the motor car soon after became a status symbol that rivaled the home.

Fred Hirsch, in his 1976 book *Social Limits to Growth*, coined the term *positional goods* to refer to goods that have status value. These often derive their status value from physical scarcity or social scarcity (where access is limited and demand high), so that only those with relatively high incomes can afford them. A good measure of the satisfaction deriving from consumption of positional goods comes from having something that others do not have.

Hirsch pointed out that "as the level of average consumption rises . . . satisfaction that individuals derive from goods and services depends in increasing measure not only on their own consumption but on the consumption by others as well."

As a consequence, happiness in a consumer society is more reliant on relative status than on absolute levels of wealth or consumption per se. This is because the pleasures associated with consumption may be limited and short term, but the respect of others provides a much more fulfilling pleasure. Because consumption is so shaped by cultural forces, and because people are judged in a modern consumer society by what they own, those with sufficient income to house, clothe, and feed themselves but few material possessions are likely to be far more unhappy in such a society than those in a poorer country, where having few possessions is more the norm.

The problem with attaining status through consumption (or the advantage from the point of view of marketers) is that it is a never-ending quest because the competition does not have an end point. Getting ahead is never enough because there are always others who are further ahead, and those you have overtaken might catch up and overtake you if you stand still. Positional goods become more expensive as relative incomes increase and there is more competition for them. This leads to people feeling poor despite having high incomes that can cover all their immediate needs.

It is for this reason that generalized increases in living standards offered by economic growth are less and less satisfying and the competition for status goods or positional goods becomes fiercer. Once basic needs for convenience and comfort are satisfied, no amount of growth in living standards will be able to satisfy everyone's need for positional goods, as these are necessarily scarce and will become more scarce and more unattainable as incomes in general rise.

By promoting positional goods, marketers can encourage obsolescence of desirability in products. Last year's models will always have less status than this year's models, even if they perform the same function equally well. Many industries, most notably cars and clothing, will bring out new versions of their products each year, even if they only have superficial changes, so as to encourage people seeking status to ditch their perfectly serviceable old products for new ones.

It is also the concept of positional goods that enables some brands to be sold for a premium. For example, Nike spends more money on advertising and promoting the reputation of its products than most other companies in the world. Celebrities are paid millions of dollars for their endorsement of and association with Nike products. Such endorsements are intended to confer status on the owners of Nike shoes so that consumers will pay much more for them than they cost to produce. The higher price, in itself, restricts access to the shoes and increases their positional value in poor neighborhoods.

As the affluence of a neighborhood increases, so do the scarcity and price of positional goods increase. Magazines and television programs that show off the lifestyles of the rich and famous feed the hunger for positional goods.

The more that people come to value positional goods, the less they value collective goods that everyone has access to, such as library books and national parks. This leads to a "commercialization bias," in which the market has an incentive to produce private commodities of increasingly less utility while underproducing collective goods that are essential. To profit from collective goods, the market tends to privatize them, thus restricting access to them. Ironically, the more private consumer goods we buy, the more the environment deteriorates and the more a home in a pollution-free, noise-free, uncongested location becomes a positional good because of its increasing scarcity.

**See Also:** Conspicuous Consumption; Consumer Society; Diderot Effect; Durability; Social Identity; Symbolic Consumption.

### Further Readings

Easterlin, Richard. "Will Raising the Incomes of All Increase the Happiness of All?" In *The Consumer Society*, Neva R. Goodwin, et al., eds. Washington, D.C.: Island Press, 1997.

Hirsch, Fred. *Social Limits to Growth*. London: Harvard University Press, 1976.

Packard, Vance. *The Status Seekers: An Exploration of Class Behaviour in America*. Harmondsworth, UK: Penguin, 1961.

*Sharon Beder*
*University of Wollongong*

## Poultry and Eggs

Poultry farms, in which chickens (or turkeys, ducks, or other birds raised for food) are raised for meat, eggs, and breeding, can be significant sources of pollution, and numerous concerns have been raised about the safety and ethics of meat produced in the "factory farms" (also known as "confined animal feeding operations" in the language of some legislation) that developed in the second half of the 20th century. Despite the promise of "a chicken in every pot" when Herbert Hoover was elected president in 1928, whole chickens were not a common supermarket item in the United States until the 1950s; roast chicken, like turkey or duck now, was previously a luxurious special-occasion food, perhaps served as Sunday dinner for some middle-class families, instead of a baked ham, and the cook often had to eviscerate the chicken herself.

Whole chickens were not a common supermarket item until the 1950s. Roast chicken was previously a luxurious, special-occasion food, until factory farms supplanted family farms and made chicken affordable.

*Source:* Stephen Ausmus/Agricultural Research Service/USDA

The factory farms that supplanted family farms—for which egg sales were the main revenue stream—made cheap chicken possible, while at the same time making the tough old chickens that had been the basis for many classic meals (the French *coq au vin*, or many a great-grandmother's chicken noodle soup) a thing of the past. With a seemingly limitless supply of chickens, they no longer had to be kept until their egg-laying days were over—they could be slaughtered while young, when the meat was still tender, and sold cheaply enough to be an everyday food, while leaving plenty of chickens alive for egg production. At the same time, egg production was enhanced by breeding chickens for larger yields—bigger eggs, and more of them. (From 1900 to 2000, the average number of eggs produced by a hen in a year increased from 83 to 300, and average egg size increased as well.) From the Great Depression to the end of World War II, the price of eggs dropped so low that many family farmers had little choice but to get out of the business, converting to some other farm product or selling their lands to a factory farm or one of the housing developments that spread across postwar America.

Industrial agriculture is resource-intensive, consuming vast amounts of water and energy, while increasing pollution of soil and water through animal waste products, dead animals, and other substances that are disposed of. Vitamin supplements, antimicrobial injections, and other substances are used to promote and accelerate the development of the chickens, and dominant behaviors are bred out as much as possible. Often chickens are physically restrained to keep them from attacking each other in their close quarters. They may also have their beaks removed to make it harder for them to hurt each other if they do attack. Animal waste at factory farms is required to be stored in "lagoons," which, when they leak or are damaged, have caused significant pollution problems in the past. Concentrating such waste in such large amounts also leads to larger concentrations of trace elements of arsenic and copper, which may contaminate nearby bodies of water. The environment of a factory farm also becomes unhealthy for the workers there; the U.S. Centers for Disease Control and Prevention has determined that workers on such farms may be susceptible to chronic lung disease, injuries, and infections as a result of the heavy concentration of chemicals, bacteria, and viral compounds.

Sustainable agriculture is the alternative to industrial agriculture, combining profitable farming with environmental and social responsibility. Advances in science over the last two centuries have enabled a new approach to farming. In crop farming, this means a greater understanding of soil impact and the use of fertilizers, other chemicals, and irrigation. In livestock farming, it means managing the environmental impact of the livestock—avoiding the conditions that can create an unhealthy work environment and leach poisons into the soil and water; raising healthy chickens; minimizing the use of nonrenewable resources; and responsibly managing the use of renewable resources such as water. Most sustainable poultry farms raise free-range poultry, giving the birds access to the outdoors, where they can feed on grass in addition to their feed, instead of being confined (literally "cooped up") indoors. This type of poultry farm is not only good for the chickens but is also good for the grass they are raised on, and some smaller operations raise chickens and cows together for this reason—the same pasture can be used to graze them both. So-called chicken tractors, for instance, are mobile chicken coops without floors that can be moved over a pasture, giving the chickens access to one area of grass at a time. The mobile coop can be moved along behind grazing cattle, letting the chickens forage insects from the cow manure—and in the process of doing so, spreading out the manure, encouraging fertilization of the soil. Eating the insects and larvae also reduces the population of flies that have traditionally harried cattle.

The eggs of free-range poultry are typically higher in vitamins and omega-3 fatty acids and lower in cholesterol, and the yolks are more orange than yellow, because of high concentrates of beta-carotene from the grass the chickens graze on. Despite ad claims that "brown eggs are local eggs," free-range eggs—and factory-farmed eggs—can be either brown or white. For consumers wanting to avoid eggs from battery farmed/caged hens, alternatives are available, though the eggs may cost more. Organic eggs are also increasingly available, laid by hens that have been fed on food free of animal by-products, synthetic chemicals, and antibiotics. However, the regulation and form of rearing practices that use terms such as *cage-free*, *barn-laid*, and *free-range* can vary significantly. As a consequence, it is not always easy for consumers to make fully informed choices about the poultry or eggs they wish to buy.

Sustainable poultry farming often goes hand in hand with the recently revived popularity of heritage turkeys (and, to a lesser extent, heritage chickens). Heritage (or heirloom) poultry breeds are not wild poultry but, rather, breeds of poultry that fell out of favor when factory farms took over the poultry industry. Turkeys over the last hundred years or more, for instance, have largely been bred for larger proportions of breast meat. Just as tomatoes that are bred for durability suffer in flavor and texture, turkeys that have been selected for one characteristic tend to suffer in others—and even when they do not, the result is that all turkeys from the major producers taste essentially the same. In contrast, there are a dozen different heritage turkey breeds being sold commercially, and there may be more breeds raised on small-enough family farms that there is not much public awareness of them. Heritage turkeys are generally raised on small farms, though as commercial interest in them grows, these farms are more and more often larger than the "mom and pop" operations that the term "family farm" may conjure up. In addition, these turkeys are often pastured and grow at a slower rate than factory farm turkeys do.

**See Also:** Composting; Dairy Products; Fish; Food Miles; Gardening/Growing; Genetically Modified Products; Grains; Green Food; Meat; Pesticides and Fertilizers; Plants; Slow Food.

### Further Readings

Hinrichs, C. Clare and Thomas A. Lyson, eds. *Remaking the North American Food System: Strategies for Sustainability*. Lincoln: University of Nebraska Press, 2009.
Ikerd, John E. *Crisis and Opportunity: Sustainability in American Agriculture*. Lincoln, NE: Bison Books, 2008.
Redclift, Michael. *Sustainable Development*. New York: Routledge, 1987.

*Bill Kte'pi*
*Independent Scholar*

# Poverty

Poverty refers to the condition of lacking basic human needs such as nutrition, clean water, healthcare, clothing, and shelter because of the inability to afford them. This is alternatively referred to using the terms *absolute poverty* or *destitution*. Relative poverty, in contrast, is the condition of having fewer resources or less income compared with others within a society or a country.

The United Nations Development Institute has reported that the world's richest 2 percent control more than half of all the wealth, whereas the poorest half possesses just 1 percent of the global wealth. Moreover, only a small number of developed countries account for most of the wealthiest 10 percent of the world, with just 11 countries contributing to over 90 percent of the wealthiest 1 percent of individuals. In addition, a staggering 86 percent of the global population lives in countries with a gross national product per capita of $10,000 or less, with these numbers likely to increase in the near future.

Although modern economic growth and prosperity have been achieved by the European countries, the United States, Canada, Australia, and Japan over the last two centuries, other parts of the world have lagged in achieving economic growth. Some regions of the developing countries, for example, were trapped in extreme poverty. Thus, the uneven distribution of wealth over two centuries resulted in two-thirds of the world population being poor, earning $2 or less per day.

The World Bank uses purchase-power parity to measure poverty. A person with $1 to $2 per day income is considered moderately poor, whereas a person with an income of $1 or less is regarded as extremely poor. The uneven wealth distribution is best described in a pyramid model of world population versus people purchasing power. The top 0.1 billion of the world population have annual income of more than $20,000. Approximately 1.5–1.75 billion people belong to the middle annual income range of $1,500–$20,000. The remaining 4 billion poor people earn $1,500 or less per year. According to data from the World Bank, the upper-income group constitute the citizens of the United States, Canada, Western Europe, Japan, Australia, and New Zealand. The middle-income groups live in East Asia (such as Korea and Singapore), Central Europe, the former Soviet Union, and Latin America. The moderately and extremely poor people are living in parts of South America, South and East Asia, and sub-Saharan Africa.

The 2005 UN Millennium Project report further categorized human poverty on the basis of three criteria: income poverty, relating to a lack of insufficient household income; social service poverty, including deficiencies in basic public services (e.g., education, health, water, and other services); and environmental poverty, concerning dilapidation of environmental resources needed for adequate quality of life. Linking the three criteria also yields further distinguishing between the overall "extremely poor" and "moderately poor" countries. Rather than viewing the poor in aggregate as a homogenous segment, customized programs of assistance for each group enable the international institutions and firms, as well as governments, to cater to their specific needs.

As noted earlier, three regions, namely South America, South and East Asia, and sub-Saharan Africa, are the major residences of the poor in the world. The number of extremely poor (earning less than $1 per day) has actually gone down from 1.5 billion in 1981 to 1.1 billion in 2001. The number of moderately poor (earning between $1 and $2 per day) has risen between 1981 and 2001, mostly in South and East Asia, as some of the extremely poor were able to improve their earning capacity. The economic condition of sub-Saharan extremely poor has not improved much over the same period.

Although a majority of the population in these regions lives below the poverty line, a fraction of people are extremely wealthy, and their numbers have been growing. For example, in a traditionally poor country such as India, which has 400 million people earning less than $1 a day, over 50,000 households have annual incomes of $225,000 a year—double the number earning this amount compared with three years ago, as the economy grew on average 8.1 percent during this time. Similarly, the wealthiest 10 percent in China held 45 percent of the country's wealth by the end of the first quarter of 2005. As we have seen before, wealth tends to accumulate unevenly. The people with more opportunities seem to collect more wealth unless governmental and public policies intervene. Past experience suggests that poverty reduction is directly related to an overall increase in economic growth. Long-run economic growth is possible through enhancements in both physical and human capital and technology, as well as the institutions that facilitate them.

## Poverty in the United States

Poverty, however, is not a phenomenon unique to the developing world. According to the Census Bureau, the poverty rate in the United States rose to 13.2 percent in 2008, the highest level since 1997, and a significant increase from 12.5 percent in 2007. This translates to 40 million people living below the poverty line, which is defined as an income of $22,205 for a family of four. The benchmark for ascertaining poverty levels is currently set at three times the annual cost of groceries. This does not take into account the rising medical, transportation, child care, and housing expenses or geographical variations in living costs. Neither does it consider noncash aid.

During the recession of 2008–09, the poor got poorer and the middle class lost ground. The poverty rate among Americans aged 65 years and older is nearly twice as high (18.6 percent) as the traditional 10 percent. Alarmingly, for the first time in history, 1.2 million more of America's poor are living in the suburbs than in the cities. The Center for American Progress further estimates that approximately 17 percent of children in the United States live in or near poverty, and the annual cost to the country's economy of children growing up poor, resulting in their eventual lower productivity and earnings and higher crime rates and health costs, is over half a trillion dollars.

**See Also:** Conspicuous Consumption; Disparities in Consumption; Downshifting; Frugality; Quality of Life; Resource Consumption and Usage; Secondhand Consumption.

## Further Readings

Akhter, S. and K. Daly. "Finance and Poverty: Evidence From Fixed Effect Vector Decomposition." *Emerging Markets Review*, 10/3:191–206 (2009).

Assan, J., et al. "Environmental Variability and Vulnerable Livelihoods: Minimising Risks and Optimising Opportunities for Poverty Alleviation." *Journal of International Development*, 21/3:403–12 (2009).

Bahmani-Oskooee, Mohsen and Oyolola Maharouf. "Poverty Reduction and Aid: Cross-country Evidence." *International Journal of Sociology and Social Policy*, 29/5–6:264–73.

Bastos, A., et al. "Women and Poverty: A Gender-Sensitive Approach." *Journal of Socio-Economics*, 38/5:764–73 (2009).

Cattaneo, C. "International Migration, the Brain Drain and Poverty: A Cross-Country Analysis." *The World Economy*, 32/8:1180–91 (2009).

Chong, A., et al. "Can Foreign Aid Reduce Income Inequality and Poverty?" *Public Choice*, 140/1–2:59–84 (2009).

Fritz, D., et al. "Making Poverty Reduction Inclusive: Experiences from Cambodia, Tanzania and Vietnam." *Journal of International Development*, 21/5:673–81 (2009).

Heltberg, R. "Malnutrition, Poverty, and Economic Growth." *Health Economics*, S1, 18, S77 (2009).

Kar, S. and S. Marjit. "Urban Informal Sector and Poverty." *International Review of Economics & Finance*, 18/4:631–42 (2009).

Kiernan, K. and F. Mensah. "Poverty, Maternal Depression, Family Status and Children's Cognitive and Behavioral Development in Early Childhood: A Longitudinal Study." *Journal of Social Policy*, 4/38:569–88 (2009).

Lepianka, D., et al. "Popular Explanations of Poverty: A Critical Discussion of Empirical Research." *Journal of Social Policy*, 3/38:421–38 (2009).

Lobao, L. and D. Kraybill. "Poverty and Local Governments: Economic Development and Community Service Provision in an Era of Decentralization." *Growth and Change*, 40/3:418–30 (2009).

Loix, E. and R. Pepermans. "A Qualitative Study on the Perceived Consequences of Poverty: Introducing Consequential Attributions as a Missing Link in Lay Thinking on Poverty." *Applied Psychology*, 58/3:385–402 (2009).

McWha, I. and S. Carr. "Images of Poverty and Attributions for Poverty: Does Higher Education Moderate the Linkage?" *International Journal of Nonprofit and Voluntary Sector Marketing*, 14/2 (2009).

Moses, J. "Leaving Poverty Behind: A Radical Proposal for Developing Bangladesh Through Emigration." *Development Policy Review*, 27/4:457–79 (2009).

Mwangi, E. and H. Markelova. "Collective Action and Property Rights for Poverty Reduction: A Review of Methods and Approaches." *Development Policy Review*, 27/3:307–31 (2009).

Pressman, S. and R. Scott. "Consumer Debt and the Measurement of Poverty and Inequality in the US." *Review of Social Economy*, 67/2:127–39 (2009).

Rosano, A., et al. "Poverty in People With Disabilities: Indicators From the Capability Approach." *Social Indicators Research*, 94/1:75–82 (2009).

Sachs, Jeffrey. *The End of Poverty: Economic Possibilities of Our Time*. New York: Penguin, 2005.

Stewart, K., et al. "Monitoring Poverty and Exclusion." *Journal of Social Policy*, 4/38:729–31 (2009).
Tahiraj, E. and P. Spicker. "The Idea of Poverty." *Journal of Social Policy*, 2/38:368–70 (2009).

*Abhijit Roy*
*University of Scranton*

# Pricing

Pricing in modern markets is a process. Pricing is set by a company to mark what it will receive for its products. The cost of producing goods includes the cost of raw materials, labor, management, profits for shareholders, and other factors. The real price will be the cost the company sets after it has averaged in costs for incentives, competition, discounts, and promotions.

Traders in primitive markets exchange goods and services in the belief that the exchange results in a value gain. With the development of trading, exchanges were made in terms of in-kind valuations. With the development of money and more advanced markets, prices were applied to goods and services as valuations were derived from supply, demand, quality, opportunity, and other factors. Setting prices for goods and services is necessary to facilitate trade. The process is somewhat subjective because the value of goods and services is priced according to what buyers and users believe they are worth.

Pricing in modern markets is set as a part of the financing of goods or services. The prices are set according to the nature of the product or service, its location, and how much money may be necessary to promote the sale of a product or service.

When pricing a product, sellers must decide on their pricing objective. There have been pricing strategies that involve setting the price high on a new invention to gain the maximum amount of revenue before competitors bring better and cheaper alternative products to the market. This strategy seeks to gain the maximum profit from the sale of the product.

Other concerns about pricing involve the question of the value in which potential customers hold the product or service. If the product is not in great demand, then a lower price may be set to attract potential customers. However, if the product is costly and is sold with some kind of snob appeal, then the price is set high regardless of the cost of the product. Premium-priced goods can be sold at high prices because the public believe that the high price represents quality. Pricing strategies like this appeal to value and have a strong psychological element to the value that the price represents. Companies at times have found that they actually have to charge more than needed because the buying public perceives the product as inferior if its price is lower than purchasers expect.

Demand pricing can also be used to promote the sale of goods that are currently in demand because of a fad or because they represent the latest technology or some other quality that is in demand. Christmas toys may be priced in some markets according to demand. Ticket scalpers often seek to sell tickets to sporting events with a price based on the demand of fans to see the event (such as a Super Bowl game).

Products that are sold below cost are very likely to disappear quickly from markets. However, in price wars, companies engage in competitor indexing. The price is set for products at ever-lower prices to gain market share. If a company seeks to penetrate a market, it will endure losses to be able to sell its goods or services in that market. Sometimes this means that good-quality competitors will be driven out of business by unfair competition; some kinds of predatory pricing practices have been criminalized in many jurisdictions.

Among nations, the practice of "dumping" occurs when a country subsidizes an industry and needs to keep its factories running to avoid political upheavals. Goods are brought into a market at prices that are really the result of subsidies, or even a loss to the product's country of origin, to get rid of surplus production. International trade rules have sought to prevent dumping and other tactics that seek an unfair advantage.

In general, prices are set by calculating the cost plus profit. The costs determine the price floor for the product or service. Using cost plus profit, a firm may need to decide whether it will allow discounts for volume purchases.

Another point to be considered is the price point of a product. Price points are price levels for a product at which demand for it stays relatively high. Milk may rise in price, but the desire to serve it to families and other customers will tend not to lower the price. Ultimately, customers will buy a product because it satisfies a need, is affordable, and there is a lack of viable alternatives.

Green pricing is based on the adoption of green technology or green products such as organic foods. Green energy from wind produces electricity that does not require the burning of fossil or nuclear fuels. However, there are costs. The costs include land acquisition and use, wind-generating machines, labor, operating costs, and power lines for power transmission. Many power companies in many states now have green pricing programs. In some of these, the cost of green energy is higher than the cost for traditional fuels.

The price of green energy is at the least the cost of production. However, because investors expect to gain profits from wind energy, the price for electricity generated with this technology will need to include costs for replacements and for investor returns. If the government were operating a wind farm, the price still would be the cost of production. However, taxpayers are not eager to foot the bill for such technologies if they always have to be subsidized at a loss.

Green pricing for products may be set at a higher rate even though products are in large part made from recycled materials. The savings are more often not gained by consumers but, instead, by the delay in resource extraction.

Green products and services are currently popular; however, studies have shown that many retail merchants as well as other businesses have engaged in making false or dubious claims about how green their products are. Prices are sometimes set higher for green products because the public expects to pay a premium for them. Organic foods often are priced higher to cover the cost of losses to farmers resulting from disease, weed competition, or insects. The consuming public is often unaware of the real price for such produce and believes that the higher prices reflect the higher quality of the product that is labeled as organic.

**See Also:** Advertising; Frugality; Organic; Positional Goods.

### Further Readings

Green, Richard and Martin Rodriguez Pardina. *Resetting Price Controls for Privatized Utilities: A Manual for Regulators*. Washington, D.C.: World Bank Publications, 1999.

Gualt, Teri and Sheryl Berk. *Shop Smart, Save More: Learn the Grocery Game and Save Hundreds of Dollars a Month*. New York: HarperCollins, 2008.
Heartfield, James. *Green Capitalism*. London: Skyscraper Digital, 2008.
Johnston, Barbara Rose. *Who Pays the Price?: The Sociocultural Context of Environmental Crisis*. Washington, D.C.: Island, 1994.
Jones, Van. *The Green Collar Economy: How One Solution Can Fix Our Two Biggest Problems*. New York: HarperCollins, 2009.
Roberts, Russell. *The Price of Everything: A Parable of Possibility and Prosperity*. Princeton, NJ: Princeton University Press, 2008.

*Andrew Jackson Waskey*
*Dalton State College*

# Production and Commodity Chains

Production and commodity chains are the most persistent approach to the study of the movement of goods through production, distribution, and consumption. Commodity chains conceptually organize the paths of commodities according to the Marxian notion of the division of labor along linear nodes of production between the initial points of resource extraction to the point of final consumption. The commodity chain approach originally emerged out of Emmanuel Wallerstein's world-systems theory but evolved into distinct analytical approaches in the early 1990s. In the original commodity chain approach that emerged out of political economy and world systems theory, four institutions—households, classes, peoples, and states—determine the social relations around the commodity at each node of a commodity chain. In all of the approach's varieties, the sets of nodes and linkages making up a commodity chain are approached by three major questions: the nature of the input–output structure, its territoriality (or the spatial reach of the different production and consumption activities), and its governance structure. The commodity chain approach to these three components as factors in global economic development is what distinguished it from more traditional ways of thinking about such development. Commodity chain analysis effectively shifted the focus of analyses of global development from the expansion of the territorial reach of national-level production and markets into the international sphere, toward the governance, territoriality, and input–output networks of sectors and industries, and their influence on global economic development.

Developments in production and commodity chain approaches in various directions have taken into account critiques of the original approach, as well as wider theoretical developments. The largest body of commodity chain literature centers on global commodity chains, or GCCs. GCCs, developed by Gary Gereffi and colleagues, split from Wallerstein's original conceptualization of commodity chains by focusing less on the role of the division of labor and the social reproduction of labor relations that make up the hierarchical world system and more on sectoral networks that make up global industries. The major difference between the Wallerstein-camp commodity chain analysis and the Gereffi-camp GCCs lies in the questions the researcher asks and how those questions are answered. World systems researchers approach commodity chains from a historical perspective and see globalization as rooted in the very beginnings of capitalism in the European context. GCC proponents, in contrast, see globalization as the result of increasingly integrated production systems characterized by producer-driven chains in which manufacturers directly control the entire chain, including its backward and forward linkages, and buyer-driven chains in which large companies or retailers coordinate the formation of decentralized production networks. Major debates emerging from this research focus on the governance of global commodity chains, especially agricultural commodity chains, and on the linkages between core and periphery and their spatiality. Said another way, there has been a particular emphasis on the conflicting spatial agendas of actors in the commodity chain, although research in the last decade has tended to focus less on structural and institutional-level commodity chains and contexts and more on sectoral and industry-specific chains. GCCs begin to reveal the complexity of commodity chain governance, but they break with the core/periphery analyses of world systems theory by placing a focus on changing industrial systems organization.

Alongside GCCs are other approaches to the study of the accumulation of value along production chains and networks. At the beginning of the 2000s, researchers of the global economy acknowledged the need for a common language to encompass the various terminologies used in the different approaches. The term *value chains* was chosen to cover the entire diverse body of research, although Gereffi's camp argues that value chains actually make up a separate, third category in addition to world systems–based commodity chains and GCCs. Their argument is that GCCs do not adequately isolate the specific factors that influence the governance forms in a particular sector. Global value chain (GVC) analysis is more explicitly rooted in international business literature.

The GCC and GVC approaches have gone a long way toward developing conceptualizations of the governance structures of commodity chains, as well as how these structures relate to sectoral, national, and regional economic development and the reproduction of core–periphery dynamics. There have also been recent efforts to integrate external factors, such as culture, environment, and others, into commodity chain analyses. The "commodity circuits of culture" approach starts with the assumption that production is consumer-driven and that the contextual meaning of a commodity attaches it to specific places of production, consumption, and distribution. This approach pays attention to the meanings attributed to commodities in specific phases of production, distribution, and consumption but tends to fall within the trap of the overly relativistic side of poststructuralism by understating the role of political economic power dynamics and overstating the role of consumers. This critique of commodity circuits underlines the need to stay in proximity to some ideas informed by Marxism—such as political economy—as a way to be able to continue to identify mechanisms of power and emergent structures such as class in the economy. A commodity network approach works toward achieving this. It is built on the commodity chain tradition but integrates thinking from cultural studies, actor-network theory, and commodity conventions theory. This approach allows for multiple tools for the identification of structures and mechanisms in an analysis of commodity linkages but follows a poststructural conceptualization of space in allowing the commodity itself to define the space of analysis, rather than it being predetermined, preventing the privileging of one space of commodity movement over another.

Some critiques of commodity chain analysis shed light on the limitations of the approach. The critique that commodity chains overgeneralize food systems and neglect local cultural factors, for instance, reflects wider critiques of structural Marxist social theories in general. Related to this, commodity chain analysis also falls short of integrating the informalities of social relations in commodity networks, instead focusing on formal relationships. It also fails to take into account nontraditional linkages that form around alternative commodity chains (such as the collaborative relationships that often occur among development nongovernmental organizations, producer cooperatives, and Fair Trade coffee companies, for instance); these fall outside the linear logic of a commodity chain approach because they involve actors that are not directly involved in the commodity chain of custody. Despite the relative exclusion of nontraditional linkages that make up alternative commodity chains, the commodity chain metaphor in its broadest sense has enabled consumer activists and social and environmental justice organizations to emphasize the connections between producers and consumers, making explicit the social and material practices that link them. Unveiling the politics of connection has for such groups been seen as a precursor to instigating change in corporate practices and in the policies of states and regulatory agencies. Making the commodity chain and its governance explicit for consumers has also been a feature of marketing and promotion surrounding Fair Trade, organic, humane, and environmentally friendly commodities. However, it must be said that recent developments in the commodity networks approach go a long way toward integrating both informal and external nodes and links into analyses.

**See Also:** Commodity Fetishism; Consumer Behavior; Final Consumption; Green Consumer; Organisation for Economic Co-operation and Development (OECD); Regulation.

### Further Readings

Bair, Jennifer. "Global Capitalism and Commodity Chains: Looking Back, Going Forward." *Competition & Change*, 9/2 (2005).

Gereffi, G. and M. Korzeniewicz, eds. *Commodity Chains and Global Capitalism*. Westport, CT: Greenwood, 1994.

Hartwick, E. R. "Towards a Geographical Politics of Consumption." *Environment and Planning A*, 32/7:1177–92 (2000).

Hopkins, T. and E. Wallerstein. "Patterns of Development of the Modern World-System." *Review*, 12 (1977).

Hughes, Alex and Suzanne Reimer, eds. *Geographies of Commodity Chains*. New York: Routledge, 2004.

Raynolds, Laura T. "Consumer/Producer Links in Fair Trade Coffee Networks." *Sociologia Ruralis*, 42/4 (2002).

*Heather R. Putnam*
*University of Kansas*

# Product Sharing

Product sharing, an in-between of public and private ownership, is one strategy of minimizing ecological impact. Rather than presuming that individual ownership (where each entity owns their "own," according to their wants) is the best system of consumer acquisition and distribution, product sharing aims to make joint use a viable and desirable alternative.

Product sharing has five main benefits to people and the environment:

- Product sharing reduces the number of items produced:
- It may expand the number of users, and may provide access to those who could not otherwise afford use of the product.
- As costs are minimized, product sharing potentially gives people the possibility to select more efficient and durable products, rather than going with the cheapest option. (This can also influence the design of products to match up to these need standards.)
- With joint use, product users are no longer individually responsible for the storage, repair, or disposal of the item. This saves space, lends to recommended maintenance, lengthens durability, requires that chemicals be disposed of properly, and generally removes some of the burdens of ownership.
- Product sharing has some social benefits in its communal, club-joining aspect that can link sharers to other participants. Even in situations such as Zipcar (see below), where the marketplace has removed the requirement of interpersonal relationships, people are proud of their affiliation and claim new identities and allegiances.

Of course there are also several risks to product sharing. Organizationally, there is always the risk that you will be left without in a time of need. Sharing requires trust in the (informal) community of sharers or for (formal) market organized sharing, at minimum, trust in the company organizing joint use. Whether formal or informal, someone needs to be trusted to keep track of using, repairing, and making the item ready and available for another's use on a timely basis. Additionally, most sharing arrangements are temporary.

When a joint use organization dissolves, so does access to the product, as residents of downtown San Diego, California, learned when Zipcar bought out the company Flexcar, and ended a car-sharing program they had organized their lives around. Culturally, product sharing may violate an individual's sense of security of the "free market evac plan" or the ability to individually buy one's way out of problems. In addition, people have to get past their cultural concerns with worn, used, and personal attachment to their possessions, according to R. Belk. It may in fact be easier to share something not previously owned as opposed to sharing the reliable car you have had for years.

Finally, sharing is a social process. Socially, product sharing may require social interaction in situations where people have previously been able to avoid interaction and possible confrontation. Bad sharing experiences, or even a fear of them, could strain human relationships, lead to bad social reputations, or even foster retaliatory property damage.

Some goods are more easily shared than others. Product sharing is a limited strategy and only works for unique items, with particular uses, in certain contexts, and at specific scales. For example, researchers have noted that durable expensive products (like washing machines, lawnmowers, and cars) that are used at irregular intervals are prime products for sharing. Often, access to a central location or willing participants can be an deterrent to product sharing, and may lead to exclusivity or limited applicability of product sharing.

Even though sharing is a concept most people are taught at a young age, it remains a rare consumer strategy. Product sharing is not renting (a one-time experience), but rather requires an informal contract or formal club membership. It is not borrowing (a neighborly negotiation of something owned), which requires a different type of social relationship, nor are buying used goods or swapping (e.g., clothing swaps) instances of product sharing. The idea of joint use is that all members use the product; this is different than an old item passed along to new users when the original owner is done with it.

Currently, there are two types of product sharing: self-organized collaboration at the community level, and market organized sharing. In the case of self-organized community sharing, a group of people jointly purchases and uses a product. A classic example is a neighborhood tool lending library. With market-based sharing schemes, a business provides the service of managing the complexity of coordinating joint use, maintenance, and storage. Car-sharing companies like Zipcar or Flexcar are prominent examples.

In sum, product sharing is joint use of a product, organized through formal or informal agreements, where products are jointly used, maintained, and stored. Of all of the various strategies of green consumerism, product sharing may be the least prone to greenwashing. Rather, product sharing pools both resources and community to buy less goods of better quality and durability, and may give society new ideas of how to sustainably organize and limit consumption.

**See Also:** Consumer Behavior; Consumer Culture; Consumerism; Consumer Society; Ecological Footprint; Frugality; Green Consumer; Green Design; Lifestyle, Sustainable; Social Identity; Sustainable Consumption.

### Further Readings

Belk, Russell. "Why Not Share Rather Than Own?" *The Annals of the American Academy of Political and Social Science*, 611:126–40 (May 2007).

Benkler, Y. "Sharing Nicely: On Shareable Goods and the Emergence of Sharing as a Modality of Economic Production." *Yale Law Journal*, 114/2:273–358 (2004).

Jones, Van. "Environmental Justice Advocates Symposium: Is It All Talk?" Keynote Address, 2007. http://lawlib.lclark.edu/podcast/?p=255 (Accessed December 2009).

Mont, O. "Innovative Approaches to Optimizing Design and Use of Durable Consumer Goods." *International Journal of Product Development*, 6/3–4:227–50 (2008).

Ruth, Todd. "Life After Flexcar. Car Sharing in San Diego." http://carsharingsd.blogspot.com/2008/01/life-after-flexcar.html (Accessed December 2009).

Schor, Juliet B. *The Overspent American: Why We Want What We Don't Need*. New York: Harper Perennial, 1998.

*Alison Grace Cliath*
*California State University, Fullerton*

## Psychographics

Psychographic constructs have been used by marketers to help describe the consumers of products and services since the early to middle part of the 20th century. The typical motivation to use such variables is to flesh out the description of consumers beyond what typical demographics can accomplish. That is, although demographics can provide information about the age, gender, income, and family size of consumers, psychographics can go beyond what might be a relatively sterile demographic description and potentially provide insights about motivations. Given that environmental concerns would be expected to relate to a person's attitudes and values, psychographics can be helpful in understanding individuals' propensity to engage in environmentally responsible behaviors. In addition, understanding consumer motivations allows marketers and public policy actors to tailor persuasive communications aimed at changing attitudes and behaviors.

## Definition and Use of Psychographics

There have been numerous definitions of what constitutes psychographic variables, with some of these definitions being restrictive in terms of what is included as psychographics, and others being more broad and inclusive. Marketers often consider psychographic and lifestyle variables to include, but distinguish between, psychological constructs (psychographic variables such as personality traits and values) and behavioral constructs (lifestyle variables such as activities in which a person may engage). Therefore, psychographics and lifestyle variables relate to the things a person may believe and how they may behave.

Psychographics and lifestyle variables can be distinguished from demographics, which typically describe characteristics of populations or groups. Some demographic variables are characteristics over which the individuals of a group have little or no control (e.g., age, race, and gender), whereas others relate to specific attainments (e.g., educational level, household size, and household income). Therefore, demographics do not describe how people behave or what they may believe, although demographics have been shown to correlate with some psychographic and lifestyle variables.

Researchers typically consider psychographics to include a people's personal values (e.g., truthfulness, self-respect, security), their internalization of cultural-level value orientations (e.g., individualism, activity orientation), personality traits (e.g., locus of control, tolerance for ambiguity, warmth, self-reliance), and general attitudes (e.g., attitudes toward political and social issues). Lifestyle variables may include activities and interests (e.g., gardening, reading, media usage).

Personal values have a rich history in the social sciences and humanities. In recent decades, value inventories have been developed to help quantify the importance of different values to individuals. Personal values include constructs such as the importance of self-fulfillment, warm relationships with others, and self-respect. Most notably, Milton Rokeach and Lynn Kahle have developed such inventories that have been extensively used by social scientists and marketers. Cultural value orientations are broader than personal values and can be conceptualized as basic beliefs that cultures hold about the relationship of the people and their world. Cultural value orientations include such concepts as individualism and collectivism (the extent to which a culture values the individual or the group), activity orientation (the extent to which a culture values activity or reflection), and the relationship with nature (the extent to which a culture believes that people have mastery over nature, live in harmony with nature, or are controlled by nature). Although cultural value orientations are broad beliefs that a culture holds, these can be measured at the individual level, in that such analysis would represent the individual internalization of the cultural beliefs.

Personality traits are relatively enduring aspects of what makes up a person. They are typically measured by personality inventories that tap several different traits, as well as scales that measure specific personality traits, such as locus of control or authoritarian personality.

Although psychographics can and have been applied to research questions in a number of ways by academics or practitioners, there are three common methodologies that are used. One of these methods is to do a psychographic and lifestyle profile of a particular group of interest. For example, a marketer could do a psychographic profile of the heavy user of a particular product. This could simply be cross-tabulations of the group of interest (e.g., heavy users of wine) with numerous psychographic variables. The researcher would then look for patterns of psychographic constructs in which the group of interest differs from other people. In the environmental domain, a researcher might do a psychographic profile of those who recycle, comparing recyclers with nonrecyclers on a number of psychographic variables.

A second approach to using psychographics is to develop psychographic segments based on a set of psychographic variables. These segments can then be considered in relation to product usage or behaviors of interest to a particular researcher. Proprietary segmentation schemes are commercially available from a number of sources; one of the most widely used is VALS, provided by SRI Consulting Business Intelligence. The VALS system comprises eight groups as a function of personality variables (and selected demographics). VALS groups include innovators, thinkers, achievers, experiencers, believers, strivers, makers, and survivors. Typically, a researcher may evaluate how these different groups vary with respect to a behavior of interest (e.g., environmentally responsible behavior).

A third approach, which is commonly used by academic researchers, is to evaluate particular psychographic variables of interest and consider their relationships to behaviors, as well as other predictor variables. This approach typically involves an analysis of the psychographic variables in the context of a complex model to understand their relation to these other variables.

## Psychographics and Environmentally Responsible Behavior

Given that psychographics relate to the psychological motivations of individuals, psychographics should be a fruitful path to understanding the environmental behavior of individuals. Indeed, a variety of psychographic constructs have been shown to relate to environmental attitudes and behavior.

At a most basic level, value orientations, personal values, and personality characteristics have been shown to relate to individuals' propensity to engage in environmentally responsible behavior. For example, the value orientations of individualism and collectivism relate to attitudes and beliefs about recycling, and these beliefs in turn affect recycling behaviors. Individualism is the extent to which a person (or culture) values the individual relative to the group, whereas collectivism involves beliefs that the group should be valued more than the individuals making up a group. Cultures vary in their beliefs with respect to these constructs, but there can be individual differences among people within cultures. Research shows that individuals who score high on collectivism tend to believe that recycling is important, which leads to more likelihood of recycling behaviors. Conversely, those who score high on individualism tend to believe that recycling is inconvenient, which leads to a lower level of recycling. Personal values have also been shown to be related to environmentally responsible behavior. In particular, self-fulfillment, a sense of accomplishment, and self-respect have been shown to relate to proenvironmental behavior.

Several personality traits have been studied as potential antecedents of environmental behavior, often with mixed results. The personality variable of "locus of control" is one personality trait that has generally shown fairly consistent results with respect to environmental beliefs and behavior. Individuals with an internal locus of control tend to believe that they have control over the events in their lives, whereas those with an external locus of control believe that they have little control over outcomes. Research has generally shown that those with an internal locus of control are more likely to engage in environmentally responsible behavior compared with those with an external locus of control. Presumably, those who feel that they have control (internal locus) believe that their environmental actions will have an effect.

Consumer segments based on the VALS system have shown important variations in how different consumer groups act with respect to the environment. Research on the environment, climate change, and recycling shows that innovators (individuals who are successful, innovative, and have high self-esteem) and thinkers (individuals who are motivated by ideals) are more likely to engage in environmentally responsible behavior compared with other VALS segments, such as achievers (individuals who are goal-oriented and value stability and predictability), who tend to view fixing the environment as "not my problem."

In terms of general attitudes, political ones have often shown a relationship to environmental behavior. Those who are more liberal are more likely to engage in proenvironmental behaviors compared with those who are more conservative.

**See Also:** Consumer Behavior; Demographics; Green Marketing; Materialism.

### Further Readings

Kahle, Lynn and Larry Chiagouris, eds. *Values, Lifestyles, and Psychographics*. Mahwah, NJ: Lawrence Erlbaum, 1997.

Shrum, L. J., et al. "Recycling as a Marketing Problem: A Framework for Strategy Development." *Psychology & Marketing*, 11/4 (1994).

SRI Consulting Business Intelligence. http://www.sric-bi.com/VALS (Accessed July 2009).

Wells, William D. "Psychographics: A Critical Review." *Journal of Marketing Research*, 12/2 (1975).

Ziff, Ruth. "Psychographics for Market Segmentation." *Journal of Advertising Research*, 11/2 (1971).

*John A. McCarty*
*College of New Jersey*

*L. J. Shrum*
*Tina M. Lowrey*
*University of Texas at San Antonio*

## Public Transportation

*Public transportation* refers to modes of transit that are available for use by the general public and that multiple people can use at once. Public transportation is most often contrasted with modes of transit that individuals can use, mostly commonly the automobile. Public transportation may include the use of buses, subways, heavy and light rail, and in some countries, ferries. In general, transportation options that are for hire, such as taxi services, are not considered public transportation, but other public operations such as vanpool services and services for the elderly or individuals with disabilities are

Large cities tend to have heavy-rail commuter service, like this elevated train in Chicago's Loop neighborhood.

*Source:* iStockphoto

considered public transit. Many of these services are subsidized by national, regional, and local governments, and as such have come to be considered a public good. Although there are many reasons individuals choose to use public transportation options, a recent trend is to do so to be green or to decrease one's carbon footprint.

The growth of the highway and road systems in the last decades caused more and more individuals to become increasingly dependent on transportation to get to work, entertainment activities, and shopping. Approximately 80 percent of the eligible population in the United States chooses to drive a car, or at least to obtain a driver's license. The history of individuals' reliance on personal transportation starts in 1916, when Congress created the Bureau of Public Roads and began matching funds with states for highway development. The Federal Highway Act of 1956 included earmarked revenue from a gasoline tax to provide funding for highway development that could be passed to the states. As a result, most state and local governments began to construct highways and roads.

Although the growth of the highway system resulted in many benefits for U.S. citizens, such as convenience of travel and the ability to pursue the "American Dream" of owning a house in the suburbs with a white picket fence, there were also a number of negative consequences associated with this trend. Some of these include the decentralization of cities, the high cost of road and highway maintenance for state and local governments, and an increase in the number of vehicle accidents. There were also environmental consequences, including the relationship between transportation and air quality—the most obvious indication of which was the creation of smog from vehicle emissions.

Individual vehicles are often singled out as examples of pollution creation. A number of countries and large cities monitor vehicle emissions in an attempt to control smog. In reality, emissions from automobiles are relatively minimal when compared with those of industrial polluters. It does remain the case, however, that cars are the largest single polluter because of the mass of vehicles in large metropolitan areas, in which emissions from individual automobiles add up. Pollutants from vehicles include hydrocarbons, nitrogen oxides, carbon monoxide, and carbon dioxide. Although other transit options also emit pollutants, the joint uses of these vehicles decrease the overall effect. Some studies suggest that public transportation in the United States uses approximately a little more than half the fuel required by automobiles. Some environmentalists also argue that individual automobiles emit more carbon monoxide, carbon dioxide, and nitrogen oxide than public transit options per mile.

As noted above, there are many different types of public transportation. The one most familiar to individuals in small to medium cities is bus service. A bus can transport numerous passengers, and differing bus routes can increase or decrease service based on demand.

Buses use the road and highway system that is already in place and do not require much additional infrastructure. Larger cities tend to have heavy rail commuter service and/or light rail service in addition to bus service. Trains can transport more individuals than buses can, but they do require track infrastructure and stations. Many extremely large cities—mostly overseas—have also moved toward rapid transit systems. A rapid transit system includes electric passenger railways with designated service lines. Rapid transit systems usually move fewer people at once than traditional heavy or light rail but may travel at a faster pace.

As noted earlier, federal, state, and local governments often need to subsidize public transportation. As a public good, governmental entities may want to provide transportation for people who are unable to transport themselves for financial, physical, or legal reasons. Local governments may also wish to promote business and economic growth in urban centers by making transportation convenient. Transportation usually is one of the major budget items for state and local governments. A number of local and regional authorities in the United States and abroad are seeking to make public transportation more energy efficient and nonpolluting by such measures as careful transportation planning and routing, using hybrid fuels, developing efficient rapid and light rail transport systems, and using trolley buses or electrical rail systems based on renewable energies.

Individuals choose to use public transportation for a variety of reasons, from personal convenience to an effort to save money on fuel. Regardless of motive, however, there are environmental benefits from a decrease in the amount of automobile emissions.

**See Also:** Automobiles; Carbon Emissions; Kyoto Protocol.

### Further Readings

Hanson, S. and G. Giuliano. *The Geography of Urban Transportation*. New York: Guilford, 2004.

Kutz, M. *Environmentally Conscious Transportation*. Hoboken, NJ: John Wiley & Sons, 2008.

U.S. Department of Transportation and U.S. Bureau of Transportation Statistics. *National Transportation Library*. Washington, D.C.: U.S. Department of Transportation, 1997.

*Jo A. Arney*
*University of Wisconsin–La Crosse*

## Quality of Life

From a sustainable development point of view, it is the most encompassing and less materialistic quality of life, rather than the limited standard of living, that is, together with environmental sustainability, the right yardstick to use to assess economic activity and social arrangements in general. Indeed, although economic growth is consuming, in absolute terms, more and more scarce environmental resources, there is ample evidence that past a given threshold, it no longer brings real improvement in the quality of life of the population, as measured by different indicators of objective (e.g., life expectancy or education) and subjective (e.g., reported happiness and satisfaction with life) well-being. In contrast, when asked what makes them feel good and happy, people in rich Western societies typically answer, in decreasing order of priority: partner and family relations, health, and a nice place to live, before money and financial situation.

It is therefore of the utmost importance to develop an adequate definition of quality of life, a precise identification of its main components, and reliable indicators for these components. Actually, this was exactly the research program of the social indicators movement that emerged in the 1960s in the United States and culminated with the publication in 1976 of *The Quality of American Life: Perceptions, Evaluations and Satisfactions* by A. Campbell, P. Converse, and W. L. Rodgers. Unfortunately, almost 40 years later, no consensus has been reached in the scientific arenas, or in the policy ones, on a workable definition, conception, and measure of quality of life. Even the vocabulary has not been stabilized so far; for instance, the terms *quality of life* and *well-being* are considered by some as equivalent, whereas others see significant differences between them. What is clear, however, is that both terms stand in contrast to narrow, monetary, or materialistic conceptions of the good life. Indeed, in the meaning of quality of life, both terms—*quality* and *life*—are important. The first emphasizes the fact that it is not only the number of things, relations, and satisfactions experienced that matters but also their kind, their intrinsic quality, and their worth. The second term, *life*, may signify different things; for example, that we are concerned with life taken as a whole and in all its dimensions in contrast to what is implied in the concept of "standard of living," or it can stand for "daily life," the set of mundane habits, practices, and routines that make the fabric of day-to-day existence. To illustrate, when a person explains she is trading higher earnings for a more interesting job, she expresses her preference for the quality of her work experience over of its pecuniary reward. In contrast, if she trades the same earnings for less strain, more time for children, and so on, it is the quality of her overall everyday life that she privileges.

What makes the concept so difficult to define and, still more, to measure, is the fact that it refers to a mix of objective living conditions (natural environment, neighborhood, housing, working conditions, income and wealth, political and civil rights, and so on), of functionings (being in good mental and physical health, loving and being loved, etc.), and of subjective evaluations of these living conditions and functionings. The multidimensionality and richness of everyday life, together with the mesh of objectivity and subjectivity in any process of evaluation by people of their own existence, are such that no satisfactory conception of quality of life can avoid some level of complexity. This is apparent in the two currently prevailing and most elaborated theories of quality of life: the one proposed by Amartya Sen and the one advocated by Ruut Veenhoven.

The influence of Amartya Sen's conception of well-being and quality of life is such that, nowadays, any contribution to this issue cannot but take position with respect to it. Sen's model of quality of life holds in a fourfold matrix built on two axes: a freedom–achievement axis and an agency–well-being axis. The distinction agency–well-being accounts for the fact that people can have other goals and values than the pursuit of their own well-being. They can commit themselves to collective aims, such as humanitarian causes, and so on, so that their agency achievement would involve an evaluation of the state of affairs in the light of those objects, independent of their contribution to their well-being. The distinction between achievements and freedoms, be it with respect to agency or with respect to well-being, is important because people are, in the last resort, responsible for taking or not taking advantage of the existing opportunities. The distinction is particularly significant in the well-being dimension, in which it leads to distinguishing (achieved) functionings from capability to function. In fact, Sen conceives of life as a set of interrelated doings and beings, some elementary, such as being adequately nourished, being in good health, avoiding escapable morbidity, and so on, and others more complex, such as being happy, having self-respect, taking part in the life of the community, and so on. He terms these beings and doings *functionings*. What matters for quality of life is, of course, the quality of the functionings a person manages to achieve, but at least as important is the freedom the person has in choosing them. For a given person, taking account of her personal characteristics and of the social context, not all possible functionings are always reachable. The subset of functionings a person is really free to achieve is called her capability set, and the assessment of her well-being should involve both her current, actual beings and doings (the vector of achieved functionings) and her capability set (the vector of reachable ones). If, for instance, the actual functionings are the only ones the person could ever achieve, her quality of life is deemed inferior to if she had the opportunity to choose between a large set of achievable beings and doings. To take an example, there is a large difference between being undernourished because of a lack of capabilities to command enough food and being undernourished because one has decided to starve for a religious or other reason. This boils down to giving (real, not purely formal) freedom a paramount importance in quality of life and its evaluation.

The most comprehensive and consistent alternative to the capability-functioning framework is Ruut Veenhoven's fourfold model of quality of life. Veenhoven denies that it is possible to speak meaningfully about "quality of life" at large. Instead, he suggests distinguishing between four different kinds of qualities that he calls liveability of the environment, life-ability of the person, utility of life for the environment, and appreciation of life by the person. Here again, as in Sen's model, the framework is the outcome of crossing two axes: a chances–outcomes axis and an outer–inner qualities axis. The chances–outcomes axis corresponds to the important distinction in sociology between life chances and life results, between the opportunities provided by the environment to have a good life and the way different individuals can take advantage or not of these opportunities. It is close to Sen's freedom–achievements axis. Veenhoven's second category—outer–inner—accounts for the distinction between the qualities that lie inside the individual and those that pertain to the environment. Of course, the two influence each other, as an internal quality could be revealed only under some external conditions. Let us look at the four qualities that result from these categorizations:

- The combination chances–outer corresponds to what Veenhoven calls the liveability of the environment—the opportunities and constraints one's environment presents with respect to well-being. Many empirical studies reduced overall quality of life to this single aspect of liveability of the environment.
- The life-ability of the person results from the combination chances–inner qualities. It expresses the capacity of a person to make use of the resources available in the environment. This would correspond to the "personal capabilities" in Sen's vocabulary, or to "internal resources."
- Utility of life (a combination of outcomes and outer qualities) refers to the quality of life assessed in terms of its utility for one's (social as well as physical) environment—the external worth of a life. It expresses the fact that a life must be good for more than just itself and presumes some higher values. It evokes also the "meaning of life."
- Finally, appreciation of life as the combination of outcomes and inner qualities designates the value one's life has for oneself—one's subjective assessment of its quality. It is the quality in the eyes of the beholder. This is subjective well-being properly said: the subjective assessment of one's own life. Thus, if utility of life can be considered an "other-regarding" assessment of one's life, appreciation of life is a "self-regarding" perspective.

To illustrate these categories, one can resort to a biological and ecological metaphor. In a given ecosystem, life chances for a given species depend, for one part, on the richness of the biotope or habitat (its liveability) and, for the other part, on the fitness (life-ability) of the species to that particular biotope. Fitness and a healthy habitat result in the well-being of its members (appreciation of life) and, in consequence, the growth of the species population insofar as individuals are adequately nourished, and therefore able to reproduce themselves (utility of life).

Veenhoven's framework helps analyzing another, more ad hoc, operationalization of quality of life and classifying the social indicators most commonly mobilized in their various empirical assessments at the country, area, or community level. In the liveability category, one would find the usual indicators of economic prosperity, social security, social structure openness, environmental cleanliness and safety, quality of human relations, and so on. The life-ability of individuals depends on what is more commonly called their "human capital," physical and mental health, skills and talents, and personality. Appreciation of life corresponds to happiness (emotional assessment) and satisfaction with one's life (cognitive evaluation) and its various components (job satisfaction, family life, leisure time, etc.). The choice of indicators of (external) utility of life is more challenging, as it can necessitate some touchy and risky value judgments. However, a promising candidate in that respect, from a green perspective, could be the "ecological footprint." This indicator measures at various scales, from the individual to the global level—but on an annual basis—the consumption of scarce natural resources (the share of the carrying capacity of the Earth) by human beings. It can be argued that the utility of an individual life—all other things being equal—must be inversely proportional to its ecological footprint.

In addition to these two leading frameworks, the literature (especially in journals such as *Social Indicators Research* and the *Journal of Happiness Studies*) is replete with other examples, generally less systematic—if not plainly ad hoc—definitions, conceptions, and operationalizations of quality of life. Many of them make explicit references to the concept of human (basic or not) needs. Indeed, it certainly makes sense to look at the liveability of an environment through the human-needs lens, provided the list of needs considered is comprehensive enough. Unfortunately, the problem with the needs-based approaches is that, hitherto, no consensus has ever been reached on a common list of these needs. However, the problem is not specific to the needs-approach—it exists also with respect to basic or universal capabilities or functionings.

**See Also:** Ecological Footprint; Green Consumer; Needs and Wants.

## Further Readings

Jackson, Tim. *Prosperity Without Growth? The Transition to a Sustainable Economy.* London: Sustainable Development Commission, 2009.

Layard, Richard. *Happiness. Lessons From a New Science.* New York: Penguin, 2005.

Nussbaum, Martha and Amartya Sen, eds. *The Quality of Life.* Oxford: Oxford University Press, 1993.

Sen, Amartya. *Development as Freedom.* New York: Oxford University Press.1999.

Veenhoven, Ruut. "The Four Qualities of Life. Ordering Concepts and Measures of the Good Life." *Journal of Happiness Studies*, 1:1–39 (2000).

*Paul-Marie Boulanger*
*Institut pour un Développement Durable*

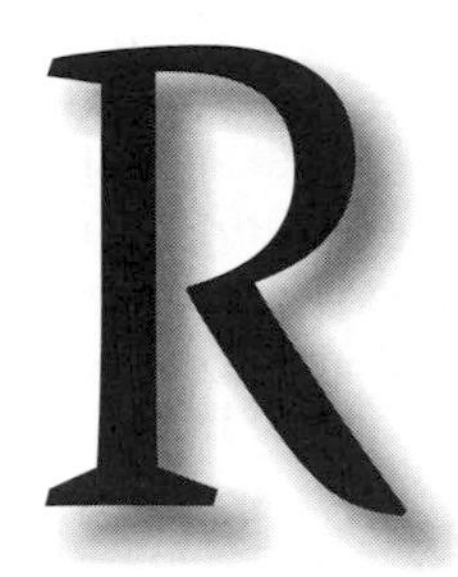

# Recyclable Products

Recyclable products include many types of plastics, textiles, glass, paper, electronics, and metals that are reprocessed from extant goods. Recyclable products are made from the same commodities as the basic materials that make up new goods. Recyclable products have many benefits for individual and societal interests. These benefits include the salvaging of resources for profit (or precaution), protecting and expanding manufacturing jobs, conserving natural resources, reducing the need for landfills and incineration of garbage, decreasing the emission of greenhouse gases that contribute to global climate change, preventing pollution caused by the manufacturing of products, and the saving of energy. Recyclable products have historically been of greater interest during times of war, scarcity, and when the process made sense economically. A variety of recyclable products have become common in the marketplace, including goods made of plastic, glass, and paper. In addition, certain commodities such as methane are becoming more available. Government planning and policy decisions can greatly promote or impede the production of recyclable products because the process of recycling has, at least heretofore, relied heavily on governmental subsidies. The main detriment of the process largely hinges on the financial burdens of reprocessing, transporting, and marketing recyclable products.

Recycling collection involves the gathering of items to be used for recyclable products, and it may involve some combination of drop-off centers, buy-back centers, and curbside collection.

*Source:* iStockphoto

The use of recyclable products predates the Industrial Revolution, as the high cost of new goods made reusing products a necessity. During the 1930s, the global economic downturn made the increased use of recyclable products desirable, as financial setbacks and resource shortages made the purchase of new products an impossibility for many. World War II, and its effect on international shipping, also made the use of recyclable products a necessity for most of the 1940s in the United States and internationally. In the postwar era, however, with its economic prosperity, pent-up consumerism, and surfeit of inexpensive goods, recyclable products faded as an alternative. Only during the 1970s, with its increased emphasis on the adoption of ecologically friendly practices, did interest in broadening public awareness about the identification and collection of recyclable products increase. During the 1980s, the amount of waste sent to landfills in the United States reached its zenith, at approximately 150 million tons (136.08 million metric tons) annually. Concern over this development spurred increased popular interest in the use of recyclable products as a means of reducing this amount. Governmental subsidies designed to increase the use of recyclable products became popular during the 1990s and continue to this day.

## Green Response and Opportunity

Recyclable products are a popular green response to environmental degradation, in part because of the potential for profit on behalf of those involved in the process. The environmental impact of recyclable products is less than that of virgin products when the two are compared over their entire life cycles. When compared with a system of production focused on the extraction of virgin materials, as well as the use of landfill and incineration to eliminate waste, recyclable products provide an economically favorable and ecologically friendly alternative. Employment opportunities exist in various stages of production for recyclable products. These stages include collection, transportation, sorting, cleaning, reprocessing, and marketing.

Collection involves the gathering of items to be used for recyclable products and can involve drop-off centers, buy-back centers, and curbside collection. Transportation of collected materials is necessary to move items to the place where they can be most conveniently separated. Sorting involves separating materials by hand, by weight, by magnets, or by color of glass—each step is necessary depending on which materials are being recycled. Cleaning is a necessary stage in the recycling process, as some materials, such as cans or bottles, may be adulterated with foreign material that would impede their refurbishment. Cleaning is the primary stage in the reuse of recyclable products that have a life span of more than one use, such as the cleaning and subsequent reuse of bottles. Reprocessing involves the remanufacturing of recyclates into different products for reuse or resale. An important yet often overlooked necessity in the viability of the recycling process is the marketing of the refurbished products.

Without a lucrative market for reprocessed or reused recyclable products, there exists little incentive to corporate or state interests to commit themselves to anything other than the least-expensive option. Policy makers, activists, and industry representatives intent on changing this paradigm often place emphasis on the environmental costs of failure to salvage or to use recyclable products. These groups argue that the ultimate cost of not using recyclable products will have long-term detrimental effects that are more expensive than the alternative. To remedy this, a system of government-sponsored financial incentives to encourage the use of recyclable products is advocated. These financial incentives include rebates, low-interest loans, start-up funds, and government subsidies.

Often the practicality of recycled materials manifests itself in the development of products not to be found in department or grocery store aisles; indeed, many recyclable materials are commonly used in the production of roads, bridges, park benches, and other civic structures one might not ordinarily consider when regarding the reprocessing of recyclable products. Many of these materials can be collected from the off-sourcing of virtually any industrial manufacturing process. Such projects have the potential to not only improve the conditions of many existing roadways and city structures but also to simultaneously save millions of dollars that would otherwise be diverted to the use of virgin materials.

## Financial Considerations

Thus far, much hesitation toward the adoption of recyclable products for use in remanufacturing has stemmed from the same financial concerns that make the recycling of other materials seem advantageous to begin with—the use of recycled steel is significantly cheaper than the processes of mining and refining iron ore to forge virgin steel. Indeed, United States laws mandate that 25 percent of all steel must be recycled. In addition, although wood pulp is a renewable resource, the energy costs associated with planting and nourishing new trees greatly outweigh those of recycling paper. Yet some recyclable products are costlier to reprocess, whereas perceived lack of demand or utility inhibits their remanufacture, thus contributing to unessential, nonbiodegradable refuse in landfills.

Seemingly unrelated political decisions, such as the prohibition and suspension of the use of industrial hemp in many parts of the world beginning in the 20th century, encouraged in part by stakeholders and lobbyists within the timber industry, can have drastic and unforeseen implications, such as an exponential contribution to unnecessary deforestation resulting from the stark contrast in the growth cycles of plants, the amount of resources they consume, and the amount of viable fibers they yield.

The use of recycled glass is considered to be greatly beneficial because, unlike the process of recycling plastics and other materials that yield lower-grade recyclable products, the quality of recycled glass remains much the same. This concept of lower-grade materials resulting from their remanufacturing is known as "down-cycling." Conversely, "up-cycling" is the process of taking materials that were previously regarded as rubbish and finding new ways to increase their value and utility.

Aluminum is an ideal material for reuse and is consequently found in many recyclable products. Recycled aluminum costs a fraction of the amount needed to create the product from raw materials and, when recycled, retains all of its original properties, allowing for infinite reuse. Recycled aluminum costs 95 percent less than products made using raw bauxite ore. Recycled aluminum requires the metal to be collected, shredded, chemically or mechanically cleaned, compressed into large blocks, and then melted down in furnaces. Solid and liquid waste from paint and other material is then filtered from the molten aluminum. Aluminum recycling has been widely practiced since the beginning of the 20th century, and the metal's increased use to produce beverage cans in the 1960s led to some of the earliest recyclable products.

Paper is another commonly used recyclable product. Over 70 percent of the newsprint sold in the United States is estimated to be recovered for reuse or export. Of this amount, nearly 60 percent is returned to paper mills for reuse. Paper is recycled through a procedure whereby any printing on or coloration of the paper is removed by a process known as "deinking." Deinking requires paper to be sorted, debaled, and pulped. When pulped, the paper is chopped, water is added, and chemicals are mixed in to remove ink and coloration. Chemicals commonly used as part of the deinking process include sodium silicate for pH control, hydrogen peroxide for bleaching, lime or calcium chloride for a calcium ion source, and fatty acid for collection of ink. After deinking, the pulp is mixed with pulp from newly harvested wood. Deinking also removes fillers, clays, and fiber fragments, which is significant as they reduce the quality of the recycled paper. Each time paper is recycled, its quality decreases, making the use of pulp from harvested wood necessary. The U.S. Environmental Protection Agency estimates that recycled paper causes 35 percent less water pollution and 74 percent less air pollution than making virgin paper. Recyclable paper products are popular, although recent changes in municipal recycling programs have made it more difficult for paper manufacturers to produce recycled paper. Because many municipalities have moved toward "single stream" recycling, in which various waste products are collected in a single compartment and not sorted, paper mills have been forced to spend more to procure a clean, appropriate source for pulping purposes.

Today even the methane gas released by decomposing landfills may be harvested for energy production. Although most methane is produced from natural gas fields, it can also be obtained via "biogas," a gas produced by the biological breakdown of natural matter in the absence of oxygen. Biogas can be made by fermenting organic matter, such as manure, feedstock, sludge, and solid waste under anaerobic conditions; that is, without air. Methane can also be extracted from coal deposits or produced by electrolysis from common atmospheric gases and hydrogen. Methane also is important for use in electrical generation, as it is often used as fuel for gas turbines or steam generators. Compared with other hydrocarbon fuels, such as petroleum or coal, methane produces less carbon dioxide for each unit of heat produced.

## Participation Needed

Widespread participation in municipal recycling programs is necessary to make these programs remunerative. To that end, municipalities often pass ordinances that prohibit the needless scrapping of recyclable products, but these ordinances often go unenforced, especially in large urban areas, where it is most difficult to encourage participation in recycling programs. In industrialized, Western nations throughout Europe and North America, recycling rates have increasingly improved to around 30 percent to 40 percent of disposed solid wastes. However, the preservation of recyclable products is virtually nonexistent in other parts of the world, and especially in developing nations.

An examination of recyclable products' economic benefits is complicated, insofar that certain externalities, or those impacts that affect individuals not involved in a transaction, are often ignored. When externalities are ignored, prices of recyclable products fail to reflect the full costs or benefits in production or consumption of that product. Although municipalities, for example, can incinerate trash less expensively than a recycling program, this ignores the positive externalities that stem from recycled products. Positive externalities that result from recycled products include decreased air pollution and greenhouse gases from incineration and reductions in hazardous waste leaching from landfills, energy consumption, waste, and resource consumption. Taxes and subsidies are often needed to internalize externalities for business concerns, which can otherwise ignore the costs borne by society as a whole. Exploring recyclable products' economic benefits as related to all parties improves decision making.

Recyclable products are not without their critics. The majority of products manufactured were not designed with recycling in mind. As a result, the costs of transforming used goods into recyclable products may be greater than using virgin materials to produce the same item. Certain recyclable products marketed as "green" also fail to consider the complete environmental ramifications of the product. For example, although cardboard food packaging may be more easily recycled than plastic, it may result in more food spoilage and increased transportation costs because it is heavier, and thus more costly to ship. Similarly, the push to use recycled paper products rather than virgin alternatives often ignores that tree farmers plant more trees than they cut down. The oft-repeated statement that "A green economy provides jobs" also ignores the numbers of forestry industry, paper mill, and other workers laid off as a result of recyclable products. To remedy these conflicts, greater emphasis on sustainable design is needed for recyclable products, and indeed all goods and services, to maximize the opportunities for and efficiencies of recyclable products.

Sustainable design is an environmentally driven philosophy that explores the design of physical objects, the built environment, and services. Sustainable design demands that the design of physical objects, the built environment, and services comply with the principles of ecological, economic, and social sustainability. To achieve these ends, every recyclable product, and the packaging it requires, needs a complete closed-loop cycle to be mapped out for each component it contains. By doing this, recyclable products and their components can be designed to return after use to the natural ecosystem or to be recycled indefinitely.

Human societies must develop adequate, practical methods that provide for the commonplace reuse and recycling of previously extracted and manufactured material resources. The dawn of the Industrial Age has bestowed on humanity the many significant conveniences associated with technological advancement. However, this evolutionary milestone has come at the price of the commoditization and devaluation of many common possessions that humans had previously been forced to reuse. The damages we continue to wreak on the natural biosphere as a result of our irresponsible overextraction of nearly every known resource will, in the end, cost humanity far more to cope with than the burdens associated with developing adequate recycling systems. Industries, as well as the individual, must reassess firmly entrenched attitudes toward their consumption of resources long enough to comprehend the urgency of the environmental situation. The longer the avoidable, unnecessary hyperextraction of the Earth's natural resources goes unchecked, the more devastating the realization will be that these practices are completely unsustainable. Societies around the globe can reduce human suffering and the squandering of resources through better stewardship of resources used for products. Recyclable products can play an important role in this change.

**See Also:** Advertising; Biodegradability; Consumer Culture; Ecological Footprint; Green Marketing; Lifestyle, Sustainable; Production and Commodity Chains; Regulation; Sustainable Consumption.

## Further Readings

Braungart, M. and W. McDonough. *Cradle to Cradle: Remaking the Way We Make Things*. New York: North Point, 2002.

Chen, Y. J. and J. Sheu. "Environmental-Regulation Pricing Strategies for Green Supply Chain Management." *Transportation Research*, 45/5:667–77 (2009).

Dahmus, J. B. and T. G. Gutowski. "What Gets Recycled: An Information Theory-Based Model for Product Recycling." *Environmental Science & Technology*, 41/2:7543–50 (2007).

Hall, E. J. *Garbage. Our Endangered Planet Series*. San Diego, CA: Lucent Books, 1997.
Roberts, D. "The Truth About Green Jobs." *Mother Jones,* 33/6:52–54 (2008).
Saphores, J. M., et al. "Household Willingness to Recycle Electronic Waste: An Application to California." *Environment & Behavior,* 38/2:183–208 (2006).

*Stephen T. Schroth*
*Jason A. Helfer*
*Daniel O. Gonshorek*
*Knox College*

# Recycling

*Recycling* refers to both physical and social processes that have the capacity to remake or modify materials, most often at the point of disposal. This article will therefore examine definitions of the physical recycling process before exploring the various cultural, political, economic, and social issues that make recycling such a major part of everyday life for citizens around the world.

In simple terms, recycling is the reuse of materials through some form of natural, mechanized, or industrial process. Accordingly, recycling differs from reuse, which refers to the use of materials for the same purpose without any physical modification to that product. A useful example of this distinction is the reuse and recycling of glass; the reuse of glass bottles or jars involves the refilling or repacking of the same, unmodified glass object with either the same or different products. Although the trend is waning somewhat, numerous countries have relied on the reuse of glass bottles for delivering milk and storing beer or other liquids. In contrast, recycling glass involves a natural, mechanical, or often industrially based process of modification: glass is broken down, modified, and remade, either into another glass product or an entirely different object. This is perhaps what people most readily think of when the term recycling is used—a trip to the bottle bank to recycle wine or beer bottles.

Recycling is therefore a physically based process relying on some form of natural or mechanized change to remake or modify materials. However, this seemingly simple physical process is complex in a number of ways, and the rest of this article explores the various issues surrounding recycling that have made this topic a major area of study for academics concerned with green consumerism.

The first major issue that the study of recycling presents is positioned at a philosophical level and concerns the ways in which both societies and individuals view the underlying issue of waste. In debates concerning how people and societies manage their waste, academics and practitioners have derived what is termed a "waste hierarchy," which places waste reduction at the top as the most preferable form of activity and the landfilling of waste at the bottom, as the action of last resort. Within this hierarchy, reuse is placed below reduction and recycling below reuse. Accordingly, within the broader issue of waste management, recycling is the third preferred option, after reduction and reuse. Recycling is not, therefore, representative of the most environmentally sound alternative to disposal but is an option to be preferred before landfilling waste. This has led many academics and practitioners to argue that, in principle, encouraging recycling is less preferable to reducing level of consumption (and therefore the products and materials used) or the reuse of products. Such assertions are challenging because they imply that citizens should seek to consume fewer products and reuse those that they do purchase, thus altering the dominant consumer culture of capitalist societies.

## Attitudes and Practices

However, within the waste hierarchy, recycling is still the most commonly applied technique for preventing waste from being landfilled. Indeed, in wider debates on green consumption and sustainability, recycling is usually the mechanism most frequently cited as being environmentally conscious. Accordingly, recycling also has an important cultural embeddedness, which researchers need to recognize when discussing green consumption. This embeddedness is, of course, varied, and large differences in both the attitudes toward and practices of recycling occur between and within societies. A useful example of this is the way in which developed societies have come to rely on large-scale mechanized recycling as the way of collecting and altering materials (e.g., curbside collections of materials taken to a central sorting facility). However, developing countries often have efficient but smaller-scale recycling programs, which may be either formal or informal. A good example of this would be composting of food and plant wastes, which are then used as fertilizer on farmland. Such composting schemes are now becoming popular in developed nations, although the dominant means of dealing with food or plant waste is still to landfill these materials.

The cultural embeddedness of recycling is also reflected between different nations with seemingly similar characteristics. Europe provides an excellent example, as recycling rates vary among countries with similar political and economic systems. The recycling rate in Austria, for example, is just over 60 percent, whereas in Greece is it only 10 percent. These marked differences are significant because the European Union sets targets for waste management in member states.

Accordingly, recycling represents differences between nation states, which leads to another major issue with recycling—the political or policy dimension. Within most nations there are both formal and informal modes of recycling. Formal modes relate to municipal collections of recyclable materials either at the curbside or at static recycling centers. However, in most societies these schemes are voluntary, and individuals are encouraged but not forced to participate. Nonetheless, the increasing urgency with which the problem of waste is viewed in many nations is witnessing an increase in more sophisticated policies for encouraging participation, including both incentives for recycling (financial or otherwise) and penalties for contaminating recycling receptacles with nonrecyclable materials. Politically, societies vary in their tolerance of such measures, which may partly account for the varying recycling rates in regions such as Europe. However, alongside these formal, state-driven measures, there has been a recent upsurge in developed nations of community groups taking a larger role in recycling initiatives. These may be either solely focused on recycling or may be part of wider social inclusion movements. Whichever is the case, such groups are often based at the neighborhood scale and focus on local citizens collecting a range of materials from local householders. These community-based schemes are often viewed as highly effective, as they provide a community focus and sense of ownership, which is rarely the case with municipal programs.

## Economic Factors

Allied to the political dimension are economic factors, which have a major effect on the viability of recycling schemes, particularly those based at the community level. Overall, recycling in developed nations is particularly susceptible to fluctuations in global market prices for recyclates, such as paper or glass. This often means that those in charge of collecting materials have to make a choice between halting collections of materials for which there is a reliable market value or continuing collections and storing materials until the market changes. These economic fluctuations are significant because they affect the perception of citizens concerning the value of their recycling efforts.

Accordingly, the final issue that poses a major challenge to effective recycling is the social dimension. In most recycling programs, citizens are the ultimate decision makers and thus have considerable power in determining the success or failure of recycling initiatives. Research into the barriers and motivations for individuals and households to participate in recycling programs is considerable and spans work from social psychologists, sociologists, anthropologists, geographers, and many others. Although this research is complex and varied, there is broad consensus that recycling is most effectively pursued when it is made convenient and simple to understand. Indeed, simply providing a curbside recycling facility makes a major difference to recycling rates. Other factors that influence participation range from psychological factors (such as social desirability, perceived effect of one's behavior, and the satisfaction of recycling) to structural or situational concerns (such as access to facilities, social composition, and access to information). Indeed, recent scholarship has also explored the importance of exploring recycling at different scales (the household, citizen, neighborhood) alongside the importance of examining how recycling relates to both other contexts (e.g., recycling at work or on vacation) and its relation to other behaviors (e.g., how does recycling influence someone's energy or water conservation behavior?).

Recycling is an accepted and commonly applied form of environmentally responsible behavior. However, numerous challenges remain for academics and practitioners to make recycling more effective and also to ensure that recycling is not the only waste management solution that individuals and societies use to promote green consumption.

**See Also:** Composting; Consumer Culture; Environmentalism; E-Waste; Waste Disposal.

### Further Readings

Barr, Stewart. *Household Waste in Social Perspective: Values, Attitudes, Situation and Behaviour*. Aldershot, UK: Ashgate, 2002.

Davies, Anna. *Geographies of Garbage Governance: Interactions, Interventions and Outcomes*. Aldershot, UK: Ashgate, 2008.

Gregson, Nicky, et al. "Identity, Mobility, and the Throwaway Society." *Environment and Planning D: Society and Space*, 25 (2007).

Lund, Herbert. *McGraw-Hill Recycling Handbook*. Chicago, IL: McGraw-Hill, 2000.

Tudor, Terry, et al. "A Tale of Two Settings: Does Pro-Environmental Behaviour at Home Influence Sustainable Environmental Actions at Work?" *Local Environment*, 12/4 (2007).

*Stewart Barr*
*University of Exeter*

## Regulation

Government regulation has proven necessary to achieve many green goals. Although the prospect of corporate profit may encourage the development of biofuel and other alternative fuel sources, or the sales of hybrid cars, there are many areas in need of green policies and regulations where a profit motive does not exist. The benefits of preventing water, soil, and air pollution; of conserving water; of protecting biodiversity; and of preventing deforestation, are analogous to the gains of pure research compared to applied research. Although applied research—such as creating tape that has more adhesive power, or more energy-efficient refrigerators—leads to a direct benefit to the researcher's employer, the benefits of research into the human genome or the origins of the solar system are either intangible or distributed throughout the populace without ownership.

Similarly, the benefits of keeping water clean are felt by all of society and the ecosystem as a whole, not just those who take the time and effort to do so. However, such benefits in both cases do lead to tangible benefits later—it was the pure research of physicists like Albert Einstein that made the applied research of the atomic bomb possible. Over the years, the private sector has greatly reduced its commitment to pure research, feeling that they were not gaining profit from it; the government, to protect the scientific interests of the country, had to step in and increase public-sector funding of such research. Government intervention is similarly required when market forces fail to provide protection for the environment.

In the United States, a variety of federal agencies are empowered to regulate and enforce the regulation of various issues regarding the environment. The most obvious and most visible is the U.S. Environmental Protection Agency (EPA), which was established in 1970 and is the principal body responsible for U.S. environmental policy. Though pesticides had been regulated in the past, the EPA provided the federal government with the structure to regulate all pollutants. Some of its powers are delegated to state and tribal governments, but the agency itself is large, with 17,000 employees at 10 regional offices, 27 laboratories, and various program offices, as well as the headquarters in Washington, D.C.

## Energy Star

In 1992, the EPA began the voluntary Energy Star program, which sets standards of energy efficiency for a variety of products. The program has since been adopted by Canada, the European Union, Japan, Australia, and New Zealand. Although Energy Star specifications vary from product to product—including almost 50,000 products—in general, they call for a 20 percent to 30 percent improvement in energy efficiency compared with standard versions of the product. Energy Star products include lighting, computers and electronics, and major appliances, and the program Website maintains a list of "payback periods." For instance, it takes three years for the savings resulting from energy efficiency to accrue enough for an Energy Star washing machine to "pay for itself." The average life of a washing machine is of course much longer. The idea is to inform the public of the long-term personal gains of energy conservation, even when the initial cost is higher. Because Energy Star is voluntary, it has been spared much of the politicization of the agency; the program is effective only for so long as its "brand" represents real gains and meaningful levels of energy efficiency, and because of agency and industry efforts to preserve that truth, it has been a considerable success. In recent years, Energy Star standards have been extended to new homes, which can be made more energy efficient through tight construction and proper insulation, energy-efficient air conditioning and heating, and special windows and water heaters. As of 2008, more than 10 percent of new U.S. homes were Energy Star compliant, a promisingly high number. Separate ratings are used for commercial buildings, and specific standards have been developed for banks, courthouses, acute care hospitals, children's hospitals, hotels/motels, schools, offices, dormitories, retail stores, supermarkets, refrigerated warehouses, nonrefrigerated warehouses, municipal wastewater treatment plants, automobile assembly plants, cement plants, and corn refineries.

Despite the general success of the program, its implementation in the United States hit some stumbling blocks in 2006, when ratings on more than 20 product categories were found to be faulty. In response to complaints from states, a federal court required the ratings to be reviewed and updated—an ongoing process that is expected to be completed in the summer of 2011. In many cases, the problem is with the way the tests are run to determine energy usage; for instance, some refrigerators are tested with the automatic ice makers disabled, which is not congruent with their real-world usage.

The WaterSense program launched in 2006 is similar to the successful Energy Star, being a voluntary program that sets standards of water efficiency for products such as toilets, faucets, and irrigation equipment. Water is something Americans in most parts of the country have taken for granted, as droughts and water rationing having historically been limited to particular regions. With the growing problem of drought in the southeast, and a more informed concern for water usage in general, water conservation may finally become the public concern that energy conservation has.

Under the Safe Drinking Water Act, the EPA sets the safety standards for 160,000 public water systems in the country and regulates injection wells to protect underground water sources. Bottled water, because it is sold as a beverage, is regulated by the U.S. Food and Drug Administration.

The EPA is responsible for testing vehicles for fuel economy, and those test results are the only ones manufacturers are permitted to report. The testing methodology includes a 12-year review cycle of test procedures because of the observation in the early 21st century that drivers experienced substantially lower fuel economy than the EPA tests reported, a result of changes in driving since the tests were originally written—for instance, the original test used a highway driving speed of 48 miles per hour, never exceeded 60 miles per hour, and accelerated slowly. The most recent version of the test has been changed to account for 21st-century drivers' higher speeds, particularly on the highway, and generally more aggressive style of driving—as well as cold weather driving and driving with the air conditioner running. Vehicles of model year 2008 and later are expected to have much more accurate fuel economy ratings as a result.

## Politics and Criticism

Though half of its large personnel is scientists, engineers, and other technical experts, the EPA has been increasingly criticized for having become politicized. Ridiculed in the 1970s and the 1980s by the Right as hippies and meddlers who wanted unreasonable standards met for mythical or impractical gains, in the 1990s and 2000s the EPA began to rely more on its administrative personnel when forming policies and specific standards. In the 21st century, for instance, the EPA limits for particulates, carbon monoxide, lead, nitrogen oxides, and ground-level ozone have all been changed, attracting the ire of the scientific community. The agency has decreased its interaction with the scientific community in general, and sometimes acts as though practical concerns—the usual negotiations of lobbyists, business interests, and politicking—are more important than hard scientific data.

In 2007, the EPA was sued by 17 state governments, led by California, for refusing to allow the states to raise fuel economy standards for cars sold or driven in those states. This was one move credited to the agency's shift in focus as they lessened the involvement of the scientific community; many saw the apparent forcing on these states of lower standards, by an agency created to raise environmental standards, as pandering to the weakening U.S. auto industry. Furthermore, later that year, then-administrator Stephen Johnson gave his okay to an agency document that declared that climate change had become a severe-enough threat to jeopardize public welfare. Entering this document into the public record would require the agency to enforce mandatory global warming regulations—a move that the federal government has long avoided. Specifically to prevent the agency from being politically manipulated, federal law states that this decision—evaluating whether climate change is a public danger, and therefore requires regulation—is left to the agency, not to other bodies or officials in the government. Nevertheless, when the White House was emailed the document, they refused to open it, as doing so would enter it into the public record. Months later, in July 2008, Johnson issued a revised report that removed the clause declaring climate change a public threat.

An April 2008 report of the Union of Concerned Scientists revealed that out of 1,600 EPA-employed scientists surveyed by detailed questionnaire, 40 percent reported political interference in their work and that the amount of interference was greater in 2008 than it had been five years earlier. The area of research with the greatest number of complaints was that of determining the carcinogenicity of chemicals used in food and consumer products like makeup.

The Consumer Product Safety Commission, an independent federal agency, was created by the 1972 Consumer Product Safety Act to protect consumers from "unreasonable risks of injuries associated with consumer products," a duty that sometimes overlaps with green concerns. For instance, the commission banned lead-based paint for residential use in 1978 and has subsequently outlawed lead toys for children.

Among numerous recent acts and strategies being enacted at the U.S. state level, New York passed the Green Jobs/Green New York Act in late 2009 to fund energy audits and energy-efficient retrofits for residential, small business, and nonprofit property owners. New York hopes to reduce fossil fuel dependence, lower housing costs, support community development and create green jobs to sustain and enhance the state economy. Connecticut has appointed task forces to study energy scarcity and sustainability and create a state solar energy strategy. Maine passed an act to increase the availability of solar and wind power with a rebate program, and California has dozens of green legislation issues that were recently enacted or are still in the proposal stage, on issues ranging from greenhouse gas emissions, building zero net energy buildings, and providing job training to expand the workforce needed to implement renewable energy, solar panel manufacturing, energy-efficient structure retrofitting, environmental cleanup, and pollution control.

**See Also:** Automobiles; Biodegradability; Carbon Credits; Carbon Emissions; Carbon Offsets; Certification Process; Environmentalism; Government Policy and Practice (Local and National); Green Politics; International Regulatory Frameworks; Water.

### Further Readings

Carter, Alan. *A Radical Green Political Theory*. London: Routledge, 1999.
Dobson, Andrew. *Green Political Thought*. London: T & F Books, 2009.
Radcliffe, James. *Green Politics: Dictatorship or Democracy?* New York: Palgrave Macmillan, 2002.
Torgerson, Douglas. *The Promise of Green Politics: Environmentalism and the Public Sphere*. Durham, NC: Duke University Press, 1999.

*Bill Kte'pi*
*Independent Scholar*

# Resource Consumption and Usage

In the context of expanding populations and growing economy, resources have to be consumed/used on a sustainable basis. The environment is the basis for resources: sources of supply or available means, whether natural or manmade, that enhance the quality of life. The environment around us can be considered as source and sink—it is fundamental for economic activity and human well-being. However, its ability to perform these functions is being irreversibly destroyed/degraded by our ruthless overexploitation. Three common themes (often inseparably linked) underlie all environmental and resource problems: population density, distribution, and growth (as well as demand), poverty issues (distribution, accessibility, and entitlement), and politics and control over resources.

Hence, the following questions on the relationship between environmental sustainability and economic growth are of key concern when dealing with resource consumption and usage:

- Can the contribution of the environment/resources to human welfare and to the human economy be sustained?
- How can economic growth be sustained without causing environmental/resource degradation and ecosystem collapse?

This article deals with current status, resource policies, and sustainable consumption issues for resources.

## Current Status of Resources

Natural resources are currently under increasing pressure, threatening environmental and human health and development. Water shortages; soil exhaustion; loss of forests; air, soil, and water pollution; and degradation of coastlines afflict many areas. As the world's population grows, improving living standards without destroying the resource base/environment is not only a local and regional issue but also a global challenge.

Most of the developed countries currently consume resources much faster than they can regenerate. Most developing countries with rapid population growth face the urgent need to improve living standards. Seventy-five percent of energy resources are consumed by 25 percent of the population in developed countries. They also consume more than 70 percent of mineral resources (copper, steel, aluminum, etc.), 75 percent of cars, 75 percent of newsprint, timber, and so on. Furthermore, 70 percent of carbon is emitted by this population. One American child requires more than 30 times as much resources as an Indian child. Decarbonization requires shifts in energy policy, dramatic technical progress, and major changes in consumption patterns of the rich. Paradoxically, the poor aspire to follow the path taken by the rich tomorrow. Three-quarters of the poorest families live in rural areas and still depend in large measure on natural resources for their existence; they remain vulnerable and their future insecure.

The ecological footprint concept is very relevant when we consider the issue of increasing resource consumption. It is the area of productive land and water—global hectares ("fair earth share" value)—ecosystems require to produce the resources that a population consumes and to assimilate/absorb the wastes that the population produces, wherever on Earth that land and water may be located over a certain time period (usually a year).

Because human impact on ecosystems/resources is a function of population, affluence, and technology, reducing the global footprint will require a combination of population stabilization and reductions in consumption and waste levels, which can be facilitated by the development of lower-impact technologies.

Global footprint or "Earthshare" in 2001 was 2.2 global hectares/capita, although the bio-capacity was only 1.80. In 2003, humanity's ecological footprint exceeded the Earth's biocapacity by over 20 percent, and in 2006 by 40 percent. Thus, the ecological footprint is essentially a dynamic concept because of population change and composition, technology change (in the supply chain), changes to the amount of resources consumed, and changes in the types of resources consumed.

Resource intensity is a measure of the efficiency of resource use and is a calculation based on the unit cost of resources (e.g., coal, water, materials) necessary for the production, processing and disposal of a unit of a commodity. It is used in green or sustainability accounting and is often expressed as units of resources used per unit of Gross Domestic Product (GDP). A high value suggests a resource-intensive production, suggestive of significant environmental costs in converting the energy, raw materials, and inputs into a good or service. It can be calculated as units of resource expended per unit of GDP. Resource efficiency is concerned with the efficient use of resources as inputs to production. Calculating intensities and efficiency not only enables resource use comparisons, but can be used as a basis for using resources more effectively and ultimately more sustainably.

- resource use (RU) = resource intensity (RI)
- value added (VA) = resource efficiency (RE)

Resource/pollution intensity can be consistently defined across various scales.

- For a company, it is resource use per value added of the company.
- For a product, it is life cycle resource use per unit cost of the product.
- For an industry, it is total resource use divided by total value added in the industry.
- For a nation, it is total resource use divided by gross domestic product:

$$RI(A + B) = (RU[A] \times RU[B])/(VA[A] \times VA[B]).$$

Resource intensity can be calculated for, and compared among, individual processes, facilities, consumption activities, households, regions, industry sectors, and nations. Resource productivity/ecoefficiency/energy/pollution intensity can be calculated as follows:

Resource productivity = $/(tons, GJ, kWh) or $/tons emissions

Resource or Ecoefficiency = tons output/(tons, energy input) or energy output/energy input

Resource/Energy/Pollution intensity = inverse of productivity and efficiency

## Sustainable Production and Consumption

The emphasis of sustainable production is on the supply side of the equation, focusing on improving environmental performance in key economic sectors such as agriculture, energy, industry, tourism, and transport. Sustainable consumption addresses the demand side, looking at how the goods and services required to meet basic needs and to improve quality of life—such as food and health, shelter, clothing, leisure, and mobility—can be delivered in ways that reduce the burden on the Earth's carrying capacity. It is common to think of production and consumption as discrete stages in a product's life cycle chain, with production (an industrial activity) preceding consumption (a domestic activity). However, production and consumption are inextricably interwoven. All production consumes resources and energy: to produce something requires that something must be consumed.

The use of goods and services that respond to basic needs and bring a better quality of life while minimizing the use of natural resources, toxic materials, and emissions of waste and pollutants over the life cycle, so as not to jeopardize the needs of future generations, is called "sustainable consumption." Policy makers in both developed and developing countries have a crucial role to play in stimulating the recognized need for "fundamental changes in the way societies produce and consume." The concept of sustainable consumption and production (SCP) encompasses this need. There are basic questions we can ask ourselves. Are there enough resources to support our ever-growing economies? Do the increasing amounts of waste and emissions exceed the carrying capacity of the environment? Are we using resources in a sustainable manner?

The answers boil down to fundamental changes that are required in the way we produce and consume resources. This leads to the following three basic strategies for sustainable use of natural resources: decoupling, or reducing negative impacts of resource use in a growing economy; improving resource efficiency; and focusing on key economic sectors.

By decoupling resource from economic growth, we get "more value per kilogram"; by decoupling environmental impact from resource use, we obtain "less impacts per kilogram." It is increasingly evident that current economic growth and development patterns cannot be sustained without significant innovation in both the supply (production) and demand (consumption) sides of the market. Decoupling economic growth from environmental impact will require producers to change design, production, and marketing activities, and consumers will need to provide for environmental and social concerns—in addition to price, convenience, and quality—in their consumption decisions.

Among the several internationally accepted policy agendas for achieving SCP, the following are important: Agenda 21 Chapter 4 (1992), Earth Summit II (1997), Human Development Report (1998), United Nations (UN) Guidelines for Consumer Protection (1999), Commission for Sustainable Development (1999), UNEP GEO-2000, and WSSD: 10-year Framework of Programs (2002). Although some countries have dedicated SCP policies and programs, in other countries their SCP-related commitments may form part of broader sustainable development and/or national economic programs. For instance, Rio Declaration Principle 8 deals with the duty of the states to reduce and eliminate unsustainable patterns of production and consumption, and Agenda 21 Chapter 4 has identified that the major cause of continued deterioration of the global environment is the unsustainable patterns of consumption and production.

SC requires strategic policy framework, business development, consumer engagement, and social discussion. This necessitates multistakeholders' dialogue that would bring people together across the sectors: governments, industry, civil society, scientists, and academic researchers with a cross-sectoral approach. This would bridge the gap between different issues and dimensions of SCP across sectors, space, and time.

Resource consumption and environmental impact during production, use, disposal, and so on are predetermined by the product design. Hence, ecodesign (from "cradle to grave" or "cradle to cradle") is one key to SCP. The following are essential ecodesign tools:

- Environmental description of a product: quantitative and qualitative details.
- Assess environmental impacts of a product (along the whole life cycle).

- Record and analyze production processes: I/O, ERA, SEA, ISO 14000, EMAS, and so on.
- Consider stakeholder requirements systematically for design/innovation.
- Develop improvement strategies from process, product, and stakeholder requirements.
- Evaluate the improvement ideas to deliver clear options for a green product concept.

The desirable change needs to be guided by public policy—the market cannot deliver it by itself (because of externalities). Markets are driven principally by prices (also by public perception/reputation, etc); public policy must systematically use prices to drive markets in the desired direction (greater resource efficiency, reduced emissions, more dynamic innovation/new industries). Regulation is often necessary or complementary to policy based on market incentives (e.g., demand-side management where markets fail) but does not achieve dynamic efficiency (innovation), and a key public policy is green tax and budget reform.

## Resource Consumption and Usage: Future Perspectives

Resource consumption and usage can be sustained by adopting integrated natural resource management, which aims to augment social, physical, human, natural, and financial capital. It does this by helping solve complex real-world problems affecting natural resources in ecosystems. Its efficiency in dealing with these problems comes from its ability to do the following:

- Empower relevant stakeholders
- Resolve conflicting interests of stakeholders
- Foster adaptive management capacity
- Focus on key causal elements (and thereby deal with complexity)
- Integrate levels of analysis
- Merge disciplinary perspectives
- Make use of a wide range of available technologies
- Guide research on component technologies
- Generate policy, technological, and institutional alternatives

The achievement of SCP requires significant technological innovation—to optimize the design, development, production, distribution, sale, and marketing of products and/or services—as well as changes in social behavior, most notably in terms of changed consumption patterns.

The current production and consumption patterns of the wealthy developed world cannot be replicated in the rapidly industrializing nations of the developing world. The morally defensible desire of developing countries to enjoy the (more resource-intensive) consumption patterns of the developed world, and the reluctance of those in the North to reduce their increasing material consumption, presents significant political challenges. In the developing country context, policy makers are faced with the challenge of meeting basic needs (by increasing economic development and consumption), while at the same time reducing the associated environmental burden. If everyone in developing countries were to live like U.S. citizens, we would need four planet Earths!

This requires the development and implementation of policy measures that decouple economic growth from environmental pressures and that prevent the rebound effect in terms of which growing consumption outstrips technology improvements and efficiency gains. SCP requires innovative approaches, both socially and technologically, that enable developing countries to eradicate poverty while respecting the natural constraints to consumption and production. A pragmatic, far-sighted set of indicators can be of significant support to decision makers in developing countries, helping to identify national risks and opportunities in time and to promote sustainable growth and development of their economies.

**See Also:** Consumer Culture; Electricity Usage; Energy Efficiency of Products and Appliances; Environmentalism; Environmentally Friendly; Ethically Produced Products; Fuel; Green Consumer; Green Consumerism Organizations; Lifestyle, Sustainable; Organic; Overconsumption; Production and Commodity Chains; Recyclable Products; Recycling; Regulation; Simple Living; Sustainable Consumption; United Nations *Human Development Report 1998*; Waste Disposal.

## Further Readings

Food and Agriculture Organization of the United Nations (FAO). *Conflict and Natural Resource Management*. Rome: FAO, 2000.

Natcher, D. and C. Hickey. "Putting the Community Back Into Community-Based Resource Management: A Criteria and Indicators Approach to Sustainability." *Human Organization*, 61:350–63 (2002).

United Nations Environment Programme (UNEP). *SCP Indicators for Developing Countries: A Guidance Framework*." Paris: UNEP, 2008.

World Resources. *Roots of Resilience—Growing the Wealth of the Poor*. United Nations Development Programme, United Nations Environment Programme, World Bank, and World Resources Institute, 2008.

*Gopalsamy Poyyamoli*
*Pondicherry University*

# S

## Seasonal Products

Conventionally, *seasonal products* referred to merchandise specific to a holiday season—items and associated colors and fragrances sold during well-established times of year: heart decor in February, pastel baskets in spring, beach imagery in July, foliage motifs in autumn, and cinnamon scents in December. Within the rubric of green consumerism, however, *seasonal products* adopts a more ecologically minded—and thus literal—meaning. Rather than referring to the time of consumption, the seasonal aspect of a product has come to describe the time of its production. Usually, seasonality within green parlance refers to foods and agricultural products—products directly reliant on, and thus related to, the various seasons of the year.

Globalization of the agrifood industry has allowed for rapid and regular international transport of fresh produce. This mobility was first facilitated by refrigeration, and then by an alteration of the produce itself—from agribiodiverse and organic (though not yet certified as such) to the uniformity of monocrop produce, bred and sprayed to withstand mechanized harvests, long-term refrigeration, transport, and shelf life. In North America, mass production allowed for relatively (or falsely cheap) Thanksgiving asparagus, Christmas fruit salad, and Valentine's Day chocolate-dipped strawberries.

Over the past few generations, this agricultural transformation has resulted in a marked shift in aesthetics. What began as luxury became necessity. After World War II, the ability to purchase nonlocal, nonseasonal foods became a marker of status, conferring an aura of personal wealth, and ultimately of societal progress. Now, however, postindustrialized communities, defined as individualized consumers, have come to expect the same array of fruit and vegetable options in January as they do in June, and they depend on daily snacks comprising fresh produce grown around the globe.

This presumption is predicated on the deliberate erasure of the time, place, mode, and means of produce's production. The produce aisle of the modern supermarket presents its goods as the context-less, place-less, timeless (nonseasonal, nondecaying) product of name brand, multinational corporations. Photogenic, shiny uniformity thus becomes the standard by which fresh produce is judged, not the subtleties of its taste, ripeness, texture, nontoxicity, or nutritional quality, nor the social and ecological consequences of its production and distribution.

Large-scale seasonal eating requires the patience to wait for such fruits as watermelon to be in season.

*Source:* Keith Weller/Agricultural Research Service/USDA

Green consumerism, in contrast, seeks to address this spatial and temporal disconnect, with the local foods movements working to redress the former and the seasonal products phenomenon the latter. As the essential counterpart to the celebrated attribute of local, seasonality constitutes and necessitates a shift in expectations and appetite. For instance, consumers and nutritionists have come to value—and recognize—only a small handful of ironically healthy fruits (apples and bananas) and vegetables (lettuce and tomatoes). Large-scale seasonal eating, however, would require a decidedly more exploratory and creative consumptive paradigm on the part of the masses, wherein eaters appreciate more "exotic" wild or locally adapted foods and varieties, from kohlrabi to dandelion leaves, as well as the edible parts of otherwise familiar foods, from beet greens to garlic scapes to pumpkin seeds. It would also require collective patience to wait diligently for melon season and an investment in researching local and seasonal sources of vitamins and minerals available in each place, around the year.

The twin goals of locality and seasonality emerge from environmental, social, political, and physical health concerns. The slow food movement, born in response to the global fast food trend and industry, was one of the first to champion seasonality as the crux of gourmet food. Since then, elite chefs have begun to agree on developing menu prototypes reflecting respective local growing seasons and declaring freshly harvested foods as the best-quality fodder for fine cuisine.

Those on the local/seasonal food public front line also seek to frame eating seasonally as primarily an environmental justice issue. Often, although not absolutely, heirloom or organic, seasonal food has come to represent the delicacy of sustainability—the quality-of-life component of ecological responsibility. Buying seasonally necessarily means buying locally, thereby reducing the "food miles" of a product, or the distance it has traveled from field to plate (the average being 1,500 miles.) Offering local and seasonal foods, as well as drinks, flowers, textiles, and crafts, expresses ecological commitment, in that less product travel and refrigeration translates into lower carbon emissions, and less packaging means less waste.

Moreover, in purchasing from nearby producers, a larger percentage of the price paid goes to the farmer herself. This supports local agricultural economies, which then conserve local farmlands. Farmland conservation staves off ecologically disruptive seas of concrete, as well as the "green concrete" of shallow-rooted lawns and golf courses—all of which exacerbate floods, droughts, and runoff, which in turn diminish local air and water quality, and arguably the rural landscape itself.

In addition to their ecological and social import and improved taste and texture, seasonal foods also have substantial health benefits. The vitality of ripeness confers higher and more absorbable vitamins and minerals and, clearly, lacks preservatives. In addition, many traditional knowledge traditions show that often, nutrients offered by freshly harvested foods align with common nutritional needs at that time of the year. For instance, after a long winter without fresh produce, the intricate and powerful micronutrients, vitamins, minerals, and antioxidants found in spring greens are particularly nourishing to the immune system, detoxifying the major organs and the blood stream.

Despite the green consumer push to transition the agrifood system and general public's taste toward seasonally grown foods, obstacles remain. Chief among them are the problem of winter, especially in regions with short growing seasons, and the high price in general of fresh produce. Small-town farmers markets may offer an array of locally grown sweet corn, beans, and zucchini during the northern hemisphere summer. However, the winter months may only offer expensive bananas and bland Red Delicious apples from Wal-Mart and gas stations.

Another obstacle is the unfavorable ratio of growers to eaters. Another ubiquitous, surprising sustainable-agriculture statistic is that 2 percent of U.S. citizens make a living farming, and the majority of them grow nonedible crops: field corn bred for animal feed or corn syrup, soy bred for processed foods, cotton, or hay. The challenge to eat seasonally is hampered by the fact that so little food is currently grown for immediate consumption.

Moreover, the highly centralized distribution lines supplying food for chain restaurants, grocery stores, and supermarkets and the vast networks of school, business, hospital, and prison cafeterias are built to handle highly processed, chemically preserved, uniform foods with indefinite refrigerator and shelf life. They remain stubbornly unconducive to the vicissitudes, diversity, perishability, and tender ripeness of freshly harvested produce.

Accordingly, green-food growers and consumers who advocate seasonal eating are working to establish, cultivate, and connect alternative distribution channels, such as farmers markets, community-supported agriculture, natural food co-ops, and niches of grocery store produce aisles. Some are also embarking on growing their own seasonal foods. The National Gardening Association has declared 2009 the Year of the Vegetable Garden and estimates that 7 million more American households will plant vegetables, fruits, or herbs in 2009, raising the total to 43 million households—an increase of 19 percent from 2008.

Overall, seasonal products—exemplified within green consumerism as seasonal foods—now constitute a growing and timely phenomenon within sustainability movements.

**See Also:** Food Additives; Green Food; Markets (Organic/Farmers); Slow Food; Vege-Box Schemes; Vegetables and Fruits.

## Further Readings

Bendrick, Lou. *Eat Where You Live: How to Find and Enjoy Fantastic Local and Sustainable Food No Matter Where You Live*. Seattle, WA: Skipstone Press, 2008.

Natural Resources Defense Council. "Eat Local." http://www.nrdc.org/health/foodmiles/?gclid=conm7/rnjk14CFSUsawordGj3how (Accessed November 2009).

Petrini, Carlo. *Slow Food Nation: Why Our Food Should Be Good, Clean, and Fair*. New York: Rizzoli Ex Libris, 2007.
Sustainable Table. "Eat Seasonal." http://www.sustainabletable.org/shop/eatseasonal/ (Accessed November 2009).

*Garrett Graddy*
*University of Kentucky*

# Secondhand Consumption

If consumption is understood as a process of selecting, purchasing, using, maintaining, and disposing of goods and services, secondhand consumption can be understood as the consumption of goods that have already been through this cycle and reached the point of disposal. Essentially, it is the process through which a consumer good enters a second cycle of consumption. It involves a range of consumer practices and takes place in a range of social spaces including (but not limited to) charity shops, retro-retailers, flea markets, swap meets, car boot sales, and eBay. Unlike a lot of first-cycle consumption, it occupies a mundane location in everyday life, insofar as it typically involves the exchange of ordinary goods in ordinary locations.

The second cycle of consumption carries traces of the first, such as mileage on a car, wear and tear on clothing, or scribbling in the margins of a book.

*Source:* iStockphoto

Nevertheless, it is an interesting topic that requires a completely different understanding of consumption and exchange. For example, in first-cycle consumption, money serves to abstract the goods and services being exchanged from the individuals undertaking the transaction, just as prices provide an external measure of the item's value. In contrast, the second cycle of consumption is haunted by the first, insofar as an item is likely to carry traces of its previous use (miles on a car, wear and tear on clothing, scribbling in the margins of a book), meaning that the exchange—and more important, the value—needs to be mediated by the individuals involved. Similarly, the symbolic meaning of goods exchanged, and indeed, the meaning accorded to the practice of secondhand consumption, needs to be mediated in ways that are different from the first cycle of consumption. Here, it is important to note that secondhand consumption is a heterogeneous activity that is characterized by multiple practices and meanings.

To begin, it can be understood as a type of consumption that marks differences, displays identities, and makes one's culture and tastes visible. For example, in the case of buying retro clothing or antique furniture, individuals are required to have a stock of knowledge that is often a manifestation and marker of social status. Indeed, an antique chair or a vintage leather jacket carries certain connotations that a chair from IKEA or a jacket from the Gap does not. Here, the meaning of secondhand consumption is defined by its opposition to first-cycle consumption, insofar as it reflects individuality and distinction, as opposed to the homogenizing effects of mass consumption. In contrast, secondhand consumption can also be understood in terms of capturing economic value, as opposed to the creation of symbolic value. For example, in the case of shopping at car boot sales, flea markets, and charity shops, the emphasis is often on getting a bargain. Indeed, secondhand consumption is very much caught up with the virtues of thrift, but even here the practice can assume different meanings. For example, shopping at a charity shop because one is poor and has no choice but to save as much money as possible is very different than shopping at, say, a flea market because one derives pleasure from getting something for less than the going rate. The emergence of new technologies, most notably eBay, has added another layer of meaning to secondhand consumption. For example, there are some individuals who would never buy secondhand goods at a car boot sale on the grounds that doing so implies a lack of money, standing, and respectability. These same individuals may be perfectly happy to purchase them on eBay because, in this context, the same goods do not carry the same connotations.

Of course, secondhand consumption is not just about thrift and the work that this entails. To the contrary, it can be understood as something that is fun or a leisure activity in its own right. For example, it can provide an opportunity to simply "shop around" in different, possibly more interesting, spaces of consumption in the search for interesting bric-a-brac that may or may not actually be purchased, just as it can provide the opportunity to unearth something unusual, a one time treat for oneself or gift for a loved one. Similarly, secondhand consumption can be seen as a pleasurable activity that allows for the subversion of ordinary relations of exchange. For example, secondhand consumption often involves a more personal form of contact between buyers and sellers than is the case in first-cycle consumption, just as it opens the possibility of haggling over prices or exchanging other goods.

Finally, there is a tacit assumption that secondhand consumption has a natural affinity with green consumption, insofar as, at its most basic level, it is concerned with reuse and using things to their fullest potential. There is good reason to make connections between this and a wider critique of consumer culture, especially in terms of overproduction, overconsumption, and waste. Similarly, secondhand consumption is often caught up with a wider set of practices such as shopping at independent retailers (for what they represent, not just what they sell), local markets, and food cooperatives, as well as buying Fair Trade products or food that is otherwise environmentally friendly (organic, low food miles). Viewed as such, secondhand consumption can be understood as more than simply a disposition for reuse, insofar as it forms part of a wider ethical consumption milieu. That said, Nicky Gregson and Louise Crewe's seminal work on secondhand consumption has demonstrated that it represents less of an alternative to—and critique of—consumerism than it might initially appear. They demonstrate how secondhand consumption creates more consumer choice and enables persons to get more for less, such that it ultimately entails more consumption rather than less. Nevertheless, it is worth noting that many secondhand consumers are also those who dispose of their own first-cycle goods into the secondhand arena. For example, it is very common for items with limited scope for use in the first cycle of consumption (such as books or children's clothing) to be passed on so that others can benefit from them and, of course, to avoid waste. Here, the emphasis is on redistribution, as well as on environmental benefits, and it suggests that there is still scope for secondhand consumption to provide a powerful critique of consumer culture.

**See Also:** Consumer Ethics; Fair Trade; Frugality, Lifestyle, Sustainable; Sustainable Consumption.

### Further Readings

Gregson, Nicky and Louise Crewe. *Second Hand Cultures*. Oxford: Berg, 2003.
Hawkins, Gay and Stephen Muecke. *Culture and Waste*. New York: Rowman & Littlefield, 2003.
Jackson, Peter, et al., eds. *Commercial Cultures: Economies, Practices, Spaces*. Oxford: Berg, 2000.

*David Evans*
*University of Manchester*

## Services

The service industry, sometimes called the tertiary sector of the economy, consists primarily of the sale of "intangible goods." Employing a greater part of the workforce than the primary (agriculture and mining) or secondary (manufacturing) sectors, the service industry is the source of soft-sector employment such as the hospitality industry, restaurants, healthcare and hospitals, real estate, legal and other professional services, education, and the media. "Service industry" and "service sector" are to be understood differently than "service economy," which is a model of doing business in which service is emphasized even though the core of the business is the sale of physical goods. Because the service sector does not involve factories or fertilizers, it is often overlooked in discussing environmental issues and the green revolution. Much of its consumption is less obvious to the general public, who may not stop to think about how much water is involved in washing their dishes after a dinner out, or the environmental impact of air-conditioning a casino in a hot climate. The environmental toll can in fact be substantial, and efforts have been made, particularly in recent years, to become more efficient and conscious of resources usage.

More and more service-sector companies are developing a specifically green focus, often advertising themselves as such.

### Hospitality

The hospitality industry encompasses hotels, restaurants, bars, clubs, casinos, and spas. A typical green issue in this area is the use of water. Hotels use an average of about 150 gallons of fresh water per occupied room per day; some use much more. Even apart from intentional usage, small amounts of waste from dripping faucets or inefficient toilets add up considerably when multiplied by the numerous bathrooms in the hotel. In an effort to conserve water, some hotels provide financial incentives for employees to help meet water-reduction goals, which encourages the cleaning staff to reduce the amount of water they use when cleaning. Bathroom fixtures can be replaced to reduce inefficiency caused by leaky plumbing, and to update them to current usage standards. Older toilets used at least three gallons per flush, whereas contemporary toilets use about half of that. Washing machines can be programmed to use shorter rinse and prewash cycles, and water flow controls can be installed on shower heads (though low water pressure in the shower is a common area of guest complaint).

Since 1996, the 12,000-member American Hotel & Motel Association's Good Earthkeeping program has encouraged hotels to conserve water by limiting the laundering of towels and linens. Hotels participating in the program leave cards in the rooms that guests can use to indicate that they would be happy with housekeeping straightening the room and making the bed, but not replacing the towels and sheets. A similar program conducted through the Holiday Inn chain has expanded to 500 participating members from the 82 who joined when it was launched in 1995.

Throughout the hospitality industry, urinals in men's restrooms can be replaced with waterless urinal systems, which come in various designs. Most use a trap insert using a layer of lighter-than-water sealant liquid that floats to the top of the collected urine. Converting a single urinal from the conventional style to a waterless design saves as many as 50,000 gallons of water a year.

The grounds of a hotel or restaurant can be a considerable source of water use because of the irrigation needs of the greenery. Efficient irrigation systems can be installed, using timers, and wasting water can be avoided by hosing down the sidewalks and driveways. Mulch around shrubbery and other plants reduces evaporation and discourages weed growth. Furthermore, the landscaped area can be redesigned entirely with water efficiency in mind. In Las Vegas, the dry, desert climate required the MGM Grand Hotel to use thousands of gallons of water every day to keep its lawns green until management redesigned the landscaping with plants appropriate to the region.

A number of the items used in hotels can be made with recycled or green materials. Many green hotels use recycled paper tissues and provide in-room recycling receptacles, while providing cleaning staff with environmentally safe cleaning products, using organic cotton for their linens (the pesticides used in the production of conventional cotton are a significant environmental hazard), and installing clean air-filtration systems that do not rely on chemical-laden air fresheners. Green hotels also often offer sustainable menus, with local ingredients when possible.

The Environmental Protection Agency maintains the Energy Star program, a set of product-type-specific standards for energy efficiency. Green hotels can apply to the program to have the energy efficiency of their buildings evaluated under the specific standards set for hotel construction. Similarly, the U.S. Green Building Council maintains the Leadership in Energy and Environmental Design (LEED) Green Building Rating System, a set of standards for environmentally sustainable construction. LEED buildings are designed to use resources more efficiently, from the use of renewable resources and water-efficient landscaping to improved ventilation and the use of low-emitting materials. The LEED standards have been criticized for emphasizing responsible use of fossil fuels more than the use of alternative fuels, a criticism that may affect future versions of the standards. At the moment, there is no specific set of standards for newly constructed hotels, as there is for some other kinds of new construction; the LEED standards for Existing Building Construction do include hotels.

## Restaurants

In some parts of the United States, restaurants are marketing themselves as "green restaurants." These restaurants are typically housed in Energy Star–rated energy-efficient buildings and offer sustainable food, almost always with an emphasis on local ingredients (and especially local family farms and sustainable food businesses). Menu transparency is a hallmark of the green restaurant, noting the provenance of all the major ingredients—who raised the cattle the steak came from, who grew the potatoes, and where those farms are. Herbs and many vegetables may be grown on-site, or the restaurant may own or have a relationship with a farm. Some restaurants even have on-site apiaries, raising their own bees for the honey.

In addition to the green hotel innovations that are applicable to restaurants—such as recycling and the use of low-flow plumbing and waterless urinals—green restaurants can also use low-flow spray nozzles for washing dishes and recycled containers for take-out. Waste food can be composted. Grease traps and kitchen hoods must be properly maintained to prevent overflow into sewer or storm drain systems.

The green restaurant movement predates most other sectors of green business. Alice Waters founded Chez Panisse in 1971 with an emphasis on seasonal, sustainable, local ingredients and a strong advocacy of farmers markets. The Chez Panisse Foundation, founded on the restaurant's 25th anniversary in 1996, is a charitable organization that sponsors such initiatives as the Edible Schoolyard program, which teaches elementary and middle school students how to grow their own food.

## Hospitals

One of the newest types of green construction is the green hospital. Conventional hospitals can easily spend $20 million a year on utilities, as they are far more energy intensive than office buildings of equivalent size, and ground-up redesigns offer rich possibilities for greater efficiency. The $442-million University Medical Center at Princeton, New Jersey, expected to finish construction in 2011, is a prime example. The building faces south, letting sunlight into the rooms, and sensors adjust the level of electric lighting based on natural light levels. More sensors pick up the carbon dioxide level in the room and from that deduce the number of people in the room and adjust the temperature accordingly. There are no lawns in need of fertilizers and pesticides—instead, the hospital is surrounded by a 32-acre park on the shore of the Millstone River, with a bike path and indigenous plants. Solar panels provide electricity, and a collection system recycles rainwater.

Other green initiatives in use by some hospitals include using recycled jeans instead of fiberglass for building insulation and recycled steel in the construction of the building. The vast roof spaces of hospitals are ideal for either solar panels or rooftop gardens. The constant cleaning of hospital floors means that just switching the janitors' mops from cotton to more efficient microfiber saves a hospital thousands of gallons of water a year. For health reasons, some hospitals had long ago switched to mercury-free lightbulbs and nontoxic cleaning supplies, and recycling programs have long been in place to raise money for toys for children's wards. An innovation dating from the 1990s is the biodegradable bedpan—flushable after use, made of recycled telephone books and beeswax, and popular with patients because it is not cold to the touch like the conventional metal pan.

LEED certification is available for hospitals. The points-based system rewards the use of local building materials, improved ventilation, rooftop gardens, and environmentally conscious paints and adhesives, as well as energy-efficient construction. Hospitals generate a good deal of toxic and medical waste, and part of being green means responsibly processing those, rather than burning them, as has conventionally been done. The first hospital to be awarded LEED Platinum certification, the highest level of the LEED standards, was the Dell Children's Medical Center of Central Texas, in Austin. In addition to the above innovations, the center saves 1.7 million gallons of water a year by using dual-flush toilets and low-flow plumbing. The heating and cooling system and power plant is 75 percent more efficient than the coal-fired plants traditionally used by hospitals, and the steam used by the power plant's combustion chamber cools the hospital's water. Almost 10 tons of material are recycled every month. Green hospitals produce much more recyclable waste than the general public would think—not only cardboard, metal, and plastic but also alcohol, linens, batteries, used cooking oil, and packing peanuts.

As with green hotels, green hospitals have the option of offering sustainable menu options. Some have begun growing their own vegetables and fruits and even have farmers markets on hospital property. Four hundred and fifty healthcare entities in the United States have signed a food pledge to buy local and organic food whenever possible.

## Spas

As places of luxury and indulgence, spas can often also be places of wastefulness and excessive energy consumption. This need not always be the case. Ecofriendly spas have been around for a long time, but the green revolution has placed them even more prominently in the spotlight of the spa/resort industry. Catering to environmentally conscious clientele, green spas feature picturesque water-efficient landscaping—typically made up of indigenous plants rather than lawn—and prioritize recycling and the use of postconsumer materials, while offering their services in ecofriendly buildings. Food and spa treatments are usually organic and sustainable (and may often be vegetarian), while activities revolve more around running, bicycling, swimming, or martial arts than resource-intensive exercise equipment.

Many spa resorts are located on properties with a good deal of plant life and woodlands, and some spas have begun purchasing carbon offsets to balance the impact of the transportation guests use to visit. Power may be provided in part by solar panels or wind farms.

**See Also:** Ecotourism; Final Consumption; Green Homes; Public Transportation; Swimming Pools and Spas; Slow Food.

### Further Readings

Brower, Michael and Warren Leon. *The Consumer's Guide to Effective Environmental Choices: Practical Advice From the Union of Concerned Scientists*. New York: Three Rivers, 1999.

Garlough, Donna, et al., eds. *The Green Guide: The Complete Reference for Consuming Wisely*. New York: National Geographic, 2008.

Rogers, Elizabeth and Thomas Kostigen. *The Green Book*. New York: Three Rivers, 2007.

*Bill Kte'pi*
*Independent Scholar*

# Shopping

Shopping can be understood as a very public manifestation of global capitalism and modern consumer culture. If consumption is to be understood broadly as the selection, purchase, use, maintenance and disposal of goods and services, shopping can be understood as a distinctive form in which the focus is on pleasure, leisure, and choice. Shopping emerges in a monetary economy in which the institutions and spaces exist for truly global systems of production and distribution to intersect with the individual consumer. Although shopping is synonymous with the high levels of consumption in modern societies, it is a mistake to think of it simply as consumption for the masses. Rather, it is more useful to think of shopping in terms of the high levels of individual consumption that stem from the apparent insatiability of individual needs and wants. Academic interest in shopping can be traced back to Walter Benjamin's account of the Parisian shopping arcades of the 19th century, and consequent engagement has tended to conceive of it in terms of the spectacle, with a particular focus on archetypical sites such as shopping malls. Perhaps unsurprisingly, a popular view of shopping is one of it being a trivial and frivolous activity through which money is spent frivolously and (self) indulgently in the pursuit of hedonistic desires. Furthermore, in this framing, shopping serves to abstract individual consumers from the world around them such that they cannot engage critically with consumer culture or global capitalism. Compelling as this view is, it is problematic insofar as it obscures the disagreement that exists between those who have turned their attention to this rather extraordinary type of shopping and empirical evidence that demonstrates how the majority of shopping that people do is far more mundane, routine, and ordinary.

There is a strong tradition of theorizing shopping in terms of industrial capitalism's ability to manipulate individual consumers to serve its own ends. For a start, Karl Marx discussed how the fetishization of commodities served to conceal the exploitative relations of their production, such that individuals could be pacified and distracted by the acquisition of goods instead of questioning their working conditions. Similarly, it has been suggested that shopping is geared toward the pursuit of "false" needs and that these are created—under the aegis of wants and desires—to persuade people to purchase things that they do not really need. The idea here is that shopping never really manages to deliver the satisfaction that it promises, leaving the individual feeling hollow and empty with only the promise of yet more shopping to fill the void. Finally, it has been argued that the choices we appear to have while shopping are merely an illusion, insofar as products only differ to the extent that they are targeted at different categories of shopper. For example, from a functional point of view, there is very little difference between a sports car and a small hatchback, and yet they are differentiated by any number of symbolic characteristics that reflect existing social hierarchies. Through shopping (and displays of taste), we reproduce these hierarchies and service the interests of the dominant social groups.

Taken together, this implies that shopping is a manifestation of consumer culture's ability to position individuals as cultural dopes and restrict individual choice and freedom. Nevertheless, there is a completely different tradition of theorizing shopping that recognizes the symbolic capacity of shopping and celebrates the creative freedoms it affords. For example, Mike Featherstone has highlighted how shopping enables persons to create an identity or design a lifestyle for themselves. The emphasis here is on genuine choice and the opening up (or breaking down) of social structures, such that "anyone can be everyone" and persons can mark their individuality. On the question of "false" needs, such an approach would question the extent to which social theorists can proclaim a need to be "false." Even if we agree with their evaluation, it is highly likely that those to whom they attribute this "false" need may not experience or even acknowledge it as such. Similarly, it is problematic to suggest that wants and desires are somehow insignificant when there are no grounds for thinking of wants and desires as innately trivial or flippant. One could, in fact, easily claim that wants and desires are simply "higher-order" needs that transcend the needs of biological survival.

Finally, there exists considerable disagreement as to whether or not shopping can really make people happy. For example, there is a good deal of popular literature documenting the inability of shopping to bring about happiness, the gist of which suggests that shopping is very good at providing short-lived pleasures but less good at providing long-term happiness. The idea here is that shopping is concerned with having material goods and possessions, which obscures a focus on being a happy individual who is living a meaningful life. Again, this is problematic insofar as shopping can be understood as an important activity through which persons give meaning to their lives. Of course, there are those who would be concerned by the handing over of this process to commercial agencies, just as there are those who would stress that shopping fails to deliver the meaning that it promises. Nevertheless, Grant McCracken has made the case that the ability of shopping to provide meaning lies precisely in its ability to satisfy wants and desires. Similarly, shopping is increasingly being understood as an important realm of social interaction and experience that structures everyday life beyond the isolated act of purchasing things. Indeed, it is widely held that shopping is an activity that involves a range of practices such as experiencing public spaces, "hanging out," and wandering around. As such, it can be understood as a leisure activity in its own right that does not necessarily involve the actual purchase of goods and services.

Although there is clearly disagreement among these understandings of shopping, they all start with the idea that shopping is something rather extraordinary and interpret it through the language of meaning, identity, desire, symbolic communication, and ideology. Although these approaches are useful, they only tell half the story, insofar as the type of shopping that they account for is not typical of the type of shopping that people actually do. For example, Daniel Miller's brilliant empirical study of shopping demonstrates that it is rarely undertaken in an abstract social context, in the pursuit of pleasure, or as an end in itself. Rather, it takes place in less glamorous social spaces, is geared toward the routine provisioning of household items, and is an expression of love, care, and devotion to the family grounded in sacrifice. The idea here is that shopping is work, not leisure. Similarly, empirical studies of High Street shopping and shopping in regional shopping centers have revealed shopping to be less about individual lifestyle choices (whether freely chosen or based on "false needs") and more about wider social structures such as gender relations and the family, generational differences, and racial inequalities. Finally, there are spirited defenses of shopping in which it is positioned as an activity of considerable significance in modern life and one that is worthy of serious academic study. For example, Mary Douglas has suggested that shoppers should be understood less as persons who—for whatever reason—respond to prices and fashion and more as rational coherent individuals who make culture visible through their shopping practices. Similarly, Daniel Miller has argued that the type of shopping that he describes should be understood less as the realm of private decision making and more as an activity with profound moral and political consequences.

**See Also:** Consumer Culture; Diderot Effect; Leisure and Recreation; Overconsumption.

### Further Readings

Douglas, Mary and Baron Isherwood. *The World of Goods*. London: Routledge, 1996.
Falk, Pasi and Colin Campbell, eds. *The Shopping Experience*. London: Sage, 1997.
McCracken, Grant D. *Culture and Consumption: New Approaches to the Symbolic Character of Consumer Goods and Activities*. Bloomington: Indiana University Press, 1990.
Miller, Daniel. *A Theory of Shopping*. New York: Cornell University Press, 1998.
Shields, Rob, ed. *Lifestyle Shopping: The Subject of Consumption*. London: Routledge, 1992.

*David Evans*
*University of Manchester*

## Shopping Bags

The debate regarding shopping bags and their (mis)use, particularly regarding the relative merits of the two principal "contenders," plastic and paper, is heated and likely to become more so. In addition, recently a further complexity has been introduced in the form of the reusable fabric bag.

To convey purchases home, plastic shopping bags in 2008 accounted for something like 95 percent of the grocery (supermarket) and convenience store market (up from only 80 percent in 2003, as then estimated by the American Plastics Council) in the United States, meaning an annual consumption of 99 to 100 billion bags. Despite this popularity being based on their convenience, strength, ease of availability, and comparative cheapness (approximately one cent for plastic compared with four cents for paper), plastic shopping bags have come in for a huge amount of negative commentary, even if the situation is far from straightforward. The statistics are impressive, if not totally agreed to.

- Worldwide, between 500 billion and one trillion plastic bags (so not just of the shopping variety) are consumed each year (that is more than one million per minute), and up to 3 percent of these enter the litter stream. Of the total, something like 380 billion relate to the United States, and only 1 or 2 percent of these are recycled.
- In the United Kingdom, 17 billion plastic bags are provided by supermarkets each year, resulting in 60,000 tons of plastic going to landfill.
- Plastic bag litter kills thousands of marine creatures every year when they mistake it for food or get entangled in it. Estimates produced by the Blue Ocean Society for Marine Conservation suggest in excess of one million birds and 100,000 marine mammals become casualties annually.
- An estimate from the United Nations Environment Programme suggests there are 46,000 pieces of plastic litter in every square mile of the world's oceans.
- In a vortex of ocean currents called the Northern Pacific Gyre, 1,000 miles off the coast of California, is an area at least the size of Texas (some estimates suggest up to twice the size) containing a churning mess of plastic debris.
- Even in landfills, the bags can take up to 1,000 years to break down, and in the breaking-down process they release toxins that contaminate soil and water, meanwhile blocking natural flows through the soil of oxygen and water. The alternative of incineration leads to both toxins and carcinogens being released into the air.

- Plastic bag manufacture uses a nonrenewable resource (usually cited as petroleum, although in reality more likely to be natural gas).
- Plastic bags have been cited as choking sewer systems (Philadelphia) and, beyond that, leading to floods in Bangladesh and Mumbai through blocked storm drains (400 dead resulting in the latter case in 2005).
- Conventional plastic bags do not biodegrade easily and merely fragment endlessly. It was estimated that the stomach of a cow that died in New Delhi contained fragments from no fewer than 35,000 plastic bags.

This is (part of) the background to the case against plastic shopping bags and has encouraged both states/counties and whole countries to totally ban, or at least attempt to restrict, their prevalence. Thus, San Francisco and Oakland, California, have outlawed plastic bags in large grocery stores and pharmacies should they not be compostable, and even the obvious alternative, paper bags, can only be used if they boast 40 percent or more recycled content. Los Angeles plans to follow suit in 2010, and Baltimore, Maryland; Santa Monica, California; New Haven, Connecticut; Boston, Massachusetts, and others have all considered bans, although the results have, at best, been mixed—the plastic bag lobby, comprising plastics manufacturers who point to their industry's nearly half a billion dollar contribution to the economy, has in many cases mounted a spirited, and effective, counterattack.

Elsewhere, South Australia has become the first Australian state to put a complete ban on plastic shopping bags

Interestingly, however, the "Ban the plastic shopping bag" campaign has been one of those rare examples of the developing/recently developed world leading the developed, with North America and most of Europe lagging behind. Thus, a number of Indian states have outlawed plastic bags (Ladakh back in the early 1990s), as have Taiwan, Rwanda, Bhutan, Thailand, Papua New Guinea, and Zanzibar. Hong Kong mounted a "No Plastic Bag Day" that saw a 40 percent drop in demand; Singapore held a "Bring Your Own Bag Day," resulting in 100,000 plus bags being conserved and over 20,000 reusable bags being sold in that single day. In China, retailers can be fined for supplying "thin" bags.

On the more positive side, it has been indicated that, although largely intended for single use, plastic shopping bags often find alternative application as trash/rubbish bin bags and for cleaning up after canines. It has thus been maintained that banning supermarket plastic bags would simply cause an increase in demand for more special purpose ones. In addition, the more benign environmental alternative potentially suggested by paper bag usage might not be that simple. The Film and Bag Federations (a subsidiary of the Society of the Plastics Industry) has suggested that compared with paper grocery bags, the plastic variety consume 40 percent less energy, generate 80 percent less solid waste, produce 70 percent fewer atmospheric emissions, and release up to 94 percent fewer waterborne wastes. Paper bags are also more bulky and heavier than plastic, with obvious effects on their transportation and consequent fossil-fuel use implications. So it is not all just a case of paper bags coming from a sustainable resource.

So what are the alternatives?

- If they have not banned plastic shopping bags completely, many North American cities have at least commenced implementing a fee for their use (frequently 5 cents). This is based on the Irish initiative of "PlasTax" instituted in 2002, which led to a 90 percent drop in consumption in the first year and raised money (3.5 million euros in just five months) to fund environmental projects.

- Direction can be provided by the major retailers themselves. Wal-Mart is committed to reducing its global plastic shopping bag waste by 33 percent by 2013, IKEA by 50 percent in the year from April 2008.
- Produce bags are available that are more rapidly biodegradable. J. Sainsbury in the United Kingdom has introduced home compostable bags that are 100 percent biodegradable, and further developments should see a shopping bag equivalent being available.
- Encourage local community developments, along the lines introduced by BBC camerawoman Rebecca Hosking after her experience in the Pacific of seeing birds and mammals choking to death on plastic.
- Probably the best solution of all is to encourage the use of high-quality reusable shopping bags, made from renewable resources and using environmentally benign production, which can make hundreds of trips to and from the local supermarket. This can be reinforced by stores simply declining to offer free plastic bags.

**See Also:** Biodegradability; Environmentalism; Green Consumer; Recycling.

### Further Readings

Australian Government, Department of the Environment, Water, Heritage and the Arts. "Plastic Shopping Bags—Analysis of Levies and Environmental Impacts." http://www.environment.gov.au/settlements/publications/waste/plastic-bags/analysis.html (Accessed July 2009).

Hosking, Rebecca. *Ban the Plastic Bag: A Community Action Plan*. Bristol, UK: Alastair Sawday, 2008.

Wisconsin Department of Natural Resources. "Frequently Asked Questions About Plastic Shopping Bags." http://dnr.wi.gov/org/aw/wm/recycle/issues/plasticbagsfaq.htm (Accessed July 2009).

*David G. Woodward*
*University of Southampton School of Management*

## Simple Living

Simple living (also referred to as "voluntary simplicity") is a lifestyle that abhors materialistic tendencies and the pursuit of wealth and luxurious consumption. It primarily minimizes the notion that "more is better" for possessions. This is distinguishable from poverty, as simple living is a voluntary lifestyle choice. It is also different from asceticism, which also generally promotes living simply and refraining from luxury and indulgence, yet not all proponents of voluntary simplicity are ascetics.

Such lifestyle choices have traditions that stretch back to Asia, preached by leaders such as Buddha, Lao-Tse, and Confucius, and were also heavily stressed in both Greco-Roman culture and Judeo-Christian ethics. Epicureanism, based on Athens philosopher Epicurus, which flourished from the 4th century B.C.E. to the third century C.E., was based on the premise of an untroubled life as a paradigm of happiness. He believed the troubles attributed to the maintenance of an extravagant lifestyle outweigh the pleasure of partaking in it. He, therefore, recommended that one should maintain what is necessary for happiness, bodily comfort, and life at minimal cost, while all things beyond what is necessary for these should either be tempered by moderation or completely avoided.

Various religious groups like the Amish, Harmonic Society, Mennonites, Shakers, and Quakers have for centuries practiced lifestyles in which some forms of materialism and technology are excluded for religious or principled reasons. For example, there is a Quaker belief called Testimony of Simplicity that recommends that one live his or her life in a simple manner. Various notable individuals over the course of history throughout the world have been known for their austere lifestyle, such as Francis of Assisi, Ammon Hennacy, Leo Tolstoy, and Mohandas Gandhi. Henry David Thoreau, an American naturalist and author, in his book *Walden*, advocated a life of simple and sustainable living. Henry Stephens Salt, an admirer of Thoreau, popularized this idea in Victorian Britain. Subsequent British advocates of the simple life included Edward Carpenter, William Morris, and members of the Fellowship of New Life.

People choose simple living for a variety of reasons, such as spirituality, health, increasing quality time with family and friends, reducing their ecological footprint, reducing stress, personal taste, or frugality. In many ways, voluntary simplicity is a manner of living that is outwardly simpler and inwardly richer. It reduces the need for purchased goods or services and, by extension, reduces their need to sell their time for money. During the holiday season, one can also perform alternative giving or just relax and try to improve their quality of life. Although many individuals seek to buy happiness, materialism very frequently fails to satisfy and may even increase the levels of stress in life.

## Green Consumerism and Simple Living

Some associated sociopolitical goals aligned with the simple living movement include conservation, social justice, and sustainable development. Prominent among proponents of these goals was George Lorenzo Noyes, a naturalist and mineralogist, who was also known as the "Thoreau of Maine." He advocated a wilderness lifestyle, advocating through his creative work a simple life and reverence for nature. In the 1920s and 1930s, the Vanderbilt Agrarians of the southern United States advocated a lifestyle and culture centered on traditional and sustainable agrarian values, as opposed to progressive urban industrialism, which has dominated the Western world. At this time, there are eco-anarchist groups in the United States and Canada promoting simplistic lifestyles. In the United Kingdom, the Movement for Compassionate Living has over the last three decades promulgated the vegan message and promoted simple living and self-reliance as a remedy against the exploitation of humans, animals, and the Earth. Another grassroots awareness campaign in the same country, National Downshifting Week, encourages participants to "slow down and green up," contains suggestions for adopting green eco-friendly policies and habits, and develops corporate social and environmental responsibility at work.

Many green political parties often advocate simple living as a cornerstone of their platforms. In terms of policy, they reject genetic modification and nuclear power and other technologies they consider to be hazardous. Their values of simplicity are reflected in their reduction of natural resource usage and environmental impact. Ernest Callenbach refers to it as the "green triangle" of ecology, frugality, and health. Such "Earth-centered" spirituality includes strengthening the community and simple living.

## Simple Living and Sustainability

The relationship between sustainability and simple living has been discussed in several publications in the decades since the United Nations Conference on the Environment in 1972. Notable among these books are B. Ward et al., *Only One Earth*; Donella Meadows et al., *The Limits to Growth;* and E. F. Schumacher's *Small Is Beautiful: Economics as If People Mattered.* Some have even proposed reducing food miles, or the number of miles a given food item or its ingredients have traveled between farm and the table, to reduce their carbon footprint by eating locally.

In recent years, many have proposed the idea of "simple prosperity" as a consequence of sustainable lifestyle. It is important for one to think about whether long commutes and hedonic consumption are good for us in the long run. Are people happier when they pursue wealth and other materialistic pursuits? Simple living is the opposite of the quest for affluence, and as such, someone who is practicing simple living is less concerned about the quantity, and more about the quality, of life, and in general the preservation of traditions and nature.

James Robertson, in his book *A New Economics of Sustainable Development*, has proposed that a move to sustainability will need a broad shift of emphasis from raising incomes to reducing costs. His specific principles are as follows:

- Systematic conservation of resources and the environment as a basis of environmentally sustainable development
- Systematic empowerment of people (instead of making and keeping them dependent) as the basis for development of human capital
- Transformation from a "wealth of nations" model of economic existence to an ecological, sustainable, and decentralized one-world model of coexistence
- Restitution of political and ethical factors to a central place in economic life and thought
- Respect for qualitative and feminine values, not just quantitative and masculine ones

Simple living advocates have differing views on the role of technology—some are strong critics of modern technology, whereas others see the Internet as critical to simple living in the future because it facilitates sustainable practices via the reduction of an individual's carbon footprint through telecommuting and less reliance on paper. Furthermore, other technologies such as wind and water turbines can be used to make a simple lifestyle within mainstream culture easier and more sustainable.

Many have criticized the role of advertising in encouraging a consumerist mentality. Advocates of voluntary simplicity recommend eliminating or reducing television viewing as a key ingredient in simple living. Other media such as print or community radio are seen as viable alternatives. Given the modern challenges of having to live with wars, economic meltdowns, the ongoing threats of terrorism, and environmental degradation, it is imperative to focus on "re-valuation" and reprioritizing our needs and focusing on a simple living lifestyle.

**See Also:** Advertising; Automobiles; Conspicuous Consumption; Downshifting; Green Communities; Lifestyle, Sustainable; Materialism; Overconsumption; Poverty; Quality of Life; Social Identity; Sustainable Consumption; Super Rich.

### Further Readings

Atkinson, Robert D. "Building a More Humane Economy." *The Futurist*, 40/3:44–49 (May/June 2006).

Baird, Susan. "Self-Empowerment: Your First Step Toward Excellence." *Journal for Quality and Participation*, 17 (December 1994).
Bekin, Caroline, et al. "Defying Marketing Sovereignty: Voluntary Simplicity at New Consumption Communities." *Qualitative Market Research*, 8/4:413–29 (2005).
Brown, Kirk Warren and Tim Kasser. "Are Psychological and Ecological Well-Being Compatible? The Role of Values, Mindfulness, and Lifestyle." *Social Indicators Research*, 74/2:349–68 (2005).
Craig-Lees, Margaret and Constance Hill. "Understanding Voluntary Simplifiers." *Psychology & Marketing*, 19/2:187–210 (2002).
Etzioni, Amitai. "Voluntary Simplicity: Characterization, Select Psychological Implications, and Societal Consequences." *Journal of Economic Psychology*, 19/5:619–43 (1998).
Huneke, Mary E. "The Face of the Un-Consumer: An Empirical Examination of the Practice of Voluntary Simplicity in the United States." *Psychology & Marketing*, 22/7:527–50 (2005).
Robertson, James. "The New Economics of Sustainable Development: A Briefing for Policy Makers." Report for the European Commission, 2005.
Segal, Jerome. "Achieving the Good Life." *Dollars and Sense*, 224 (July/August 1999).
Shaw, D. and C. Moraes. "Voluntary Simplicity: An Exploration of Market Interactions." *International Journal of Consumer Studies*, 33/2:215–23 (2009).
Shaw, Deirdre and Terry Newholm. "Voluntary Simplicity and the Ethics of Consumption." *Psychology & Marketing*, 19/2:167–85 (2002).
Shi, David. *The Simple Life*. Athens: University of Georgia Press, 2001.
Weisser, Cybele. "The Quest for the Simple Life." *Money*, 35/8:134–38 (August 2006).

*Abhijit Roy*
*University of Scranton*

# Slow Food

Most people are familiar with the maxim "you are what you eat," credited to the 19th-century German philosopher Ludwig Feuerbach. The meaning of this phrase (that the food one eats has a bearing on one's mental and physical health) has since been distorted to capture the idea that the way we (as a society) eat reflects the values and behaviors of society more generally. Food, as with many other things in modern society, is now often produced, prepared, and eaten in haste, part of a so-called cult of speed. Global food marketing offers an impressive range of food choices that can be consumed with little effort. It is this global fast food culture and the productivist food systems that go with it that the slow food movement is rebelling against. This article explains the motivations and beliefs of slow food and its geographic spread and core initiatives and reviews studies and critiques related to the movement.

Slow food is a consumer movement established in Italy in 1986, when McDonald's opened a branch in the famous Piazza di Spagna in Rome. The opening of the restaurant raised concerns that traditional eating habits might be threatened by Americanized fast food. In protest, the food writer Carlo Petrini gathered chefs, authors, and journalists together for a meeting to discuss the best means of countering the spread of fast food in Italy. From this, slow food was launched. The slow food manifesto is simple: a movement for the protection of the right to taste. It promotes the following:

- Fresh, local, seasonal produce
- Sustainable farming practices
- Skills and artisanal food production
- Traditional recipes, passed down through the generations
- Eating slowly with family and friends

Slow food exists alongside other initiatives that are also calling for more sustainable, territorially embedded forms of food production and marketing, spurred on by the increasing persistence of "food scares" in the food chain and a concomitant debate about the relationship between food, diet, and health, as rates of obesity and heart disease continue to rise. This includes the burgeoning market for "natural" and "organic" foods and countermoves to return to localism, regional foods, and real cooking.

The slow food movement is strongly consumer orientated. The movement believes that food should be about pleasure and social integration. Slow food is also underpinned by the notion of "ecogastronomy," which promotes the idea that eating well can, and should, go hand in hand with protecting the environment. Taking the snail as its symbol, the central objective of slow food is to decelerate the food consumption experience so that alternative forms of taste can be reacquired. It is the antithesis of fast food culture and consumption formats such as McDonald's, which impose a standardized model on each restaurant and serve food through one formula, the "Speedee Service System." The first edition of the movement's magazine, *Slow*, explicitly sought to oppose the spread of McDonald's and other fast food chains.

The movement also has spatial motivations: it wishes to embed food in territory and bring consumers closer to these foods, reasserting also the natural bases of food production (e.g., seasonality) and the role of cultural context (e.g., culinary skills, tacit knowledge). It is a celebration of cultural connections surrounding local cuisines and traditional products, targeting discerning consumers to heighten their awareness of "forgotten" cuisines and the threats that they face. This has been very successful. In Italy, for example, over 130 delicacies have been saved, including purple asparagus from Albenga, Ligurian potatoes, and lentils from Abruzzi. The prime concern of the movement is thus for "typical" or "traditional" foods, although increasingly recognizing that some regional foods are disappearing because they are too embedded in local food cultures.

Although the roots of the movement lie in Italian food cultures, slow food also promotes "typicality" further afield. In 1989, it launched itself as an international movement and has now spread to more than 50 countries, with 78,000 members worldwide. Membership is strongest in Europe, as one might expect, given the history of indigenous cuisines there, but the movement is also growing in the United States (with a membership of 8,000+ and growing), and even in Japan, where fast food culture is endemic. At a regional/local level, groups are organized into slow food convivia (or consumer clubs). These clubs are central to the movement in terms of enabling local distinctiveness and typicality. Activities include the following:

- Promoting the slow food ethos of "good, clean, and fair food for their area"
- Building relationships with local producers and restaurants
- Raising awareness about and working to protect traditional foods in their area
- Organizing tastings and seminars around key food policy issues
- Conducting tasting workshops with different groups, including schoolchildren

The local representation and activities of convivia mean that slow food operates in a variety of places, including remote areas. In the United Kingdom, for example, there are currently over 50 local groups. These groups are diverse in size, location, and the focus of their activities, ranging from small groups in outlying rural areas such as Cornwall and Ayrshire to metropolitan convivia with a membership of several hundred. The leader and committee of each slow food group are elected by the members, and their activities reflect the members' interests and the nature of the local area.

Education is a critical element of the slow food mandate. The movement has a number of initiatives designed to educate consumers about food and food tasting. These include

- the Salone del Guston (Hall of Taste), the world's largest food festival, which takes place in Turin every two years;
- the "Ark of Taste" project, which aims to identify and catalog artisan products and regional dishes that are in danger of being lost unless promoted;
- the "Week of Taste," an educational program implemented in Italian schools since 1992, which turns classrooms into workshops and kitchens to involve pupils in supervised tasting and cooking experiences;
- a quarterly magazine (*Slow*) published in five languages, as well as various food and wine guides and educational books—for example, in 1998, *Talking, Doing, Tasting*, which was designed for teachers and parents in Italian schools, with experiments to teach children how to enjoy and appreciate their food; and
- opening its own University of Gastronomic Sciences at Pollenzo, near Bra, in 2004, at which students study the history and sensual character of food, as well as the science of food.

Slow food has clearly expanded since its inception to become a significant international movement. One of the strongest symbols of growing consumer interest in this movement is the rapid growth and popularity of farmers markets. In the United States there are now more than 3,000 markets, with an annual revenue of $1 billion. The ambitions of the movement have also expanded over time. For example, it now has a manifesto on biotechnologies that argues that the effects of genetically modified organisms on taste and the organoleptic range of cooked dishes and foodstuffs are underestimated. In 1999, slow food also established the slow city movement, which started as a group of 32 Italian towns and cities but has since expanded; the aim is to preserve local cultural traditions and promote slow living.

There is a growing body of social science work related to slow food, as well as related debates about localism, alternative food networks, and sustainable consumption. Examples include sociological examinations of slow food eating experiences as "practical aesthetic" theoretical accounts of the ethical orientation of convivium, analyses of the social economy of slow food, and various general accounts describing the historical and geographical development of the movement. There are also critiques emerging about the movement and who and what it represents. Some suggest, for example, that slow food, similar to other quality food economy initiatives, only serves a minority element of the populace, a "gourmet club for affluent epicureans." It has also been accused of "defensive localism," which can be elitist and reactionary. Slow food products are clearly very expensive in some instances, and the fine-dining element may well marginalize some groups of society. However, others argue that these criticisms need to sit alongside the movement's strong educational goals. Slow food is also viewed as "alternative" and firmly against global capitalism. Research suggests this does not reflect reality, as many artisanal products (e.g., Parmesan cheese) rely on international markets, something that the movement's founder describes as "virtuous globalization." Other work suggests that a potential mismatch may be apparent between many slow food members, who are keen to celebrate the pleasures of eating and drinking local produce and share homespun philosophies, and the movement itself, which has much wider ambitions to spread the "slow living" gospel, articulated also in slow city and slow sex movements.

**See Also:** Conspicuous Consumption; Consumer Culture; Consumer Society; Markets (Organic/Farmers); Seasonal Products; Sustainable Consumption.

### Further Readings

Honoré, Carl. *In Praise of Slow*. London: Orion, 2004.

Jones, Peter, et al. "Return to Traditional Values? A Case Study of Slow Food." *British Food Journal*, 105/4–5 (2003).

Miele, Mara and Jonathan Murdoch. "The Practical Aesthetics of Traditional Cuisines: Slow Food in Tuscany." *Sociologia Ruralis*, 42/4 (2002).

Petrini, Carlo. *Slow Food: The Case for Taste*. New York: Columbia University Press, 2001.

Petrini, Carlo. "The Slow Food Manifesto." *Slow*, 1 (1986).

Petrini, Carlo. *Slow Food Nation: Why Our Food Should Be Good, Clean, and Fair*. New York: Rizzoli Ex Libris, 2007.

Schlosser, Eric. *Fast Food Nation: The Dark Side of the All-American Meal*. New York: Houghton Mifflin, 2001.

Slow Food. http://www.slowfood.com (Accessed May 2009).

*Damian Maye*
*Countryside and Community Research Institute*

## Social Identity

Social identity is a theory developed primarily by Henri Tajfel and John Turner to understand the psychological basis of intergroup discrimination and preference. The theory attempts to identify the conditions under which members of one group discriminate in favor of the in-group to which they belong and against other out-groups. There are four critical elements of this theory:

- *Categorization:* We label ourselves (and others) into categories such as ethnic groups, hobbies, and so on, which helps stereotype these individuals.
- *Identification:* We associate with specific groups (our in-groups), which helps to reassure our self-esteem.
- *Comparison:* We compare our affiliated groups with others, typically biasing in favor of our groups.
- *Psychological Distinctiveness:* We want our identity to be both distinct from and positively compared with other groups.

According to social identity theory, individuals have several "selves" that correspond to a variety of group memberships. The "level of self" is contingent on different social contexts that trigger people to think, feel, and act on the basis of personal, family, or national factors. The theory is concerned with both psychological and sociological aspects of group behavior. An individual typically has "multiple" social identities.

## Scope and Application of Social Identity Theory

Social identity theory has a considerable effect on several fields, including social psychology. It has been tested in a wide range of settings, including prejudice, stereotyping, negotiation, and language use. The theory also has implications for the way individuals deal with social and organizational change. In one study, subjects were assigned to groups randomly, as meaninglessly as possible. They were found to display in-group favoritism, even in the most minimum circumstances.

A related application of social identity theory relates to the concept of organizational identification, or an employee's identification with his or her firm. Higher organizational identification has been related to lower turnover, reduced absenteeism, improved performance, and increased organizational citizenship behaviors. Other studies report higher incidences of extra role behavior, such as helping other members and offering innovative suggestions. A stronger organizational identification often mirrors one's self-concept—hence, to create and enhance a positive self-image, individuals are likely to seek and identify with a high status organization.

## Social Identity in Economics

In the field of economics, the notion of social identity has been applied to the principal–agent model. The main findings illustrate that when agents consider themselves to be insiders, they maximize their identity by exerting a high effort level compared with the prescription behavior. Alternatively, if they consider themselves outsiders, they will require a higher wage to compensate for their loss for behavior differences with prescription behaviors.

Other tests have shown that when individuals are matched with in-group members, they will be more likely to have a "charity concern" and will be less likely to have "envy" concern. Others have shown similar results, but in the case of negative reciprocity, people were shown to more likely get revenge when they get negative reciprocity from in-group members in sequential games.

Although individuals might identify themselves as green consumers, engaging in distinctive consumption practices designed to lessen impacts on the environment, there is some debate around the implications of this. Author and environmental activist George Monbiot, in an article published in the British *Guardian* newspaper in 2007, suggests that a green social identity promoted by marketers and activists may be a middle-class phenomenon, with individuals purchasing green and ethical products but not in fact reducing their overall consumption. In fact, he suggests that socially identifying with and practicing green choices (despite encouraging people to think more widely about environmental challenges) may be depoliticizing if it is reduced to the atomistic identity choices of individuals.

**See Also:** Automobiles; Consumer Activism; Green Communities; Lifestyle, Rural; Lifestyle, Suburban; Lifestyle, Sustainable; Shopping; Symbolic Consumption.

### Further Readings

Akerlof, George A. and Rachel E. Kranton. "Economics and Identity." *Quarterly Journal of Economics*, 115/3:715–53 (2000).

Aspara, Jaakko, et al. "A Theory of Affective Self-Affinity: Definitions and Application to a Company and its Business." *Academy of Marketing Science Review*, 1 (2008).

Chattaraman, V., et al. "Identity Salience and Shifts in Product Preferences of Hispanic Consumers: Cultural Relevance of Product Attributes as a Moderator." *Journal of Business Research*, 62/8:826–38 (2009).

Cheng, G., et al. "Reactions to Procedural Discrimination in an Intergroup Context: The Role of Group Membership of the Authority." *Group Processes & Intergroup Relations*, 12/4:463–71 (2009).

Collins, J., et al. "Why Firms Engage in Corruption: A Top Management Perspective." *Journal of Business Ethics*, 8/1:89–108 (2009).

Coomaraswamy, Radhika. "Identity Within: Cultural Relativism, Minority Rights and the Empowerment of Women." *The George Washington International Law Review*, 34/3:483–513 (2002).

Cornwell, Bettina T. and Leonard V. Coote. "Corporate Sponsorship of a Cause: The Role of Identification in Purchase Intent." *Journal of Business Research*, 58/3:268–76 (2005).

Dalton, Maxine A. "Social Identity Conflict." *MIT Sloan Management Review*, 45/1:7–8 (2003).

Downing, Stephen. "The Social Construction of Entrepreneurship: Narrative and Dramatic Processes in the Coproduction of Organizations and Identities." *Entrepreneurship Theory and Practice*, 29/2:185–204 (2005).

Dutton, J. E. and J. M. Dukerich. "Keeping an Eye on the Mirror: Image and Identity in Organizational Adaptation." *Academy of Management Journal*, 34/3:517–54 (1991).

Goldberg, Caren B. "Applicant Reactions to the Employment Interview: A Look at Demographic Similarity and Social Identity Theory." *Journal of Business Research*, 56/8:561–71 (2003).

Güth, W., et al. "Determinants of In-Group Bias: Is Group Affiliation Mediated by Guilt-Aversion?" *Journal of Economic Psychology*, 30/5 (2009).

Haslam, Alexander S. *Psychology in Organizations: The Social Identity Approach*. Thousand Oaks, CA: Sage, 2001.

Hirst, G., et al. "A Social Identity Perspective on Leadership and Employee Creativity." *Journal of Organizational Behavior*, 30/7:963–70 (2009).

Kleine, R., et al. "Transformational Consumption Choices: Building an Understanding by Integrating Social Identity and Multi-Attribute Attitude Theories." *Journal of Consumer Behaviour*, 8/1:54–63 (2009).

Knippenberg, V. E. A. "Organizational Identification After a Merger: A Social Identity Perspective." *British Journal of Social Psychology*, 41:233–52 (2002).

Lauring, J. "Rethinking Social Identity Theory in International Encounters: Language Use as a Negotiated Object for Identity Making." *International Journal of Cross Cultural Management: CCM*, 8/3:343–52 (2008).

Llewellyn, Nick. "In Search of Modernization: The Negotiation of Social Identity in Organizational Reform." *Organization Studies*, 25/6:947–68 (2004).

Mael, F. A. and B. E. Ashforth. "Alumni and Their Alma Mater: A Partial Test of the Reformulated Model of Organizational Identification." *Journal of Organizational Behavior*, 13/2:103–23 (1992).

Mana, A., et al. "An Integrated Acculturation Model of Immigrants' Social Identity." *Journal of Social Psychology*, 149/4:450–73 (2009).

Marin, Longinos and Salvador Ruiz. "'I Need You Too!' Corporate Identity Attractiveness for Consumers and The Role of Social Responsibility." *Journal of Business Ethics*, 71/3:245–60 (2007).
Marin, L., et al. "The Role of Identity Salience in the Effects of Corporate Social Responsibility on Consumer Behavior." *Journal of Business Ethics*, 84/1:65–78 (2009).
Mehra, Ajay, et al. "At the Margins: A Distinctiveness Approach to the Social Identity and Social Networks of Underrepresented Groups." *Academy of Management Journal*, 41/4:441–52 (1998).
Montbiot, George. "Eco-Junk." *Guardian* (July 24, 2007). http://www.monbiot.com/archives/2007/07/24/eco-junk/ (Accessed July 2009).
Smidts, A., et al. "The Impact of Employee Communication and Perceived External Prestige on Organizational Identification." *Academy of Management Journal*, 1/29 (2001).
Tajfel, Henri and John Turner. "An Integrative Theory of Intergroup Conflict." In *The Social Psychology of Intergroup Relations*, William G. Austin and Stephen Worchel, eds. Monterey, CA: Brooks-Cole, 1979.
Taylor, Donald and Fathali Moghaddam. "Social Identity Theory." *Theories of Intergroup Relations: International Social Psychological Perspectives*, 2nd Ed. Westport, CT: Praeger, 1994.

*Abhijit Roy*
*University of Scranton*

# Solid and Human Waste

Solid and human waste currently contribute significantly to greenhouse gas (GHG) emissions and consume valuable land acreage for requisite system infrastructure while decreasing surrounding property market values, polluting drinking water sources, and demanding billions of dollars in annual taxpayer support.

Solid waste within the United States can be divided into several categories: 7.6 billion tons of industrial solid waste, consisting of municipal sludge, industrial nonhazardous waste, agricultural waste, oil and gas waste, mining waste, and hazardous waste, is generated from industrial and commercial facilities annually, according to a government estimate calculated in the 1980s that has not since been updated. Undoubtedly, this number has risen significantly in the past two decades. In 2007, the residential, business, and institutional economic sectors generated over 254 million tons of municipal solid waste (MSW), including durable goods, nondurable goods, containers and packaging, food wastes, and yard trimmings. The average person generated approximately 4.6 pounds of waste per day. As a consequence, municipal solid waste makes up roughly 2 percent of overall waste generated annually, although the U.S. Environmental Protection Agency (EPA) claims that MSW can make up as much of 20 percent of the waste in a given area. Industrial and commercial sectors, including the building construction industry, make up a large portion of MSW within both cities and suburbs. Building construction and demolition contribute an average 68 percent of MSW to landfills annually—roughly 172 million tons. Americans currently recycle or compost 33.4 percent of their waste, 12.6 percent is burned at combustion facilities, and 54 percent is contained in landfills. This figure applies

Waste collection costs typically represent between 40 and 60 percent of a community's solid waste management system costs. Some states are eliminating twice-a-week pickup to cut expenditures.

*Source:* iStockphoto

only to MSW. Rates of recycling for industrial solid waste are estimated to be much lower, although there is currently no official monitoring to verify the industrial waste recycling rate. The EPA estimates the total cost of municipal waste disposal to be $100 per ton, costing taxpayers $25.4 billion in 2007, not including the loss in property value of surrounding sites or the public health and environmental costs of landfill facilities. The number of landfills is continually decreasing, while the amount of waste generated is increasing, forcing the import and export of waste between states based on available land and resources, thereby increasing transportation emissions and the overall financial burden on taxpayers.

The composition of MSW comprises the remains of the life cycle energy of manufactured goods. The typical life cycle consists of the extraction and processing of raw materials, product manufacturing, material and product transportation to the marketplace, consumer use, and waste management. Thus, strategies for waste reduction should focus initially on source reduction, then recycling and composting, and finally, landfill waste reclamation and waste combustion. This order is determined by greatest reduction of waste, GHG emissions, and financial cost. Significant financial and GHG emissions savings can be attained through relatively simple, affordable means.

Source reduction can be achieved through sustainably motivated design, manufacturing, and product and materials use to reduce quantities of waste. Source reduction has the greatest potential for reducing GHG emissions by minimizing energy-related carbon dioxide emissions from the raw material acquisition and manufacturing processes and eliminating emissions and waste material from waste management.

Recycling is the sorting, collecting, and processing of materials to manufacture and sell as new products. Recycling is the second-best method of reducing GHG emissions by reducing energy-related carbon dioxide emissions in the manufacturing process and avoiding emissions from waste management. Source reduction and recycling can reduce manufacturing emissions, increase forest carbon storage, and avoid landfill methane emissions.

The majority of residential MSW is organic waste. Food scraps constitute 11 percent of MSW, yard trimmings 12.1 percent, and paper 32.7 percent, all of which can be composted to provide fertilizer, heat building water supplies, or provide energy-generation sources. Composting ranges in ease based on a given site's urban typology, as rural sites are generally easier to establish compost piles in than urban sites, which may require centralized composting facilities to be effective on a municipal scale.

Collection costs typically represent between 40 and 60 percent of a community's solid waste management system costs and can be easily reduced by eliminating twice-a-week pickup, which is already being practiced in some states. Substituting one day of waste pickup with compost and yard trimming pickup has been proven to be an effective strategy that reduces landfill waste quantities, transportation costs, and GHG emissions, while increasing consumer support and involvement in composting programs.

The net GHG emissions from MSW combustion are lower than landfill emissions, reduce land acreage necessary for waste storage, and replace fossil fuel–generated electricity from utilities, thus reducing GHG emissions from the utility sector and landfill methane emissions. Landfills can reduce their emissions through gas recovery systems.

Building construction and demolition waste can be reduced up to 90 percent, as an increasing number of sustainable construction projects have shown, through the incorporation and reuse of salvaged materials in building projects, actively recycling waste on-site, and designing for easy deconstruction. Because of the value of salvaged materials and avoided disposal costs, deconstruction can cost less than demolition, generates less dust and noise pollution, provides minimally skilled workers an opportunity for training and integration into the building trades, and can also reduce the disruption of a site's soil and vegetation.

Human waste consists of human urine and feces, and its current disposal methods pose significant environmental problems, provide a route for disease transportation, and are largely responsible for much of the world's water pollution. Buildings currently use 13.6 percent of all potable water, or 15 trillion gallons per year. Toilet flushing makes up 30 percent of residential and 63 percent of office water use. In areas with combined sewer infrastructure, raw sewage is released into major waterways and public drinking water supplies during periods of high rainfall. Treating wastewater is energy and chemical intensive, requires financing and land from taxpayers, and adds chemicals to the local water supply while failing to remove pharmaceuticals from wastewater that are introduced through the collection of urine waste streams. This is a growing pandemic that can increase human cancer rates and alter hormone levels and has been found to alter the sex and gene composition of local fish. By reducing the quantity of human waste treated at these facilities, the financial, electrical, and land burden on the taxpayer is diminished by reducing the size of treatment facilities and waste transportation infrastructure, and fewer chemicals, sewage pollutants, and GHGs are introduced into the local environment. Effective solutions include composting, dual-flush and vacuum-flush toilets, waterless and low-flow urinals, and low-flow sinks, faucets, and showerheads. Composting toilets can also be used as a fertilizer source if properly maintained, as well as urine from urinals, which is a good source of nitrogen.

Human and solid wastes and their transportation, treatment, and storage infrastructure constitute large financial burdens on U.S. taxpayers, significantly contribute to GHG emissions and water and land pollution, and have detrimental health and environmental effects. The majority of these waste sources come from the industrial and commercial sectors, which are held largely unaccountable and are not currently motivated to reduce their waste stream. Federal regulation, some of which has already been legislated but is not being practiced, needs to be instituted to rein in and reduce the financial, environmental, and health burden on U.S. taxpayers. A truly sustainable waste management strategy would be to reduce or eliminate virgin material use and to hold manufacturers accountable for their products' waste and GHGs, thus promoting sustainable strategies such as product upcycling, biodegradable packaging, and reducing or eliminating a majority of today's products' toxic contents.

**See Also:** Biodegradability; Composting; E-Waste; Heating and Cooling; Packaging and Product Containers; Pharmaceuticals; Recycling; Waste Disposal.

### Further Readings

Kreith, Frank and George Tchobanoglous. *Handbook of Solid Waste Management.* New York: McGraw-Hill, 2002.

McDonough, William and Michael Braungart. *Cradle to Cradle.* New York: North Point, 2002.

U.S. Environmental Protection Agency. "Solid Waste Management and Greenhouse Gases: A Life-Cycle Assessment of Emissions and Sinks." http://www.epa.gov/climatechange/wycd/waste/SWMGHGreport (Accessed May 2009).

U.S. Geological Survey. "Estimated Use of Water in the United States in 2000." http://pubs.usgs.gov/circ/2004/circ1268/index (Accessed May 2009).

*Giancarlo Mangone*
*University of Virginia*

# Sports

In recent years, sports organizations have made a more concerted effort toward sustainability and green living in response to economic crisis, consumer interest, and unstable energy costs. A 2008 survey found that about three-quarters of professional sports teams were implementing, developing, or actively planning a sustainability plan for their team and/or stadium. Both Chicago and Tokyo included comprehensive sustainability plans in their unsuccessful bids to host the 2016 Olympics, further demonstrating the growing interest in sustainability in the sporting and athletic world. One especially intriguing part of Chicago's plan was to recycle stadium seats, after the close of the Olympics, into wheelchair seats—a plan that could certainly be adopted by other parts of the sporting world.

As of 2009, the NBA's Atlanta Hawks and Miami Heat, baseball's Washington Nationals, and hockey's Atlanta Thrashers became the first four professional teams to play in LEED-certified venues. Shown here is the Miami Heat's downtown Miami venue, the American Airlines Arena.

*Source:* Marc Averette/Wikipedia

As of 2009, the National Basketball Association teams the Miami Heat and the Atlanta Hawks, the Major League Baseball team the Washington Nationals, and the National Hockey League team the Atlanta Thrashers became the first four professional sports teams to play in Leadership in Energy and Environmental Design (LEED)-certified venues. The LEED Green Building Rating System was developed by the U.S. Green Building Council, a nonprofit organization, in 1998; LEED standards go above and beyond mandatory building codes, emphasizing a healthy working environment (avoiding "sick building syndrome"), energy efficiency, and sustainable construction practices. One criticism of LEED is that, although focusing on energy efficiency, it comes at the problem from a fossil-fuels perspective, with too little encouragement for alternative energy sources.

For instance, these venues use compact fluorescent lightbulbs and water-conservation methods to cut down on the typically massive utility bills ($3 million a year in the case of Miami's American Airlines Arena) associated with operating a sports stadium. Atlanta's Philips Arena also uses carpeting made of recycled materials.

From the start of spring training in the 2008 season, Major League Baseball has been advising its 30 teams, in collaboration with the Natural Resources Defense Council, on environmental stewardship and sustainable resource management, with each team being asked to develop a plan with clearly defined goals for reducing waste and achieving greater sustainability and energy efficiency. Beginning with the 2010 season, teams will issue regular sustainability reports to track their progress.

The National Football League (NFL) environmental program has no current plans to involve the individual teams. Instead, it aims to offset the impacts of the Super Bowl, the NFL Kickoff, and the Pro Bowl—events with huge attendance and equally huge waste and resources usage. The NFL works with the host cities of these events on carbon offsets, use of local renewable energy capabilities when possible, and recycling of as much waste as possible. The Philadelphia Eagles have led the way in green sports, actively improving the sustainability of their stadium and training facilities since 2003, including the use of solar energy and even the recycling of cooking fat to make biodiesel. In 2005, Philadelphia hosted the first carbon-neutral NFL game and remains ahead of the curve in sustainable sports.

The National Basketball Association conducts Green Week every April (beginning in 2009), using recycled basketballs, organic cotton shirts, and online auctions to benefit the Natural Resources Defense Council.

Professional soccer player Natalie Spilger founded GreenLaces in 2008, a nonprofit athletic sustainability program that works with after-school programs and children's sports leagues to educate children on environmental issues, while also bringing attention to professional athletes who are active in environmentalism.

A summer 2009 Association for the Advancement of Sustainability in Higher Education survey revealed that 44 percent of the colleges in the National Collegiate Athletic Association Football Bowl Subdivision (formerly known as Division I-A) had athletic departments that considered sustainability a high priority, compared with 72 percent of institutions that considered it a priority for the university as a whole. Few respondents were aware of whether or not their university was part of the American College & University Presidents Climate Commitment, and only 7 percent reported a personal involvement in the greening of their department. Less than half of the departments kept track of greenhouse gas emissions generated by their operations, and although energy efficiency and recycling were frequently reported as examples of departments' green initiatives, natural/local food (which one might think could coincide with an athletics departments' nutrition concerns) was the least frequently reported. When released, the Association for the Advancement of Sustainability in Higher Education's commentary on the survey did not suggest that athletics departments should independently pursue their own sustainability goals apart from their institution's but, rather, that the results showed a generally low awareness of sustainability issues and the implementation of sustainability programs in these athletic departments, not only in contrast with the rest of the institution but in contrast with professional sports organizations.

Despite this apparent low awareness, some schools were notably ahead of the curve. The University of Florida, the University of Georgia, the University of Notre Dame, and the University of Wisconsin–Madison all hosted carbon-neutral football games in the 2008 season, specifics of which included planting trees to offset the emissions of game-day activities and travel. Recycling initiatives were widespread, and the University of Colorado at Boulder has announced a plan to recycle or compost at least 90 percent of the waste generated at its home games.

**See Also:** Bottled Beverages (Water); Consumer Behavior; Electricity Usage; Leisure and Recreation; Needs and Wants; Water.

### Further Readings

Brower, Michael and Warren Leon. *The Consumer's Guide to Effective Environmental Choices: Practical Advice from the Union of Concerned Scientists*. New York: Three Rivers, 1999.

Garlough, Donna, et al., eds. *The Green Guide: The Complete Reference for Consuming Wisely*. New York: National Geographic, 2008.

Lindsay, Iain. "Conceptualizing Sustainability in Sports Development." *Leisure Studies*, 27/3 (July 2008).

*Bill Kte'pi*
*Independent Scholar*

## Supermarkets

Today, many supermarkets and local groceries have developed a new consciousness of what it is to be green and carbon neutral. When sourcing their food products or household merchandise—and when operating their businesses—supermarkets play a critical role in providing consumers with green choices. In general, the markets for green branding and green products have increased significantly over the past few years, not only in developed but also in developing countries. Not surprisingly, the world market for organic foods has, according to the International Federation of Organic Agriculture Movements (IFOAM), doubled between 2000 and 2006. In China, for example, supermarkets have almost doubled their floor space for organic goods.

The increased demand for green products or ecoproducts correlates with a growing awareness of environmental protection and a rising concern for healthy eating. Green products are very much about the personal values of consumers. Environmentally friendly companies will not only have the opportunity to satisfy their customer needs, but can also capture the willingness of consumers to spend more on green products in their own interests. As demonstrated in a recent green brands survey conducted by marketing company WPP, consumers worldwide agree that green products are more expensive than regular products, and that consumers in all of the surveyed countries expressed their intention to spend more money on green products. Almost two thirds of Chinese and Indian consumers said that they plan to spend more on green products.

When discussing green products, organic foods and the Slow Food movement must also be mentioned. The term *organic food* encompasses foodstuffs that are being produced with or without limited use of conventional nonorganic pesticides, insecticides and herbicides. What determines organic food is in general regulated: content needs to be certified and respectively labeled by recognized authorities and institutions. The Fair Trade movement—which is concerned with providing better prices for goods, improved working conditions, increasing local sustainability, and fair terms of trade for farmers and workers in developing countries—is often linked with organic food. In addition, the Slow Food movement attempts to preserve and promote local and traditional food products and local cuisine; to encourage organic farming; and to educate consumers about the dangers of fast food.

But the question is: can supermarkets provide sufficient green food options? It seems that many supermarkets are still lagging behind when it comes to satisfying increased consumer demand for environmentally friendly food options. Frequently green products are not clearly labeled, and it is unclear if the consumer knows exactly what constitutes a green product. Very often, consumers become confused by the variety of labels that declare the products to be "bio," organic, and/or green. Again, the definition of green is relatively vague. Some supermarkets may attempt to become overly green, for example, being energy-efficient and producing less waste. Increasingly, more and more government regulations require this form of approach. Providing green choices for consumers, however, has yet to be regulated.

In the interim, it is up to the supermarket chains themselves to decide whether or not they see these green choices as profitable. Clearly, governments will play a critical role in the future when it comes to convincing supermarkets to offer more green products and educate their consumers. Considering the fact that ecofood products usually have a higher profit margin (based on the assumption that green consumers are often willing to pay more), promoting green products can benefit the supermarkets as well. However, supermarkets and retailers, who plan to target green consumers, have to come to terms with the fact that even green consumers do not always buy green products. So far, green consumers are still a relatively small consumer segment. Also, due to smaller economies of scale, many green or environmentally friendly options are relatively expensive.

There is little doubt that the demand for green products has significantly increased in recent years. This is reinforced by the growing number of health food stores and retailers of natural and organic foods, such as Whole Foods market. Today, traditional supermarkets tend to offer more seasonal and organic foods than ever before. Slowly changing food traditions, along with a general rising awareness of the carbon footprint of imported products, are resulting in a higher appreciation of local and regional foods.

In a recent brands survey, consumers' environmental beliefs and behaviors in seven countries were investigated. The survey, released in mid-2009, revealed that consumer perceptions of and attitudes toward green products vary in different cultural contexts. The research showed that consumers in the United States, the United Kingdom, Germany, and France appear to have similar views, whereas Brazilian, Indian, and Chinese consumers differ in regard to their views about green products, green brands, and green companies. In China and India, the reputation of a company to be green or environmentally friendly appears to greatly directly influence consumer behavior. Thus, a company that puts effort into recycling, efficient use of energy, proper waste management, and that also offers green products will be sustainably competitive.

**See Also:** Green Consumer; Green Food; Green Marketing; Organic; Slow Food.

### Further Readings

Alter, Lloyd. "Organic Sales in UK Supermarkets Get a Healthy Boost." http://www.treehugger.com/files/2006/03/organic_sales_i.php (Accessed December 2009).

Dunn, Colin "UK Supermarkets Told to 'Green Up Their Act.'" http://www.treehugger.com/files/2009/11/uk-supermarkets-told-to-green-up-their-act.php (Accessed December 2009).

Lawrence, Felicity. "Shoppers Give Organic Sales a Healthy Boost." http://www.guardian.co.uk/business/2006/mar/30/supermarkets.food (Accessed December 2009).

Seth, Andrew and Geoffrey Randall. *Supermarket Wars: The Future of Global Food Retailing*. New York: Palgrave Macmillan, 2005.

*Sabine H. Hoffmann*
*American University of the Middle East*

## Super Rich

In recent years, there has been increasing debate about the growth of income and wealth inequality both between and within industrialized and developing nations. Particularly in countries like the United States and the United Kingdom, the income gains from growth have accrued increasingly to a new class of "super-rich." Growing inequality in the distribution of income and wealth appears to hold not merely for the world as a whole but within the world's richest countries. Inequality—not just poverty—is now recognized to entail serious social costs. A major challenge to mobilizing support for green economics is to recognize that reversing growing inequality must be part of any "new green deal."

Journalist Robert H. Frank popularized the term *super rich* in his book titled *Richistan*—a mythical country (styled on the rich in the United States) in which sumptuous lifestyles are enjoyed and whose inhabitants vie with each other for wealth and status. It is not just the persistence of poverty that is the problem, it is the growth of huge inequality. What is true for the United States seems true for the world as a whole. The World Institute for Development Economics Research of the United Nations University reported in 2006 that richest 2 percent of adults in the world owned more than half of all global household wealth; the richest 1 percent of adults alone owned 40 percent of global assets in the year 2000, and the richest 10 percent of adults accounted for 85 percent of the world's total. In contrast, the bottom half of the world adult population owned barely 1 percent of global wealth. Understanding that wealth distribution is highly unequal is an essential part of understanding why world income distribution is unequal.

How has world inequality changed in recent years? Some academic studies find the distribution of income to have worsened considerably, others that it has become more even. Although the distribution of income between countries seems to have improved somewhat over the past two to three decades—although not if fast-growing China is excluded—the absolute income gap between rich and poor countries has widened considerably, as has the gap within countries.

Among the industrialized countries of the Organisation for Economic Co-operation and Development, income inequality both between countries and within countries fell for three decades following World War II. However, during the Reagan and Thatcher years, inequality worsened markedly, particularly in the United Kingdom, where the Gini coefficient shot up from 0.23 to 0.33 in a single decade. (On the Gini scale, perfect equality is 0.00 and perfect inequality is 1.00.) The United States and the United Kingdom are today the most unequal countries in the Organisation for Economic Co-operation and Development, with Gini coefficients for household income of 0.41 and 0.34, respectively. In contrast, most European Union countries have suffered only a slight deterioration in equality in the last two generations when social transfers are taken into account, and in some countries (e.g., France), equity has improved.

Rising inequality is sometimes attributed rather loosely to "globalization"; that is, to the growingly footloose nature of international capital. The U.S. labor economist Robert Frank has coined the phrase "winner-take-all markets"; such markets are inefficient, in the sense that they pay gigantic rewards to people who are only marginally more talented than their peers. The fabulous sums paid to sports stars and media celebrities are obvious examples of how inefficient markets create inequality. Equally, the shift away from industrial production toward the financial services industry, with its culture of bonuses and stock options, helps explain the phenomenon. A further reason for rising inequality is that tax rates on income and wealth have been greatly reduced in the United States and the United Kingdom, largely negating the postwar progressivity of the fiscal system.

A widely held view is that growing inequality is no bad thing, as long as everyone is getting richer; that is, as long as the poor are being lifted out of poverty. This view has been strongly challenged. British social epidemiologist Richard Wilkinson has gathered a wealth of cross-country and intracountry data showing that high inequality is associated with poor scores on a variety of social indicators; namely, life expectancy, infant mortality, mental illness, obesity, incarceration rates, academic achievement, and several others. Consider the apparent paradox that infant mortality is higher in Harlem (New York City) than in Bangladesh, although household income is far lower in Bangladesh than in Harlem. For many social scientists such as Wilkinson, the explanation lies in the fact that the poor in Harlem occupy a lower position on the relative income scale than the average Bangladeshi. The social costs of inequality are high and rising.

The rise of the super rich has quite serious implications for green economics. For one thing, the super rich, with their multiple homes, large personal staffs, and private jets, consume fabulous amounts of energy. For another, the fact that an increasingly large proportion of the fruits of economic growth are siphoned off by top earners leaves the rest of society behind. Using the example of the United States, the ordinary, middle-class family finds it increasingly difficult to make ends meet, whereas, further down the scale, the real income of the bottom fifth of households has fallen relative to 1980. What this means is that "green taxes" (taxes on fuel, for example), which fall disproportionately on middle- and lower-income households, become harder for them to pay and are likely to meet growing resistance. More generally, economic growth in developing countries must be accompanied by greening technology, but these countries are often too poor to afford carbon taxes and new technologies. In sum, resources for greener growth must come largely from the rich.

**See Also:** Carbon Credits; Carbon Offsets; Finance and Economics; Lifestyle, Sustainable; Organisation for Economic Co-operation and Development (OECD); Sustainable Consumption.

### Further Readings

Frank, R. H. *Falling Behind: How Rising Inequality Harms the Middle Class*. Berkeley: University of California Press, 2007.

Frank, R. H. *Richistan: A Journey Through the 21st Century Wealth Boom and the Lives of the New Rich*. New York: Crown, 2007.

Irvin, George. *Super Rich: The Rise of Inequality in Britain and the United States.* Cambridge, UK: Polity, 2008.
Wilkinson, R. G. and K. E. Pickett. *The Spirit Level.* London: Penguin Books, 2009.
World Institute for Development Economics Research of the United Nations University. *The World Distribution of Household Wealth.* Helsinki: World Institute for Development Economics Research of the United Nations University, 2006.

*George Irvin*
*University of London (SOAS)*

# Sustainable Consumption

Sustainable consumption is only one facet of sustainable development, which seeks to challenge prolific patterns of consumption. This could relate to the sustainable consumption of natural resources, energy use, and consumer goods more broadly. Initiatives such as the Marrakech Process, a global multistakeholder process to promote sustainable consumption and production (SCP), are working toward a Global Framework for Action on SCP, a so-called 10-Year Framework of Programmes on SCP. This signifies that there is widespread acknowledgement that consumption has an effect on sustainability, which in turn has given rise to debates surrounding how to strive toward more sustainable and equitable patterns of consumption.

Tim Jackson, in his report for the Sustainable Development Commission, stated that the term *sustainable consumption* entered into being from Agenda 21 (a policy document that emanated from the 1992 Rio Summit), even though consumption had been an international policy issue since the 1970s. He noted that there are differing interpretations as to whether sustainable consumption means consuming differently, responsibly, or less. He also noted that to fully understand whether sustainable consumption is feasible, the rationale and history of consumption has to be considered, as well as thoroughly understanding patterns of consumer behavior, as many consumption actions are seemingly habitual.

The maxim "reduce, reuse, recycle" demonstrates the priorities of the waste hierarchy that should be observed in the quest for sustainable consumption. Where this is not feasible, research by the British Sustainable Consumption Roundtable suggests sustainable consumption techniques for businesses. For example, by ensuring that consumers are offered clear choices of environmentally responsible products or behaviors, positive incentives exist for consumers to choose more environmentally responsible products, social stimuli are needed to make environmentally responsible products more desirable, people engage in initiatives that encourage them to help each other, and governmental departments also set a good example.

The danger of encouraging people to consume less to save money may have unintended rebound effects—such as an increase in consumption in other areas. For example, consumers who are encouraged to buy new, more energy-efficient electrical goods may notice a drop in their energy bills. However, if the money saved is spent on going on holiday abroad, this may negate the environmental benefits that are created in the first place. In addition, the embedded energy inherent in products that may have been put together with components from all over the world, assembled in one country, and then shipped to another for retail has to be considered when making purportedly "sustainable" purchases. The embedded energy may also negate the energy savings of buying new products, although at this time this information is not readily available for consumer goods, nor is it factored into the price of the product.

Initiatives such as ecolabeling and reducing packaging are other techniques that can be adopted to encourage consumers to think about making more sustainable purchases. However, it is counterintuitive to the sustainable consumption debate if these initiatives provide third-party organizations with an opportunity to develop new labels that could potentially increase the scope for new product development. Recent debates in improving sustainable consumption behavior have also centered on retailers' efforts in choice editing, where set contracts and specifications exist that frame procurement choices before products get to the shelf, though retailers are reluctant to slim down their product lines, possibly out of fear of foregoing profits. Perhaps it could also be said that current consumers are not willing to accept responsibility for their own actions and their own consumption habits and may be reluctant to change unless other actors (retailers/producers) do more to encourage this.

Online community initiatives such as Freecycle demonstrate that there is some willingness among consumers to consume differently, as well as some new initiatives, such as the Oxfam Boutique stores that sell secondhand designer labels on behalf of the Charity. However, this competes with the rise in obsession with celebrity culture and magazines that chart all of the styles and items that celebrities are buying, most of which have similar, inexpensive versions available on the High Street.

It is debatable whether the term *sustainable consumption* is a contradiction in itself, especially as typical Western consumption habits are far from sustainable, and that the most sustainable option would be not to consume anything at all. It is ironic that such a concept of striving for sustainable forms of consumption has arisen out of the phenomena of globalization, itself fueled by neoliberal economics, an era of cheap oil, and deregulation alongside privatization in free-market economies. These myriad factors, coupled with persuasive advertising from big companies, have created societies and individuals that have become accustomed to profligate and acquisitive behavior.

There is acknowledgement of an attitude–behavior gap when researching consumer attitudes toward forms of sustainable consumption. Survey questions asking consumers whether they would actually pay more for "green" products suggest that they would, in principle. In practice, this does not hold, so therefore it is a complex area to attempt to regulate, for example, via policy mechanisms. In addition, governments may be keen to encourage or support businesses in stimulating sustainable consumption; however, such initiatives, such as suggesting that businesses should seek to discourage their consumers from consuming so much, would likely be highly political and unpopular in economies in which economic growth and profit are key measures of success.

Marketing concepts such as consumer sovereignty—in which "the customer is always right" or "the consumer is king"—have all contributed to the development of products that attempt to satisfy even the smallest consumer need. Exacerbating the unsustainable nature of consumption is the decline of the repair culture, partly driven by the fact that certain products are sealed and therefore cannot be repaired. Furthermore, occasionally new products are the same price, if not cheaper, than repair services. With the increasing rate of innovation and new product features, the desire for newer goods more frequently is stimulated. Therefore, sustainable consumption is inextricably linked with production, and it is difficult to look at one aspect in isolation from the other. Hence, the phrase "sustainable consumption and production" is, by now, also well established.

Furthermore, in countries where recycling facilities are available, it could be said that this provides consumers with a reason not to change their consumption, as excess waste and packaging can be recycled, and as such is considered by some to be sustainable. This could mask issues emanating from unsustainable consumption habits, as there is still waste being generated, albeit not being sent to landfills. Recycling has its limitations, as certain materials can only be recycled a certain number of times, certain products are currently nonrecyclable, and the process can be highly energy intensive. True sustainability could be achieved with closed-loop systems, but this would require a shift in culture on the part of governments, designers, manufacturers, retailers, and consumers, and a focus on sustainable production, as well as consumption.

Research by World Wildlife Fund on One Planet Living has shown that if European consumption levels were matched in every country the world over, we would need three planets worth of resources to sustain this level. The equivalent amount of resources needed for the global population to live like a North American would require five planets—clearly, such levels of consumption would be impossible to sustain. At this time, these levels are only able to continue in this inequitable fashion because of global differences in consumption levels: 20 percent of the world's population consumes 80 percent of its resources, leaving the poorest 80 percent of the planet to consume very little among themselves. However, if the prosperity in developed countries is perceived as "progress," can this be denied to other people in other countries? Added to the pressures on resources is the increasing population—forecast to reach 9 billion people by 2040—so the time in which to redress overconsumption remains crucial.

**See Also:** Advertising; Affluenza; Conspicuous Consumption; Consumer Behavior; Consumer Culture; Consumerism; Ecolabeling; Frugality; Green Consumer; Malls; Overconsumption; Pricing; Product Sharing; Resource Consumption and Usage.

## Further Readings

British Sustainable Consumption Roundtable. *I Will If You Will, Towards Sustainable Consumption.* London: Sustainable Development Commission, 2006. http://www.sd-commission.org.uk/data/files/publications/I_Will_If_You_Will.pdf (Accessed May 2011).

Jackson, Tim. "*Motivating Sustainable Consumption: A Review of Evidence on Consumer Behaviour and Behavioural Change.*" University of Surrey: Centre for Environmental Strategy, 2004.

Jackson, Tim. "Policies for Sustainable Consumption: A Report to the Sustainable Development Commission." London: Sustainable Development Commission, 2003. http://www.sd-commission.org.uk/publications.php?id=138 (Accessed August 2009).

United Nations. "The Marrakech Process." http://esa.un.org/marrakechprocess (Accessed October 2009).

United Nations Environment Programme, Division of Technology, Industry, and Economics, Sustainable Consumption & Production Branch. http://www.unep.fr/scp (Accessed October 2009).

World Wildlife Fund. "One Planet Living." http://www.oneplanetliving.org/index.html (Accessed October 2009).

*Cerys Anne Ponting*
*Cardiff University*

# Swimming Pools and Spas

Swimming pools and spas provide exercise, leisure, and pleasure for their owners and users. However, the environmental impact of swimming pools should not be underestimated—the need to sanitize and ensure that they are hygienic often entails the use of large inputs of chemicals, and maintaining a comfortable temperature requires large quantities of energy in addition to the requirement for vast quantities of clean water. This is before the environmental impacts of constructing a pool are accounted for.

## Environmental Impacts of Swimming Pools

One of the main environmental impacts of maintaining a pool is producing the energy required to heat it, especially if it is an out-of-season outdoor pool. In addition, energy is required to pump and circulate the water, and a large input of manufactured chemicals is needed to keep the pool clean.

### Pool Chemicals

In addition to the environmental impact of chemicals used to keep pools sanitary, there are also the personal impacts—some people find the strong odor of some pool chemicals, dry skin that results from swimming, and eye irritation to be sufficient reason to seek out alternatives to chlorine.

One of the issues is whether a pool is private or public. To ensure sanitary conditions with such a vast quantity of bathers, public pools are often "over-chlorinated" as a precaution. In a private pool, where tighter control over swimmers' habits and routine can be maintained, such as the adoption of sanitary practices—for example, ensuring users shower before and after entering the pool—the amount of chemicals required to keep the pool clean can often be reduced.

Pool maintenance has many environmental impacts, including energy use required to heat, pump, and circulate the water, and the large quantity of chemicals, like chlorine or bromine, needed to keep the pool clean.

*Source:* Demeester/Wikipedia

Traditionally, pools are treated with either chlorine products or bromine. Bromine smells less strong than chlorine and can also be used at higher temperatures, making it a favored option for spas.

A "salt" system is often viewed as an alternative to chlorine; however, the device in a salt system is a chlorination device that produces chlorine from the salt. Pools using a salt system that are then backwashed into freshwater streams can cause big problems for the local environment. Some reduced chlorine options are as follows:

- Ozone gas can be used to disinfect water and also break down organic matter. Ozone is used with a greatly reduced quantity of chlorine. The ozone is generated on-site using electricity to power the process, and as such can be energy intensive.
- Ionized copper and silver slowly disinfect water. A device called an ionizer is fitted with metal electrodes that wear away as electrochemical processes are used to deposit the metal ions into the water.

Some zero-chlorine options are as follows:

- Ultraviolet light can be used to help sanitize pool water, with the water passing through a sealed chamber, where it is exposed to a bright ultraviolet light source.
- Copper sulphate, approved for use as a drinking water additive in Europe, often sold under the trade name PristineBlue, is an alternative to pool chemicals, with the benefit that waste pool water can then be used to water plants.
- Persulphates are an option that can be used instead of chlorine. Although they break fats and oils down effectively and kill bacteria, they are less effective at killing algae.
- Polymeric biguanides are a three-stage process, consisting of a treatment that kills bacteria, followed by a second treatment that kills algae, which is finally followed by a hydrogen peroxide treatment that breaks down oils and fats.

## Swimming Pool Heat Losses

One of the main ways that energy is "lost" from swimming pools is through heat loss. Appropriate measures can be put in place to ensure that heat loss is minimized, saving energy and reducing the running costs of the pool.

The main ways that heat is lost from a swimming pool are as follows:

- Evaporation losses (approximately 51 percent): Swimming pools lose a lot of heat as water evaporates from the pool. This can be reduced in practice by fitting covers to the pool when not in use to prevent the hot water from evaporating from the surface.
- Convective losses (approximately 17 percent): Through the heat transfer process of convection, pools lose heat. A cover over the water surface goes some way to reducing heat lost through convection.
- Radiative losses (approximately 27 percent): Some heat is lost from the pool by radiation; reflective insulating covers can be used to reradiate this heat back to the pool when the pool is not in use.
- Conductive losses (approximately 4 percent): Pools lose a relatively small amount of energy through conduction of heat through the walls and floor of the pool. This can be eliminated to some extent during the design stage of the pool by incorporating insulation into the walls and floor of the pool to reduce heat transfer.
- Top-up water losses (approximately 2 percent): Some water is inevitably lost from the pool as a result of splashing and water carried on the bodies of people entering and exiting the pool. Heat is required to preheat the replacement water used to "top up" the pool.

## Heating Pools in an Environmentally Sound Manner

For heated pools, there are a number of options with reduced environmental impact that can be used to provide heat in a less carbon-intense manner than direct fossil fuel–powered or electrical heating

One option is solar pool heating. A solar collector array harnesses the sun's energy to heat water. Often, very simple unglazed collectors will provide adequate heating performance for many applications; however, evacuated tube collectors can also be used. A consideration is that any solar collectors used for pool heating will have to withstand the use of any pool chemicals.

A geothermal heat pump uses an external source of energy to "pump" heat from one side of the heat pump to another. You can think of the technology in terms of your refrigerator-freezer, which pumps heat from within the insulated box to the outside, where it can be conducted away. For swimming pool applications, heat can be pumped from coils of tubing embedded in the surrounding earth in a closed loop to heat the pool water.

Fuel cell combined heat and power technology is another potential avenue for pool heating, which lends itself more readily to larger installations, such as public swimming pools. Some high-temperature fuel cells can reform natural gas internally and produce heat and electrical power as output products. This approach was used in Woking Park Leisure Centre—the United Kingdom's first fuel cell–heated swimming pool. Heat from the fuel cell is used to heat the pool's water and also to produce heat for heating, cooling, dehumidification, and air-conditioning (using absorptive chillers). The fuel cell produces 1.2 MW of electrical energy and 1.6 MW thermal energy.

## Natural Swimming Pools

Natural swimming pools dispense with the clear blue water and manufactured sides in favor of a more ecological approach, with self-cleaning biological systems comprising microorganisms and aquatic plants taking the place of expensive, and irritant, pool chemicals.

The water in a natural pool is clear and attractive; however, unlike a conventional chemically dosed swimming pool, it is not sterilized, as it has to sustain a range of plants, organisms, and invertebrates.

The concept revolves around a two-pool system. One pool is used for swimming, and the other contains the plants that are used to filter and regenerate the water. These can either be isolated from each other, connected by piping, or integrated into a single pool with a division, with the deeper portion of the pool being reserved for swimming and a shallow portion containing the plants.

The swimming area is usually between just over one to two meters deep and is kept free of plants to permit ease of swimming. The regeneration section of the pool relies on the roots of the plants and the microorganisms that grow on these plants, acting as oxygenators: floating aquatic plants, and shallow and deep marginals.

The bog plants deal with green water that is caused by algal formation. Algae forms in the presence of nitrogen, potassium, and phosphorus elements in the water, which, combined with sunlight and (often warm) water, provide an ideal environment for rapid formation of algae blooms. Nitrogen enters the pond as ammonia, a constituent of urea. Potassium and phosphorus are also present as the product of protein breakdown. All three can be broken down by bacterial aerobic and anaerobic digestion processes. Anaerobic bacteria thrive in conditions that are above 46 degrees F.

**See Also:** Green Consumer; Leisure and Recreation; Positional Goods; Water.

### Further Readings

Littlewood, M. *Natural Swimming Pools: A Guide to Building*. Agri-Media, 2008.

Littlewood, M. *Natural Swimming Pools: Inspiration for Harmony With Nature*. Atglen, PA: Schiffer, 2005.

Todd, Nancy Jack and John Todd. *From Eco-Cities to Living Machines: Principles of Ecological Design*. Berkeley, CA: North Atlantic Books, 1994.

*Gavin D. J. Harper*
*Cardiff University*

# Symbolic Consumption

*Symbolic consumption* refers to the meanings conveyed by goods or other consumables, such as entertainment, leisure activities, cultural practices, and group membership in a broader social group. The term reminds us of the socially assigned meaning to goods, often tied to desirable attributes for certain identifiable groups in specific contexts. For example, symbolic consumption could refer to a woman's wardrobe, where she "dresses for success" and displays an identifiable business attire style on the city streets of New York. In the past 40 years, during a period of significant increases in consumed goods across the world, many scholars have addressed the symbolic or meaning-making properties of what people purchase, use, and display. Symbolic consumption has been addressed primarily in consumer studies, but also in anthropology, sociology, economics, marketing, human ecology, and social psychology. The term is used to theoretically explore the relational dimensions of what people buy and to probe what individuals wish to convey by what they buy and why consumption of particularly personal goods has accelerated.

The term *symbolic consumption* originated with Thorsten Veblen's coining of the term *conspicuous consumption*, or the display of goods to signify status and wealth. Conspicuous consumption can signify status in a materialist society where one's command over their work, home, and leisure environment suggests greater control over their life. This enviable position is produced because of social comparison, in which people often asses how well off they are relative to others. Similarly, the concept of positional consumption suggests that once material needs are met, people seek to "position" themselves relative to others. They do this through the display of goods that are scarce, situating them ahead of the crowd. Symbolic consumption can allow one to advertise their social position, and perhaps wealth and power, and maintain class distinctions in society. Social anthropologists have explained this need to display what one consumes as part of evolutionary psychology, in which individuals try to establish themselves within a "pecking order," or status hierarchy. Status hierarchies play an important role in the social organization of rights and access to resources. Those who are high in the status hierarchy generally have better access to financial and physical resources, potential mates, friends and family, and information.

Symbolic consumption can be carried out through a variety of social practices, such as trying to match one's furniture with their newly renovated house, in which the pursuit of matching can be referred to as the "Diderot effect," or trying to convey proficiency by having all the specialized goods that are associated with a recreational activity, such as downhill skiing and all its accessories, even though the consumer may ski infrequently. Symbolic consumption might also occur through elaborate gift-giving in gifts that are given to convey the provider's generosity, taste, and wealth, for example. By recognizing the symbolic nature of some consumer goods, researchers have also explored how goods once thought to be a luxury, such as a freezer, become commonplace and assumed a necessity. Advertising plays a key role in symbolic consumption, planting and reinforcing new notions of what conveys desirable qualities displayed to others, and normalizing the possession of certain goods, such as a cell phone.

Symbolic consumption encompasses a broader meaning to include how consumption allows one to continually maintain or re-create one's self-concept by the relational message one conveys to others, which may entail indicators of wealth, status, stylishness, popularity, fertility, potency, fidelity, creativity, bravery, rebellion, talent, irreverence to convention, belongingness to a group, and so on. Many goods associated with symbolic consumption are cars, clothing, homes, and more specifically, a set of distinct tastes for one's physical looks, food, wine, furniture, vacation places, and entertainment. Given that societies are always changing and new products continually become available, symbolic consumption is always changing, allowing people to repeatedly reinforce—and in some cases reinvent—their identity. Symbolic consumption is most often applied to consumer behavior in nations where shopping is a mainstream recreational activity and is more individualistic in its value orientation. Consumer cultures breed social exchanges in which people are more likely to communicate their status through display of consumer goods than through other material cultural markers of religion, ethnic group, caste, or tribe.

Symbolic consumption studies may also recognize that people's purchasing behavior may be less about what they purchase and more about the opportunity to be around others who purchase similar things. For example, a shopper may go to a farmers market regularly, even though she needs nothing, to support local farmers, see others who share a priority to shop for locally grown or made goods, and feel a communal association with others who wish to see the same kind of society evolve. Several scholars have also studied the role of symbolic consumption in thwarting thrift, purchasing high-quality and durable goods, and reusing fixable items; that is, how symbolic consumption can promote wasteful, impulsive, or excessive consumption. Recent advocacy around "sustainable consumption," for example, is trying to redefine the meaning of many kinds of consumption as symbolizing poor quality, offering only temporary gratification, and portending harmful additions to the Earth's ecosystems through the garbage it creates. Symbolic consumption is embedded in all cultures, but the meanings conveyed by consuming are continually open to revision.

**See Also:** Conspicuous Consumption; Consumer Culture; Consumerism; Consumer Society; Diderot Effect; Overconsumption; Positional Goods; Social Identity.

### Further Readings

Douglas, Mary and Baron Isherwood. *The World of Goods*. London, Routledge, 1996.

Huddart Kennedy, Emily and Naomi Krogman. "Toward a Sociology of Consumerism." *International Journal of Sustainable Society,* 1/2:172–89 (2008).

McCracken, Grant D. *Culture and Consumption: New Approaches to the Symbolic Character of Consumer Goods and Activities*. Bloomington: Indiana University Press, 1990.

Shipman, Alan. "Lauding the Leisure Class: Symbolic Content and Conspicuous Consumption." *Review of Social Economy,* 62/3:277–88 (2004).

*Naomi T. Krogman*
*University of Alberta*

# TAXATION

Taxation originated as a means to finance the activities of government, and that still remains its primary purpose in most of the world. However, in recent years, taxation also has been used to discourage undesirable behavior or encourage socially desirable activity. In some cases, these policies take the form of a "user fee," in which the taxpayer is charged for specific activities, such as a charge to dump garbage in a landfill. These fees should serve as a deterrent that would discourage the activity or behavior. In addition, these fees may be designed to pay, at least in part, for the cost of the activity to the government. In other cases, a reduction in taxes is given to encourage desired activities or behavior. An example would be a deduction from income for mortgage interest paid, which serves to encourage home ownership. Elements of all these approaches are found in green taxation.

Green taxation, also known as environmental taxation or ecotaxes, most likely originated in 1920. Economist A. C. Pigou suggested that the problem of externalities could be addressed by imposing a tax equal to the cost of the environmental damage caused by the activity. An externality is an effect imposed on someone who is not a party to the activity in question. Frequently, it is an unintended consequence. For example, the effects of pollution are felt by those who did not cause the pollution, whereas those causing the pollution do not bear the cost of what they caused. Pigou would tax the polluter for the costs caused by the pollution.

The central aim of taxation in a green economy would be to make prices reflect true costs, thus internalizing the externalities. Studies in a number of countries have found that replacing existing taxes on employment, income, and profits (goods) with taxes on energy use (bads) yield at least three advantages: better overall national economic performance, higher levels of employment, and an improved environment. These taxes are seen as enabling society to achieve environmental goals in the most efficient way. Such a tax structure facilitates a move to a more sustainable economy, in addition to encouraging innovation and the development of new technology. Finally, with revenues being realized from the environmental taxes coupled with an improved economy, the level of other taxes can be reduced. In essence, taxing the "bads" has the desired result of moving toward a cleaner environment, with the added benefit of better economic performance. When taxes on employment and income are reduced, the result is an economic boost.

Austria, Denmark, Finland, Germany, the Netherlands, Sweden, and the United Kingdom have taken this environmental tax reform approach, but the United States has followed a different path. The United States has used tax credits and deductions for targeted activities that have positive environmental effects. By and large, the United States has not embraced the approach of taxing environmentally damaging activities. Revenues from environmentally related taxes in the United States have been about 3.5 percent of total tax revenues, compared with 7 percent in Organisation for Economic Co-operation and Development countries, including highs of 10 percent in Denmark and 16 percent in Turkey.

## The "Polluter Pays" Approach

When people or firms create pollution, they are imposing costs on others. Environmental taxation may adopt a "polluter pays" principle, under which those causing the pollution are made to pay for the costs or damages that they cause. This penalty approach gives firms incentives to improve efficiency and environmental quality. As they are allowed to choose the best economic alternative available to them—pay the tax or reduce pollution—the tax must be sufficiently large if it is to achieve its goal of a cleaner environment. The other approach to environmental taxation is the incentive approach. This approach frequently gives tax exemptions or deferrals to companies that undertake environmentally desirable activities. Tax credits for the purchase or installation of energy-efficient property such as hybrid vehicles or insulation in homes are two such examples.

The first step in environmental taxation should be to remove subsidies that damage the environment. Tax preferences for extraction of minerals, including oil, encourages the mining activity and offers no incentive to research alternative energy sources. In the United States, the exclusion of vehicles over 6,000 pounds from the luxury vehicle depreciation limitations encourages the purchase of large, inefficient vehicles. The mortgage interest deduction has been faulted in this regard, as it encourages people to build bigger homes or acquire second homes. It has been estimated that as much as $500 billion in tax subsidies is given to environmentally destructive industries annually. The next step is to impose taxes on activities that are considered environmentally damaging. From a policy and practical standpoint, it is desirable to implement these two steps simultaneously, as part of a comprehensive environmental tax reform.

Congestion pricing is a concept that has been used in several countries, most notably for highway use. This tax levies a charge for the use of roads that receive heavy use. On heavily traveled urban roads, this takes the form of a fee to drive on designated roads. In some cases, the additional charge is assessed for the privilege of traveling less congested, designated "fast lanes." A second approach is to charge for driving within certain congested areas, often referred to as "cordon" areas. Singapore and Stockholm charge the congestion fee every time a user crosses into the cordon area, and London charges vehicles a daily fee for unlimited access. Indications are that congestion taxes are effective, as traffic volume reductions of up to 30 percent have been reported. In addition to less congestion, air pollution, and noise, traffic accidents have also decreased.

Congestion pricing has potential in other areas such as airports, ports, and waterways. Variable-rate parking meter systems have been discussed for certain large cities. The entire concept is not unlike pricing structures found in telephone and electric utilities, subways, and airlines, among others.

## "Feebates"

Another, broader approach to environmental taxation has been described as the "feebate" approach. This taxation scheme contains dual elements. The first element levies additional taxes, or fees, on less sustainable products such as sport utility vehicles. The other side of this approach uses these funds to grant credits or other subsidies on more sustainable alternatives such as hybrid electric vehicles. Small changes in corporate tax rates can radically change return on investment for capital projects. If a credit or reduced tax rate is applied to environmentally beneficial activities, these proposals become more attractive. Conversely, a surtax could be imposed on environmentally damaging projects. This could result in an economic stimulus, as companies become increasingly invested in environmentally sustainable projects.

Germany enacted its ecotax in three steps. The 1998 legislation taxed electricity and petroleum at variable rates based on environmental considerations. Renewable sources of electricity were not taxed. In 1999, they adjusted taxes to favor efficient conventional power plants, whereas the 2002 statute increased taxes on petroleum. As these taxes were enacted, income taxes were reduced proportionally, so that the total tax burden was constant. Several nations have introduced differentiations in their car registration taxes to encourage the purchase of environmentally friendly vehicles.

Opposition to environmental taxes frequently focuses on the fact that these taxes are regressive, hurting those in the lower economic brackets more than those whose income is larger. However, an environmental tax package, by reducing income or sales taxes, can eliminate any regressitivity. In addition, the poor benefit from the reduction in environmental harm.

In conclusion, five advantages are cited for environmental taxes: they can provide incentives for behavior that protects or improves the environment, they can enable environmental goals to be reached in the most efficient manner, they can help signal structural environmental changes needed for a more sustainable economy, they can encourage innovation and the development of new technology, and revenue from environmental taxes can be used to reduce other taxes, reducing distortions and raising efficiency in the economy. One disadvantage would be short-term displacements that occurred in the economy as jobs and companies were eliminated. Second, unlike the current tax systems in place, environmental taxes would need the coordination and cooperation of all nations, or the benefits realized by those enacting these taxes would be diminished. A third disadvantage would be continuing environmental damage by those who sought to circumvent the system.

**See Also:** Environmentalism; Environmentally Friendly; Green Consumer; Regulation; Sustainable Consumption.

### Further Readings

Durning, Alan Thein, et al. *Tax Shift: How to Help the Economy, Improve the Environment, and Get the Tax Man off Our Backs*. Seattle, WA: Northwest Environmental Watch, 1998.

Gale, William G. "The Case For Environmental Taxes." Washington, DC: Brookings Institution, 2005. http://www.brookings.edu/opinions/2005/0721taxes_gale.aspx?P=1 (Accessed February 2009).

"A Primer on Green Taxation, *Alternatives Journal*, 1996. http://www.articlearchives.com/government/public-finance-taxes-taxation/12957661.html (Accessed February 2009).
Robertson, James. "Eco-Taxes." *New Internationalist,* 2005. http://www.netint.org/Issue278/taxes.htm (Accessed February 2009).
Snape, John and Jeremy de Souza. *Environmental Taxation Law: Policy, Contexts and Practice.* Aldershot, UK: Ashgate Publishing, 2006.

*John L. Stancil*
*Florida Southern College*

# TEA

Tea is the leaf of an evergreen plant of the Camellia genus, more specifically, *Camellia Sinensis*. It has historically been grown in Asia, including Myanmar, southeast China, Laos, Cambodia, and Vietnam. It still grows there, as well as in many areas in India and Japan. Tea plantations sprang up in India when the British had them planted in the mid-1800s to supply their habit of drinking tea. Some of the environmentally negative aspects of tea production are that it can cause soil degradation and a loss of flora and fauna habitat when open land is converted to profitable tea crops. Tea cultivating land is often located in remote areas, which tend to be home to the highest biodiversity. Local deforestation can also occur due to the tea drying process, which involves roasting the leaves over a wood fire.

Tea consumption is growing worldwide, possibly due to increased media reports touting its healthful effects.

*Source:* Vanderdecken/Wikipedia

The tea plant has thick leaves that are dark green in color and have a strong, thick stem. Its white or pink blooms emit a delicate fragrance. There are about 200 different species of the tea plant around the world, but all teas—black, green, white, and oolong—come from the same plant. Tea thrives in a hot, moist climate. The processing of the leaves determines the difference in the type of tea. Tea consumption is growing worldwide, most likely as a result of increased health reports in recent years. All tea in the United States is imported. Therefore, selecting tea that promotes the health of the environment and people is important. Tea is handpicked and gathered in wide baskets that tea pickers wear on their backs, ensuring that the best leaves are collected. In Asia, the tea-picking season begins in spring and continues until August. In Africa, tea picking occurs all year.

The quality of tea varies greatly, and most tea that is prepackaged in bags or packets is often of lower quality than tea that is produced as loose, dried leaves. All tea contains caffeine in varying amounts. The differences in black, green, white, and oolong teas are in the way the leaves are picked and processed. Where and how the tea plant is grown also influences its taste and aroma.

Tea leaves are 80 percent liquid content, so they must be dried to be sold. How and to what degree the leaves are dried determines the result. For black tea, the leaves are spread out on mats for sun-drying. Alternatively, currents of warm air might be blown over the leaves. Then the leaves are rolled, allowing fermentation and oxidation to begin. During the fermentation phase, humid air blows over the leaves and they collect oxygen. Over a period of about three hours, they begin to blacken and take on flavor. A second drying takes place so that the oxidation stops and the product is shelf-stable.

To make green tea, the leaves are sun-dried on bamboo trays for a few hours. Then the leaves are roasted or stirred in hot pans to evaporate additional moisture. The leaves are hand-rolled and then go through a second drying in pans. Often they are rolled once more to achieve their final shape. Green tea does not go through a fermentation phase, so there is no oxidation, and the leaves retain their delicate green color and flavor.

White tea comes from tea buds from the top of the plant, which are picked before they ripen and open. The closed buds are sun-dried. The delicate leaves are barely processed to preserve the original taste of the tea plant. If they are steamed, it is done immediately so that there is no oxidation. White tea is produced in small quantities and is generally very expensive.

Oolong tea has an almost identical process as black tea, except that it has a shorter fermentation stage and is often called semifermented. The flavor of oolong tea is slightly weaker than that of black tea.

Tea can be purchased in bulk, bags, or loose in packages, which affects the taste of the brewed product. Most loose-leaf tea is of higher quality than bagged tea, which you can observe by examining the leaves—which should be larger than what you see in tea bags—and by inhaling the aroma. Traditional tea bags work well, however innovative biodegradable pyramid-shaped bags have been developed to allow for better water circulation. This will result in a tea that is closer in taste to the preferred method of making tea—by steeping loose leaves in a pot.

Black tea contains the most caffeine of all the brews, but only half as much per cup as coffee. Green tea is the oldest form of tea, while white tea contains the least caffeine. Tea contains antioxidants, and some studies have shown that green tea is highest in antioxidants, whereas others report that white tea is highest. The reported benefits of drinking tea include a reduction in the incidence of various cancers, heart disease prevention, cholesterol-lowering effects, and possibly aiding weight loss.

According to TransFair USA, "Tea is one of the fastest growing Fair Trade Certified product categories, with Fair Trade Certified tea imports increasing an unprecedented 187 percent in 2005." Choosing organic and/or Fair Trade tea will more positively benefit the environment and people. According to TransFair USA, "Fair Trade promotes the use of sustainable farming methods that are safer for humans and the environment." Many tea concerns are made up of small-scale farmers. There are now more than 70 Fair Trade tea estates and small-scale producer groups, located in 11 Asian, African, and Latin American countries. There, wages and working conditions are monitored. Fair Trade premiums help to elevate communities by enhancing education and healthcare programs.

**See Also:** Beverages; Coffee; Fair Trade; Organic.

**Further Readings**

Hossain, M. Afzal. "Favourable Impact of Tea Plantation." *The Daily Star* (April 25, 2009). http://www.thedailystar.net/story.php?nid=85452 (Accessed August 2009).
TransFair USA Certified Tea Companies. http://www.transfairusa.org/content/certification/licensees.php?category=Tea&include=Everyone&sort=name (Accessed August 2009).
Willson, K. C. *Coffee, Cocoa and Tea (Crop Production Science in Horticulture)*. Cambridge, UK: University Press, 1999.

*Jill Nussinow*
*Santa Rosa Junior College*

# Television and DVD Equipment

Every year, the 275 million televisions in the United States consume over 50 billion kilowatt-hours of energy and are responsible for as much as 10 percent of total household electricity usage. Televisions are a considerable source not only of energy consumption but also of toxic substances contributed to landfills. The 2009 switch to digital broadcasting in the United States led to many older analog televisions being thrown out—no longer usable without a converter or cable connection. Many of those old cathode-ray tube (CRT) television sets contained lead, among other toxins conventionally found in consumer electronics. However, for all their bulkiness, the truth is that CRT televisions were, on the whole, more energy efficient than the flat screens that have replaced them.

Flat screens come in three varieties: LCD, plasma, and rear-projection. Rear-projection is generally reserved for the largest sizes, 50 inches and up, and in that size range they are more efficient than the other two. At other ranges, LCDs are more efficient. Plasma televisions are inherently resource intensive—the sharper their resolution is, the more energy they use because each pixel is illuminated independently, whereas other televisions use a single light source to illuminate every pixel on the screen. For that reason, there is a greater difference in energy use between two plasma televisions of different sizes and a pair of LCD or rear-projection televisions. The average plasma television consumes more energy than a refrigerator—which, because of its resource demands and "always on" nature, is otherwise the most consumptive appliance in the average household.

DVD players were introduced in the late 1990s, and by the 21st century had almost completely replaced the VHS player. This technology switch resulted in a large number of VHS players being relegated to landfills or exported as e-waste to developing countries. Manufacturers are attempting to make DVD players more environmentally friendly, including reducing their energy use, eliminating the use of brominated flame retardants, using lead-free solder, eliminating halogenated substances in their manufacture, and using less packaging.

The Environmental Protection Agency's Energy Star certifies products of various types that meet optional standards of energy efficiency. The Energy Star 4.0 standards for televisions and combination television/DVD players go into effect on May 1, 2010, to be followed by Energy Star 5.0 on May 1, 2012. The efficiency standards vary according to the size of the television, but in general they require the television set to be 30 percent more efficient than average sets of its type. As for the Energy Star 3.0 standards—which include requirements that will be carried forth in 4.0, 5.0, and subsequent standards—Energy Star–certified televisions are required to limit the power use in both on-mode and standby. The energy use of televisions and home entertainment equipment that are plugged in but not in active use is a common area of consumption that consumers remain unaware of—similar to many other appliances, televisions "leak" power. Energy Star televisions must also demonstrate their efficiency while in use. It should be noted that realistically speaking, no plasma television should be considered green compared with an LCD counterpart, and a big television, no matter how efficient, will not be a more environmentally conscious choice than a smaller one.

One of the concerns about flat screen televisions has been the use of nitrogen trifluoride in their production. Used to clean the manufacturing equipment, nitrogen trifluoride is a greenhouse gas 17,000 times more potent than carbon dioxide, but unregulated by federal and international bodies. Though a cleaner alternative has been developed—pure fluorine, with zero global warming impact—most manufacturers have not yet adopted it. Among those who have are LG and Toshiba-Matsushita.

In the near future, manufacturers are expecting to mass-produce televisions using light-emitting diodes (LEDs), which promise to be significantly more energy-efficient than current designs. Until then, LCDs seem the most ecofriendly choice. Television manufacturer Vizio introduced a line of green LCD televisions in 2008 that consume half as much power as standard televisions of the same size, and Philips announced a 42-inch set that uses about 75 watts—as much as many 30-inch televisions.

Televisions, as well as computer displays, are significant sources of electronic waste, containing nontrivial amounts of lead, mercury, and other hazardous substances as well as gold, platinum, copper, and silver in amounts that are worth harvesting. The export of electronic waste from the United States to other countries has become a major environmental concern and human health hazard, and the dismantling of electronics often presents an occupational health hazard because of the exposure to toxins within. More and more companies, particularly those appealing to green consumers, are manufacturing their products in such a way that they are more easily disassembled at the recycling stage, limiting toxicity for the disassembler and helping to encourage recycling. Sony and LG both offer numerous drop-off sites for recycling their electronics, operated by Waste Management Recycle America.

In the fall of 2009, the California Energy Commission proposed a new regulation to mandate better energy efficiency in televisions sold in the state. The move was prompted in part by the recent popularity of bigger and bigger televisions, and by California's perennial energy shortages, as well as the state's historical concern for energy efficiency and environmentalism. The proposal required a 33 percent improvement in energy efficiency in all sets sold beginning January 1, 2011, and a 50 percent improvement starting January 1, 2013. The utility companies immediately supported the move, although public opinion was mixed, in part because of headlines emphasizing that no existing plasma televisions over 40 inches would meet the requirements. Although some manufacturers balked, others spun the story to their advantage, claiming their projections had them meeting those goals anyway. Given the size of the California market, it is unlikely any manufacturer would sell one type of televisions in that state and another in the rest of the country, so the regulation would essentially raise the standards for the whole country.

**See Also:** Audio Equipment; E-Waste; Green Consumer; Leisure and Recreation; Materialism.

**Further Readings**

Brower, Michael and Warren Leon. *The Consumer's Guide to Effective Environmental Choices: Practical Advice From the Union of Concerned Scientists*. New York: Three Rivers, 1999.

Garlough, Donna, et al., eds. *The Green Guide: The Complete Reference for Consuming Wisely*. New York: National Geographic, 2008.

Rogers, Elizabeth and Thomas Kostigen. *The Green Book*. New York: Three Rivers, 2007.

*Bill Kte'pi*
*Independent Scholar*

# Tools

The category of tools can be usefully divided into subdomains: domestic utensils, hand tools, power tools, communications tools, machine tools, and so on. A commonsense understanding of a tool is that of a device that can be used to affect some kind of change on a material. A broader definition would stress the inclusion of social tools: networks of people and social technologies (e.g., public libraries) that operate on a large, societal scale and that enable users to effect successful actions through their use. Green approaches to tools consider the multitude of tools of different kinds and their effects at different scales of action.

The private ownership of underused tools makes little sense. Most domestic power tools, like this leaf blower, are used for perhaps 10 percent of the time that they are designed to last.

*Source:* iStockphoto

Critical attention has been drawn to a number of features and effects of familiar, domestic utensils and hand tools. As with most other consumer goods, there has been a marked proliferation, through duplication, of simple tools. Functionally identical tools are increasingly styled to appeal to lifestyle sensibilities. Nuances of color, shape, surface decoration, packaging, and advertising are the bases of such differentiation. The gendering of tools such that functionally very similar devices are offered in men's and women's versions is also noted.

Related to the above is the overdesign of functionally simple tools. Most of this overdesign results from the differentiation already considered, such as when styling predominates over ease of use. Function-driven design and manufacture of tools has somewhat given way to the form-driven imperatives associated with consumerism.

Increasing tool differentiation is also apparent. The range and configuration of tools with very similar functional purposes has gradually expanded. For example, there are many different types of devices to remove corks from wine bottles, each working in a different way and employing differing technological approaches, and each one offered as the solution to the problem of cork removal. Such differentiation is not new, but whereas in the past model differentiation predominated (e.g., offering a range of pocketknives with a family resemblance but differentiated through the functions on offer—more blades and so on), nowadays conceptual differentiation is emphasized as manufacturers offer a range of families of devices with similar functions (penknives and multitools and pocket assistants).

Tools are also becoming increasingly specialized. First, many traditional hand tools have been redesigned as power tools, and this technological elaboration is usually the basis of a promise of increased functionality and ease of use. Although often welcome, such technological elaboration is often unnecessary. For example, powered screwdrivers can easily overtighten screws, damaging both the head of the screw and the medium into which the screw is being driven. Second, specialization is apparent as new tools with novel, specific functions are developed. These tools are often matched to new products such that there is a seeming symbiosis between the two. For example, many electronic consumer goods are built from carcasses to which access is only possible through the use of specific tools. There is a strong suspicion that the development of both new fastening devices and their associated tools is a superfluous, though profitable, aspect of product design and development. A corollary of such specialization is the discouragement of adapted use or functional improvisation in relation to tools; witness the number of gadget catalogs offering very specific solutions to problems seemingly discovered or invented for the sole purpose of offering the solution.

In short, tools are as prone to the same imperatives of consumer product culture as any other commodity. However, a series of concerns relating to the social aspects of tools have also been raised and have been increasingly central to green approaches. First, the mass, private ownership of underused tools makes little social sense. Most domestic power tools are used for perhaps 10 percent of the time that they are designed to last. The rest of the time they are depreciating and usually deteriorating objects of great potential utility. The growth of what Ezio Mazini terms *enabling solutions*—tool libraries, tool banks, and the microleasing of tools—are seen as alternatives to mass, duplicated ownership. Second, the sheer amount of tools produced has a serious associated material impact on the environment. This is especially the case in terms of raw material extraction, production, and disposal, although, because of widespread tool idleness, not so much of use. Tool collection, repair, refurbishment, and redistribution charities can be regarded as a further enabling solution to such problems.

Such remedies, with their stress on neighborhood and community organization, sociality, collaboration, self-organization, and environmental benignancy, can themselves be seen to be tools. These initiatives can be positively allied with what Ivan Illich has termed *convivial tools*; that is, social devices that enable individuals and groups to use them in a manner of their choosing and toward variable and adaptable ends. Illich promoted the potential of the public library and telephone systems, as well as the rejection of the consumerist developments of utensils and tools highlighted above. It is in these frameworks, which integrate individual tools and technologies with social networks and initiatives, that the green response to tools is most fertile.

**See Also:** Consumerism; Durability; Garden Tools and Appliances; Green Design; Overconsumption.

**Further Readings**

Illich, Ivan. *Tools for Conviviality*. London: Marion Boyars, 1973.

Manzini, Ezio and Jégou François, eds. *Collaborative Services: Social Innovation and Design for Sustainability*. http://www.sustainable-everyday.net/main/?page_id=26 (Accessed April 2009).

Papanek, Victor. *Design for the Real World*. London: Thames and Hudson, 1984.

*Neil Maycroft*
*University of Lincoln*

# Toys

Toys are highly profitable consumer items worldwide. As a multibillion dollar industry, the production and distribution of children's toys is a competitive global market. As such, toy manufacturers may often forgo the costly collection of health and safety data on their products to maintain a competitive edge over other toymakers. This lack of safety testing puts children at risk for exposure to a variety of toxic chemicals. Moreover, children's toys, in addition to containing chemicals that are harmful to human health and the environment, often use materials that are unsustainable, with plastics being the most common. However, toys do not have to be unsustainable hazards—many manufacturers are switching to safer materials and production practices that may better assure the well-being of the environment and our children's health.

As a multibillion dollar industry, the production and distribution of children's toys is a competitive global market. Children's toys often use materials that are unsustainable, with plastics being the most common.

*Source:* iStockphoto

## Role of Regulatory Agencies in Assessing Children's Toy Safety

In the United States, the federal government does not require the full testing of chemicals before they are added to consumer products, making toy manufacturers less likely to conduct safety testing on specific chemical ingredients before their use. The federal law presumed to regulate individual chemical ingredients, the Toxic Substances Control Act (TSCA), became law in the early 1970s, giving the Environmental Protection Agency the task of collecting data on the thousands of chemicals used in commerce. Unfortunately, this regulatory task has been a large undertaking, and the Environmental Protection Agency has managed to collect data on fewer than 200 of the more than 62,000 chemicals registered for use since TSCA's inception in the 1970s. Therefore, toy manufacturers have very few restrictions on the types of chemicals they may use in children's products. The restrictions that are in place are limited to five types of chemicals, all of which are universally seen as highly toxic substances: halogenated chlorofluoroalkanes, polychlorinated biphenyls, dioxins, hexavalent chromium, and asbestos.

The Consumer Product Safety Commission (CPSC) is the U.S. federal agency tasked with regulating the safety of consumer products, including children's toys. Although the majority of CPSC restrictions deal with the use of the toy, rather than the materials it is made of, the Consumer Product Safety Improvement Act of 2008 imposes new requirements on the manufacturers of a variety of consumer products, including toys, and also requires some new testing procedures and reduced acceptable exposure levels for some chemicals (such as lead). However, the outcomes of this legislative act are still unclear.

The unwieldy nature of U.S. regulatory mechanisms and their failure to respond quickly to perceived toxic threats means that some toxic chemicals inevitably end up in children's toys. Lead, bromine, and polyvinyl chloride (PVC) are hazardous chemicals that have been found in children's toys in recent years, often to the detriment of children's health and safety.

## Chemicals of Concern

When found in toys, lead is often used as pigmentation in paint, plastics, or rubber. Widely recognized as a neurological and developmental toxin, lead has been restricted in children's toys and, according to CPSC policy, should not be present in children's products at any level higher than 300 parts per million.

Bromine is a chemical component of flame retardants and is often found in children's toys that are treated with flame-retardant chemicals. Exposure to brominated flame retardants is linked to both neurological and developmental defects. These flame retardants are also bioaccumulative and persistent in the environment. PVC is a common plastic that is often a material used in producing children's toys. To make PVC a flexible material, chemicals called phthalates are added during production. Phthalates have been linked to reproductive health problems, and some types of phthalates may be carcinogenic.

A group of environmental health organizations have created a database called HealthyToys.org that has tested more than 5,000 children's toys. The coalition of groups involved in the project found that a variety of popular children's products contained toxic chemicals at levels that pose a threat to children's safety.

## Safer Toys and Sustainability

Children's toys do not have to be hazardous products that pose a threat to health or to the environment. Regulations and manufacturing practices can be altered and, in many instances, are already being changed.

A number of U.S. states have enacted their own state-level policies to keep toxic chemicals out of children's products. From PVC and bisphenol-A to lead and mercury, many states have adopted policies that are more protective than those at the federal level. In addition, a variety of environmental and healthcare organizations are pushing for federal reform of policies such as TSCA and agencies such as the CPSC.

Along with this, a number of toy manufacturers are working to make their products safer and more sustainable. The use of bio-based plastics and inks made of corn, soy, and other natural materials is becoming a common practice among some manufacturers, and the use of alternative materials like bamboo and organic cottons is also gaining popularity. Importantly, the use of alternative and more sustainable materials in toys is still a niche market, serving the needs of informed and affluent customers. To promote the safer production and consumption of all toys, both regulatory and manufacturing approaches to toy production must be altered.

**See Also:** Consumer Activism; Regulation; Sustainable Consumption.

### Further Readings

Consumer Product Safety Commission. "Consumer Safety." http://www.cpsc.gov (Accessed September 2009).

Healthy Stuff. "Healthy Toys." http://www.healthystuff.org (Accessed September 2009).

Lowell Center for Sustainable Production at the University of Massachusetts. "State Chemicals Policy Database." http://www.chemicalspolicy.org/chemicalspolicy.us.state.database.php (Accessed September 2009).

*Amy Lubitow*
*Northeastern University*

# U

## UNITED NATIONS *HUMAN DEVELOPMENT REPORT 1998*

Every year since 1990 the United Nations Development Programme publishes a reliable source that provides an alternative perspective on critical issues for human development worldwide. Human development means the enlargement of all human choices, and in this case, it refers to improving consumer choices respecting human life. Every annual report presents agenda-setting data and analysis. The report tries to focus international attention on issues and policy options to meet the challenges of economic, social, political, and cultural development.

The 1998 report investigates the 20th century's growth in consumption. World consumption has expanded at an unprecedented pace over the 20th century, with private and public consumption expenditures reaching $24 trillion in 1998—twice the level of 1975 and six times that of 1950. In 1900, real consumption expenditure was barely $1.5 trillion. The report states that the benefits of this increase in consumption have been badly distributed, leaving a gap between poor and rich countries. Furthermore, ever-expanding consumption puts strains on the environment—emissions and wastes that increase pollution—and growing degradation of natural resources that undermines the capability of future generations to take advantage of them. The report demonstrates that the rising pressures of conspicuous consumption can turn destructive, reinforcing exclusion, poverty, and inequality.

Consumer choices should be turned into a reality for all. Consumption per capita has increased steadily in industrial countries (about 2.3 percent annually) over the past 25 years, spectacularly in East Asia (6.1 percent). Nonetheless, these developing regions are far from catching up to levels of industrial countries. Consumption growth has been slow or stagnant in others. The average African household today consumes 20 percent less than it did 25 years ago. The poorest 20 percent of the global population have been left out of the consumption explosion. Over a billion people are deprived of basic consumption needs, and of the 4.4 billion people in developing countries, nearly three-fifths lack basic sanitation. Almost a third of these people have no access to clean water. A quarter do not have adequate housing. A fifth have no access to modern health services. A fifth of these children do not attend school beyond grade five. A child born in the industrial world adds more to consumption and pollution over his or her lifetime than do 30–50 children born in developing countries, about a fifth of whom do not have enough dietary energy and protein. Nutrition deficiencies are even more widespread. Worldwide, 2 billion people are anemic, including 55 million in industrial countries.

## Disparities

The global population is projected to be 9.5 billion in 2050, with more than 8 billion in developing countries. This population will require three times the basic calories consumed today to be adequately fed—the equivalent of about 10 billion tons of grain a year. Twenty percent of the world population in the highest-income countries account for 86 percent of total private consumption expenditures, and the poorest 20 percent for 1.3 percent. More specifically, the richest fifth consumes 45 percent of all meat and fish, the poorest fifth 5 percent. The richest fifth consumes 58 percent of total energy, the poorest fifth less than 4 percent. The richest fifth has 74 percent of all telephone lines, the poorest fifth 1.5 percent. The richest fifth consumes 84 percent of all paper, the poorest fifth 1.1 percent. The richest fifth owns 87 percent of the world's vehicle fleet, the poorest fifth less than 1 percent.

The growth in the use of mineral resources has considerably slowed in recent years, and fears that the world would run out of such nonrenewable resources such as oil have proven false. New reserves have been discovered, and the growth of demand has slowed. Energy efficiency also has improved, thanks to technological advance and recycling of raw materials. The per capita use of basic materials such as steel, timber, and copper has stabilized in most Organisation for Economic Co-operation and Development countries.

The concern of the international community is the pollution and waste being created that exceed the world's capability to absorb and convert them. The use of mineral fuels is emitting gases that change the ecosystem. Annual carbon dioxide emissions have quadrupled over the past 50 years, and global warming has been proven to be a serious problem. The consequences of this phenomenon are permanent floods of large areas, an increase in the frequency of storms and droughts, the extinction of some species, and the spread of infectious diseases.

The consumption of freshwater has almost doubled since 1960, and the marine catch has increased fourfold. Wood consumption, both for industry and for household fuel, is now 40 percent higher than it was 25 years ago. These elements show the growing deterioration of renewable resources such as water, soil, forests, and fish. Twenty countries already suffer from water stress, having less than 1,000 cubic meters per capita a year, and water's global availability has dropped from 17,000 cubic meters per capita in 1950 to 7,000 today. A sixth of the world's land area—nearly 2 billion hectares—is now degraded as a result of overgrazing. Forests worldwide, which bind soil and prevent erosion, regulate water supplies, and help govern the climate, are shrinking. Since 1970, the wooded area per 1,000 inhabitants has fallen from 11.4 square kilometers to 7.3.

The fifth of the world population in the highest-income countries accounts for 53 percent of carbon dioxide emissions, whereas the poorest fifth account for just 3 percent. Brazil, China, India, Indonesia, and Mexico are among the developing countries with the highest emissions. However, having large populations, their per capita emissions are still relatively small—3.9 metric tons a year in Mexico and 2.7 in China, compared with 20.5 metric tons in the United States and 10.2 in Germany. The consequences of global warming will be devastating for many poor countries—with a rise in sea levels, Bangladesh could see its land area shrink by 17 percent. The 132 million people in water-stressed areas are predominantly in Africa and parts of the Arab states, and if present trends continue, their numbers could rise to 1–2.5 billion by 2050.

Deforestation is concentrated in developing countries. Over the last two decades, Latin America and the Caribbean lost 7 million hectares of tropical forest; Asia and sub-Saharan Africa lost 4 million hectares each. Most of this deforestation has taken place to meet the demand for wood and paper, which has doubled and quintupled, respectively, since 1950.

## Seeking Solutions

The world community has been active in environmental problems that directly affect poor people. Such areas include desertification, biodiversity loss, and exports of hazardous waste. For example, the Convention on Biological Diversity has near-universal signature, with over 170 parties. The Convention to Combat Desertification has been ratified by more than 100 countries. However, the deterioration of arid lands—a major threat to the livelihoods of poor people—continues, and there are other immediate environmental concerns for poor people, such as water contamination and indoor pollution, which have yet to receive serious international attention. Globalization is integrating consumer markets around the world and opening opportunities, but it is also creating new inequalities and new challenges for protecting consumer rights. Globalization is integrating not just trade, investment, and financial markets. This has two effects—economic and social. Economic integration has accelerated the opening of consumer markets with a constant flow of new products.

Every year some 100 countries prepare national human development reports, assessing their present situations and drawing conclusions about actions to achieve more human patterns of development. Most of these plans have analyzed needs in the critical areas of education, health, and employment—often linking them with opportunities for generating resources from reduced military spending. In the poorer countries, many priorities in consumption still need to be addressed. Increases in consumption should be planned and encouraged, but with attention to nurturing the links, to making sure that the increases contribute to human development, and to avoiding extremes of inequality.

**See Also:** Carbon Emissions; Conspicuous Consumption; Consumer Society; Demographics; Environmentalism; Sustainable Consumption.

### Further Readings

Anand, Sudhir and Amartya Sen. *Consumption and Human Development: Concepts and Issues*. Oxford: Oxford University Press, 1997.

Khor, Martin. *Globalization, Income Distribution, Consumption Patterns and Effects on Human and Sustainable Development.* San Francisco, CA: International Forum on Globalization, 1997.

Prescott-Allen, Robert. *Consumption Patterns, Ecosystem Stress and Human Development.* Washington, D.C.: Island, 1997.

United Nations Development Programme. *Human Development Report 1998*. http://hdr.undp.org/en/reports/global/hdr1998 (Accessed August 2009).

*Michele Caracciolo di Brienza*
*Graduate Institute of International and Development Studies, Geneva*

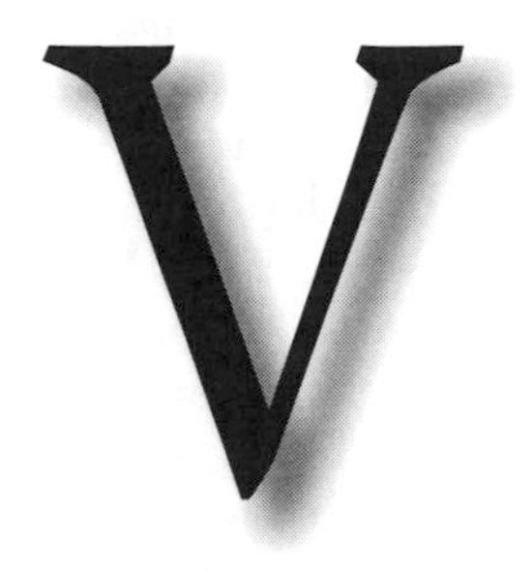

# Vege-Box Schemes

Vege-box schemes (veggie-box or vegetable-box schemes) are agreements between growers and consumers or marketers of fresh produce. The produce, usually fresh and locally grown, is delivered to a fresh produce stand, a larger fresh foods store, or directly to individual customers. It is viewed as a green enterprise using less energy than commercial farming or importing produce.

In many places, especially in the urban areas of the United States and many other industrialized countries, the demand for fresh food has been increasing. In previous decades, grocery chain stores with displays of virtually blemish-free fruits and vegetables was a great attraction to customers, many of whom had never visited a farm or seen fresh vegetables in a garden. Many customers abandoned smaller fruit stands or neighborhood vegetable and fruit peddlers, who once drove through neighborhood streets in horse-drawn wagons or trucks selling their wares. These types of peddlers and small neighborhood stores had been common from the time of the Great Depression through the Korean War.

Since the prosperity of the 1960s, most urbanites and suburbanites shopped at convenient chain grocery stores for vegetables and fruit. The development of industrial farming turned many fruits and vegetables into commodities that were uniformly the same and sold year-round without the limitations of seasonal availability. Complaints that produce, for example, tomatoes ripened in methane gas chambers, did not taste the same as produce that was naturally grown led to consumers seeking better-quality goods elsewhere.

In addition, as the general public became aware of the vast quantities of chemicals used as fertilizers, pesticides, and herbicides, many grew uncomfortable with the possible negative health consequences. Organic gardens and farms grew in popularity in response. In addition, the growth of huge agricultural corporations using a limited variety of hybrid seeds sparked an interest in preserving heirloom seeds—the seeds of plant varieties that are no longer readily available in mainstream commercial vegetable markets.

In many instances, from the 1980s onward, young people from prosperous backgrounds in urban and suburban areas moved to small towns and farms in search of a quieter lifestyle. To make a living, many turned to organic farming or to specialty-produce farming, especially using heirloom seeds. Urban and suburban families were seeking fresh vegetables just hours out of a garden, rather than those sold in a store days after being picked on an industrial farm or imported from a Third World country.

Vege-box schemes have been developed to fill the desires of customers, usually urban and wealthy, for locally grown fresh foods. In general, vege-box schemes are arrangements with local farmers to deliver an assortment of vegetables that are available as they ripen. A great many vege-box schemes provide organically-grown vegetables directly to the consumer from the producer. These schemes have been growing in popularity in a number of countries. In Great Britain, they are estimated to number well over 600.

Vege-box schemes usually operate by subscription. The customer generally contracts the farmer or gardener to deliver fresh vegetables or fruits every week or every other week. If the farmer also produces goods such as honey, preserves, pickles, meat, or diary products like cheese or goat milk, arrangements can be made for delivery of these products as well.

Organic farms and produce sold as organic must be certified by a government agency or by a trade association before it can be marketed as "organic." To be considered organic, it must be grown without the use of chemical fertilizers, pesticides, and herbicides. Many environmentally conscious people support the use of traditional fertilizers, such as cow manure, and the use of weed and insect management programs by using plants that discourage weeds, or planting crops at a certain time of the year to avoid a weed season, or to incorporate plants such as marigolds in their crops to discourage insects.

Produce purchased through organic vege-box schemes is usually praised by customers as superior in flavor to commercially grown produce. Customers of vege-box schemes usually pay a premium, and many vege-box-scheme farms allow or even encourage potential customers to visit the farm, although the usual arrangement is to deliver the produce to the home, a cooperative, or to a public space where a local market is held.

**See Also:** Markets (Organic/Farmers); Organic; Seasonal Products; Vegetables and Fruits.

### Further Readings

Coleman, Eliot and Sheri Amsel. *The New Organic Grower: A Master's Manual of Tools and Techniques for the Home and Market Gardener*. White River Junction, VT: Chelsea Green, 1995.

Decoteau, Dennis R. *Vegetable Crops*. Upper Saddle River, NJ: Prentice Hall, 1999.

Knott, James Edward and Donald N. Maynard. *Knott's Handbook for Vegetable Growers*. New York: John Wiley & Sons, 1988.

Lee, Andy W. and Patricia Foreman. *Backyard Market Gardening: The Entrepreneur's Guide to Selling What You Grow*. Buena Vista, VA: Good Earth, 1995.

Stark, Tim. *Heirloom: Notes From an Accidental Tomato Farmer*. New York: Broadway Books, 2009.

*Andrew Jackson Waskey*
*Dalton State College*

## Vegetables and Fruits

Much of green consumerism entails consumption of healthy and sustainably grown foods. Perhaps considered the healthiest of foods, vegetables and fruits are often, however, grown, packaged, and distributed in ecologically unsustainable manners. Moreover, even

Fruits and vegetables are upheld as linchpins of good health, while the organic, non–genetically modified, and locally grown types are often heralded as icons of sustainability.

*Source:* Bob Nichols/Agricultural Research Service/USDA

the most emblematic of healthy foods—apples, bananas, spinach, corn—have recently come under scrutiny as carriers of highly toxic residual pesticides, controversial transgenetic alteration, and even diseases, such as *Escherichia coli* and *Salmonella*. Despite—or because of—these environmentally and socially unsustainable aspects of conventional vegetable and fruit production, green consumers have begun to value ripe, fresh, organic, and/or local produce even more by investing heavily in purchasing and even growing it. Concurrently, as industrialized diets move deeper into the health quagmire (and vicious cycle) of highly processed food, drink, and pharmaceutical medication, vegetables and fruits become even more recommended, prescribed, and lauded. Across the board, fruits and vegetables are upheld as linchpins of good health, and the organic, non–genetically modified, and local among them are often heralded as icons of sustainability.

Biologically speaking, the term *vegetable* refers to the edible part of a nonwoody plant cultivated particularly for that edible part. Many herbaceous plants are related as family clusters, such as nightshades, allium, cucurbits, and brassica. Meanwhile, fruits are the ripened, swollen ovary of a flower after the ovules inside have been fertilized. Fruits contain the seeds of any flowering plants and can be divided into three main subclasses: those with seeds in the flesh, such as pears, citrus, melons, and berries; those containing pits (drupes), such as cherries, peaches, and plums; and dry fruits (dry drupes), such as certain nuts, beans, and peas.

On closer botanical inspection, however, this technical distinction blurs. Because a vegetable is by definition the edible part of any herbaceous plant (be it the root of the carrot plant, the floret of the broccoli plant, the seed of the bean, the bulb of the garlic, or the leaf of the spinach), and fruits are the seed-containing part of any flowering plant, the two are not mutually exclusive. Tomatoes, peppers, pumpkins, and cucumbers fit both vegetable and fruit classifications, and somewhat counterintuitively, olives, sugar peapods, avocados, coffee, and almonds would all technically count as fruits. *Fruit* serves as a more botanically specific term, whereas *vegetable* remains more a grocer or chef classification (though this fruit definition begs the question of the oxymoron of seedless fruit).

Despite the arbitrary difference and contested definitions of vegetables and fruits, the categories remain relatively firm in mainstream, culinary, public health, and horticultural parlance. Vegetables usually contain complex carbohydrates, whereas fruits offer mostly simple carbohydrates with more readily accessible sugars (fructose). In many traditional cuisines, different fruits and vegetables have respective properties, often associated with heating or cooling attributes, and their various combinations enhance nutrient absorption and benefit. According to Ayurveda, the ancient Indian medical lineage, for instance, vegetables and fruits should not be mixed, and most fruits should be eaten separately, because each requires particular (and often conflicting) digestive enzymes.

The defining feature of fruits and vegetables, however, remains their renowned health benefits. Both contain an impressive array of vitamins, minerals, micronutrients, fiber, and antioxidants. Green-colored vegetables, such as spinach and kale, have long been revered as superfoods for their multifaceted nutritional content—which nutritionists try to identify and quantify and vitamin supplements aim to mimic. Such greens serve as crucial sources of iron, calcium, and B-vitamins, and beans and legumes give fiber and folate, potatoes ensure potassium and vitamin C, and orange- and red-colored vegetables supply substantive doses of vitamin A (among many other benefits). Fruits also contain extraordinary nutritional properties, boosting immunity, improving digestion, and cleansing internal organs and the bloodstream. Studies regularly document the details of how sustained consumption of vegetable and fruit substantially reduces the risk of type 2 diabetes, cardiovascular disease, obesity, and hypertension and helps prevent or improve survival of colon, prostate, breast, and other prevalent cancers.

Nevertheless, not every apple confers the same healthful properties. Industrialized production of fruits and vegetables has altered, and often decreased, their overall absorbable nutrition. This is primarily a result of the decline in nutrient levels in, and thus overall quality of, soils. The nutrient composition of the soil supplies the nutrient composition of the fruits and vegetables themselves. Long-term application of chemically enhanced fertilizers on monocrop production without rotation or fallow periods depletes soils of nitrogen, potassium, phosphorus, calcium, and magnesium. The synthetic versions of these minerals work to ensure a harvest but fail to replicate the natural complexity of noneroded topsoil, and thus fail to transfer those high-quality nutrients to the crops themselves.

The industrialization of agriculture has also changed the seeds used, from vast biodiverse arrays of heirloom vegetables and heritage fruits to post–World War II use of hybrid seeds to the contemporary crisis of agri-biodiversity decline and adoption of genetically engineered varieties. The hybrid corns of the green revolution and the crops of its offspring—the gene revolution—share aims of increased productivity through an agri-food system resistant to herbicide sprayings and conducive to mechanized harvest, global transport, and extended shelf life. The vegetables and fruits engineered in these contexts have been bred accordingly.

Health benefits of fruits and vegetables have also suffered under industrialized agriculture with the increased use of insecticides, herbicides, and fungicides—some of which have been found to cause cancers, skin conditions, neurotoxicity, and birth defects, thereby posing risks to both the farmers who apply them and the eaters who consume them.

Nevertheless, it is generally agreed that the assets of fruits and vegetables outweigh their hazards. U.S. Food and Drug Administration–based nutrition standards suggest five daily servings of fruit and vegetables, though the majority of the U.S. population does not meet this quota. Reasons for this include lack of accessibility and lack of affordability. Those living in "food deserts" do not have the financial and/or logistical means to find and purchase fresh produce regularly. Not as expensive as meats, dairy, or fish, fruits and vegetables have nevertheless increased in expense and are now—even when nonorganic—considerably and problematically costlier than processed counterparts.

U.S. Food and Drug Administration recommendations have been criticized, however, as simplistic—cartoon versions of health with detrimental ecological, social, and even health consequences. This generic conception of healthy produce engenders shiny red apples, heavily sprayed and highly waxed for indefinite produce-aisle sheen; bouquets of summer berries and tropical pineapples at Christmas; and conflict bananas grown under notorious working conditions in Central America. Moreover, when green consumerism becomes sheer rhetoric—for example, in school lunch cafeteria marketing—tater tots suffice for vegetables and high-fructose corn syrup–laden grape drinks count as fruit servings.

The rallying cry of green consumers is a reclamation of fruits and vegetables from overpuffed, tasteless look-alikes, but also from the periphery of the plate in general. Vegetables have graduated from their secondary status as a paltry side dish to the front-and-center meat dish—anchor of any postwar meal, be it breakfast, lunch, or dinner. A rise in vegetarianism and even veganism—consuming no animal products whatsoever—has opened a vegetable- and fruit-based culinary universe in restaurants and cuisines, wherein the myriad variations of beans and spices alone supply protein and nutrients, but also a satisfyingly broad array of tastes and textures.

Overall, however, green consumerism in general has revived collective respect and demand for the organic and/or local fruit or vegetable, in all its seasonal glory. The subsequent rise in household and community gardens and miniorchards, home preservation, farmers markets, the organic and local phenomena, and preventative medicine all point to fresh fruits and vegetables as being pivotal for human and ecological health.

**See Also:** Food Miles; Gardening/Growing; Organic; Pesticides and Fertilizers; Seasonal Products.

## Further Readings

Ballister, Barry. *The Fruit and Vegetable Stand: The Complete Guide to the Selection, Preparation and Nutrition of Fresh and Organic Produce*. New York: Overlook, 2007.

The Chopra Center. "Ayurveda: The Science of Life." http://www.chopra.com/ayurveda (Accessed November 2009).

Coleman, Eliot, Barbara Damrosch and Kathy Bray. *Four-Season Harvest: Organic Vegetables From Your Home Garden All Year Long*. New York: Chelsea Green, 1999.

U.S. Department of Agriculture. http://www.MyPyramid.gov (Accessed November 2009).

*Garrett Graddy*
*University of Kentucky*

# Waste Disposal

Globally, the amount of waste from human activities is increasing. In 2006, the amount of municipal solid waste (MSW) generated worldwide was 2.02 billion tons. Some estimates predict global MSW will increase by 37.7 percent from 2007 to 2011. This enormous amount of waste requires appropriate disposal mechanisms. This article discusses some of the options available for waste disposal.

The simplest definition of waste includes all items no longer useful to humans. According to the U.S. Environmental Protection Agency, "solid waste means any garbage, or refuse, sludge from a wastewater treatment plant, water supply treatment plant, or air pollution control facility and other discarded material, including solid, liquid, semisolid, or contained gaseous material resulting from industrial, commercial, mining, and agricultural operations, and from community activities." However, the definition of waste is subjective, as waste to one community could be a resource to another community. Nevertheless, such a definition of waste helps in creating legal statutes and laws for management purposes.

The major sources of waste are construction and demolition, agriculture and forestry, mining and quarrying, industries, municipal solid waste, and energy production. Moreover, the common types of waste are household waste, commercial waste, e-waste, hazardous waste, industrial waste, construction waste, biomedical waste, agricultural waste, and universal waste. Broadly, wastes are of three types: hazardous, nonhazardous, and inert. Because of their environmental and health hazards, waste disposal is a major concern in many nations. Rich nations have effective and efficient disposal systems, whereas poor nations do not have an established system to manage the waste. In many developing nations, open dumping is very common.

## An Age-Old Problem

Historically, waste disposal was done as long ago as 500 B.C.E. in Athens, Greece. Disposal of waste became challenging with an increasing population. The Industrial Revolution of 1750–1850 increased the movement of population from rural areas to urban areas, which subsequently generated more waste. Governments created new laws to manage this increasing waste. In the United Kingdom, a law in 1297 required residents to keep the front of their homes clean, and the Public Health Act of 1875 required local authorities to remove and dispose waste.

In the United States, one of the early legislations for waste disposal was the 1795 law by the Corporation of Georgetown, which prohibited waste disposal on streets. In 1885, the first municipal incinerator was introduced in Allegheny, Pennsylvania. In the early 1900s, nearly 79 percent of U.S. cities were collecting garbage. This waste was mostly dumped into lakes and rivers and in "less desirable" neighborhoods. A study on waste disposal techniques between 1899 and 1924 found the following methods: dumping on land, farm use, dumping in water, burning, reduction, multiple methods, and no methods.

The dumping of waste in water declined after World War I because of global concerns. Likewise, burning became expensive. Some cities mixed waste with other materials and used it in construction projects like roads and levies. Some even fed the waste to swine. However, the huge amount of waste meant that a city required 75 pigs to dispose of one ton of garbage per day. In the United States and other industrialized nations, availability of cheap land made landfill the best option to manage the waste.

After World War II, environmental and health concerns from waste disposal increased. Incidents like that at Love Canal, New York, in 1977 compelled nations to adopt effective strategies to manage their waste. For example, the European Union's strategy on waste management uses a concept of "hierarchy of waste management." Within the hierarchy, waste reduction is at the top, followed by reuse, recycling and composting, energy recovery, and landfill—the least desirable option. In the United States, the Resource Conservation and Recovery Act in 1976 amended the Solid Waste Disposal Act of 1965 and set national goals to protect human health and the environment from hazards of waste disposal, to conserve energy and natural resources, to reduce the amount of waste, and to ensure waste is managed in an environmentally sound manner. The Resource Conservation and Recovery Act banned all open dumping of waste and mandated strict controls over the treatment, storage, and disposal of hazardous waste.

## Modern Solutions

The different waste disposal methods used today are recycling, waste reduction, waste reuse, landfill, incineration, pyrolysis, gasification, combined pyrolysis-gasification, composting, anaerobic digestion, and phytoremediation. Regardless of the methods used, the basic mechanism in waste disposal involves collection, separation, and disposal. Collection consumes about 50–70 percent of the total cost of waste disposal. Collection cost depends on the number of collection points, type of containers, size of the vehicle, frequency of collection, and transport cost to final destination. The different types of containers used in the United States to collect MSW are plastic or paper bags, 901-volume metal cans, 350-5001 volume plastic wheeled carts, and 6001-300001 volume containers. The collected waste is then transported by trucks, trains, and barges to transfer stations, where it is compacted and sent for treatment or disposal.

Waste reduction is economically and environmentally beneficial to society and businesses, as it saves energy, space, and transportation and administrative costs. Waste reduction practices are simple and very effective and can be practiced at all levels. Waste reuse is another step toward waste reduction.

Recycling involves the collection, separation, cleanup, and processing of wastes into reusable marketable products. The common recyclable materials are paper, textiles, plastics, glass, metals, wood, and tires. The advantages of recycling include reducing use of virgin raw materials, saving energy, reducing emissions, and contributing to mitigation of climate change. Recycling depends on a secure and stable supply of waste materials, a suitable system for the collection and then transportation of waste, a reliable materials separation and cleanup mechanism, and a secure and stable market for the new products

Landfill is the most common waste disposal method used in many nations. The popularity of landfill over other methods is the result of its low cost and the wide variety of wastes that can be disposed in it. Other advantages include restoring landscape, energy recovery, and socioeconomic benefits. In addition, other disposal options also require landfill as their final destination to dispose residues. Disadvantages of landfills include meeting environmental and health regulations, leakages, bad odor, traffic, and noise and air pollution. There are 3,091 active landfills and over 10,000 old landfills in the United States. In 2001, about 56 percent of the MSW was landfilled in the United States. However, a study of Organisation for Economic Co-operation and Development nations for 1995–2000 found that the share of landfill as the primary method of waste disposal decreased from 64 percent to 58 percent.

The different types of landfills include hazardous landfills, nonhazardous landfills, inert waste landfills, controlled flushing bioreactor landfills, and entombment dry landfills. Landfill disposal occurs in the following steps:

1. Site selection and assessment for proximity to waste, road access, affect on local environment, and geological and hydrological stability.
2. Development of a landfill has to consider landform profile, site capacity, waste density, waste settlement, material requirements, drainage, and operational procedures. Key to the design of a landfill is the containment barrier system at the base, sides, and cap of the landfill.
3. Landfill capping is done after the final waste has been deposited. This is done to prevent rainwater and surface water from percolating into the site and generating lecheate, and to block the release of a and prevent entry of air, which would disrupt the anaerobic biodegradation process.
4. The final step of landfill disposal is its completion and restoration. The completed landfill site should be physically, chemically, and biologically stabilized and no longer pose risks to the local environment. The stabilization can take decades to complete and requires responsibility and liability from the operator. The closed site can be developed for agriculture, woodland, conservation, and various infrastructures.

Incineration is the burning of combustible materials in the waste to produce heat, water vapor, nitrogen, carbon dioxide, and oxygen. Some advantages of incineration are that it can be carried out near the point of waste collection, waste is reduced to ash products (which is 10 percent of original volume and 30 percent of original weight), it produces no methane, and it can generate energy. Disadvantages include high cost, steady supply of products, pollution, and that the residue requires further disposal (a landfill). The most common incineration system is mass burn incineration, which can burn 10–50 tons of waste per hour. Other types include fluidized bed, cyclonic, pyrolytic, rotary kiln, rocking kiln, cement kiln, and gaseous incinerators. A typical incineration of MSW includes waste delivery, a bunker and feeding system, a furnace to burn the waste, heat recovery, emissions control, energy recovery, and management of wastewater and ash residues.

Pyrolysis is a thermal process that degrades organic waste at 400–800 degrees Celsius, in the absence of oxygen, into carbonaceous char, oil, and combustible gases. The by-products, because of their high energy content, can supply the energy requirements of the plant. The different technologies used for pyrolysis are fluidized bed, fixed-bed reactors, ablative pyrolysis, rotary kilns, entrained flow reactors, and vacuum pyrolysis.

Gasification is also a thermal process, but unlike pyrolysis, this uses oxygen during the process. Heated oxygen is reacted with carbon present in the waste to produce gas, ash, and tar products. Gasification is conducted at high temperatures of 800–1400 degrees Celsius.

Combined pyrolysis-gasification is a thermo-chemical processing of waste, which uses both pyrolysis and gasification to dispose the wastes. However, during the process pyrolysis and gasification are done separately. This combined process helps in disposing of a wide range of waste materials. Globally, there are 100 combined pyrolysis-gasification process plants operating.

Composting is a biodegradation process that is best suitable for organic waste from garden and households. Typically, it takes about four to six weeks to stabilize the product. Composting can be done even in the backyards of homes. The degraded product (compost) is added to soil as an organic fertilizer. The only difficulty in composting is the separation of nonorganic waste before putting it into the compost pit. In developing nations, composting helps to meet the fertilizer needs of many subsistence-based farmers. One example of a large-scale composting program is the Barcelona Composting Scheme in Spain. This site composts about 10,700 tons of waste annually from 55,000 households (137,000 people).

Anaerobic digestion is similar to landfill disposal but differs in two ways. First, it controls the process, which means all the gas produced during degradation is used. Second, the process is complete within a few weeks. Anaerobic digestion is conducted in an enclosed and closely controlled reactor. The different types of anaerobic digestion systems are dry continuous digestion, dry batch digestion, leach-bed process, wet continuous digestion, and multistage wet digestion.

Phytoremediation uses certain plant species like poplar and sunflower to absorb chemicals present in contaminated soils. One example of phytoremediation is in the recovery of abandoned metal mines and coal mines.

The challenges for nations are in disposing hazardous waste, such as radioactive waste. For example, in the United States, there are 1.7 trillion gallons of contaminated groundwater, 2,000 tons of radioactive spent nuclear fuel, 4,000 facilities, and 40 million cubic meters of contaminated soil and debris. To clean all the nuclear waste it will cost $212 billion and take 70 years. Some areas of conflicting values regarding the disposal of nuclear waste are effectivity, protection of future generations, restoration of resources, and decision-making process.

Waste disposal is a complex process. Regardless of the methods used, the waste disposal system should consider the following criteria: visual impacts, air emissions from incinerators, landfill sites, water discharges, ash discharges, human health, fauna and flora, noise, costs, traffic, socioeconomic, social justice, and land use and cultural heritage. Large-scale waste treatment and disposal projects require large-scale investments. The best option to dispose of waste is to reduce waste, which is cheap and effective.

**See Also:** Biodegradability; Composting; E-Waste; Recycling; Solid and Human Waste; Water.

### Further Readings

Market Research. *Global Waste Management Market Assessment 2007*. Hampton, UK: Key Note, 2007.

Melosi, M. V. *Garbage in the Cities: Refuse, Reform and the Environment*. Pittsburgh, PA: University of Pittsburgh Press, 2005.

Power, M. S. *America's Nuclear Wastelands: Politics, Accountability, and Cleanup*. Pullman: Washington State University Press, 2008.

U.S. Environmental Protection Agency. "Wastes." http://www.epa.gov/osw/index.htm (Accessed May 2009).

Vaughn, J. *Waste Management: A Reference Handbook*. (Contemporary World Issues). Santa Barbara, CA: ABC-CLIO, 2009.

Williams, P. T. *Waste Management and Disposal*. West Sussex, UK: John Wiley & Sons, 2005.


*Krishna Roka*
*Penn State University*

## WATER

Depleted sources of fresh water, severe droughts, and water contamination from both industry and consumers have combined to reduce the availability of good water, resulting in an escalating water crisis throughout the world.

*Source:* Amy Smith/Natural Resources Conservation Service

More than 70 percent of the Earth's surface is covered by water, the majority of which is salt water in the oceans and seas. Despite this seeming overabundance, only a small portion of this water is suitable for drinking. Merely 3 percent of all water, according to author Gary Null, is freshwater, which is the only type safe for human consumption.

With global warming shrinking our polar ice caps and glaciers—which together provide over 75 percent of planetary freshwater—and a global population well in excess of 6 billion people, freshwater is fast becoming a precious commodity, and the public has already begun to feel the effects of its scarcity in rising water prices. In 2007, water prices rose 27 percent in the United States, 45 percent in Australia, 50 percent in South Africa, and 58 percent in Canada. The solution to the water problem will likely involve carefully limiting water use in the home and in industry, recycling and reusing "grey water," and alertly guarding and cautiously replenishing our freshwater sources in lakes and reservoirs.

Less than 3 percent of water is available for human consumption because of the chemical fouling of so many freshwater sources with substances that render their water unfit for human consumption. The resulting pollution has rendered regions all over the world with far less access to clean water supplies than just 50 years ago. In 1994, 54 percent of Africa's population did not have access to sanitary water, along with 20 percent of the people in Latin America and the Caribbean. The vast majority in Asia—over 80 percent of its population—also do not have access to clean drinking water. Overall, more than a billion, mostly poor people lack clean drinking water supplies. That's close to one-sixth of the population of the Earth. Although the poor suffer the most from increasing water scarcity because they cannot afford to develop new sources, developed industrial countries such as the United States are also feeling its effects.

A typical issue in the coming water crisis is described by spokesman Jeff Ruch of Public Employees for Environmental Responsibility, who notes that in Tallahassee, Florida, government officials proposed to cut off the flow of water from Lake Okeechobee into the Caloosahatchee River. Residents of south Florida say this would increase salinity in the system and thus degrade their drinking water. Examples such as this are just the beginning. If moves to conserve freshwater are not made soon, access to clean drinking water will soon be a problem for populations in every country, rich and poor.

Several factors have contributed to these critical water shortages, but the biggest are severe droughts and water contamination. Droughts can have serious health and environmental consequences. During drought periods, failing crops spread famine as underground aquifers recede and reduce access to drinking water. With crop failure, livestock start to die, and soon afterward people begin to starve. The eventual results threaten entire regional populations with malnutrition and starvation. Since the 1970s, severe droughts have killed 24,000 people per year. Many of the victims resided in Third World countries, which have limited access to drinking water supplies and experienced food shortages when world food prices peaked in 2008.

Contaminated water has also contributed to the growing water crisis. Indiscriminate dumping of toxic wastes by industry into lakes and rivers has resulted in much of the water supply containing high levels of chemicals like mercury. These chemicals render the water undrinkable in many areas of the world and are so toxic that they can cause severe birth defects. As a result, people develop diseases like dysentery and can often die from the lack of clean, uncontaminated water. Hydrologists are already implementing "blended" systems to reclaim wastewater and recycle it for home and industrial use.

## Conservation

One of the ways in which relatively wealthy consumers can combat the escalating water crisis is to conserve as much as possible. In the United States alone, consumers waste billions of gallons of water a day in their homes. Americans flush more than 5 billion gallons of water down the toilet every day—3.5 billion more gallons than necessary. Longer than necessary daily showers waste even more water, up to 3 billion gallons a day, more than 1.5 billion gallons of which is unnecessary. On the basis of these numbers, the United States can be seen as a major part of the problem, using more than 8 billion gallons a day, much of it wasted.

If such wasteful practices continue, soon only the richest of the rich will be able to afford clean, un-recycled, fresh drinking water. The United Nations Educational, Scientific and Cultural Organization reports that demand for water will continue to increase at an even faster rate, soon resulting in significantly higher prices for a basic necessity of life.

There are many ways consumers can conserve water in their homes without dramatically altering their lifestyles. One is to turn off the faucet when brushing your teeth. This can save two to three gallons of water per minute. Guidelines of the U.S. Environmental Protection Agency also advise consumers to take shorter showers instead of baths and turn off the water while soaping up their bodies and shampooing their hair. The average shower consumes up to 21 gallons per minute; shorter showers can save five to seven gallons per minute. Another way to conserve water at home is by not prerinsing dishes before putting them in the dishwasher, and running that appliance only when it is full. These simple, rather painless, adjustments in habits can lead to savings of 20 gallons per dishwashing load: over 6,500 gallons annually. Consumers should also periodically check for toilet leaks—a waste that can account for more than 14 percent of water used in the home. Mandating such conservation and combining it with new water technology can save billions of gallons of water annually.

## Technology

Improved consumer water consciousness may be the cheapest way to save the most water, but it is not the only way consumers can contribute to water conservation. With technology progressing faster than ever before, there are plenty of devices that consumers can install in their homes to save more. More than 35 models of high-efficiency toilets are on the U.S. market today, some of which use less than 1.3 gallons per flush—less than the U.S. government's required standard of 1.6 gallons. Starting at $200, these toilets are affordable and can help the average consumer save hundreds of gallons of water per year.

Appliances officially certified as most efficient are tagged with the Energy Star logo to alert the shopper. Washing machines with that rating use 18 to 25 gallons of water per load, compared with older machines that use 40 gallons. High-efficiency dishwashers save even more water. These machines use up to 50 percent less water than older models, and some use less than two gallons of water per wash.

Another consumer approach to water conservation is the use of a water purifier or filter to reduce the presence of foreign substances and chemicals in tap water. Filters absorb, modify, or trap pollutants through a medium. One type of medium works by mechanically trapping pollutants in an ultrafine sieve. Another employs a process called "adsorption," in which pollutants are trapped within the medium's microscopic pores. Top-quality carbon filters remove 80 percent of organic chemicals, such as chlorine and radon, rendering the water fully safe.

Depleting sources of freshwater, increasingly severe droughts, and continuing water contamination by both industry and consumers have combined to reduce the availability of good water, resulting in an escalating water crisis throughout the world. Hundreds of millions of people have very limited access to acceptably clean water, and many of them are becoming sick and dying because of it. Consumers in wealthier and water-blessed countries of the world, such as the United States, Canada, and Russia, can make a significant difference with fairly little effort by conserving water in their homes and using water purifiers to remove chemicals from their drinking water.

**See Also:** Beverages; Ecological Footprint; Green Homes; Lawns and Landscaping; Swimming Pools and Spas; Waste Disposal.

### Further Readings

Brooks, S. Michael. "Green Leases and Buildings." *Probate & Property,* 22/6 (November 2008).

*Consumer Reports.* "How to Save Water." *Consumer Reports,* 55/7 (July 1990).

*Consumer Reports.* "Water Savers: 50 Ways to Live With Less." *Consumer Reports,* 67/6 (June 2002).

Dy, Charlene. "How to Stop Being a Drip." *Newsweek,* 150/71 (August 13, 2007).

*Environmental Health Perspectives.* "Water Prices Surge." *Environmental Health Perspectives,* 115/6 (June 2007).

Ingram, Colin. *The Drinking Water Book: The Complete Guide to Safe Drinking Water.* Berkeley, CA: Ten Speed Press, 1991.

L'Homme, Cristina. "Three Scenarios." United Nations Educational, Scientific and Cultural Organization (UNESCO), March 2000.

Morris, Robert. *The Blue Death, Disease, Disaster, and the Water We Drink.* New York: HarperCollins, 2007.

Null, Gary. *Clearer, Cleaner, Safer, Greener: A Blueprint for Detoxifying Your Environment.* New York: Villard Books, 1990.

Public Employees for Environmental Responsibility (PEER). http://www.peer.org (Accessed April 2009).

U.S. Environmental Protection Agency (EPA). "How to Conserve Water." *The Water Encyclopedia: Hydrologic Data and Internet Resources,* 3rd Ed. Washington, D.C.: EPA, 2007.

*Anthony R. S. Chiaviello*
*Michael A. Zamora*
*Luis Saravia*
*University of Houston–Downtown*

# Websites and Blogs

Triggered by the seminal contributions by Vance Packard (1957) and John Kenneth Galbraith (1958), the shift of sovereignty from consumers to producers has been attributed to the producers' domination of mass communication channels. However, modern digital media are not dominated by a few key communicators, and thus enable a voice for the people again. With increasing attention devoted to these media, consumers recover part of their sovereignty by gaining both "share of voice" and "share of attention." Websites and blogs are the main instruments for consumers to strengthen their communication abilities.

According to Ralf Wagner (2009), Websites are defined as representations of individuals, groups, or organizations within the World Wide Web that are accessible by a uniform resource locator (or URL), using a browser. The Websites can be

- static—presenting text and audiovisual content according to a predefined navigation system—or
- dynamic—customizing the presented content according to both visitors' profiles and visitors' actions while browsing the site.

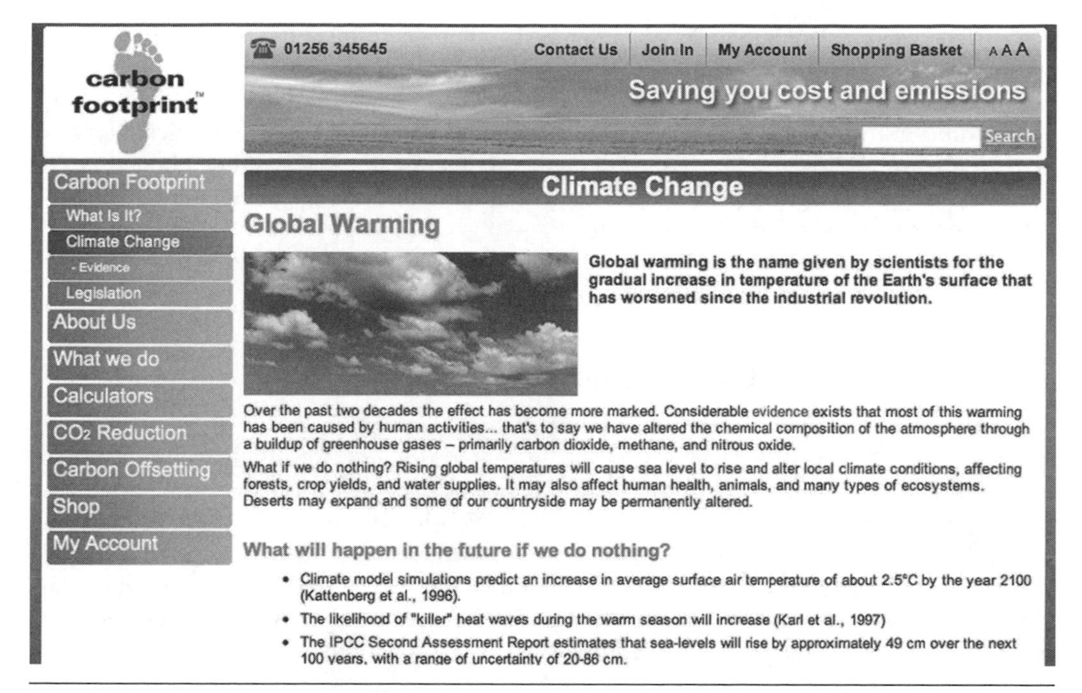


One example of an informative blog from an NGO is the Britain-based carbonfootprint.com, which allows consumers to calculate their own carbon footprint and learn tips on how to reduce it.

*Source:* carbonfootprint.com

Website vendors point to other Websites by incorporating hyperlinks to these Websites. This makes up a network of information. The links can be followed by human visitors and Web bots, which feed the search engines with information about Websites. Because most Web search engines use the number of in-links and the quality of the in-linking Websites for the ranking of Websites, the net citizens control the visibility of Websites and reinforce one another's visibility.

Blogs, or weblogs, are digital diaries on the Web. Consumers can share not only their experiences when using products and services but also their opinions and concerns about issues such as energy production and consumption, for example. Unlike Websites, blogs have typically been associated with individuals—nowadays, however, more and more organizations and companies are setting up blogs to communicate with their stakeholders or to promote their brands and labels—who make up a network by adding other blogs to their blog roll. Although blog rolls and links between Websites rely on the same hypertext technology, they have a different quality. As links can spread throughout the Website, they are typically embedded in a specific context. In contrast, the relations consolidated in blog rolls are highlighted by the fact that they are assigned to the blog roll. The blogger is also free to add links to other blogs or Websites when assembling the content. Moreover, bloggers usually provide one another with a ping as notification that they refer to other blogs, thus supporting community building within the blogosphere.

Blogging is becoming more popular than setting up Websites because some providers offer bloggers maintenance of personal blogs free of charge. Moreover, sophisticated tools disburden bloggers from any programming tasks: blogging allows the participant to focus on the content, whereas maintaining Websites requires technical knowledge of navigation design, file hosting, and so on.

In addition to conventional blogs, special blog types like videoblogs and microblogs are also gaining in popularity. The latter are made up of short messages and can be fed in and read not only by using computers with Internet access but also via mobile devices like cellular phones or personal digital assistants.

This allows for a new kind of opinion leadership by grassroots journalism, which is relevant for both consumerism and the discussion of green media and the intersection of both topics.

## Use of Websites in Consumerism

Websites are operated by companies and other organizations merely to promote products and to provide product-related information and entertainment. Although Websites need to be explored interactively, they attract a higher degree of attention in comparison with television commercials or printed ads. However, Websites are not a dialogue medium per se. Consumer dialogues are hampered by contact forms or shift the consumer dialogue to other media such as e-mails or phone calls. Similar obstacles need to be overcome when consumer unions and nongovernmental organizations set up their Websites. Focusing on green consumerism, Websites are used for four related functions:

1. *Communication Channel:* Websites are used to inform about ecological issues. This approach is used primarily by nongovernmental organizations and other organizations. A typical example of informing and facilitating is the issue of the carbon footprint (http://www.carbonfootprint.com), which allows consumers to calculate their own carbon footprint and learn tips and tricks on how to reduce it. The company itself is an ecofriendly consulting agency.
2. *Donating Channel:* By incorporating payment functions (e.g., PayPal or AlertPay), Websites can facilitate donating directly or indirectly by providing online forms through which visitors to the Website can submit their credit card details and make their donations. Donating may support a charitable organization or an individual suing for consumer rights.
3. *Sales Channel:* When Websites are used to sell ecofriendly products, even small and niche vendors offer their products to many (potential) customers. This channel has helped reduce the power of the conventional retail chains, which frequently refuse to add green products to their product assortments or flagrantly hamper the open diffusion of green products by promoting competing nonecological products and services.
4. *Delivery Channel:* Digital products, such as software, music, videos, and books can be offered for download on a Website. If a consumer buys these products online and uses them without creating a hard copy, a number of natural resources are preserved. CDs and DVDs, in particular, have a short life cycle, and their production, distribution, and disposal cause substantial environmental damage.

## Use of Blogs in Consumerism

Blogs and microblogs are purely dialogue oriented and challenge companies because they need to interact individually with interested stakeholders, who raise questions and post comments and opinions. These efforts are not sufficient and are only a precondition for successful corporate communications. In contrast, consumers are an interactive cloud of communicators with similar interests and, often, similar opinions. Consumers in such networks frequently support one another and share information and ideas in an open manner. Companies and governmental organizations are reduced to being one of many communicators. Therefore, blogs and microblogs are thought to reinforce a democratic structure in the markets, but individuals have a different importance for communication within the virtual social network. Martin Klaus and colleagues (2009) outline three quantitative assessments of an individual's relevance to the communication process:

1. The degree of centrality is deemed to be the dimension of possible communication activity within the network. The more relations to other individuals a person has, the higher is the probability of direct communication to other individuals or organizations.
2. The closeness centrality quantifies an individual's independence from other individuals. The higher the centrality of an individual's blog or microblog is, the more direct the connections are from other individuals linked to the blog. Thus, a person is less dependent on another person or organization if there are many others close by.
3. The betweenness centrality assesses individuals' opportunities for controlling the communication process. Assuming the information flow usually takes the shortest path with a network structure, the betweenness becomes relevant. If many shortest distances of the communication network run over a person's blog, it has a high influence on the network communication process.

Despite these inequalities, blogs enhance consumerism substantially. Usually, the more-relevant bloggers do not accentuate their position but, instead, aim to support persons with a weaker virtual voice. If any member of the community is affronted by vendors, for example, he feels his complaints are not taken seriously, the community supports the individual by spreading the information and adding comments. A famous case is the attempt by the blogger Vincent Ferrari to cancel an AOL account, in which the user posted an audio-file, recording his interactions with a call center agent in 2006, and more recently the attempt of Facebook to change the terms of service clauses. The coverage of such information and the consumer's actions are not restricted to an audience of blog readers because the mass media eagerly seize on such stories, and thus boost the consumer's voice.

The crowd of consumers makes up an implicit set of principles ruling the marketplace. This social mind guides all kinds of judgments with green and sustainable market behavior. Vendors of all kinds of goods and services are well advised to use these principles for designing new offers and related marketing campaigns.

Not only the number of interactions, but also accepted response delays, define a new communication environment: Web 2.0. Web 2.0 usually becomes immediately aware of current issues. A good example of this is Twitter, which nowadays is quoted in all mass media news channels. Because Web 2.0 connects people from all over the world, a flood in China is immediately no longer a news item far, far away but something that threatens the lives of people you know and talked with just a week ago.

Green consumerism affects almost all markets, but the key areas of interest can be systematized as follows:

- *Food:* In addition to the obvious advantage of allowing city dwellers the "luxury" of organic food, there are many consumers who have their own blogs in which they share their ideas about food and experiences with various vendors. These blogs cover the preparation of food (recipes) and recommend the best places in a town to buy organic food (e.g., http://www.organicangels.com/blog). Some blogs even specialize in reviews of restaurants.

- *Health and Childcare:* In this area of interest, blogs and consumer forums outnumber companies' Websites. However, vendors' blogs are usually well integrated in the community because they do not restrict their topics to their own offers but, rather, promote different kinds of news for healthy and green living. The blog Keetsa (http://keetsa.com/blog/), for example, provides an "eco-friendly and green blog search engine."
- *Clothing:* A well-known example of an online community of ecofriendly consumers is the ThreadBanger network (http://www.threadbanger.com/blog). This community merges people who aim to create and discover their own style through alternative forms of fashion. They discuss recycling, up-cycling, and refashioning and share their passion for DIY, fashion, and crafting.
- *Energy Production and Consumption:* Blogs related to these topics frequently interact with blogs focusing on related topics such as green tourism, innovative car engines, climate change, and the like (e.g., http://ecoiron.blogspot.com). Here, nongovernmental organizations frequently make up information hubs.

## Usage of Microblogs in Consumerism

Microblogs are similar to blogs but are restricted with respect to technical features for the users, particularly the number of characters for a message and the length of videos uploaded to the platform. The major advantage of microblogs is that information can be added or retrieved not only by using the Internet but also via various kinds of mobile devices such as cellular phones or personal digital assistants.

The current most prominent microblog is Twitter, which allows for 140 characters per entry. Links and photos may also be included. Green consumerism communicators in the Twitter community can be categorized in the following groups:

- Green citizen activists, including globalwarming, greenlivingidea, and LighterFootstep
- Green blog twitters, who mainly announce updates of complementary Websites/blogs
- (Inter)national environmental preservation organizations, such as the National Wildlife Federation (at http://twitter.com/NWF)

If companies want to start using Twitter to become involved with green consumers, they can systematically learn about the "green Twitterers" out there by using lists (e.g., http://www.mnn.com/earth-matters/energy/blogs/the-great-green-twitter-follow-parade#comment-1860).

Twitter, which is the fastest information exchange medium, is suited to spreading warnings and coordinating consumerist action in the nonvirtual world. The Twitter service enables the consolidation of topics by hashtags. Hashtags are a community-driven convention for adding extra context and metadata to tweets. A glance at the top 10 hashtags gives a clear indication of which topics are the most popular on Twitter at any given moment. The hashtag #ecomonday, which was created April/May 2009, was initially only a call for all "Green tweople" to list their accounts—a concept very similar to #followfriday. The hashtag #ecomonday quickly evolved from recommending people to follow to recommending specific Websites and blogs.

## Energy Consumption

Maintaining Websites or blogs needs electricity for users' and servers' devices. For instance, the Web- and Web 2.0–related information technology infrastructure burns up 2 percent of the total electricity in Germany. Each retrieval using a search engine is estimated to consume 11 watts, in addition to the energy the users' devices consume. Because blogs and microblogs are becoming an increasingly relevant referrer to Websites, they may reduce the Web traffic generated by search engine usage. However, Websites and blogs consume less than unsolicited e-mails. Nevertheless, simply by using dark colors (similar to http://godark.us) in Websites and blogs, energy consumption could be reduced substantially.

**See Also:** Advertising; Consumer Activism; Consumer Society; Electricity Usage; Green Marketing.

### Further Readings


Dennis, Alan. *Networking in the Internet Age.* Hoboken, NJ: Wiley, 2002.

Galbraith, John Kenneth. *The Affluent Society.* Boston, MA: Houghton Mifflin, 1958.

Gant, Scott. *We're All Journalists Now: The Transformation of the Press and Reshaping of the Law in the Internet Age.* New York: Free Press, 2007.

Kadirov, Djavlonbek and Richard Varey. "Marketplace Wisdom and Consumer Experience: Redefining Sustainability." *Proceedings of the ANZMAC 2005 Conference.* Perth, Australia, 2005.

Packard, Vance. *The Hidden Persuaders.* New York: McKay, 1957.

Shiffman, Denise. *The Age of Engage: Reinventing Marketing for Today's Connected, Collaborative, and Hyperinteractive Culture.* Austin, TX: Hunt Street Press, 2008.

Wagner, Ralf. "Customizing Multimedia With Multi-Trees." In *Encyclopedia of Multimedia Technology and Networking*, 2nd ed., M. Pagani, ed. Hershey, PA: IGI Global, 2009.



*Ralf Wagner*
*University of Kassel*


## Windows

Windows can have an enormous effect on a consumer's energy bills. Experts predict that 25 to 30 percent of heat loss comes through the window and window frame. Therefore, buying energy-efficient windows may cost more at the time of initial purchase, but they will pay for themselves over the windows' lifetime. Windows in a structure have direct contact with the elements without any insulation. Poorly constructed windows can let out cool air in summer and warm air in winter. Thus, windows can be the single most important element when it comes to heating costs or cooling costs. In winter, windows can let in sunlight to warm the house through natural means, and letting in natural light in wintertime also can improve mood and fight against seasonal affective disorder.

To get the most appropriate windows, one must understand the needs implied by the climate in which one resides. For example, the United States (excluding Alaska and Hawaii) is broken down into seven zones based on the average temperature. The coldest areas in northern Minnesota, Wisconsin, and North Dakota are a number seven. The farther south the location, the lower the number. The area around Miami is coded as number one. These codes are very important for choosing the type of window and how much sunlight is desired passing through the window (remember, if the sunlight goes through the window, it warms up the house—bad in houses in the south, good in places in the north).

The windows themselves can come in a single pane, double panes, or triple panes. The space between the panes can be empty or filled with argon or krypton gas, which can act as insulation to keep warm air in the house and cold air outside the house in winter. A great deal of heat loss can also occur where the frame meets the window(s), as there is really no way to get a perfect fit. Window frames can be made of aluminum, wood, vinyl, or fiberglass. Windows can have glazing or tinting put on them to reduce the amount of sunlight coming through the window. Window coatings also can help heat (longer waves of sunlight) and visible transmission of light occur while blocking out shorter waves of sunlight that might fade fabrics in the room.

Windows are the part of the house that have direct contact with the elements without any insulation. Poorly constructed or old windows can let out cool air in summer and warm air in winter.

*Source:* iStockphoto

The effectiveness of windows is measured in four ways: the U-factor, the solar heat gain coefficient, visible transmittance, and air leakage. The U-factor measures how much heat is lost through the window. The U-factor is expressed as 1/R-factor. The R-factor measures thermal resistance in insulation materials. In U-factor measures, the lower the number, the less heat is lost through the window. For example, a single, clear pane of glass has a U-factor of one, whereas a window of two panes of glazed glass has a value of 0.5 (think of this as two panes of glass that provide double the insulation). The solar heat gain coefficient measures how much the sun heats the room through the glass. Think of this type of heating as if one stood near a warm stove. The coefficient is expressed in numbers from 0 to 1. So, those in a cooler climate would want a higher number (because more heat would pass through the window).

The visible transmission number tells how much visible light is getting through the window. Visible transmission is expressed in a number between 0 and 1. The higher the number, the more light gets through (this could affect heating). For example, a single pane of glazed glass would be around 0.69, and a double-glazed piece of glass would be lower (because the glazing lets less light through the window). Of course, air leakage measures how much air gets through the window, which includes the frame. Air leakage is also expressed in a number between 0 and 1. The lower the air leakage number, the better the window. Remember, the more leakage in summer, the more air conditioning is leaking out of the house.

Obviously, to get the most sunlight (or less), it is best to work with the Earth's natural rotation. Because the sun rises in the east and sets in the west, windows on these sides of the house will get more sun. So a different type of window might be preferable in these areas. This fact is especially important in the northern United States, where a homeowner wants the natural sunlight to warm the house as much as possible in the winter. and thereby reduce energy costs.

Energy-efficient windows can control condensation, or the water that develops on the glass (especially around the edges where the glass meets the frame). The less water on the glass, the less amount of energy is transmitted through the glass. This results in a cost savings to the consumer.

**See Also:** Green Homes; Insulation.

## Further Readings

Efficient Windows Collaborative. http://www.efficientwindows.org (Accessed September 2009).

Fisette, Paul. "Understanding Energy Efficient Windows." http://www.amesc.org/assets/files/Energy-Efficient_Windows.pdf (Accessed September 2009).

U.S. Department of Energy (DOE). "Energy Efficient Windows" (January 2000). http://www.energycodes.gov/implement/pdfs/lib_ks_energy-efficient_windows.pdf (Accessed September 2009).

*Charles R. Fenner, Jr.*
*State University of New York at Canton*

# Green Consumerism Glossary

## A

**Acute Exposure:** A single exposure to a toxic substance that may result in severe biological harm or death. Acute exposures are usually characterized as lasting no longer than a day, as compared to longer, continued exposure over a period of time.

**Administered Dose:** In exposure assessment, the amount of a substance given to a test subject (human or animal) to determine dose-response relationships. Since exposure to chemicals is usually inadvertent, this quantity is often called potential dose.

**Adulterants:** Chemical impurities or substances that by law do not belong in a food or pesticide.

**Alternative Fuels:** Substitutes for traditional liquid, oil-derived motor vehicle fuels like gasoline and diesel. Includes mixtures of alcohol-based fuels with gasoline, methanol, ethanol, compressed natural gas, and others.

**Animal Dander:** Tiny scales of animal skin, a common indoor air pollutant.

## B

**Back Pressure:** Pressure that can cause water to backflow into the water supply when a user's wastewater system is at a higher pressure than the public system.

**Backyard Composting:** Diversion of organic food waste and yard trimmings from the municipal waste stream by composting them in one's yard through controlled decomposition of organic matter by bacteria and fungi into a humus-like product. It is considered source reduction, not recycling, because the composted materials never enter the municipal waste stream.

**Biological Magnification:** Refers to the process whereby certain substances such as pesticides or heavy metals move up the food chain, work their way into rivers or lakes, and are eaten by aquatic organisms such as fish, which in turn are eaten by large birds, animals, or humans. The substances become concentrated in tissues or internal organs as they move up the chain.

**Body Burden:** The amount of a chemical stored in the body at a given time, especially a potential toxin in the body as the result of exposure.

**Bottle Bill:** Proposed or enacted legislation that requires a returnable deposit on beer or soda containers and provides for retail store or other redemption. Such legislation is designed to discourage use of throwaway containers.

**Buy-Back Center:** Facility where individuals or groups bring recyclables in return for payment.

## C

**Carcinogen:** Any substance that can cause or aggravate cancer.

**Chemical Stressors:** Chemicals released to the environment through industrial waste, auto emissions, pesticides, and other human activity that can cause illnesses and even death in plants and animals.

**Child Resistant Packaging (CRP):** Packaging that protects children or adults from injury or illness resulting from accidental contact with or ingestion of residential pesticides that meet or exceed specific toxicity levels. Required by FIFRA regulations. Term is also used for protective packaging of medicines.

**Chronic Effect:** An adverse effect on a human or animal in which symptoms recur frequently or develop slowly over a long period of time.

**Commingled Recyclables:** Mixed recyclables collected together.

## D

**Demand-Side Waste Management:** Prices whereby consumers use purchasing decisions to communicate to product manufacturers that they prefer environmentally sound products packaged with the least amount of waste, made from recycled or recyclable materials, and containing no hazardous substances.

**Dermal Toxicity:** The ability of a pesticide or toxic chemical to poison people or animals by contact with the skin.

**Detergent:** Synthetic washing agent that helps to remove dirt and oil. Some contain compounds that kill useful bacteria and encourage algae growth when they are in wastewater that reaches receiving waters.

**Distillation:** The act of purifying liquids through boiling, so that the steam or gaseous vapors condense to a pure liquid. Pollutants and contaminants may remain in a concentrated residue.

## E

**End User:** Consumer of products for the purpose of recycling. Excludes products for reuse or combustion for energy recovery.

**Environmental Exposure:** Human exposure to pollutants originating from facility emissions. Threshold levels are not necessarily surpassed, but low-level chronic pollutant exposure is one of the most common forms of environmental exposure.

**Environmental Tobacco Smoke:** Mixture of smoke from the burning end of a cigarette, pipe, or cigar and smoke exhaled by the smoker.

## F

**Finished Water:** Water is "finished" when it has passed through all the processes in a water treatment plant and is ready to be delivered to consumers.

**Food Chain:** A sequence of organisms, each of which uses the next, lower member of the sequence as a food source.

**Food Waste:** Uneaten food and food preparation wastes from residences and commercial establishments such as grocery stores, restaurants, produce stands, institutional cafeterias and kitchens, and industrial sources like employee lunchrooms.

**Fuel Economy Standard:** The Corporate Average Fuel Economy Standard (CAFE) effective in 1978. It enhanced the national fuel conservation effort imposing a miles-per-gallon floor for motor vehicles.

## G

**Garbage:** Animal and vegetable waste resulting from the handling, storage, sale, preparation, cooking, and serving of foods.

**Gasahol:** Mixture of gasoline and ethanol derived from fermented agricultural products containing at least nine percent ethanol. Gasohol emissions contain less carbon monoxide than those from gasoline.

**Global Warming:** An increase in the near surface temperature of the Earth. Global warming has occurred in the distant past as the result of natural influences, but the term is most often used to refer to the warming predicted to occur as a result of increased emissions of greenhouse gases. Scientists generally agree that the Earth's surface has warmed by about one degree Fahrenheit in the past 140 years. The Intergovernmental Panel on Climate Change (IPCC) recently concluded that increased concentrations of greenhouse gases are causing an increase in the Earth's surface temperature, and that increased concentrations of sulfate aerosols have led to relative cooling in some regions, generally over and downwind of heavily industrialized areas.

**Grey Water:** Domestic wastewater composed of wash water from kitchen, bathroom, and laundry sinks, tubs, and washers.

## H

**Halogen:** An incandescent lamp with higher energy-efficiency that standard lamps.

**Heptachlor:** An insecticide that was banned from some food products in 1975, and in all of them in 1978. It was allowed for use in seed treatment until 1983. More recently it was found in milk and other dairy products in Arkansas and Missouri where dairy cattle were illegally fed treated seed.

**Homeowner Water System:** Any water system that supplies piped water to a single residence.

**Household Hazardous Waste:** Hazardous products used and disposed of by residential as opposed to industrial consumers. Includes paints, stains, varnishes, solvents, pesticides, and other materials or products containing volatile chemicals that can catch fire, react or explode, or that are corrosive or toxic.

## L

**Lead (Pb):** A heavy metal that is hazardous to health if breathed or swallowed. Its use in gasoline, paints, and plumbing compounds has been sharply restricted or eliminated by federal laws and regulations.

**Lifetime Exposure:** Total amount of exposure to a substance that a human would receive in a lifetime (usually assumed to be 70 years).

## M

**Maximally (or Most) Exposed Individual:** The person with the highest exposure in a given population.

## N

**Nitrilotriacetic Acid (NTA):** A compound now replacing phosphates in detergents.

**Non-Transient Non-Community Water System:** A public water system that regularly serves at least 25 of the same nonresident persons per day for more than six months per year.

## O

**Office Paper:** High-grade papers such as copier paper, computer printout, and stationary almost entirely made of uncoated chemical pulp, although some ground wood is used. Such waste is also generated in homes, schools, and elsewhere.

**Organotins:** Chemical compounds used in anti-foulant paints to protect the hulls of boats and ships, buoys, and pilings from marine organisms such as barnacles.

**Other Glass:** Recyclable glass from furniture, appliances, and consumer electronics. Does not include glass from transportation products (cars, trucks, or shipping containers) and construction or demolition debris.

## P

**Passive Smoking/Secondhand Smoke:** Inhalation of others' tobacco smoke.

**Personal Measurement:** A measurement collected from an individual's immediate environment.

**Pest Control Operator:** Person or company that applies pesticides as a business (an exterminator); usually describes household services, not agricultural applications.

**Pesticide Tolerance:** The amount of pesticide residue allowed by law to remain in or on a harvested crop. The EPA sets these levels well below the point where the compounds might be harmful to consumers.

**Point-of-Use Treatment Device:** Treatment device applied to a single tap to reduce contaminants in the drinking water at the faucet.

## R

**Raw Agricultural Commodity:** An unprocessed human food or animal feed crop like raw carrots, apples, corn, or eggs.

**Redemption Program:** Program in which consumers are monetarily compensated for the collection of recyclable materials, generally through prepaid deposits or taxes on beverage containers. In some states or localities legislation has enacted redemption programs to help prevent roadside litter.

**Reuse:** Using a product or component of municipal solid waste in its original form more than once; that is, refilling a glass bottle that has been returned, or using a coffee can to hold nuts and bolts.

## S

**Sanitary Water (also known as Grey Water):** Water discharged from sinks, showers, kitchens, or other nonindustrial operations, but not from commodes.

## T

**Teratogen:** A substance capable of causing birth defects.

**Teratogenesis:** The introduction of nonhereditary birth defects in a developing fetus by exogenous factors such as physical or chemical agents acting in the womb to interfere with normal embryonic development.

**Theoretical Maximum Residue Contribution:** The theoretical maximum amount of a pesticide in the daily diet of an average person. It assumes that the diet is composed of all food items for which there are tolerance-level residues of the pesticide. The TMRC is expressed as milligrams of pesticide/kilograms of body weight/day.

## V

**Vulnerability Analysis:** Assessment of elements in the community that are susceptible to damage if hazardous materials are released.

## W

**Waste:** 1. Unwanted materials left over from a manufacturing process. 2. Refuse from places of human or animal habitation.

*Sources:* U.S. Environmental Protection Agency (http://www.epa.gov/OCEPAterms), U.S. Energy Information Administration (http://www.eia.doe.gov/tools/glossary)

# Green Consumerism Resource Guide

## Books

Anand, Sudhir and Amartya Sen. *Consumption and Human Development: Concepts and Issues*. New York: Oxford University Press, 1997.

Barr, Stewart. *Household Waste in Social Perspective: Values, Attitudes, Situation and Behaviour.* Aldershot, UK: Ashgate, 2002.

Brower, Michael and Warren Leon. *The Consumer's Guide to Effective Environmental Choices: Practical Advice from the Union of Concerned Scientists*. New York: Three Rivers, 1999.

Burke, Cindy. *To Buy or Not to Buy Organic. What You Need to Know to Choose the Healthiest, Safest, Most Earth-Friendly Food*. Boston, MA: Marlowe, 2007.

Carter, Alan. *A Radical Green Political Theory*. London: Routledge, 1999.

Cato, Molly Scott. *Green Economics: An Introduction to Theory, Policy and Practice*. London: Earthscan Publications Ltd., 2009.

Charles, Daniel. *Lords of the Harvest: Biotech, Big Money, and the Future of Food*. Cambridge, MA: Perseus Publishing, 2001.

Dobson, Andrew. *Green Political Thought*. London: T&F Books, 2009.

Duran, S. C. *Green Homes: New Ideas for Sustainable Living*. New York: Collins Design, 2007.

Durning, Alan Thein, Yoram Bauman, and Rachel Gussett. *Tax Shift: How to Help the Economy, Improve the Environment, and Get the Tax Man off Our Backs*. Seattle, WA: Northwest Environmental Watch, 1998.

Frank, R. H. *Falling Behind: How Rising Inequality Harms the Middle Class*. Berkeley, CA: University of California Press, 2007.

Garlough, Donna, Wendy Gordon, and Seth Bauer, eds. *The Green Guide: The Complete Reference for Consuming Wisely*. New York: National Geographic, 2008.

Gasper, Des. *The Ethics of Development*. Edinburgh, UK: Edinburgh University Press, 2004.

Goleman, Daniel. *Ecological Intelligence: How Knowing the Hidden Impacts of What We Buy Can Change Everything*. New York: Broadway Books, 2009.

Hayden, Dolores. *Building Suburbia: Green Fields and Urban Growth, 1820–2000*. New York: Pantheon Books, 2003.

Heartfield, James. *Green Capitalism*. London: Skyscraper Digital Publishing, 2008.

Hecht, Joy E. *National Environmental Accounting: Bridging the Gap Between Ecology and Economy.* Washington, DC: Resources for the Future, 2005.
Illich, Ivan. *Tools for Conviviality*. London: Marion Boyars Ltd, 1973.
James, S. and T. Lahti. *The Natural Step for Communities: How Cities and Towns Can Change to Sustainable Practices*. Gabriola Island, British Columbia, Canada: New Society Publishers, 2004.
Jones, Van. *Green Collar Economy: How One Solution Can Fix America's Two Biggest Problems*. New York: HarperCollins Publishers, 2008.
Kasser, Tim. *The High Price of Materialism*. Cambridge, MA: MIT Press, 2002.
Kutz, M. *Environmentally Conscious Transportation*. Hoboken, NJ: John Wiley & Sons, 2008.
Lee, Andy W. and Patricia Foreman. *Backyard Market Gardening: The Entrepreneur's Guide to Selling What You Grow*. Buena Vista, VA: Good Earth Publications, 1995.
Littlewood, Michael. *Natural Swimming Pools: Inspiration for Harmony With Nature.* Atglen, PA: Schiffer Publishing, 2005.
Makower, Joel and Cara Pike. *Strategies for the Green Economy: Opportunities and Challenges in the New World of Business*. New York: McGraw-Hill Companies, 2008.
Melosi, M. V. *Garbage in the Cities: Refuse, Reform and the Environment*. Pittsburgh, PA: University of Pittsburgh Press, 2005.
Murty, M. N. *Environment, Sustainable Development, and Well-Being: Valuation, Taxes, and Incentives*. Oxford: Oxford University Press, 2009.
Null, Gary. *Clearer, Cleaner, Safer, Greener: A Blueprint for Detoxifying Your Environment.* New York: Villard Books, 1990.
Ottman, J. A. *Green Marketing: Opportunity for Innovation*. New York: BookSurge, Inc, 2004.
Papanek, Victor. *Design for the Real World*. London: Thames and Hudson Ltd., 1984.
Peattie, K. *Environmental Marketing Management: Meeting the Green Challenge*. Philadelphia, PA: Trans-Atlantic Publications, 1995.
Power, M. S. *America's Nuclear Wastelands: Politics, Accountability, and Cleanup*. Pullman, WA: WSU Press, 2008.
Prescott-Allen, Robert. *Consumption Patterns, Ecosystem Stress and Human Development.* Washington, DC: Island Press, 1997.
Radcliffe, James. *Green Politics: Dictatorship or Democracy?* New York: Palgrave Macmillan, 2002.
Rogers, Elizabeth and Thomas Kostigen. *The Green Book*. New York: Three Rivers, 2007.
Snape, John and Jeremy de Souza. *Environmental Taxation Law: Policy, Contexts and Practice*. Aldershot, UK: Ashgate Publishing, 2006.
Soderlind, Steven D. *Consumer Economics*. New York: M.E. Sharpe, 2001.
Torgerson, Douglas. *The Promise of Green Politics: Environmentalism and the Public Sphere.* Durham, NC: Duke University Press, 1999.
Twitchell, J. B. *Living It Up: Our Love Affair With Luxury*. New York: Columbia University Press, 2002.
Uno, Kimio and Peter Bartelmus, eds. *Environmental Accounting in Theory and Practice*. New York: Springer-Verlag, 2002.
Vaughn, J. *Waste Management: A Reference Handbook*. Contemporary World Issues. Santa Barbara, CA: ABC-CLIO, 2009.

Wagner, Sigmund. *Understanding Consumer Behavior*. New York: Routledge, 2003.
Williams, P. T. *Waste Management and Disposal*. West Sussex, UK: John Wiley and Sons, Ltd., 2005.
Wilson, Alex, Mark Piepkorn, and Nadav Malin, eds. *Green Building Products: The Greenspec Guide to Residential Building Materials*. Gabriola Island, British Columbia, Canada: New Society Publishers, 2008.

## Journals

*American Journal of Economics and Sociology* (John Wiley & Sons)

*Business Strategy and the Environment* (John Wiley & Sons)

*Confronting Consumption* (MIT Press)

*Ecological Economics* (Elsevier)
*Environmental Management* (Springer)
*Environmental Science and Technology* (American Chemical Society)
*Environment and Behavior* (SAGE Publications)

*International Journal of Consumer Studies* (John Wiley & Sons)
*International Journal of Sociology and Social Policy* (Emerald)

*Journal of Consumer Behavior* (John Wiley & Sons)
*Journal of Consumer Policy* (Springer)
*Journal of Environmental Economics and Management* (Elsevier)
*Journal of Environmental Management* (Elsevier)
*Journal of Happiness Studies* (Springer)
*Journal of Industrial Ecology* (John Wiley & Sons)
*Journal of Organizational Behavior* (John Wiley & Sons)
*Journal of Social Policy* (Cambridge Journals)
*Journal of Socio-Economics* (Elsevier)

*Leisure Studies* (Taylor & Francis)
*Local Environment* (Taylor & Francis)

*Resource and Energy Economics* (Elsevier)
*Review of Social Economy* (Taylor & Francis)

*Strategy and the Environment* (John Wiley & Sons)

## Websites

Acres U.S.A.
www.acresusa.com

Adbusters
www.adbusters.org

Center for a New American Dream
www.newdream.org

Environmental Defense Fund
www.edf.org

Green America
www.greenamericatoday.org

GreenerChoices
www.greenerchoices.org

GreenMuze
www.greenmuze.com

Healthy Stuff
www.healthystuff.org

LocalHarvest, Inc.
www.localharvest.org

Natural Resources Defense Council
www.nrdc.org

One Planet Living
www.oneplanetliving.org

Organic Consumers Association
www.organicconsumers.org

Public Employees for Environmental Responsibility
www.peer.org

Sustainable Furnishings Council
www.sustainablefurnishings.org

UN Environment Programme
www.unep.org

U.S. Consumer Product Safety Commission
www.cpsc.gov

U.S. Department of Agriculture
www.usda.gov

U.S. Environmental Protection Agency
www.epa.gov

U.S. Green Building Council
www.usgbc.org

Whole Building Design Guide
www.wbdg.org

# Green Consumerism Appendix

## Affluenza

http://www.pbs.org/kcts/affluenza

This is a Website for a Public Broadcasting System (U.S. television) program on consumerism, which popularized the term *affluenza*, and explores the costs of American consumer culture, both to the environment and to humans and their social relationships. It takes a humorous approach to the situation, using cartoons to illustrate points, including several quizzes to "diagnose" affluenza, suggestions for "treatment," and a timeline titled "Consuming Moments in American History," tracing the role consumerism has played from the 17th century to today. The site also includes a teacher's guide with activities and background information keyed to the program for several subjects, including: mathematics, economics, social studies, business, and language arts. It also provides a downloadable viewer's guide, and suggestions for holding a viewing party and leading a group discussion. There is also a list of further resources on general questions of economics, consumerism, and marketing, including books, periodicals, organizations, and Websites.

## Biotechnology (Genetically Modified Foods) and Nanotechnology

http://www.who.int/foodsafety/biotech/en

This Website, created by the World Health Organization (WHO), brings together numerous resources on biotechnology, including genetically modified (GM) food. It includes a general overview of biotechnology and food-related nanotechnology (research and development on very small particles, of 1–100 nanometers), and a document with 20 questions and answers on GM food that is downloadable in several languages. There is also a section on the resolution of the 53rd World Health Assembly on GM food, several reports of expert consultations with the Food and Agriculture Organization of the United Nations, a downloadable report of a 2002 WHO-commissioned study on the health implications of GM food, documents from the Codex Intergovernmental Task Force on Foods Derived from Biotechnology, a list of meetings related to biotechnology and GM foods, and a number of downloadable publications relating to the safety of GM food.

## Center for Sustainable Destinations

http://www.nationalgeographic.com/guides/travels/sustainable

The Center for Sustainable Destinations (CSD), part of the National Geographic Society's Research, Conservation, and Exploration Division, is dedicated to creating and supporting geotourism. Geotourism is defined as tourism that is not only environmentally sustainable, but also enhances the geographic character of a place, including its environment, culture, aesthetics, heritage, and the welfare of its residents. The Website provides information for both the individual traveler and the travel professional. For the former, it offers general tips about being a responsible traveler, as well as links to information about organizations offering ecotours, advice about differentiating good from bad ecotours, and a list of organizations involved in sustainable tourism. For travel professionals, it provides links to organizations concerned about geotourism policies and industries, and governmental and academic sites related to geotourism. The site also includes a map with links demonstrating how the principles of geotourism have been applied in different destinations around the globe.

## eCycling

http://www.epa.gov/waste/conserve/materials/ecycling/index.htm

This Webpage, maintained by the U.S. Environmental Protection Agency (EPA), is a resource for information about eCycling, defined as the recycling of electronic products such as televisions, computers, and cell phones. It includes basic information about why discarded electronic products require special consideration, and guides to reusing, donating, and recycling electronic equipment, including information on where to find a local, government, or manufacturer donation program. It also has information about U.S. laws regarding recycling and disposal of hazardous wastes (such as cathode ray tubes), statistics about the amount of consumer electronic equipment discarded each year and the amount that is recycled, information about certification programs for electronics recyclers, and a list of links to further resources about recycling and electronic waste.

## Fair Trade Federation

http://www.fairtradefederation.org

The Fair Trade Federation (FTF), created in 1994, is part of the global fair trade movement, with a particular focus on North American organizations committed to fair trade. This Website is a resource for information about fair trade, for both consumers and businesses, including the philosophy and history behind the movement, and its current practice. It includes a searchable directory to help consumers locate members of the Fair Trade Federation by geographic location, name, or type of business, as well as a list of trade shows that include fair trade companies. The Website also includes information for people who want to get involved in the movement, information about starting a fair trade business, a list of fair trade tour operators, a list of job opportunities in the field, information about World Fair Trade Day, and many other resources.

## Green Homes

http://www.epa.gov/greenhomes

This Website, created by the U.S. Environmental Protection Agency (EPA), brings together information about environmental and energy issues for renters, homeowners, and homebuilders. It suggests solutions for reducing energy use and protecting health. Basic information is presented on the definition of green building, the history of green building in the United States, and research on green building, along with statistics on the benefits of green building. It also explains the different components of green building, including: energy efficiency and renewable energy, water efficiency, environmentally preferable building materials and specifications, waste reduction, toxics reduction, indoor air quality, and sustainable development. Links to relevant EPA programs and outside information are provided with information also accessible by building type (such as home, retail, or school). Links are also provided for funding opportunities in the United States from many sources, including: national, local, and state governments, industry organizations, and nonprofits.

## Greener Choices: Products for a Better Planet

http://www.greenerchoices.org/home.cfm

This Website, run by the nonprofit Consumers Union, offers independently researched, noncommercial information to consumers about environmentally friendly products. It includes consumer information in five product areas—appliances, cars, electronics, food, and home and garden—as well as offers tools such as energy calculators, and information about more general topics such as global warming and electronics waste disposal. Within each product category, the Website includes general articles (such as "20 Free Ways to Save Energy") as well as reports on specific classes of products similar to that provided in the subscription magazine *Consumer Reports* (also published by Consumers Union). For instance, the section on air conditioners includes general advice about cooling a home and choosing an air conditioner, a calculator to determine the size required (considering factors such as geographic location, room size, and the number and direction of windows), and the energy efficiency of specific models. The Website also includes a searchable interface to help consumers evaluate a broad range of claims made on product labels (such as vegan, organic, and fair trade certified) by identifying the certifying organization, the meaning of the label, and products certified under that label.

*Sarah Boslaugh*
*Washington University in St. Louis*

# Index

*Note*: Article titles and their page numbers are in **bold.**